SOIL ANALYSIS

BOOKS IN SOILS AND THE ENVIRONMENT

edited by

E. A. Paul
Department of Plant and Soil Biology
University of California
Berkeley, California

Organic Chemicals in the Soil Environment, Volume 1, edited by C. A. I. Goring and J. W. Hamaker

Organic Chemicals in the Soil Environment, Volume 2, edited by C. A. I. Goring and J. W. Hamaker

Humic Substances in the Environment, by M. Schnitzer and S. U. Khan

Microbial Life in the Soil: An Introduction, by T. Hattori

Principles of Soil Chemistry, by Kim H. Tan

Soil Analysis: Instrumental Techniques and Related Procedures, edited by Keith A. Smith

Also in the Series

Soil Biochemistry, Volume 1, edited by A. D. McLaren and G. H. Peterson

Soil Biochemistry, Volume 2, edited by A. D. McLaren and J. Skujiņš

Soil Biochemistry, Volume 3, edited by E. A. Paul and A. D. McLaren

Soil Biochemistry, Volume 4, edited by E. A. Paul and A. D. McLaren

Soil Biochemistry, Volume 5, edited by E. A. Paul and J. N. Ladd

Additional Volumes in Preparation

SOIL ANALYSIS

Instrumental Techniques and Related Procedures

Edited by KEITH A. SMITH
The Edinburgh School of Agriculture
Edinburgh, Scotland

MARCEL DEKKER, INC. New York and Basel

Library of Congress Cataloging in Publication Data
Main entry under title:

Soil analysis.

(Books in soils and the environment)
Includes bibliographical references and indexes.
Contents: Atomic absorption and flame emission spectrometry / Allan M. Ure – Ion-selective electrodes / Oscar Talibudeen and Michael B. Page — Continuous-flow and discrete analysis / Keith A. Smith and Albert Scott — [etc.]
1. Soils—Analysis. I. Smith, Keith A., [date]. II. Series
S593.S742 1983 631.4'1 82-17965
ISBN 0-8247-1844-5

MARCEL DEKKER, INC.
270 Madison Avenue, New York, New York 10016

Current printing (last digit):
10 9 8 7 6 5 4 3 2

PRINTED IN THE UNITED STATES OF AMERICA

Preface

The last twenty years have seen the rapid development of instrumental methods of analysis. These have virtually replaced the traditional methods and have made it possible to undertake on a routine basis some determinations that could not be attempted previously.

The rate of adoption of instrumental techniques has been very uneven in different scientific fields and has been heavily influenced by the size of budgets, by traditions, and even by legislation prescribing that a particular analysis be carried out in a particular way. However, many changes have occurred in the last few years, and much instrumentation that previously might have been found only in industrial or medical laboratories is now commonplace in agricultural and environmental laboratories in which analysis focuses mainly on soils and plant material.

In spite of these developments the research soil scientist or the analyst working in governmental or commercial soil testing services generally has to make do with books covering traditional methods of analysis and specialist monographs on individual instrumental techniques. The latter are usually not written with soil or plant analysis specifically in mind, and are often inadequate in their consideration of the particular problems arising from the nature of the materials being analyzed. This book attempts to fill the gap between the two extremes, by covering the principles of the major instrumental techniques currently of relevance to soil analysis, together with discussions of sample preparation and matrix problems and critical reviews of applications in soil science and related disciplines. It will, I hope, be of value to scientists who wish to make a critical evaluation of available techniques in order to select the most appropriate ones for their needs, whether in terms of speed, cost, sensitivity, or other criteria, and also to the teachers and students of postgraduate courses in soil chemistry and soil analysis.

The principles underlying the various analytical techniques described by the contributors to this volume will continue to apply, irrespective of changes in the actual instrumentation available. However, the pace of development of this instrumentation has dramatically increased with the advent of microprocessor-based control systems. This phenomenon has occurred since this book was first conceived, and many significant innovations have appeared even since individual chapters were written and rewritten. It is therefore appropriate to supplement the information given in individual chapters with some mention here of recent developments in microprocessor control to illustrate the potential of this new approach. It is a safe bet that within a year or two there will be little analytical instrumentation commercially available that does not have some element of microprocessor-based control of operations or data processing facilities.

In the latest generation of gas chromatographs and liquid scintillation counters, for example, the operating parameters are set up not by operation of switches or dials on control panels, but by an interactive computer program; that is, the operator responds to questions displayed on a screen (or printer roll) by typing in appropriate answers via a keyboard. Invalid responses are not accepted, and, if no response is made, the system may automatically enter a preset stored value. After each acceptable response, the next question is posed and the process repeated. A "library" of standard programs may be available, or, alternatively, programs specifically tailored to fit a unique requirement may be written. Microprocessor-controlled instruments will normally be able to print out a list of all the operating parameters contained in a particular library program or those entered via a keyboard.

Many instruments (including those with some degree of automation) currently in service have simple analog output devices such as chart recorders, or digital meter displays or printers. Any further processing of data, whether it be colorimetric absorbance, chromatographic peak height, or electrode voltage, requires the intervention of the user, either through manual calculation or through transferring the data to a computer. In contrast, a major feature of the latest instruments is the ability to perform such functions as storing standard readings, calculating response curves and then calculating sample values, subtracting backgrounds, and calculating means and standard deviations of replicate samples. Furthermore, day-to-day or month-to-month variation in instrumental performance can be automatically monitored by comparison of fresh standard readings with earlier ones stored in the memory. Where further computer processing or data filing is necessary, there is a major reduction in the quantity of data to be handled, as a result of the initial processing within the microprocessor-based instrument, and the computer-compatible output of this instrument avoids manual handling of the data with the associated risk of transcription error. In many laboratories where existing instruments devoid of such facilities as those described have years of useful life in them, microprocessor-based attachments are now being constructed to up-grade the instruments at a fraction of the cost of complete replacement.

There has been a dramatic "leveling up" of features available in instruments from different branches of analytical science. We now see the humble pH meter equipped with the sort of sophisticated circuitry that was previously limited to complex instruments such as pulse height analyzers or mass spectrometers. Before long we may expect to reach an equilibrium stage where no analytical instrument is manufactured without the incorporation of some microprocessor-based facilities; but until then the instrumentation available in any one field is likely to change from month to month, rather than year to year. This is a small price to pay for what is, by any criterion, a major revolution in chemical analysis. The versatility, convenience, and improved quality control of the new generation of instruments is to be wholeheartedly welcomed. All we users need now is the funds with which to reequip ourselves!

I would like to offer my thanks to the other contributors to this volume for their efforts, to Gordon Finnie and Jackie Calder for producing many of the illustrations, to Ruth Easton for checking the manuscript, and, not least, to the members of my family for their tolerance while this project struggled toward completion.

Keith A. Smith

Contributors

John M. Bremner Department of Agronomy, Iowa State University, Ames, Iowa

Peter A. Cawse Environmental and Medical Sciences Division, Atomic Energy Research Establishment, Harwell, Oxfordshire, England

David J. Eagle Agricultural Development and Advisory Service, Ministry of Agriculture, Fisheries and Food, Cambridge, England

Anthony Haystead* Hill Farming Research Organisation, Bush Estate Penicuik, Midlothian, Scotland

Edward J. Jewell Agricultural Development and Advisory Service, Ministry of Agriculture, Fisheries and Food, Cambridge, England

John L. O. Jones Agricultural Development and Advisory Service, Ministry of Agriculture, Fisheries and Food, Cambridge, England

Victor Middelboe Physics Laboratory, The Royal Veterinary and Agricultural University, Copenhagen, Denmark

Michael B. Page Soils and Plant Nutrition Department, Rothamsted Experimental Station, Harpenden, Hertfordshire, England

Leonard Salmon Environmental and Medical Sciences Division, Atomic Energy Research Establishment, Harwell, Oxfordshire, England

Albert Scott Central Analytical Laboratory, The Edinburgh School of Agriculture, Edinburgh, Scotland

Keith A. Smith Soil Science Department, The Edinburgh School of Agriculture, Edinburgh, Scotland

**Present affiliation*: Ministry of Agriculture and Fisheries, Ruakura Soil and Plant Research Station, Hamilton, New Zealand

Roger S. Swift* Soil Science Department, The Edinburgh School of Agriculture, Edinburgh, Scotland

M. Ali Tabatabai Department of Agronomy, Iowa State University, Ames, Iowa

Oscar Talibudeen Soils and Plant Nutrition Department, Rothamsted Experimental Station, Harpenden, Hertfordshire, England

Allan M. Ure Department of Spectrochemistry, The Macaulay Institute for Soil Research, Aberdeen, Scotland

Carolyn Wilkins† Soils and Plant Nutrition Department, Rothamsted Experimental Station, Harpenden, Hertfordshire, England

**Present affiliation*: Department of Soil Science, Lincoln College, Canterbury, New Zealand

†Former affiliation

Contents

1

Atomic Absorption and Flame Emission Spectrometry

ALLAN M. URE The Macaulay Institute for Soil Research, Aberdeen, Scotland

I. INTRODUCTION

Since the independent realization in 1955 by Walsh [1] and by Alkemade and Milatz [2] of the analytical potential of atomic absorption spectrometry in flames, and the pioneering work of Walsh and his colleagues [3,4] in designing a practical atomic absorption instrument, there have been almost two decades of research and development. The first applications of the technique were to soils, plant ash, and other agricultural materials [5-7], while the use of solution samples, nebulized into a flame, derives from the earlier use of flame emission spectrometry largely by Lundegårdh [8], again in an agricultural context. The history of these developments and earlier work has been comprehensively reviewed [9]. At present, atomic absorption spectrometry is established as the principal analytical technique for elemental analysis in agricultural and environmental laboratories.

Atomic emission spectrometry using flame photometers with photoelectric detectors was of considerable importance in the 1950s and 1960s, but its use, except for the determination of the alkali metals and the alkaline earths, has to a large extent been superseded by atomic absorption spectrometry. More recently, however, there has been a revival of interest in atomic emission spectrometry, due to the development of high temperature sources such as the nitrous oxide/

acetylene flame [10-12], various types of high-frequency plasma [13], and, in particular, radio-frequency inductively coupled plasma (ICP) sources [14-18]. This interest stems not only from the high sensitivity of the ICP sources: It is due also to the increasing realization of the advantages of simultaneous multielement analysis offered by atomic emission and atomic fluorescence spectrometry. Atomic absorption spectrometry remains essentially a single-element technique.

This account will be concerned mainly with atomic absorption spectrometry (AAS) and its use and potential in the analysis of soils and other environmental materials, but reference will be made where appropriate to atomic emission spectrometry (AES) and to atomic fluorescence spectrometry (AFS). In this connection it is important to realize that most commercial atomic absorption spectrometers, using conventional atomic absorption burners, make excellent flame photometers and an emission mode of operation is usually provided. Most instruments can be adapted for AFS but specialized flames or electrothermal atomizers will usually be required. Initial discussion of the principles of AAS will be in terms of flame atomizers, but electrothermal atomization using graphite tube furnace and carbon-rod techniques will also be considered, together with other nonflame methods of analysis.

II. BASIC PRINCIPLES

In its simplest sense atomic absorption spectrometry makes use of the fact that free atoms of an element absorb light at wavelengths characteristic of that element and that the extent of the absorption is a measure of the concentration of these atoms in the light path. The production of atoms from chemical compounds requires a source of energy such as a flame which can vaporize and dissociate the sample compounds into the gaseous elemental state in which *atomic absorption* of radiation can take place. In favorable circumstances, this light energy absorbed by the atoms can subsequently be re-emitted as light of characteristic wavelengths whose intensity can form the basis for analysis by *atomic fluorescence* spectrometry. The function of the flame, therefore, in atomic absorption and fluorescence spectrometry, is to produce atoms from the sample compounds, i.e., to act as an *atomizer*. For atomic emission spectrometry, on the other hand, additional energy is required from the flame to excite the emission of atomic spectra, again characteristic of the element, whose intensity is not only a function of the atomic concentration in the flame but also of the temperature of the flame. The energy required for the production of atomic emission spectra increases as the wavelength decreases, and practical chemical flames such as air/acetylene, and even nitrous oxide/acetylene, are of little value for elements whose analytical wavelengths are lower than about 250 nm. Atomic absorption and fluorescence spectrometry, on the other hand, do not have this limitation and are

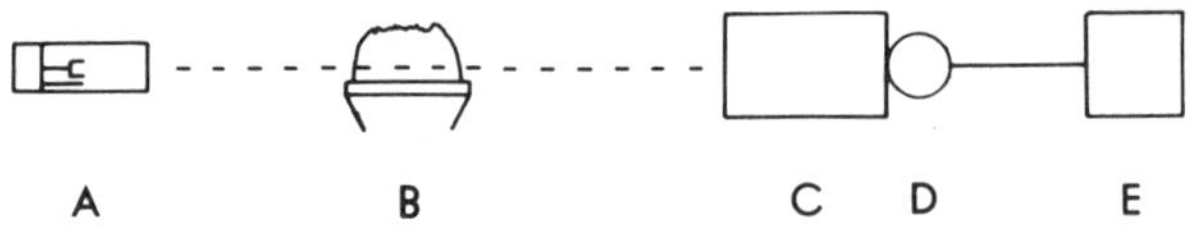

Figure 1 Schematic diagram of a single-beam atomic absorption spectrometer comprised of a hollow-cathode lamp, A; a burner/nebulizer, B; a monochromator, C; a photomultiplier detector, D; and an output device, E.

successfully used even at wavelengths below 200 nm. The extremely high temperatures attained by the radio-frequency plasma sources, as distinct from chemical flames, are sufficient to produce emission spectra with excellent sensitivity for elements whose spectral lines lie well below 200 nm.

The basic principles of atomic absorption spectrometry are illustrated in Fig. 1, which shows, schematically, the basic components of a simple single-beam instrument. A hollow-cathode discharge lamp, A, whose cathode is made of the element (copper, for example) that is to be determined, emits a spectrum characteristic of the cathode element (copper). This light is passed through the atomizer in the form of a flame, B, into which a fine mist of a solution of the analyte (copper) is sprayed by a pneumatic nebulizer. Atoms of copper are formed from the sample mist by thermal processes in the flame and absorb some of the light from the hollow-cathode lamp at the wavelength of an absorbing (resonance) copper line. The light passed through the flame is received by a monochromator, C, set to accept and transmit radiation of this wavelength. By means of the adjustable, ganged entrance and exit slits the bandwidth of the monochromator is made narrow enough to reject lines of other wavelengths emitted by the hollow-cathode lamp. The light emerges from the monochromator exit slit and falls on the photocathode of a photomultiplier detector, D, and an output current, proportional to the incident light intensity, is amplified, processed electronically, and finally presented to a readout device, E, such as a moving-coil meter, a pen recorder, or digital display.

If the light intensity measured in the absence of copper atoms in the flame is I_0 and the intensity when copper atoms from a nebulized sample solution are present is I_T, then

$$\%T = \text{percentage transmission} = \frac{100 \times I_T}{I_0} \qquad (1)$$

$$\%A = \text{percentage absorption} = \frac{100 \times (I_0 - I_T)}{I_0} \qquad (2)$$

$$\text{Absorbance} = \log \frac{100}{\%T} = 2 - \log \%T \qquad (3)$$

The atomic concentration in the flame is proportional to the measured absorbance and from this the element concentration in the sample solution can be found by standardization (see Sec. IV.B).

The function of the monochromator in atomic absorption spectrometry is merely to isolate the required line from the remainder of the emitted hollow-cathode lamp spectrum, which is usually simple. The effective bandwidth of the absorption measurement is determined by the line width (ca. 0.001 nm) of the spectral lines emitted by hollow-cathode lamps. It is this excellent monochromation which contributes to atomic absorption spectrometry its remarkable freedom from spectral, superpositional interference.

The situation is quite different in atomic emission spectrometry where isolation of the desired line from the emitted spectrum is dependent on the monochromator resolution, which cannot approach the equivalent resolution of atomic absorption spectrometry.

III. THE ATOMIC ABSORPTION SPECTROMETER

A. Introduction

The block diagram of a simple, single-beam, atomic absorption spectrometer, Fig. 1, serves to illustrate the basic components of the measuring system whose various functions are discussed below. An example of a commercial single-beam instrument (Varian Techtron AA6) is shown in Fig. 2, in which these basic components are also indicated.

Since the atomic absorption measurement near the detection limit is the measurement of a small reduction in the initially large signal from the hollow-cathode lamp, this detection limit is determined largely by the stability of the lamp output and by variations in the sensitivity of the detector and electronics. One method of optimizing the detection limit is to use averaging output devices such as pen recorders or signal integrators. Another approach is to use a double-beam instrument in which the light from the hollow-cathode lamp is split into two beams, one passing through the flame and the other, the reference beam, passing outside the flame. The atomic absorption measurement made, by a common detector, is in this case the ratio of the two beam intensities so that lamp and detector noise and instability are considerably reduced. In addition, instrument warmup time, normally about 15 min with a single-beam spectrometer, is virtually eliminated. However, because of the requirement to divide the light beam for double-beam operation, the available light energy at the detector is reduced and this can present some small disadvantage. Double-beam instruments can perhaps more readily incorporate background correction continuum sources (see Sec. VI) but some instruments may sacrifice double-beam operation in the process. Double-beam instruments are necessarily more complicated and their maintenance more difficult but,

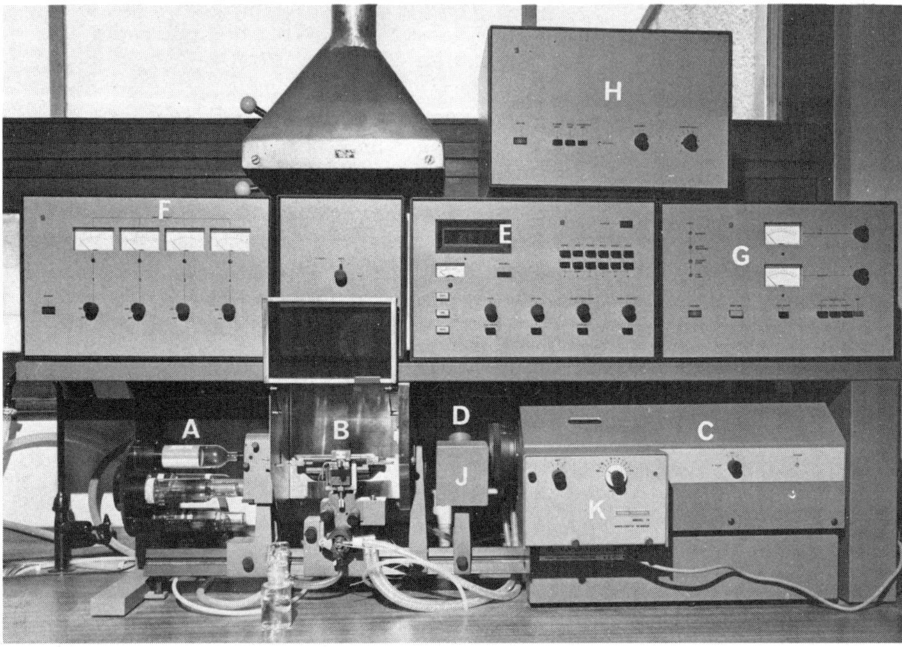

Figure 2 Commercial single-beam atomic absorption spectrometer (Varian Techtron AA6) showing: hollow-cathode lamp, A; burner/nebulizer assembly, B; monochromator, C; photomultiplier, D; digital indicator output device, E; hollow cathode lamp power supply for turret of four lamps, F; automatic gas control unit, G; background correction unit, H. Also shown are the chopper, J, in position for atomic emission measurements, and wavelength scanning attachment, K, also for atomic emission spectrometry.

with modern solid state electronic technology, this disadvantage is a minor one. In practice, similar performance and detection limits are obtainable by both types of instrument, and in choosing an instrument this difference is not necessarily the dominating criterion.

The desirability of attempting to minimize the principal remaining sources of instability, i.e., the production of atoms by the nebulizer/flame combination, has led to the introduction of two-channel double-beam instruments which have a common light source and nebulizer/burner assembly but two monochromators and measuring channels. This not only allows two elements to be determined simultaneously by atomic absorption (or atomic emission) but, by using one channel for an internal standard element whose concentration is known or fixed, and the other for the element being determined (the analyte), a measurement of the ratio of the absorbance (or measured concentrations) of the two elements can be made. If the two elements are chosen so that the internal standard element and the analyte behave similarly in the flame this ratio measurement compensates for fluctuations in nebulizer performance and in flame conditions. This internal standard technique, well known in emission spectrometry, requires the incorporation into sample and standard solutions of an internal standard element. The choice of a suitable internal standard, however, is often limited in practice, but where this is feasible considerable gains in precision can be obtained [19].

B. The Light Source

Considering, one by one, the components of the atomic absorption spectrometer we turn first of all to the hollow-cathode discharge lamp (Fig. 1,A), almost universally employed as the light source, and illustrated in Fig. 2,A. The lamp is filled at low pressure with a rare gas, generally argon or neon, whose ions carry an electrical discharge between the wire anode and the hollow cylindrical cathode which is made of the element whose spectrum is required, e.g., of copper for copper analysis. As a result of bombardment of the inner surface of the hollow-cathode by the filler-gas ions, copper, in this example, is sputtered into the discharge where its atoms are excited to emit their characteristic spectrum; this emerges through the glass, UV-transmitting glass, or quartz end window. Hollow-cathode lamps are stable and reliable and have lives often guaranteed by the manufacturer for six months or some 5000 mA hr of operation.

For a few elements, especially the alkali metals, metal-vapor arc discharge lamps are somewhat superior to hollow-cathode lamps. They are less convenient to use, partly because they require operation at reduced current to keep the emitted spectral-line widths narrow, but chiefly because they cannot readily be fitted into most commercial instruments. They are only available for the more volatile elements such as the alkali metals, Cd, Hg, and Zn, and have found their chief use in atomic fluorescence where their high intensity is desirable and line broadening is less disadvantageous.

Electrodeless discharge lamps (EDL) excited by microwave-frequency sources have also been developed for atomic absorption and fluorescence spectrometry [20-26] for a wide range of elements. For many elements, however, stable EDL operation requires the use of heated chambers [27] thermostated at temperatures of several hundred degrees Celsius. Their chief virtue lies in the very high intensity, which has given them a principal role in atomic fluorescence spectrometry, where the intensity of the fluorescence signal is proportional to the intensity of the exciting radiation. In commercial instrumentation, however, they have not found much favor as atomic absorption sources. Similar devices operated at radio frequencies [28], more suitable for routine use, are also available from some manufacturers. In atomic absorption EDLs are mainly of use for the volatile elements arsenic and selenium, for which hollow-cathode lamps are not particularly satisfactory.

C. The Atomizer

The flame atomizer illustrated in Fig. 1,B, is the most common method of producing atoms from the elements in the sample, but in recent years this has been supplemented by a variety of nonflame devices, most of which use electrothermal methods of atomization (see Sec. VII). The flame atomizer consists of a burner in which the flame is produced and a nebulizer to convert the sample solution into a fine mist or aerosol which is fed to the flame. The nebulizer is usually a pneumatic device operating on the principle of a scent or paint spray, although the less convenient ultrasonic nebulizer has been used occasionally because of its greater efficiency. In the flame a complicated process involving desolvation of the solution droplets and dissocation of the chemical compounds of the resultant clotlet, results in the production of gaseous atoms of the constituent elements. While over the years a great variety of burners and nebulizers has been used [29,30] the burners now almost universally employed are of long-slot design burning premixed fuel and oxidant gases and fitted with a pneumatic nebulizer. A typical burner and nebulizer assembly is shown in Fig. 3. Sample solution is aspirated through the sample capillary by the pneumatic nebulizer operated by the oxidant gas and projected, as a spray of fine droplets, into the spray chamber where the larger droplets are intercepted or broken up by the flow spoiler. The surplus solution eventually drains out to waste. The oxidant and fuel gas are mixed in the chamber and carry the sample aerosol to the burner head and into the flame. Although pneumatic nebulizers only succeed in presenting less than 10% of the aspirated sample solution to the flame, alternative, more efficient methods, such as ultrasonic nebulization [31], have not found wide acceptance despite much development work. Heated spray chambers [32,33], or a preheated air supply to the nebulizer [34], have shown improved sensitivity but have not generally been adopted mainly because of memory effects. Automatic sampling devices are available from most manufacturers which sequentially insert the nebulizer capillary into the samples of a batch.

The type of flame is of critical importance and several criteria govern the choice. If the flame temperature is too low, atomization will be incomplete since the flame cannot supply sufficient energy to dissociate the compounds in the sample. In addition, the degree to which atomization occurs will be a varying function of sample composition, giving rise to matrix, interelement, or anionic interference effects in the atomic absorption measurement. If, on the other hand, the flame

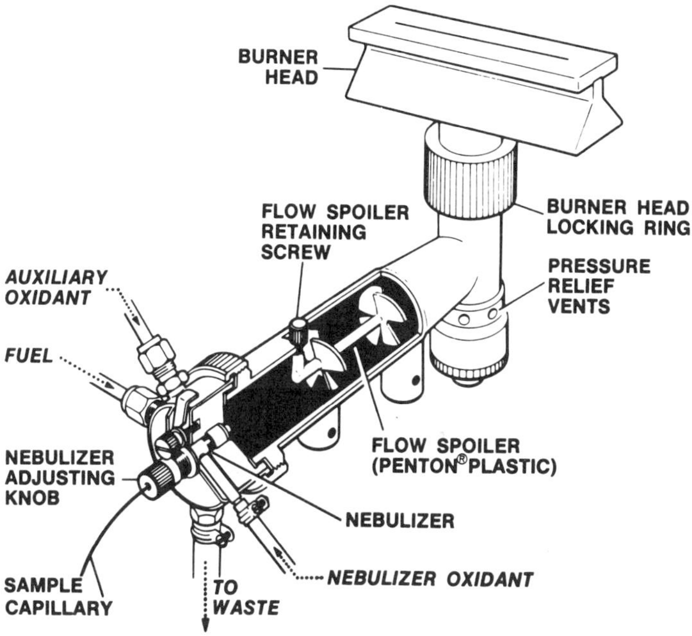

Figure 3 Diagrammatic illustration of a burner/nebulizer assembly. (From the Perkin Elmer Model 306 Absorption Spectrophotometer, Order No. L333, courtesy of Perkin Elmer Ltd., Buckinghamshire, England.)

temperature is too high, the atoms formed may be ionized to a sufficient extent to reduce the number of atoms present; this number will be dependent not only on the flame temperature but on the composition of the sample, since interelement competitive and ionization suppression reactions can occur. The selection of the flame is therefore a compromise between achieving complete atomization (high-temperature flame) and avoiding ionization (low-temperature flame).

For most elements and sample types the air/acetylene flame represents the best compromise. Even for the easily ionized alkali metal elements, sodium and potassium, the relative freedom from matrix or anion interference effects offered by this flame often makes it the best choice since ionization effects incurred can usually be avoided by the addition of an excess of another easily ionized (alkali) element. For rubidium and cesium, however, lower temperature flames such as air/propane may be advantageous, provided they are used with standards whose matrix composition is similar to that of the average sample.

For elements such as Al, Ba, Be, Ca, Mo, Si, Ti, etc., which form refractory oxides or other compounds not completely dissociated in the air/acetylene flame, the higher temperature nitrous oxide/acetylene flame [10] must be used with a fuel-to-oxidant ratio in excess of the stoichiometric proportions, to obtain reducing conditions.

Various other specialized flames are in use, of which only a few can be mentioned here. A multislot burner, the Boling burner [35], is commercially available, which, because the wider cross section of the flame more completely fills the optical path, is much less critical both in its positioning and in its fuel and oxidant settings. It is also less subject to clogging when sample solutions with a high content of dissolved solids, such as EDTA extracts of soils, are aspirated. Separated flames offer considerable improvements in signal to noise ratio and detection limits, particularly in atomic emission spectrometry. In these flames the inner combustion zone is separated from the outer diffusion flame by a silica tube [36], plates [37], or by a curtain of argon or nitrogen [38] which prevents the access of atmospheric air necessary for the completion of this secondary combustion. They have also been used routinely in atomic absorption spectrometry for the determination of Cu, Mn, and Zn in EDTA extracts of soil [39]. In addition to their value for emission spectrometry, separated flames seem likely to offer useful atomizers for atomic fluorescence spectrometry [40,41]. The cool argon/hydrogen flame has been used for atomic fluorescence spectrometry [42] and for the determination of alkali metals [43]. It has useful applications in the determination of arsenic and selenium by the volatile hydride method (see Sec. VIII.B), a method which obviates matrix effects, which are normally large in this flame [43]. The low background of these flames in the ultraviolet is of importance in the determination of arsenic and selenium, for which short wavelength lines are employed. Similar nitrogen/hydrogen flames have been studied [44].

The nitrous oxide/butane/propane flame [45] is of limited application, but of considerable value for the determination of calcium in the presence of phosphate. This flame is hot enough almost to eliminate this interference although aluminum interference is only reduced. Because of its docility this flame requires only the precautions usual for air/acetylene, although normally it is used with 50 mm long high-temperature burners. This is in contrast to the necessity for stringent operational safety procedures with the nitrous oxide/acetylene flame.

Mention might also be made of the use of mixtures of a powdered rock sample and a solid fuel powder [46] which are ignited and burned, without the need for fuel gases, in the optical path of an atomic absorption spectrometer - a system which could find application in portable, prospecting, or field instruments. In general, however, samples for flame atomic absorption, emission, or fluorescence spectrometry must be in solution form and although reasonably successful attempts have been made to aspirate and analyze suspensions or slurries of solid powders in liquids [47,48], matrix interference effects and other difficulties do not usually commend the method.

D. The Monochromator and Detector

Light from the hollow-cathode lamp is passed through the flame to the entrance slit of the monochromator (Figs. 1,C and 2,C) by a simple optical system generally consisting of two lenses, one on each side of the flame. As outlined above, the function of the monochromator in atomic absorption is to separate the resonance line used from other lines emitted by the hollow-cathode lamp while the desired wavelength and bandwidth is chosen by an appropriate setting of the monochromator.

Many instruments incorporate a wavelength scanning facility (Fig. 2,K), which allows emission spectral lines to be displayed on a recorder together with the background emission for which correction can thus be made.

The monochromator receives light from the hollow-cathode lamp through the flame together with the light emitted from the flame. In the atomic absorption mode, it is almost essential that the signal arising at the detector from flame emission be rejected, and only that originating from the hollow-cathode lamp be accepted. In most commercial instruments this is achieved by modulating the hollow-cathode lamp output intensity so that the signal from the lamp can be separated electronically from the unmodulated flame emission signal and the latter rejected. Strongly emitting species such as the alkali metals will still generate a noise signal at the input of the detector amplifier, and a degradation of the atomic absorption signal-to-noise ratio may occur. The emission-generated noise signal can be minimized by using a narrow monochromator slit to minimize the accepted spectral bandwidth. To permit an instrument incorporating such a modulation system to be used for the measurement of atomic emission from the flame, a mechanical, rotating-sector light-beam chopper (Fig. 2,J), synchronized to the detector, must be inserted between the flame and the monochromator to modulate light emitted by the flame.

The detector used almost universally is a photomultiplier tube (Figs. 1,D and 2,D), whose current output corresponds to the intensity of the light falling on its photocathode. This feeds the amplifiers and output devices which display the measured signals. Although the photomultiplier usually fitted covers a wide band of wavelengths from the ultraviolet to the near infrared, one could be selected to provide maximum sensitivity in the ultraviolet, visible, or red region of the spectrum.

E. Output Devices

The simplest output devices consist of a moving-coil meter or a pen-recorder displaying percentage transmission (%T), from which the absorbance, A, can be calculated. An example of a pen-recording of the transmission for a series of standard solutions for the determination

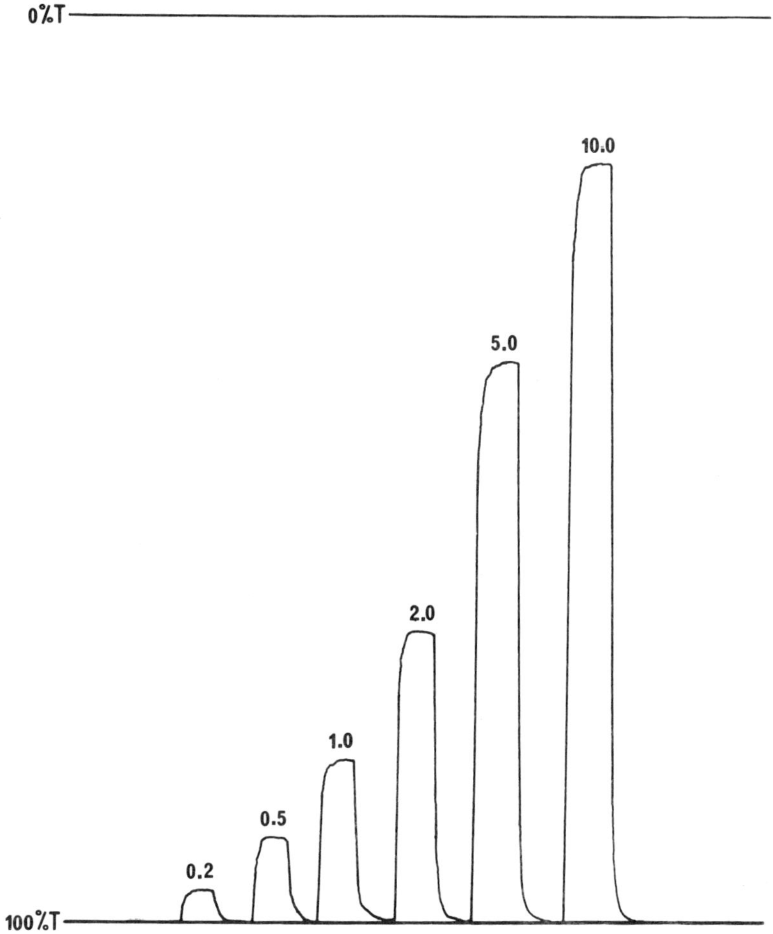

Figure 4 Pen recording of measurements of standard sample solutions. Manganese, 0-10 μg ml^{-1} in 0.05 M EDTA solution.

of manganese in EDTA soil extracts is given in Fig. 4. A graph of absorbance versus concentration, i.e., the standard curve, is obtained as shown in Fig. 5. The absorbance versus concentration graph is essentially linear over most of the range with only a slight and quite typical curvature at high concentrations. Most instruments incorporate a logarithmic amplifier, which provides a direct readout of absorbance values, and this, with provision for curve linearization,

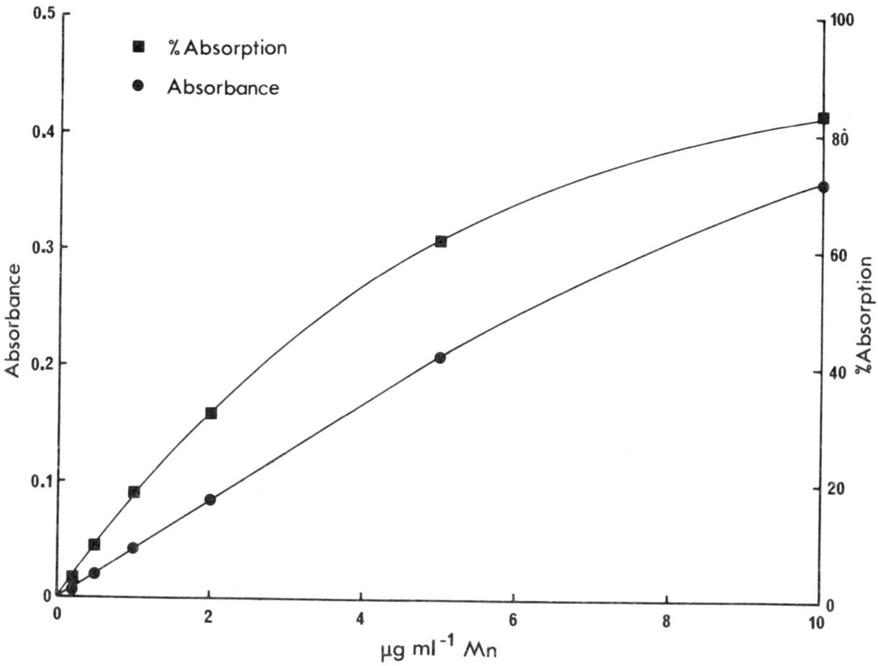

Figure 5 Standard curve, absorbance versus concentration, for manganese in 0.05 M EDTA solution; also shown is the nonlinearity of the graph of % absorption versus concentration.

forms the basis for displaying outputs directly in concentration terms, using standard solutions for calibration. Operationally, this process of standardizing the instrument to read concentration directly has been greatly facilitated by the incorporation into recent instruments of a microprocessor, or by interfacing a programmable calculator to the instrument. The latter approach is more flexible and gives the user control of the programming of the calculator [49,50].

Provision is usually made for *scale expansion*, which improves the reading accuracy for small absorbance, and also for integration of the signals. For the rapid pulse signals from electrothermal atomizers, there are fast recorders and facilities for peak height and peak area measurement. Printout of absorbance or concentration, and digital punched-tape output for computer processing, make automated or semi-automated instrumentation possible for atomic absorption [51,52], particularly where the requirements are for the analysis of a large number of samples for only a few elements.

IV. OPERATIONAL REQUIREMENTS

A. General Considerations

For each elemental analysis the best wavelength must be selected and the instrumental conditions must be optimized. These include the choice of flame type, fuel and oxidant flow rates, burner height, and slit width. Most manufacturers supply "cookbooks" of methods, and these form a good starting point for establishing operating conditions.

The laboratory itself must be clean and sources of contamination [53] studiously guarded against, especially where trace element analyses are concerned. There is always the possibility of contamination by constituents of structural building materials. These can be calcium from plaster work, transition metals from stainless steel and paints, and Ba, Cd, Pb, Sn, and Zn in particular from plastics. This last also applies to the use of plastic laboratory ware; PVC and polystyrene may contain these and other metals as water-soluble stearates. Natural grade (not colored) polyethylene and polypropylene are the preferred plastic materials for containers of standard and sample solutions. Borosilicate glass laboratory ware is equally useful except for the alkalis at low concentrations, but soft glass containers should be avoided. The problem is particularly severe when very sensitive methods, such as electrothermal atomization (ETA), are used [54]. Contamination occurs not only in the laboratory but also at the sampling stage, and should be of concern to the analyst since a contaminated sample is not worth analyzing [53].

B. Standardization

The usual method of analysis entails the preparation of an analytical working curve, by measuring the absorbance of a series of standard solutions of the elements of interest and plotting the graph of absorbance versus concentration, as illustrated in Fig. 5. The element concentration in the sample solutions can then be obtained from this graph by interpolation.

V. ANALYTICAL CAPABILITIES

A. Comparison of Atomic Absorption, Emission, and Fluorescence Spectrometry

As has already been indicated, atomic absorption must be regarded as a single-element technique and, although several multi-element instruments have been described and reviewed [55], they have so far had little influence on commercially available equipment. Both atomic emission [56] and atomic fluorescence [57] are capable of simultaneous multielement analysis and, while the development of multielement atomic fluorescence spectrometers has not proceeded far, direct reading spectrometers using dc arc plasma-jet, microwave plasma, and in particular inductively coupled radiofrequency plasma (ICP) sources are available commercially for the simultaneous determination of elements in solution by atomic emission methods. In addition to an extremely high temperature, the ICP source offers detection limits which are often superior to those yielded by atomic absorption, and particularly so for refractory elements such as molybdenum (detection limits of a few nanograms per milliliter). This atomic emission technique has attractive possibilities for the direct analysis of soil extracts, making unnecessary the preconcentration techniques often required for atomic absorption analysis. In view, however, of the complexity of spectra produced by this source, the degree of expertise required will be considerably greater than for atomic absorption spectrometry and a good deal of exploratory work on its application has still to be done before the applied laboratory can make full use of it.

The limits of detection by flame AES, AAS, and AFS have been compared in an excellent review of AFS as an analytical technique [58]. AFS shows significant advantages over AAS and AES in detection limit for elements such as Ag, Bi, Cd, Hg, Te, Tl, and Zn. The advantages of AFS, however, are more obvious where nonflame techniques are used, since these avoid the quenching effects of the flame gases and combustion products. The subject of AFS has been comprehensively treated [41]. The place for AFS as a sensitive multielement technique may well have been usurped by the advent of the ICP AES technique, and the continued lack of a good range of stable intense excitation sources for AFS.

B. Element Coverage: Sensitivity and Detection Limits

More than 60 elements can be determined by atomic absorption methods, and the extent to which this wide range of element coverage can be realized in soil analysis depends first of all on the sensitivity of atomic absorption spectrometry for the elements of interest. An indication of the possibilities of soil analysis for different minor and trace elements, judged solely on their detection limits, is given in Table 1. It should be remembered, however, in considering Table 1, that other factors such as matrix or interference effects have a major influence on the viability of a particular analysis and that the quoted detection limits are for single-element solutions under optimum conditions. In many cases the lowest concentration that can usefully be determined in practical sample solutions will be greater, by a factor of 5 or 10, than these detection limits.

The major soil constituents Al, Ca, Fe, K, Mg, Na, and Si have not been included in Table 1, since atomic absorption methods are more

Table 1 Contents of Various Elements in Soils and Their Detection Limits by AAS

Element	Total content		Extractable content[c] (normal range; $\mu g\ ml^{-1}$ in extract)		Detection limit ($\mu g\ ml^{-1}$ in solution)	
	Soil: typical conc.[a] ($\mu g\ g^{-1}$)	Analysis solution[b] ($\mu g\ ml^{-1}$)	0.05 M EDTA	0.5 M acetic acid	Flame[d]	ETA[e]
Ag	0.1	0.0001			0.002	0.000005
As	6	0.006			0.11	0.00006
B	10	0.01			2.0	
B[g]				0.005-5[f]	2.0	
Ba	1000	1			0.02	0.00004
Be	6	0.006			0.0007	0.000001
Cd	0.5	0.0005			0.0007	0.000004
Cd			<0.002-0.06	<0.0003-0.008		
Co	15	0.015			0.007	0.00003
Co			<0.01-0.8	<0.001-0.05		
Cr	100	0.1			0.005	0.000005
Cr			0.02-0.8	<0.0003-0.03		
Cu	20	0.02			0.002	0.000008
Cu			0.06-2	<0.001-0.08		
Ga	25	0.025			0.038	
Ge	1	0.001			0.038	
Hg	0.1	0.0001			0.16	
Li	50	0.05			0.0015	
Mn	800	0.8			0.002	0.000004
Mn			1-20	0.1-2.5		
Mo	2	0.002			0.03	0.00006

Mo			<0.006-0.2	<0.0001-0.0008		
Ni	50	0.05			0.008	0.000025
Ni			0.04-1.0	0.003-0.1		
Pb	20	0.02			0.15	0.00003
Pb			0.2-2	<0.00005-0.1		
Rb	100	0.1			0.002	
Sc	8	0.008			0.025	
Se	0.5	0.0005			0.25	
Sn	3	0.003			0.031	0.00006
Sn			<0.004-0.2	<0.0005-0.005		
Sr	300	0.3			0.002	0.00001
Sr				0.005-0.25		
Ti	4000	4			0.05	0.00033
Ti			0.1-2	<0.003-0.03		
V	100	0.1			0.05	0.00015
V			0.04-1	<0.001-0.03		
Y	40	0.04			0.11	
Zn	50	0.05			0.001	0.000001
Zn			<0.6-4	<0.05-0.75		
Zr	400	0.4			1.0	

[a]Ref. 59.
[b]$LiBO_2$ fusion, HNO_3 dissolution, 100 mg soil per 100 ml solution, i.e., 1000 × dilution.
[c]Calculated from data in Ref. 59, EDTA extract, 15 g soil per 75 ml extract, i.e., 5 × dilution; acetic acid extract, 20 g soil per 800 ml extract, i.e., 40 × dilution.
[d]Varian Techtron AA6. Manufacturer's data.
[e]Instrumentation Laboratory 555 Furnace atomizer. Manufacturer's data.
[f]Hot water extraction; 20 g soil per 40 ml extract, i.e., 2 × dilution.
[g]Solid line indicates that determination is possible at "normal" levels on sensitivity criteria; dashed line indicates that determination is possible at "excess" levels on sensitivity criteria.

than adequately sensitive for their determination as total contents in soils and, with the exception of silicon, in soil extracts. The determination of these elements is discussed in Sec. X, and applications of AAS to major elements in soil extracts are comprehensively listed in the annotated bibliography by Kalra (see Sec. XII).

Table 1, column 2, shows typical, total soil contents [59], together with the contents of the corresponding analysis solutions (column 3) prepared, for example, by a lithium metaborate fusion/nitric acid dissolution procedure (see Sec. X.A) which, typically, involves a dilution of the contents from soil (100 mg) to solution (100 ml) of about 1000 times. The dilution will depend on the dissolution procedure but will seldom be much less than 500 times and may often exceed 1000 times. Also listed in Table 1 (columns 4 and 5) are normal content ranges for some elements extractable by 0.05 M EDTA and by 0.5 M acetic acid, which are in common use for assessing extractable or plant-available contents. These contents are calculated from data in Ref. 59. The last two columns list the detection limits obtained in aqueous solutions by atomic absorption spectrometry using flame and electrothermal atomization (ETA) methods, respectively. Full horizontal lines from element to total or extractable contents in Table 1 indicate that flame atomic absorption offers sufficient sensitivity for a particular element without preconcentration. Dashed lines indicate that the sensitivity is not sufficient for the full range of contents expected, but may be adequate for some samples and some purposes, e.g., detection of excess levels.

The determination of the total soil contents of the minor and trace elements Ba, Cr, Li, Mn, Rb, Sr, Ti, and Zn are indicated in Table 1 to be within the range of flame AAS methods without preconcentration. Examples of the determination of total contents of these elements are given in Sec. XI. The determination of the total contents of Be, Co, Cu, and Ni is marginal, and the remainder of the elements are insufficiently sensitive and require the use of preconcentration techniques, the employment of special methods (see Sec. VIII), or electrothermal atomization techniques. The detection limits for the latter technique (Table 1) are often some two orders of magnitude lower than flame limits.

Table 1 also shows that, on sensitivity criteria, it should be possible by flame atomic absorption spectrometry to determine the extractable contents of Cd, Co, Cr, Cu, Mn, Ni, and Zn in EDTA and acetic acid extracts, Pb, Ti, and V in EDTA and Sr in acetic acid extracts. This also illustrates the fact that the extractant is often chosen not only for its assessment of availability but also to maximize the solution content of the element to bring it within the scope of an analytical technique. For many of the other elements and other extracting solutions, preconcentration or the use of special methods such as ETA will be essential.

The characteristic concentration (often referred to in AAS as *sensitivity* or *reciprocal sensitivity*) is defined as the concentration of an element (measured in micrograms per milliliter) that gives a signal corresponding to 1% absorption, i.e., to an absorbance of 0.0044. This value, used in conjunction with detection limits, allows some assessment of the viability of an analytical technique for a particular application.

The detection limit is usually defined as the element concentration which provides an absorption equal to twice or three times the standard deviation of the background (zero absorption).

C. Accuracy and Precision

For many trace element analyses the obtainable precision of AAS procedures is in the region of ±1-5%, while with special procedures and sophisticated instrumentation precisions of better than 1% can be obtained in some cases. The accuracy depends on the extent to which interference effects can be neglected or overcome and, in complex solutions such as those used for the determination of total contents of soils, it may be desirable to match standards and samples for matrix composition if the maximum accuracy is to be achieved. The precision of flame measurements has recently been discussed [60] and it has been shown that for each element a concentration (often some 100 times the characteristic concentration) exists for which the precision is a maximum [61]. The accuracy of AAS methods for soil analysis has been assessed by comparison with neutron activation analysis [62].

VI. INTERFERENCE EFFECTS

A. Spectral Effects

The relative freedom of AAS from superpositional spectral interference effects has already been commented on and this largely accounts for the fact that simple, single-element solutions can often be used as standard solutions without rigorous matching of matrix composition.

B. Physical or Matrix Effects

Physical effects such as viscosity of the sample solution, the presence of organic solvents, or high salt contents can also produce interference effects. This can usually be overcome by dilution, by matching the standard solution matrix composition to that of the sample, or by using the method of standard additions [63].

C. Ionization Effects

The extent to which ionization interference occurs depends on the ionization potential of the element being determined and on the flame temperature. An element with a very low ionization potential, such as the alkali metal cesium, can be ionized to the extent of 80% in the air/acetylene flame, while lithium, with a much higher ionization potential, is only some 5% ionized. In high-temperature flames, such as that of nitrous oxide/acetylene, cesium is completely ionized [64] and most elements are ionized to some extent. This effect can be countered by the addition of an excess of an easily ionized element such as potassium to suppress the ionization of the element being determined. For very easily ionized elements such as cesium it may be necessary to use a flame of lower temperature.

D. Background Effects

Interference of this type arises in flame atomic absorption where the concentration of major elements such as Al, Ca, Mg, and Na is high. The effect is most pronounced at shorter wavelengths. This interference is due to a scattering of light by dried salt particles and results in an apparently enhanced absorption. In some instances molecular absorption can produce a similar interference effect. This type of interference can be removed by separating the element from the interfering species by a technique such as solvent extraction. Alternatively the error can be corrected for by repeating the determination at an adjacent wavelength, using a nonabsorbing line and thus measuring and subtracting the contribution made by scattering or molecular absorption to the original measurement. Most instruments now incorporate a deuterium or hydrogen lamp in a background corrector. These lamps produce a continuous emission, at wavelengths shorter than about 350 nm, which is used to measure the scattering or molecular absorption light loss. This is then automatically subtracted from the original measurement of atomic absorption plus background absorption. Such correctors can correct for a light loss equivalent to an absorbance of about 2 but when the interference is large compared with the atomic absorption signal, the precision of the background-corrected analysis can be significantly degraded.

The use of a background corrector is essential for work with electrothermal atomizers such as the graphite furnace, since the furnace itself retains the smoke and molecular species produced during atomization of the sample, as well as carbon particles emitted from the rod, all of which can contribute a light-scattering error. It should also be available for use with the boat technique (see Sec. VIII.A).

Other sophisticated background correction systems include the use of the Zeeman effect produced by an applied magnetic field [65-68].

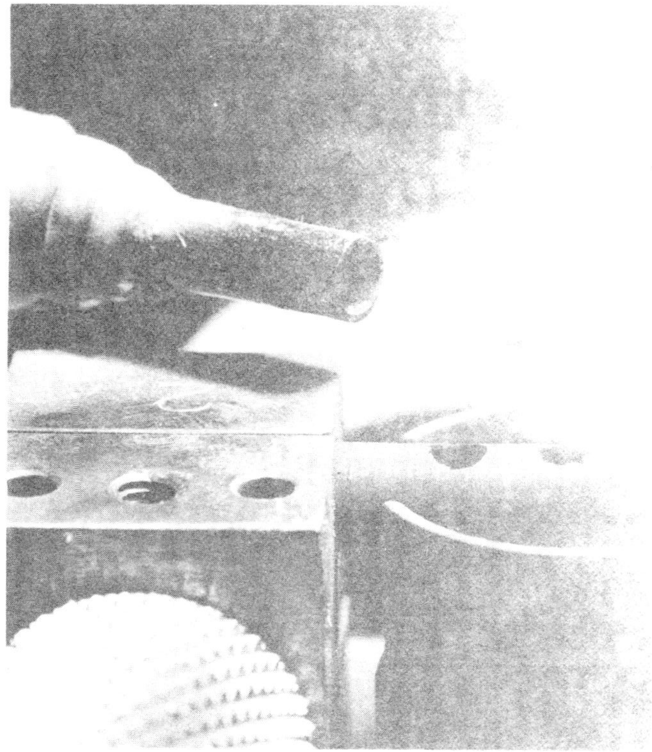

Figure 7 Carbon-rod atomizer for the determination

[illegible] ments discussed [84]. Some of [illegible] thermal atomizers such as the [illegible] filament atomizers using, for ex[illegible] wire [86-89]. The characteristic[illegible] carbon rod [91] atomizers have be[illegible] ly. The advantages of graphite furnace a[illegible] atomizers compared with flame atomizers can [illegible] as follows:

1. Smaller (microliter) samples are required.
2. Higher sensitivity and lower detection limits (see Table 1) are possible as a consequence of longer atomic lifetimes, complete sample consumption (cf. 5-10% for burner nebulizers), and absence of dilution by flame gases.
3. Viscous liquids and even solid samples can be used.
4. The absence of flame background can give a better signal-to-noise ratio and may also allow measurement in the ultraviolet region.
5. For some elements, particularly those forming refractory oxides, the different mechanism of atomization, namely, reduction of oxide to element by hot carbon, gives more effective atomization than a flame.

The principal disadvantage is that the nonspecific light losses caused by smokes and vaporized matrix material are very much greater in furnaces than in flames, and the use of a hydrogen or deuterium background correction facility is almost mandatory. In view of the difficulty in ensuring that the background correction optical path is identical with the atomic absorption light path, much of the element-specific nature inherent to flame atomic absorption spectrometry is lost. This means in effect that each type of sample and each analyte presents a separate problem of method development. This is in striking contrast to the situation with flames. The problem of matrix effects is also much more severe than in flames although the occurrence is usually specific to the particular analyte element and matrix.

Electrothermal atomization is a technique which offers major advantages in those cases where flame methods are insufficiently sensitive, where sample sizes are too small for flame methods, and in those instances where the different atomization mechanism offers freedom from an interference problem occurring in flames. Flame methods are, however, likely to remain the method of first choice for many routine analyses.

Interferences can occur as a result of carbide formation in the furnace by elements such as Hf, Mo, Nb, Th, W, and Zr, which can be minimized by the use of a tantalum furnace liner [92,93]. In contrast to the flame method, solutions are best prepared, not as chlorides, but as nitrates, because the former can lead to losses of volatile elements as chlorides [94] prior to the atomization stage and to interferences by recombination of analyte atoms to form chloride complexes in the vapor phase [95,96]. Many other interferences have been described [97-104] including the effect of H_2O_2 and various acids [105]. Interference effects may be so severe that resort may still be ncessary to a solvent extraction separation procedure. It may also be necessary, because of complex interference effects, to make use of standard addition techniques.

With carbon rod atomizers, the use of hydrogen to replace argon and form a hydrogen flame around the rod reduces the interference effects [106]. Pyrolytic carbon coated furnaces have shown not only an improved performance for elements such as molybdenum and vanadium but also a longer life [91].

VIII. SPECIAL TECHNIQUES

A. Boat Technique

Into a small boat of tantalum a small volume (less than 1 ml) of sample solution is pipetted and dried near the atomic absorption spectrometer flame before being introduced into the center of the flame. There the analyte is atomized within a few seconds, producing an atomic absorption signal in the form of a discrete pulse [107]. As the optimum temperature achieved will not exceed 1200°C this method is only suitable for easily atomized elements such as Ag, As, Bi, Cd, Hg, Pb, Se, Te,

Tl, and Zn, but for these elements improvements in detection limits of 5-50 times can be achieved. The technique can be used with solutions too viscous for pneumatic nebulizers. Compared with the graphite furnace and carbon-rod methods, somewhat larger samples can be used with the boat technique. In a development of the boat technique, the Delves cup method [108], a nickel boat is used with an open tube of nickel or silica placed over the boat and flame to increase atomic residence times and improve the sensitivity. This method has been widely applied to the determination of lead. The use of a background corrector is essential to achieve accuracy with this technique. Comparative detection limits of flame, boat, and Delves techniques are given in Welz [109].

B. Volatile Hydride Method

The technique makes use of the fact that from elements such as As, Bi, Ge, Sb, Se, Sn, and Te, gaseous covalent hydrides can be formed by reduction in solution and passed into a flame or a furnace, where atomization occurs and an atomic absorption measurement is made. The method was introduced by Holak [110] for arsenic, developed by Dalton and Malanoski [111], and optimized in a form made available commercially by Fernandez and Manning [112] and Manning [113]. In addition to Zn/HCl other reducing agents, including $Mg/TiCl_3$ [114] and $NaBH_4$ [115], have been used and the methods for preparing stable borohydride solutions described [116]. The method has also been applied to elements of special agricultural interest such as selenium [117], but interference effects from copper and nickel, particularly, are likely to be severe in soil and plant matrices [118-120]. The effects of acids used in digestion procedures (for arsenic) have been discussed, together with the effect of the oxidation state of arsenic [121]. The technique has recently been reviewed by Robbins and Caruso [364].

C. Reduction, Aeration, and Cold-Vapor Methods for Mercury Analysis

Since the detection limit for mercury by normal flame methods is about 0.5 mg $liter^{-1}$, its determination at concentrations at the μg $liter^{-1}$ level is only possible by special methods. Currently the most important of these is the cold-vapor atomic absorption technique developed by Hatch and Ott [122], based on the formation of atomic mercury in a solution by reduction, followed by aeration of the sample solution to release the mercury vapor [123,124]. A second technique widely used for geological materials volatilizes the mercury by heating the solid sample, and determines the evolved mercury directly in the vapor phase or preferably after collection and revolatilization [125,126]. The determination of mercury by such nonflame methods of atomic absorption and fluorescence spectrometry has been comprehensively reviewed [127,128] with applications to plants, rocks, soils, and other sample types. Bibliographies on mercury in the environment are available [129-131].

D. Indirect Methods

Indirect procedures have been described for phosphorus [132-134], arsenic [135], and silicon [136] determination by forming phosphomolybdate, arsenomolybdate, or silicomolybdate and determining the molybdenum content by atomic absorption spectrometry. For the determination of sulfur the interference effects of sulfates and sulfides on the determination by AAS of calcium and iron have been used [137, 138], but more commonly sulfate has been precipitated by barium and sulfate estimated by determination of the residual, or precipitated, barium [139,140]. Fluoride has been determined by making use of the effect on magnesium atomic absorption of fluoride concentration and pH [141].

E. Molecular Emission Methods

Various techniques of molecular emission spectrometry are of potential value for the determination of trace amounts of nonmetals in agricultural samples. These include the use of a cool flame such as the nitrogen/hydrogen diffusion flame for elements such as sulfur [142,143], phosphorus [144], and the halides [145]. More recently molecular emission cavity analysis (MECA) [146] has been employed not only for the determination of nonmetals such as sulfur [147] and boron [148], but has been applied in speciation studies for the determination of different sulfur anions in mixtures [149]. A recent review of the applications of MECA has been published [150].

In the MECA technique, a small (milligram) sample is injected into a cavity at the end of a stainless steel holder and introduced into a cool (ca. 300°C) hydrogen flame, which volatilizes the sample and produces molecular species such as S_2, whose molecular emission is measured by a simple spectrometer. Depending on the combinational form, detection limits of a few milligrams of sulfur can be determined with reproducibilities of 2-3%. While, in common with most molecular emission systems interference effects occur, procedures for their elimination or reduction have been evolved.

F. Miscellaneous Methods

Flame atomic absorption spectrometers have been used as metal-specific detectors in gas chromatography [151,152] and in liquid and high performance liquid chromatography [152,153]. These, and other techniques used for purposes of speciation, have recently been reviewed [154]. The graphite furnace atomizer has also been combined with gas chromatography for the measurement of volatile alkyl compounds of arsenic and selenium [155].

Various methods of analysis for microsamples by AAS have recently been reviewed [156]. Useful techniques for the atomization in flames of microsamples have been described [157-162], in most of which a discrete sample is atomized and an atomic absorption signal pulse measured.

Emission techniques of interest in agriculture include (1) the use of the band emission spectrum of boron oxide in an air/hydrogen flame for the determination of boron [163] and (2) the isotopic analysis of nitrogen by using microwave plasma emission sources [164,165].

The use of solid samples instead of solutions for direct atomization in flames has a long history, from the early work of Ramage [166], who introduced a sample in a paper spill into the flame, to that of Govindaraju et al. [167], who used a powder sample in a cup at the end of an iron rod to introduce the sample, both of these workers using atomic emission spectrometry. Various solid sampling systems for use with AAS and AFS have recently been reviewed [168] and many examples using graphite furnace atomizers have been described [169-171]. While the use of solid samples has its own special problems of matrix effects, sample inhomogeneity, and the difficulties of microsampling, some worthwhile results have been obtained.

IX. PRECONCENTRATION AND SEPARATION

The simplest way to achieve a concentration of the element of interest, and at the same time separate matrix elements which may interfere with the analysis, is to use the method of solvent extraction. The technique is discussed in depth in books by Morrison and Freiser [172] and Cresser [173]. The element is combined with an organic reagent to form a complex which can then be extracted by an organic solvent. Concentration of 10 times or more can readily be achieved and, by adjusting the conditions, usually the pH, the extraction can be made selective for the element in question. The most common procedure uses ammonium pyrrolidene dithiocarbamate (APDC) [174] as the chelating or complexing agent and methyl isobutyl ketone (MIBK) as the extracting organic solvent. Other solvents have been used but chlorinated hydrocarbons are undesirable in flames since they inhibit combustion and also form the very poisonous gas phosgene. Chlorinated hydrocarbons such as chloroform can be used safely with furnace atomizers. Silver, Bi, Cd, Co, Cr, Cu, Fe, Hg, In, Mn, Mo, Ni, Pb, Pd, V, and Zn have all been successfully extracted at various pHs and the conditions for extraction for many metals have been studied systematically [175].

100 mg ignit...

Fuse wit...

Dissolve in HNO_3 ...

Make up to 100 ml ...

Dilute 50 ml to 100 ml with 4 % HNO_3

Add 10 ml 2.4% K solution as ion suppressor

Flame N_2O/C_2H_2

Determine: Al
Ba
(Fe)
Mn
Si
Ti

Dilute 5 ml t...
$LiBO_2$ (0.35...
Add H...

Flame N_2O/C_2H_2

Determine: Fe
Mg

*Na and K by AES, remainder by AAS.
Standard solutions are single element, made up in 0.35...

Scheme 1 Scheme for the determination of 10 elemen... fusion and nitric acid dissolution using atomic abs...

[illegible] the [illegible] contents of a soil. With some rigorous digestion procedures, e.g., nitric acid/perchloric acid digestion, the determined contents will closely approach total contents with some types of sample and for some elements. Between such digestion procedures and much less vigorous extraction procedures designed to simulate plant availability, a whole range of digestion and extraction procedures have grown up in response to local conditions and for a variety of diagnostic purposes. The factors affecting the availability of trace elements in soils have been discussed by Mitchell [218]. The use and justification of the different methods of obtaining soil solutions for trace element analysis cannot, however, be considered in detail here and reference should be made to a detailed treatment [219], to an appropriate soil chemistry textbook [178,220,221], or to references surveying this subject [222-227]. Practical extraction procedures used for soils in the United Kingdom have been described, together with analytical methods which include atomic absorption spectrometry [228, 229]. Extraction procedures for clay minerals [230] and sediments [231] have recently been evaluated.

Major nutrients such as Ca, K, Mg, and Na can be assessed in topsoils of noncalcareous nature by 0.5 M acetic acid extraction, and their concentration determined by direct analysis of the extract by atomic emission or absorption spectrometry. In calcareous soils, neutral 1 M ammonium acetate extracts can be used to determine exchangeable K,

Mg, and Na concentrations, but for exchangeable calcium a different procedure, such as leaching with LiCl/Li acetate solution and determining calcium in the leachate [229], is required. Potassium and sodium are most readily determined by flame emission methods, and this allows their simultaneous measurement [232,233]. Calcium, determined by AAS or AES, suffers pronounced depressive interference by aluminum, readily extracted from soils by acetic acid and to a lesser extent by ammonium acetate, and by phosphate, although this is generally of less importance than aluminum in soils. With an air/acetylene flame an addition of lanthanum or strontium is only a palliative for aluminum interference, and for accurate calcium analysis a nitrous oxide/acetylene flame combined with the use of lanthanum or strontium is necessary. Magnesium is also affected, but to a smaller degree, and accurate analysis can be made in an air/acetylene flame by atomic absorption spectrometry, with an excess of lanthanum or strontium added to sample and standard solutions.

In Table 1 it can be seen that the quantity of certain elements in the soil extract (e.g., cadmium in EDTA extract) may exceed the quantity in the solution prepared for a total analysis, as a consequence of the large dilution (ca. 1000 times) involved in the preparation of the latter solutions. In the case of cadmium it is, in fact, possible to determine this element in most soils in EDTA extractable form directly by AAS, whereas the direct determination of total cadmium contents is difficult in uncontaminated soils. The choice of the extractant is usually made for its ability to predict plant uptake, but where this is not the prime criterion the choice of a different extractant may bring the element within range of the AAS technique. Thus the cadmium content in acetic acid extracts of many soils will be too low for direct AAS determination, and the higher content of major elements, e.g., aluminum and calcium, in this type of extract entails a greater background absorption interference. This interference, combined with the lower atomic absorption signals for cadmium, introduces greater errors than are found in EDTA extracts for this element. In addition, the use of EDTA for cadmium, like zinc, is likely to be as good as acetic acid in predicting plant availability. For Co, Cr, and Ni, acetic acid is the usual extractant and useful measurements can be made directly on the extract, although for some soils the levels of these elements may be beyond the reach of AAS unless a small concentration is made.

At the Macaulay Institute, Cu, Mn, and Zn in EDTA extracts are routinely determined using a separated air/acetylene flame [39], with the minimum soil concentrations determined being respectively 0.05, 0.1, and 0.05 $\mu g\ ml^{-1}$. After long periods of operation (2 hr) with EDTA extracts, the burner slot may become partially salted up and require cleaning before further analysis. Multislot burners of the Boling type [35] are more tolerant of such high dissolved solid contents, and can be used to advantage in this situation with no loss of sensitivity. Acetic acid extracts can also be analyzed for these elements in the air/acetylene flame, although in the case of copper the extract may require concentration by a factor of 5-10 times before analysis. This can be performed by evaporating to dryness the 800 ml extract, followed by treatment of the residue with HNO_3 and HCl, before dissolving in 100 ml of 0.05 M HCl.

Lead, Ti, and V can all be measured in EDTA extracts. For most of the other elements in Table 1 flame methods are insufficiently sensitive, so preconcentration procedures (Sec. IX), special techniques (Sec. VIII), or electrothermal atomization techniques (Sec. VII) must be followed.

These examples are only for illustration, and for details of these and other analyses in different extracts the references in Sec. XI, the annotated bibliography of applications of AAS in the agricultural field [234], or the general works of reference, of which a few are listed in Sec. XII, should be consulted.

XI. APPLICATIONS TO THE DETERMINATION OF TRACE ELEMENTS IN SOILS AND RELATED MATERIALS, AND SOME RECENT CONTRIBUTIONS TO THE METHODOLOGY*

A. Abbreviations Used for Reagents

EDTA	Ethylenediaminetetraacetic acid
APDC	Ammonium pyrrolidinedithiocarbamate
MIBK	Methyl isobutyl ketone
NH_4Ac	Ammonium acetate
DTPA	Diethylenetriaminepentaacetic acid
NaDDC	Sodium diethyldithiocarbamate
TOTP	Tri-iso-octyl thiophosphate

B. Determination of Individual and Groups of Elements

Arsenic

Arsenic has been determined in $HClO_4$ digests of soils by the volatile hydride (AsH_3) procedure using Ar/H_2 [235] or N_2/H_2 [236] flames, by direct vaporization of AsH_3 from soils into an air/C_2H_2 flame [237], and by the hydride method using an electrically heated quartz tube furnace in place of a flame atomizer [238,239]. Benzene extracts of soil [240] have also been used. A comparative and detailed study has been made of arsenic determination by the hydride method and electro-

*References to work published 1971 et seq. For earlier work see Kalra [234].

thermal atomization in a graphite furnace [241]. Interference effects in the hydride and graphite furnace methods have been examined for arsenic and selenium [242].

Antimony

Total antimony has been determined using an air/acetylene flame after fusion with NH_4I and dissolution in HCl [243]. The hydride method has been used following a HNO_3/H_2SO_4 digestion [365].

Beryllium

A sodium carbonate fusion and HCl dissolution technique followed by a solvent extraction procedure [244] has been used for the determination of beryllium in a N_2O/C_2H_2 flame.

Boron

Only a few attempts have been made to determine boron by AAS [245] because of the technique's limited sensitivity, although a recent extraction procedure [246] makes it more feasible. The analysis of boron by a variety of methods, including spectrochemical ones, has recently been extensively reviewed [247].

Cadmium

Cadmium has been determined in water and acid digests or extracts of soils directly [248-250,366]. Lower detection limits and greater freedom from matrix and background effects in the flame are obtained with concentration by solvent extraction using dithizone [251]. The health hazard associated with the combustion of the chlorinated organic solvent $CHCl_3$ or CCl_4 is eliminated by back extraction from CCl_4 into aqueous medium by HCl [252]. APDC/MIBK solvent extraction methods have been used, with the organic phase analyzed by flame [253] or graphite furnace [254,367] atomizers. Electrothermal atomization has also been used directly for soils [368] and for dilute acid or ammonium acetate extracts of soils [255]. The carbon-rod atomizer has been used for AAS and AFS determination of cadmium in dithizone/$CHCl_3$ concentrates of acetic acid extracts of soils [78,256,257]. An AAS method has been compared and found in good agreement with a NAA method [258]. Sorption of cadmium (and lead) by clay minerals has been studied [259].

Chromium

Chromium contents of soils have been determined in the air/acetylene flame by AAS in HF/H_2SO_4 digests [260] and in HNO_3 [261], HCl [262], and NH_4Ac/KCl extracts [263]. The use of $HClO_4$ treatments is inadvisable for chromium because of the volatility of its oxychloride.

Cobalt

Cobalt can be determined by flame AAS directly in HF [264] or HCl/HNO_3 digests [265] or after concentration by evaporation of acetic acid extracts [266]. It has been determined in various extracts by solvent extraction procedures using APDC/MIBK [267], or cupferron/MIBK [263] and in MIBK or amyl acetate extracts of soil treated with ammonium hexamethylene dithiocarbamate [268]. Soil extracts concentrated by APDC/MIBK have been analyzed for cobalt by graphite furnace [269].

Copper

Since copper is to a large extent concentrated in the organic fractions of topsoils, oxidative digestion procedures can give a reasonably good estimate of total copper content in such cases; even dry ashing at 500°C followed by dissolution in HNO_3 has been used with ICP AES determination [270]. Acid/perchloric acid mixtures have been widely used with flame techniques [253,271-274] and with ICP atomic emission methods [275]. Digestion with HNO_3/HCl followed by treatment with HF has been used for flame determination of Cu (and Zn) [369]. Other procedures for total copper by flame methods include HNO_3/HCl digestion with an APDC/MIBK concentration [276] and HNO_3/H_2SO_4 digestion [277]. Total copper in peat has been determined following $HClO_4$ digestion and concentration by APDC/MIBK [278].

Extractable copper contents have been determined by flame methods in water [279,280], NH_4Ac [229,280], NH_4Ac/KCl [263], LiCl/LiAc [229], EDTA [39,229,281,282], HCl [280,283], HNO_3 [284], and in various other extracts [272,282,285-287], in some cases combined with a subsequent concentration by solvent extraction [253,288] or by an ion-exchange technique [289].

Lead

Total lead contents have been determined in HNO_3 [290,291], HF/HNO_3 [292,293], HF/$HClO_4$ [370], HNO_3/HCl [276], and $HNO_3/HClO_4$ [253] by air/C_2H_2 flame AAS, but in many cases a concentration/separation procedure is required not only for reasons of sensitivity but also because the short wavelength makes it subject, like cadmium, to background interference. Solvent extraction procedures using APDC/MIBK with citrate masking of iron [267] or NaDDC/MIBK concentration with prior removal of iron [294] have all been used. For sediments, HCl/HNO_3 [265] and HF/HCl/$HClO_4$ [295] digestions have been used with the air/C_2H_2 flame.

Extractable lead has been determined in acetic acid [291,296,297], dilute HCl [298], dilute HNO_3 [261], dilute HCl/H_2SO_4 [255], NH_4Ac [255,299], hydroxylamine hydrochloride in HNO_3 [288], and water [illegible] Manganese has also been determined in the soil solution [illegible].

Mercury

The literature on the determination of mercury by various flameless techniques is so extensive that no attempt will be made here to summarize it. Readers are referred to reviews and bibliographies [80, [illegible], 129-131] on the subject and to the outline given in Sec. VIII.C.

Molybdenum

For total molybdenum, soil solutions are prepared by digestion with $HF/HClO_4$ [315] or HF/HCl [316] and will often require the use of a concentration procedure [317] such as an MIBK [315] or Aliquat 336 [318] extraction of the thiocyanate complex, or an extraction of phosphomolybdate by Aliquat 336 in MIBK [319]. Soil extracts of molybdenum by NH_4Ac are seldom within the range of flame AAS, even with solvent extraction procedures. Molybdenum is not easy to determine with the graphite furnace because of the high temperatures required for atomization and the existence of memory effects.

Nickel

Nickel in soils has been of less interest agriculturally than cobalt, but with the increasing impact of sewage sludge application to agricultural land and the presence in some of these sludges of potentially toxic amounts of nickel, the demand for its analysis is likely to increase. It has been determined in HNO_3/HCl digestions of sediments [265] in

Strontium

Although there have been few recent references to strontium analysis in soils, methods for the analysis of soil profiles [322], NH_4Ac extracts [325], and soils reclaimed after contamination with ^{90}Sr [326] have been described.

Zinc

Total zinc contents of soils have been determined by AAS in flames following treatment with $HF/HNO_3/HClO_4$ [271,327], $HF/HNO_3/H_2SO_4/HClO_4$ [328], $H_2SO_4/HNO_3/HClO_4$ [329], $HF/H_2SO_4/HNO_3$ [245], and treatments with HF [264] or $HClO_4$ [330]. Other analyses of soils [253,308,331,332], sediments [265,302,333,334], and mine spoil [311] have been made using similar treatments. Although in general the sensitivity by flame AAS methods is more than adequate and interferences are controllable, an APDC/MIBK concentration procedure [276], nondispersive AFS [335], AES using an ICP [272], and flameless methods [330] have been employed.

Extractable zinc can readily be determined with a variety of extracting reagents, including water with an APDC/MIBK preconcentration stage [336], and without one [279,280]. Also used have been HCl [280,282,283,328,337], HAc [297,338], EDTA [282,328], and other or similar treatments [252,261,263,272,288,339]. An argon-separated air/C_2H_2 flame and AFS [57], and a mechanically separated air/C_2H_2 flame, have also been used for extracts of soils [39].

41. V. Sychra, V. Svoboda, and I. Rubeska, *Atomic Fluorescence Spectroscopy*, Van Nostrand, London, 1975.
42. C. Veillon, J. M. Mansfield, M. L. Parsons, and J. D. Winefordner, Anal. Chem., *38*:204 (1966).
43. H. L. Kahn and J. E. Schallis, At. Absorpt. Newsl., *7*:5 (1968).
44. R. M. Dagnall, K. C. Thompson, and T. S. West, At. Absorpt. Newsl., *6*:117 (1967).
45. L. R. P. Butler and A. Fulton, Appl. Opt., *7*:2131 (1968).
46. A. A. Venghiattis, Spectrochim. Acta, *23B*:67 (1967).
47. M. Kashiki and S. Oshima, Anal. Chim. Acta, *51*:387 (1970).
48. W. W. Harrison and P. O. Juliano, Anal. Chem., *43*:248 (1971).
49. Pye Unicam Anal. Bull., *1*:4 (1977).
50. W. H. Newman and W. Prinz, Z. Anal. Chem., *289*:17 (1978).
51. S. Slavin and W. Slavin, At. Absorpt. Newsl., *5*:106 (1966).
52. J. E. Poldoski, Anal. Chem., *49*:891 (1977).
53. R. O. Scott and A. M. Ure, Proc. Soc. Anal. Chem., *9*:288 (1972).
54. T. Harada, K. Fujiwara, and K. Fuwa, Bunseki Kagaku, *26*:877 (1977).
55. K. W. Busch and G. H. Morrison, Anal. Chem., *45*:712A (1973).
56. B. L. Vallee and M. Margoshes, Anal. Chem., *28*:175 (1956).
57. R. M. Dagnall, G. F. Kirkbright, T. S. West, and R. Wood, Anal. Chem., *43*:1765 (1971).
58. R. F. Browner, Analyst (Lond.), *99*:617 (1974).
59. M. L. Berrow and J. C. Burridge, in *Inorganic Pollution in Agriculture*, MAFF Tech. Bull. GFM3, HMSO, London, 1979.
60. N. W. Bower and J. D. Ingle, Jr., Anal. Chem., *49*:574 (1977).
61. J. T. H. Roos, Spectrochim. Acta, *28B*:407 (1973).
62. R. L. Brunelle, C. M. Hoffman, K. B. Snow, and M. J. Pro, J. Assoc. Off. Anal. Chem., *52*:911 (1969).
63. J. Ramirez-Munoz, *Atomic Absorption Spectroscopy*, Elsevier, Amsterdam, 1968, p. 328.
64. A. M. Ure and R. L. Mitchell, in *Flame Emission and Atomic Absorption Spectroscopy*, Vol. 3, (J. A. Deans and T. C. Rains, Eds.), Marcel Dekker, New York, 1975, p. 1.
65. H. Prugger and R. Torge, German Patent 1,964,469, filed December 23, 1969.
66. Y. Uchida and S. Hattori, Oyo Butsuri, *44*:852 (1975).
67. S. D. Brown, Anal. Chem., *49*:1269A (1977).
68. E. Grassam, J. B. Dawson, and D. J. Ellis, Analyst (Lond.), *102*:804 (1977).
69. H. Koizumi, K. Yasuda, and M. Katayama, Anal. Chem., *49*:1106 (1977).
70. B. V. L'vov, Ing. Fiz. Zhur., *11*:44 (1959).
71. B. V. L'vov, Spectrochim. Acta, *17*:761 (1961).

72. B. V. L'vov, *Atomic Absorption Spectrochemical Analysis*, Adam Hilger, London, 1970.
73. H. Massmann, Z. Anal. Chem., *225*:203 (1967).
74. H. Massmann, Spectrochim. Acta, *23B*:215 (1968).
75. R. Woodriff and G. Ramelow, Spectrochim, Acta, *23B*:665 (1968).
76. C. Weillon, in *Trace Analysis. Spectroscopic Methods for Elements* (J. D. Winefordner, Ed.), Wiley, New York, 1976, p. 155.
77. T. S. West and X. K. Williams, Anal. Chim. Acta, *45*:27 (1969).
78. A. M. Ure and M. P. Hernandez-Artiga, Anal. Chim. Acta, *94*:195 (1977).
79. S. R. Koirtyohann and C. Feldman, in *Developments in Applied Spectroscopy*, Vol. 3, Plenum Press, New York, 1964, p. 180.
80. G. F. Kirkbright, Analyst (Lond.), *96*:609 (1971).
81. A. Syty, Crit. Rev. Anal. Chem., *4*:155 (1974).
82. R. A. Woodriff, Appl. Spectrosc., *28*:413 (1974).
83. C. W. Fuller, *Electrothermal Atomization for Atomic Absorption Spectrophotometry*, Chem. Soc., London, 1977.
84. G. M. Hieftje, T. R. Copeland, and D. R. de Olivares, Anal. Chem., *48*:142R (1976).
85. J. Y. Hwang, C. J. Mokeler, and P. A. Ullucci, Anal. Chem., *44*:2018 (1972).
86. J. E. Cantle and T. S. West, Talanta, *20*:459 (1973).
87. N. T. Faithfull, Lab. Practice, *26*:946 (1977).
88. N. T. Faithfull, Lab. Practice, *27*:25 (1978).
89. N. T. Faithfull, Lab. Practice, *27*:26 (1978).
90. R. E. Sturgeon, Anal. Chem., *49*:1255A (1977).
91. R. E. Sturgeon and C. L. Chakrabarti, Prog. Anal. Atomic Spectrosc., *1*:5 (1978).
92. G. D. Renshaw, C. A. Pounds, and E. F. Pearson, At. Absorpt. Newsl., *12*:55 (1973).
93. B. V. L'vov and L. A. Pelieva, Can. J. Spectrosc., *23*:1 (1978).
94. D. A. Segar and J. G. Gonzalez, Anal. Chim. Acta, *58*:7 (1972).
95. M. Binnewies and H. Schafer, Z. Anorg. Allg. Chem., *395*:77 (1973).
96. A. Katz and N. Taitel, Talanta, *24*:132 (1977).
97. J. M. Ottaway, Proc. Anal. Div. Chem. Soc., *13*:185 (1976).
98. W. Frech and A. Cedergren, Anal. Chim. Acta, *82*:83 (1976).
99. W. Frech and A. Cedergren, Anal. Chim. Acta, *82*:93 (1976).
100. W. Frech and A. Cedergren, Anal. Chim. Acta, *88*:57 (1977).
101. J. A. Persson, W. Frech, and A. Cedergren, Anal. Chim. Acta, *92*:85 (1977).
102. J. A. Persson, W. Frech, and A. Cedergren, Anal. Chim. Acta, *92*:95 (1977).
103. K. Johansson, W. Frech, and A. Cedergren, Anal. Chim. Acta, *94*:245 (1977).

104. C. W. Fuller, At. Absorpt. Newsl., *16*;106 (1977).
105. J. F. Alder and D. A. Hickman, At. Absorpt. Newsl., *16*;110 (1977).
106. M. D. Amos, P. A. Bennett, K. G. Brodie, P. W. Y. Lung, and J. P. Matousek, Anal. Chem., *43*;211 (1971).
107. H. L. Kahn, G. E. Peterson, and J. E. Schallis, At. Absorpt. Newsl., *7*;35 (1968).
108. H. T. Delves, Analyst (Lond.), *95*;431 (1970).
109. B. Welz, *Atomic Absorption Spectroscopy*, English ed., Verlag Chemie, Weinheim, 1976, p. 42.
110. W. Holak, Anal. Chem., *41*;1712 (1969).
111. E. F. Dalton and A. J. Malanoski, At. Absorpt. Newsl., *10*;92 (1971).
112. F. J. Fernandez and D. C. Manning, At. Absorpt. Newsl., *10*:86 (1971).
113. D. C. Manning, At. Absorpt. Newsl., *10*;123 (1971).
114. E. N. Pollock and S. J. West, At. Absorpt. Newsl., *11*:104 (1972).
115. F. J. Schmidt and J. L. Royer, Anal. Lett., *6*:17 (1973).
116. J. R. Knechtel and J. L. Fraser, Analyst (Lond.), *103*:104 (1978).
117. Y. Yamamoto, T. Kumamaru, Y. Hayashi, and M. Kanke, Anal. Lett., *5*:717 (1972).
118. M. Bedard and J. D. Kerbyson, Can. J. Spectrosc., *21*:64 (1975).
119. J. Guimont, M. Pichette, and N. Rheaume, At. Absorpt. Newsl., *16*:53 (1977).
120. G. F. Kirkbright and M. Taddia, Anal. Chim. Acta, *100*:145 (1978).
121. H. K. Kang and J. L. Valentine, Anal. Chem., *49*:1829 (1977).
122. W. R. Hatch and W. L. Ott, Anal. Chem., *40*:2085 (1968).
123. Y. Kimura and V. L. Miller, Anal. Chim. Acta, *27*:325 (1962).
124. N. S. Poluektov and R. A. Vitkun, Zh. Anal. Khim., *18*:37 (1963).
125. W. W. Vaughan and J. H. McCarthy, Jr., *U.S. Geol. Surv. Prof. Paper 501D*, 123 (1964).
126. S. H. Williston and M. H. Morris, U.S. Patent No. 3,173,016 (1965).
127. S. Chilov, Talanta, *22*:205 (1975).
128. A. M. Ure, Anal. Chim. Acta., *76*:1 (1975).
129. M. E. Barnes (Ed.), *Mercury Contamination in the Natural Environment*, PB 192910, U.S. Dept. of the Interior, Washington, D.C., 1970.
130. D. Taylor, *Mercury as an Environmental Pollutant. A Bibliography*, 3rd ed., ICI Brixham Laboratory, England, 1973.
131. E. A. Jenne and W. Sanders, J. Water Pollut. Contr. Fed., *45*:1952 (1973).

132. W. S. Zaugg and R. J. Knox, Anal. Chem., *38*:1759 (1966).
133. W. S. Zaugg and R. J. Knox, Anal. Biochem., *20*:282 (1967).
134. T. R. Hurford and D. F. Boltz, Anal. Chem., *40*:379 (1968).
135. S. S. Michael, Anal. Chem., *49*:451 (1977).
136. C. Riddle and A. Turek, Anal. Chim. Acta, *92*:49 (1977).
137. G. D. Christian and F. J. Feldman, Anal. Chim. Acta, *40*:173 (1968).
138. M. Kunishi and S. Ohno, At. Absorpt. Newsl., *13*:29 (1974).
139. R. H. Loeppert and H. L. Breland, Proc. Soil Crop Sci. Soc. Florida, *32*:145 (1973).
140. O. C. Bataglia, Cienc. Cult. (Sao Paulo), *28*:672 (1976).
141. C. C. Fongi and C. O. Huber, Spectrochim. Acta, *31B*:113 (1976).
142. R. M. Dagnall, K. C. Thompson, and T. S. West, Analyst (Lond.), *92*:506 (1967).
143. K. M. Aldous, R. M. Dagnall, and T. S. West, Analyst (Lond.), *95*:417 (1970).
144. R. M. Dagnall, K. C. Thompson, and T. S. West, Analyst (Lond.), *93*:72 (1968).
145. R. M. Dagnall, K. C. Thompson, and T. S. West, Analyst (Lond.), *94*:643 (1969).
146. R. Belcher, Z. Anal. Chem., *263*:257 (1973).
147. R. Belcher, S. L. Bogdanski, and A. Townshend, Anal. Chim. Acta, *67*:1 (1973).
148. R. Belcher, S. A. Ghonaim, and A. Townshen, Anal. Chim. Acta, *71*:255 (1974).
149. M. Q. Al-Abachi, R. Belcher, S. L. Bogdanski, and A. Townshend, Anal. Chim. Acta, *86*:139 (1976).
150. A. Townshend, Proc. Anal. Div. Chem. Soc., *13*:64 (1976).
151. B. Kolb, G. Kemmner, F. H. Schleser, and E. Wiedcking, Z. Anal. Chem., *221*:166 (1966).
152. J. C. Van Loon, B. Radziuk, N. Kahn, J. Lichwa, F. J. Fernandez, and J. D. Kerber, At. Absorpt. Newsl., *16*:79 (1977).
153. S. E. Manahan and D. R. Jones IV, Anal. Lett., *6*:745 (1973).
154. F. J. Fernandez, At. Absorpt. Newsl., *16*:33 (1977).
155. G. E. Parris, W. R. Blair, and F. G. Brinckman, Anal. Chem., *49*:378 (1977).
156. W. J. Price, Chem. in Britain, *14*:140 (1978).
157. W. Slavin, *Atomic Absorption Spectroscopy*, Interscience, New York, 1968, p. 16.
158. M. S. Cresser, Anal. Chim. Acta, *80*:170 (1975).
159. D. C. Manning, At. Absorpt. Newsl., *14*:99 (1975).
160. P. D. Goulden, At. Absorpt. Newsl., *16*:121 (1977).
161. G. A. Eagle and M. J. Orren, At. Absorpt. Newsl., *16*:151 (1977).

[illegible]

[illegible] E. J. [illegible] Absorpt. News[illegible]

180. [illegible] Absorpt. Newsl. [illegible] (1964).

181. C. B. Beit, Jr. Econ. [illegible] 59:240 (1964).

182. I. Rubeska and B. Moldan, Acta. Chim. Acad. Sci. Hung., *44*:367 (1965).

183. L. R. Hossner and L. W. Ferrara, At. Absorpt. Newsl., *6*:71 (1967).

184. R. A. Isaac and J. D. Kerber, in *Instrumental Methods for the Analysis of Soils and Plant Tissue* (L. M. Walsh, Ed.), Soil Sci. Soc. Am., Madison, Wis., 1971, p. 17.

185. E. J. Perkins, J. R. S. Gilchrist, O. J. Abbot, and W. Halcrow, Mar. Pollut. Bull., *4*:59 (1973).

186. J. Bolton, J. Sci. Fd. Agric., *24*:727 (1973).

187. H. Erlenkeuser, E. Suess, and H. Willkomm, Geochim. Cosmochim. Acta, *38*:823 (1974).

188. V. I. Shcherbakov, A. V. Karyakin, L. N. Bannykh, and V. I. Lebedev, Zavod. Lab., *43*:957 (1977).

...15 (19..).
208. P. Kannaker and T. C. Hug... ...49:1485 (1977).
209. Department of the Environment, *Analysis of Raw, Potable and Waste Waters*, HMSO, London, ...
210. L. Wiklander and K. Vahtras, Geoderma, *19*:123 (1977).
211. A. Rosopulo and W. Scholl, Landwirtsch. Forsch., *31*:74 (1978).
212. R. Knechtel, K. Conn, and J. Fraser, *The Analysis of Chemical Digester Sludges for Metals by Several Laboratory Groups*, Technology Development Report EPS4-WP-78-1, Water Pollution Control Directorate, Environmental Protection Service, Ottawa, 1978.
213. P. J. Ballinger, Science & Industry (Philips), *8*:12 (1976).
214. J. C. Van Loon, J. Lichwa, D. Ruttan, and J. Kinrade, Water Air Soil Pollut., *2*:473 (1973).
215. J. C. Van Loon, *Heavy Metals in Agricultural Lands Receiving Chemical Sewage, Sludges IV, Analytical Methods for Sewage Sludge Analysis*, Environmental Protection Service, Ottawa, 1976.

267. A. Sapek, Chem. Anal. (Warsaw), *19*:687 (1974).
268. A. I. Busev, V. M. Byrko, L. A. Lerner, and V. I. Migunova, Zh. Analit. Khim., *27*:607 (1972).
269. M. J. Dudas, At. Absorpt. Newsl., *13*:67 (1974).
270. R. H. Scott and A. Strasheim, Anal. Chim. Acta, *76*:71 (1975).
271. J. Masui, S. Shoji, and K. Minami, Soil Sci. Plant Nutr., Tokyo, *18*:31 (1972).
272. F. D. Macias, Soil Sci., *115*:276 (1973).
273. H. Karim, J. E. Sedberry, and B. J. Miller, Commun. Soil Sci. Plant Anal., *7*:437 (1976).
274. D. E. Pride, J. Geochem. Explor., *7*:361 (1977).
275. R. H. Scott and M. L. Kokot, Anal. Chim. Acta, *75*:257 (1975).
276. R. Horton and J. J. Lynch, Geol. Surv. Can., Paper No. 75-1, Pt. A, *213* (1975).
277. B. D. Balraadjsing, Inst. Bodemvruchtbaarheid Rapp., No. *12* (1972).
278. A. Sapek and B. Sapek, Chem. Anal. (Warsaw), *17*:339 (1972).
279. W. H. Ko and F. K. Hora, Soil Sci., *113*:42 (1972).
280. S. J. Locascio, J. G. A. Fiskell, and F. G. Martin, J. Am. Soc. Hort. Sci., *97*:119 (1972).
281. A. Oeien and K. Gjerdingen, Acta. Agric. Scand., *22*:173 (1972).
282. M. Facchinetti, R. L. Grassi, and A. L. Diez, Agrochimica, *17*:413 (1973).
283. B. R. Singh and K. Steenberg, Plant Soil, *40*:637 (1974).
284. M. E. Varju and E. Elek, At. Absorpt. Newsl., *10*:128 (1971).
285. A. Olumuyina, E. E. Osiname, F. E. Schulte, and R. B. Corey, J. Sci. Fd. Agric., *24*:1341 (1973).
286. R. G. McLaren and D. V. Crawford, J. Soil Sci., *24*:172 (1973).
287. G. W. Wallingford, L. S. Murphy, W. L. Powers, and H. L. Manges, Soil Sci. Soc. Am. Proc., *39*:482 (1975).
288. T. T. Chao and R. F. Sanzolone, J. Res. U.S. Geol. Surv., *1*:681 (1973).
289. Zs. Horvath, K. Falb, and M. Varju, At. Absorpt. Newsl., *16*:152 (1977).
290. J. L. Seeley, D. Dick, J. H. Arvik, R. L. Zimdahl, and R. K. Skogerboe, Appl. Spectrosc., *26*:456 (1972).
291. D. D. Hemphill, C. J. Marienfeld, R. S. Reddy, and J. O. Pierce, Arch. Environ. Health, *28*:190 (1974).
292. N. I. Ward, R. D. Reeves, and R. R. Brooks, N.Z. J. Sci., *18*:261 (1975).
293. N. I. Ward, R. R. Brooks, and R. D. Reeves, N.Z. J. Sci., *19*:81 (1976).
294. B. D. Balraadjsing, Commun. Soil Sci. Plant Anal., *5*:25 (1974).

295. E. A. Crecelius and D. Z. Piper, Environ. Sci. Technol., *7*: 1053 (1973).
296. D. Briggs, Nature (Lond.), *238*:166 (1972).
297. F. Beavington, Aust. J. Soil Res., *11*:27 (1973).
298. L. D. Jordan and D. J. Hogan, N.Z. J. Sci., *18*:253 (1975).
299. J. V. Lagerwerff and D. L. Brower, Soil Sci. Soc. Am. Proc., *37*:11 (1973).
300. S. Ravikovitch and J. Navrot, Soil Sci., *121*:25 (1976).
301. D. H. Yaalon, C. Jungreis and H. Koyumdjisky, Geoderma, *7*:71 (1972).
302. M. S. Cresser and R. Hargitt, Anal. Chim. Acta, *82*:203 (1976).
303. R. J. Bartlett and C. J. Picarelli, Soil Sci., *116*:77 (1973).
304. A. Siman, F. W. Cradock, and A. W. Hudson, Plant Soil, *41*:129 (1974).
305. R. F. Farley and A. P. Draycott, Plant Soil, *38*:235 (1973).
306. O. G. Oniani and N. V. Egorashvili, Agrokhimiya, *9*:137 (1974).
307. L. B. Grass, A. J. Mackenzie, B. D. Meek, and W. F. Spencer, Soil Sci. Soc. Am. Proc., *37*:14 (1973).
308. A. Kadow, E. Rabban, and I. Rubeska, Inst. Appl. Res. Natur. Resour. Abu-Ghraib Iraq Tech. Rep., No. *2* (1970).
309. D. S. Cronan, Can. J. Earth Sci., *9*:319 (1972).
310. F. J. Dewis, A. A. Levinson, and P. Bagliss, Geochim. Cosmochim. Acta, *36*:1359 (1972).
311. P. B. Hoyt and M. D. Webber, Can. J. Soil Sci., *54*:53 (1974).
312. U. K. Misra, R. K. Blanchar, and W. J. Upchurch, Soil Sci. Soc. Am. Proc., *38*:897 (1974).
313. R. J. Gibbs, Science, *180*:73 (1973).
314. S. K. Yoon, J. T. Gilmour, and B. R. Wells, Soil Sci. Soc. Am. Proc., *39*:685 (1975).
315. C. H. Kim, C. M. Owens, and L. E. Smythe, Talanta, *21*:445 (1974).
316. J. C. Van Loon, At. Absorpt. Newsl., *11*:60 (1972).
317. D. Hutchinson, Analyst (Lond.), *97*:118 (1972).
318. C. H. Kim, P. W. Alexander, and L. E. Smythe, Talanta, *23*:229 (1976).
319. P. Dharma Rao, At. Absorpt. Newsl., *10*:118 (1971).
320. H. F. Massey, Soil Sci., *114*:217 (1972).
321. A. J. Anderson, D. R. Meyer, and F. K. Mayer, Aust. J. Agric. Res., *24*:557 (1973).
322. L. W. Murdock and C. I. Rich, Soil Sci. Soc. Am. Proc., *36*:167 (1972).
323. A. A. R. Hafey and P. P. Stout, Soil Sci. Soc. Am. Proc., *37*:572 (1973).
324. T. T. Chao, J. W. Ball, and J. M. Nakagawa, Anal. Chim. Acta, *54*:77 (1971).

325. J. M. Soileau, Agron. J., *65*:625 (1973).
326. J. V. Lagerwerff and W. D. Kemper, Soil Sci. Soc. Am. Proc., *39*:1077 (1975).
327. M. K. John, Soil Sci., *113*:222 (1972).
328. M. L. Harimho and K. Igue, Agron. J., *64*:3 (1972).
329. B. D. Balraadjsing, Inst. Bodemvruchtbaarheid Rapp., No. *11* (1972).
330. D. D. Siemer and R. W. Stone, Appl. Spectrosc., *29*:240 (1975).
331. D. D. Warncke and S. Barbir, Soil Sci. Soc. Am. Proc., *36*:39 (1972).
332. M. P. Ireland, J. Soil Sci., *26*:313 (1975).
333. A. S. G. Jones, Mar. Geol., *14*:M1 (1973).
334. G. W. Bryan and L. G. Hummerstone, J. Mar. Biol. Ass. U.K., *53*:839 (1973).
335. J. D. Norris and T. S. West, Anal. Chim. Acta, *71*:289 (1974).
336. K. G. Tiller, M. P. C. De Vries, and J. L. Honeysett, Aust. J. Soil Res., *10*:151 (1972).
337. G. A. Selevtsova, L. A. Volodina, M. A. Knyazeva, and I. A. Yanchuk, Khim Sel'sk. Khoz., *13*:755 (1975).
338. D. Purves and E. J. Mackenzie, Plant Soil, *39*:361 (1973).
339. W. H. McKee, Soil Sci. Soc. Am. J., *40*:586 (1976).
340. H. Nishita and R. M. Haug, Soil Sci., *118*:421 (1974).
341. J. G. A. Fiskell and D. V. Calvert, Soil Sci., *120*:132 (1975).
342. U. K. Misra, W. J. Upchurch, and C. E. Marshall, Soil Sci., *121*:323 (1976).
343. M. L. Tsap, Z. V. Proskura, S. P. Sheredko, and N. S. Gedz., Agrokhimiya, *5*:150 (1976).
344. L. F. Elliott, T. A. Travis, and T. M. McCalla, Soil Sci. Soc. Am. J., *40*:513 (1976).
345. J. R. Jorgenson and C. G. Wells, Plant Soil, 39:373 (1973).
346. S. Payette and P. Morisset, Soil Sci., *117*:352 (1974).
347. R. M. Carlson and J. R. Buchanan, Soil Sci. Soc. Am. Proc., *38*:113 (1974).
348. E. A. Jenne, J. W. Ball, and C. Simpson, J. Environ. Qual., *3*:281 (1974).
349. P. S. Sidhu, J. L. Seghal, M. K. Sinha, and N. S. Randhawa, Geoderma, *18*:241 (1977).
350. M. D. Webber, P. B. Hoyt, M. Nyborg, and D. Corneau, Can. J. Soil Sci., *57*:361 (1977).
351. J. L. McIntosh and K. E. Varney, Agron. J., *65*:629 (1973).
352. P. L. Searle and B. K. Daley, Geoderma, *19*:1 (1977).
353. M. J. Dudas and S. Pawluk, Can. J. Soil Sci., *57*:329 (1977).
354. S. Lian and A. Tanaka, Soil Sci. Plant Nutr., Tokyo, *18*:15 (1972).
355. D. J. Carmody, J. B. Pearce, and W. E. Yasso, Mar. Pollut. Bull., *4*:132 (1973).

356. J. A. C. Fortescue, S. A. Curtis, and E. Gawron, Can. J. Spectrosc., *18*:23 (1973).
357. D. E. Baker, Soil Sci. Soc. Am. Proc., *37*:537 (1973).
358. S. M. Griffith and M. Schnitzer, Soil Sci., *120*:126 (1975).
359. E. A. Forbes, N.Z. J. Agric. Res., *19*:153 (1976).
360. J. Yan, Ti (Ch'iu) Hua Hsueh, No. *4*, p. 291 (1975).
361. S. Yamasaki, A. Yoshino, and A. Kishita, Soil Sci. Plant Nutr., Tokyo, *21*:63 (1975).
362. K. Yasudo, H. Koizumi, K. Onishi, and T. Noda, Prog. Anal. Atom. Spectrosc., *3*:299 (1980).
363. I. Rubeska and J. Musil, Prog. Anal. Atom. Spectrosc., *2*:309 (1979).
364. W. B. Robbins and J. A. Caruso, Anal. Chem., *51*:889A (1979).
365. P. N. Vijan, Am. Lab., No. *8*:32 (1979).
366. N. I. Ward, R. R. Brooks, and E. Roberts, N.Z. J. Sci., *20*:413 (1977).
367. W. Schmidt and F. Dietl, Z. Anal. Chem., *295*:110 (1979).
368. V. N. Oreshkin, Y. I. Belyaev, G. L. Vnukovskaya, and Y. G. Tatzil, Pochvovedenie, No. 5, p. 109 (1979).
369. L. M. Shuman, Soil Sci., *127*:10 (1979).
370. L. Faitondzhiev and A. Brashnarova, Landwirtsch. Forsch., *32*:1 (1979).
371. J. Azad, G. F. Kirkbright, and R. D. Snook, Analyst (Lond.), *104*:232 (1979).
372. K. L. Liu, I. D. Pulford, and H. J. Duncan, Anal. Chim. Acta, *106*:319 (1979).

BIBLIOGRAPHY OF USEFUL REFERENCE WORKS

W. H. Allaway, "The Trace Elements in Biological Systems" in *Trace Analysis. Physical Methods* (G. H. Morrison, Ed.), Interscience, New York, 1965, pp. 67-102.

G. D. Christian and F. J. Feldman, *Atomic Absorption Spectroscopy Applications in Agriculture, Biology and Medicine*, Wiley-Interscience, New York, 1970.

D. J. David, Analysis of Soils, Plants, Fertilizers and Other Agricultural Materials, Prog. Anal. Atom. Spectrosc., *1*:225 (1978).

J. A. Dean and T. C. Rains, *Flame Emission and Atomic Absorption Spectrometry*, vol. 1, *Theory*, 1969; vol. 2, *Components and Techniques*, 1971; vol. 3, *Elements and Matrices*, 1975, Marcel Dekker, New York.

J. D. Donaldson, Ed., *Atomic Absorption and Emission Spectroscopy Abstracts*, D. H. Masek, PRM Science and Technology Agency Ltd., North Finchley, London.

J. B. Dawson, Ed., *Annual Reports on Analytical Atomic Spectroscopy*, Chemical Society, London. Issued annually; comprehensively reviews the literature for the previous year.

R. A. Isaac and J. D. Kerber, Atomic Absorption and Flame Photometry: Techniques and uses in Soil, Plant and Water Analysis, in *Instrumental Methods for Analysis of Soils and Plant Tissue* (L. M. Walsh, Ed.), Soil Sci. Soc. Am., Madison, Wis., 1971, pp. 17-37.

Y. P. Kalra, *Application of Atomic Absorption Spectrophotometry in Soil, Plant, Water and Fertilizer Analysis: An Annotated Bibliography*, Part I, Northern Forest Research Centre, Edmonton, Alberta, 1977.

M. Pinta, Agricultural Applications of Flame Photometry, in *Analytical Flame Spectroscopy* (R. Mavrodineanu, Ed.), Macmillan, London, 1970, pp. 431-478.

W. J. Price, *Analytical Atomic Absorption Spectrometry*, Heyden, London, 1972.

V. Sychra, V. Svoboda and I. Rubeska, *Atomic Fluorescence Spectroscopy*, Van Nostrand, London, 1975.

B. Welz, *Atomic Absorption Spectroscopy* (English Edition), Verlag Chemie, Weinheim, 1976.

2

Ion-Selective Electrodes

OSCAR TALIBUDEEN and MICHAEL B. PAGE Rothamsted Experimental Station, Harpenden, Hertfordshire, England

I. INTRODUCTION

The first ion-selective electrode (ISE) was invented in the first decade of this century. The active component was a glass membrane sensitive to hydrogen ions, and the electrode forms the basis of the modern pH glass electrode. Such electrodes were first used in soil and plant science more than 50 years ago. About 20 years later, glass membranes sensitive to Na^+ and K^+ ions were devised by modifying the composition of the glass, although interference by hydrogen ions has not been eliminated. In the intervening period, clay mineral membranes were extensively tested for cation analyses in solutions, but these were slow in response, not very ion-specific, and too delicate for use in routine work. Since then much attention has been given to crystalline membranes and liquid membranes sensitive to a wide range of cations and anions. This has resulted in the availability of membranes of high ion specificity and robustness under appropriate environmental conditions.

This chapter is mainly concerned with the use of ISEs in the analysis of soil extracts, but also considers briefly their use in plant analysis and in vivo plant studies. The reader is referred to the voluminous literature on the subject for details of the pH glass electrode.

II. BASIC PRINCIPLES

In a modern ISE, a passive membrane separates the internal standard and test solutions of the ions. Electrons, simple ions, or charged or neutral complexes of the test ion are transported across the membrane interfaces to extents that are proportional to the compositions of the solutions on either side of the membrane. The electrostatic potential difference (EMF) thus developed across the membrane can be measured by coupling the membrane half-cell with a standard reference electrode half-cell. This potential difference E is given by the Nernst equation

$$E = E_0 + \frac{RT}{z_i F} \ln a_i \quad (1)$$

where z_i is the valence and $a_i = \gamma_i c_i$, the activity of the ion, with γ and c the ionic activity coefficient and concentration, respectively. The theoretical slope of the E : ln a_i relationship at 25°C is $59.12/z_i$ mV for ions of valence z_i, the slope being positive for cations and negative for anions.

A. Types of Membranes

Membranes are essentially of three types, further subdivisions of which are shown in Table 1, with their principal parameters in Table 2 (see Sec. II.B).

Table 1 Types of Ion-Selective Membranes Commonly Used

Type of membrane		Composition
1. Without sites	(a) Nonporous solid	Organic crystals; plastic film impregnated neutral carrier
	(b) Nonporous liquid	Organic liquid containing neutral carrier
2. With fixed sites	(a) Nonporous solid	Glass; defect inorganic crystals; high cross-link exchange resins
	(b) Porous solid	Compacted membranes of very small particles of nonporous solids
3. With mobile sites	(a) Liquid	Ion exchangers; hydrophobic salts with dissociation and ion pairing in organic solvent
	(b) Solid	Pure crystalline halides and sulfides; lanthanum fluoride

Source: Adapted from Ref. 1.

Table 2 Composite Selectivity Parameters for the Various Types of Membranes

Ion exchange membrane	Composite selectivity parameter	Comment
Solid	$\frac{u_j}{u_i} \cdot K_{ij}$	Ion mobility opposes equilibrium selectivity
Liquid		
(1) Dissociated	$\frac{u_j}{u_i} \cdot \frac{k_j}{k_i}$	u and k depend only on the solvent
(2) Associated - poorly mobile sites	$\frac{u_j}{u_i} \cdot \frac{k_j}{k_i}$	$u_i, u_j >> u_s$
(3) Associated - highly mobile sites	$\frac{u_{js}}{u_{is}} \cdot K_{ij}$	
Neutral carriers	$\frac{K_j}{K_i}$	$u_{js} = u_{is}$

Notes

1. u refers to mobility in the membrane.
2. Subscripts i, j, s, is, and js refer to specific ion, interfering ion, neutral carrier, and the complexed ions.
3. K_{ij} $(= K_j/K_i)$ is the equilibrium constant of the reaction

$$J^+ \text{ (aq)} + IS^+ \text{ (membrane)} \rightleftharpoons I^+ \text{ (aq)} + JS^+ \text{ (membrane)}$$

4. K_i is the formation constant of the reaction

$$I^+ + X^- + S \rightleftharpoons IS^+ + X^-$$

5. k_i, k_j are the partition coefficients of the dissociated ions i, j between water and the solvent of the liquid membrane.

Source: Adapted from Ref. 2.

Solid Ion Exchange or Fixed Site Membranes

These principally consist of inorganic ionic crystals in which their manner of preparation induces specific preference for the cation or the anion, and ion selectivity depends not only on the ion exchange constant but also on the relative mobility of the specific and interfering ions in the membrane matrix, much more so than in liquid membranes. Examples of these are the Ag halide-sulfide, the LaF_3, and the glass electrodes (for H_3O^+, Li^+, Na^+, and K^+).

Liquid Ion-Exchange Membranes

Like the liquid "neutral-carrier" membranes described below, these were until recently truly liquid in that the ion-complexing exchanger was dissolved in an inert solvent and physically restrained in an inert membrane, at least temporarily, and replenished at intervals. In modern practice, the dissolved exchanger is impregnated in thermosetting plastic materials so that the so-called liquid membrane is cast at ambient temperatures as a thin film in which the solution is held for much longer periods. Such membranes are much more robust but cannot be recharged or reconditioned after the impregnated solution has been lost. The replacement film costs more but is considered to be much more reliable and long-lasting. The interfaces of a cation-selective ion-exchange membrane must be permeable to the cation but not to its co-anion, nor to the cation-selective anion dissolved in the inert solvent within the membrane (Fig. 1, [3]). The cation complex formed within the membrane is uncharged and freely mobile, in which case the selectivity constant of the complexing ion for the specific and interfer-

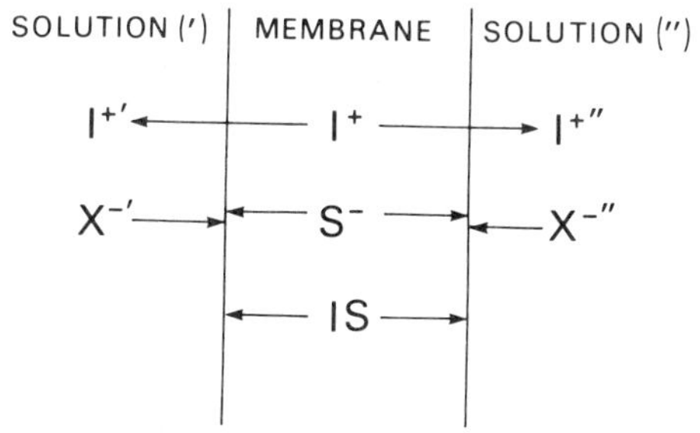

Figure 1 Ion transfer in a system incorporating a liquid ion-exchanger membrane. I^+, S^-, and X^- refer to the counter ions, the ion-exchanging site, and the co-ions, respectively. (From Ref. 2.)

ing cations determines the EMF response of the electrode to cation concentration. If, however, the cation complex dissociates fully, then the partition coefficients of the two cations between water and the solvent in the membrane are implicated.

Liquid Neutral Carrier Membranes

The ion-selective components in such membranes are macrocyclic antibiotics and cyclic polyethers in low or high dielectric constant solvents, depending on whether they are thin or thick membranes. They form a high concentration of a complex with the same charge as the complexed ion. The mechanism of the development of the potential difference is diagrammatically represented in Fig. 2, and is controlled by the equilibrium constant K_{is} of the reaction

$$I^+ + X^- + (S^*) \rightleftharpoons (IS^*)^+ + (X^*)^-$$

Neutral carriers can be tailored to bind a particular cation and hence provide a simple mechanism of high specificity.

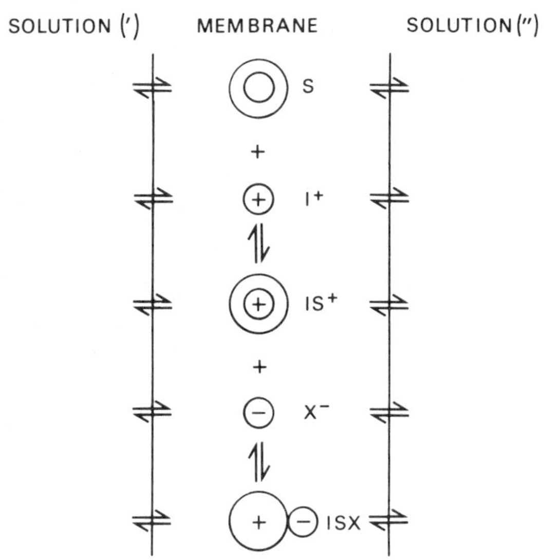

Figure 2 Ion transfer in a system incorporating a neutral carrier membrane. S, I^+, $(IS)^+$, X^-, and ISX refer to the neutral molecular carrier, the free cation, the complexed cation, the free anion, and the neutral complex, respectively. (From Ref. 2.).

B. Factors Affecting Performance

Response Time

The response time of an ISE is essentially a property of the physical structure of the electrode that might affect the time taken by the ion-sensitive part of the membrane to come to equilibrium with the solution interface, and the time taken by the ion in bulk solution to come to equilibrium with the membrane, determined by the thickness of the unstirred boundary layer between the bulk solution and the membrane surface. The rate-controlling step is then the diffusion of the ion across this layer, giving an exponential time relationship for ion exchange membranes and a square-root time relationship for neutral carrier membranes.

Interfering ions can delay the formation of the transported species that are necessary to develop a Nernst potential difference across the ISE membrane and thus affect the response time. In experiments where the kinetics of changes in ion concentration are being measured, a knowledge of the response time of an ISE is essential. Morf et al. [4] calculated theoretical response times by including a term in the Nernst relationship which was a function of all the experimental variables that could affect the development of the Nernst potential difference, e.g., rate of stirring of the bulk solution. Equation (1) was thus rewritten:

$$E = E_{i,0} + E' + \frac{RT}{z_i F} \ln a_i \tag{2}$$

where E' is the potential difference contribution produced by diffusion layers in the aqueous sample, a_i the activity of the ion in the boundary layer, and z_i its valence. From the equation subsequently derived to express the time dependence of a_i and hence the factors affecting response time, these authors drew the following conclusions. With ion exchange membranes, the boundary layer, and thus the response time, can be drastically reduced by (1) efficient stirring (or faster flow rates in continuous flow work), (2) avoiding pores and impurities on the membrane surface, or by using microelectrodes with minimal membrane surface, and (3) changing from a lower to a higher activity in the sample solution during sequential measurement. In neutral carrier membranes, diffusion of dissolved electrolyte (ligand and complex), resulting from the attainment of a steady state concentration gradient across the membrane, is the main factor governing the response time. Thus the ion extraction capacity, the resistance to diffusion, the thickness of the membrane, and the polar nature of the solvents in the membrane are all implicated.

Sensitivity of Measurements

The detection limit of an ISE is normally defined as the lowest concentration of the ion at which a finite potential difference is developed across the membrane. (More rigorously, this could be the lowest concentration at which a Nernstian, or near-Nernstian, response to concentration is measured.) Detection limits are set by the nature of the ISE, of which two examples are given below:

ISE - Types 2 and 3b (Table 1). These types consist of ionic crystal membranes in which the electron transfer is through interstitial ions or vacancies in the ionic lattice, so that the detection limit is governed by the solubility product of the ionic solid, or by the product of the ionic defects at the surface of the crystal, whichever is the larger [5]. For an AgCl membrane, for instance, the solubility product $[a_{Ag}a_{Cl}]$ sets the detection limit $a_{Ag,0}$ of the silver ion such that a Nernstian response occurs only when $a_{Ag,0}$ is appreciably higher than $[a_{Ag}a_{Cl}]^{\frac{1}{2}}$. By contrast, for an AgI membrane at whose surface the concentration of Ag defects greatly exceeds $[a_{Ag}a_{I}]^{\frac{1}{2}}$, then the detection limit of Ag is much higher than this defect Ag concentration.

ISE - Types 1a, 1b, and 3a (Table 1). In these types the active agent, or ligand (dissolved in an inert solvent in the membrane), forms a charged, or an uncharged, complex with the ion in solution. The transport of this complex within the membrane gives rise to the potential difference across it and hence the detection and measurement of the activity of the ion in the external solution. The steps leading to such transport are: the partition of ion i between the external and internal solutions and the membrane solvent, the formation of a complex between the ligand and the ion, and the solubility of this complex in the membrane solvent. These steps also apply to liquid membranes physically immobilized in a porous matrix, but when the inert solvent is replaced by an inert solid matrix, the partition and solubility steps are inapplicable. Thus, the thermodynamic constant for the formation of the complex between ion i and the ligand essentially controls the detection limit.

For most ISEs, the detection limit is around 10^{-5} M. In favorable environmental conditions in solution, ISEs of Type 2 (Table 1) can give Nernstian response down to 10^{-7} M, and some ISEs of Type 2 with favorable formation constants, to 10^{-6} M. These limits are adversely affected, however, by interfering ions and molecules. (A larger mole fraction of water molecules in concentrated aqueous solutions of ions could also prevent access of the ion to the hydrophobic ISE membrane, and thus affect the detection limit adversely.)

Selectivity

In the presence of an interfering ion j, therefore, the selectivity coefficient for ion i, K_{ij}, or the partition coefficient ratio k_i/k_j, is a more general expression of the formation constant K_i of the ligand complex with ion i. So it is convenient and adequate to consider the significance of the detection limit as a special case of the amendment of Nerstian response [Eq. (1)] by the selectivity coefficient K_{ij}.

Equations of varying complexities have been proposed and described to express this, in which terms of ion selectivity and mobility (in the membrane) are incorporated.

For solid-state membranes measuring the activity of anion Y^{z-}, typical equations for an AgX (silver halide) membrane [Eqs. (3a) and 3b)] or a mixed AgX + AgY membrane [Eq. (4)], are given by Morf et al. [5]:

$$E = E_{y,0} - \frac{RT}{zF} \ln a_y \tag{3a}$$

which describes the measured EMF for values of a_x given by

$$a_x < K^{(pot)}_{x,y} a_y^{1/z}$$

where the potentiometric selectivity coefficient, $K^{(pot)}_{x,y} = L_{AgX}/L^{1/z}_{Ag_zY}$, L representing the solubility products. The close agreement between calculated and experimental values of $K_{x,y}$ for AgX membranes (Fig. 3) confirms the validity of these relationships. At higher values of a_x, the membrane EMF is only sensitive to a_x [6] and is given by

$$E = E_{x,0} - RT \ln a_x \tag{3b}$$

The EMF of mixed-phase solid-state electrodes is sensitive to a_x and a_y, and is given by

$$E = E_{x,0} - \ln \quad a_x + K_{x,y} a_y^{1/z} \tag{4}$$

denoting that the activity of one ion can be measured at a constant activity of the other (for the Cl, Br system, see Ref. [6]).

The AgX membranes can also be used to assay cations that react with AgX, e.g, the mercuric Hg^{2+} ion [5] for which the EMF is given by

$$E = E_{Ag,0} + \frac{RT}{F} \ln (\bar{n}a_{Hg} + \alpha) \tag{5}$$

where $\bar{n}$ is the average number of X^- ions coordinated to the Hg^{2+} ion at the phase boundary and α the defect activity of Ag^+ at the membrane surface which is governed by the conditions of its preparation.

In solid-state ion exchange membranes, the low ion mobility within the membrane does not influence ion selectivity. In liquid membranes, where the complex formed with the ion within the membrane is neutral, the complex-forming ionizable species is an aliphatic amine or acid dissolved in a water-immiscible solvent. Thus, ionic mobility in the solvent, the partition coefficient of the ion between solvent and water, and the formation constant of the neutral complex are all invoked in producing the EMF.

For neutral carrier liquid membranes, measuring the activity a_i of ion i in the presence of an interfering ion activity, a_j, in the simp-

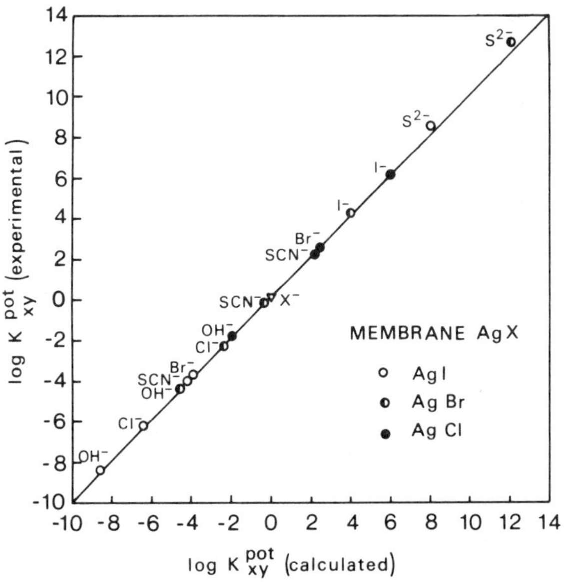

Figure 3 Comparison of the experimental and calculated values of anion selectivity of different silver halide membrane electrodes. [From Ref. 5, reprinted with permission from W. E. Morf, G. Kahr, and W. Simon, Theoretical treatment of the selectivity and detection limit of silver compound membrane electrodes, Analytical Chemistry *46*:1541 (1974). Copyright by the American Chemical Society.]

lest case of both ions with the same valence z, Morf and Simon [7] give the EMF of the ISE according to the equation

$$E = E_{i,0} + \frac{RT}{zF} \ln (a_i + K_{i,j}^{(pot)} a_j) \tag{6}$$

which is a simpler version of that for the liquid ion-exchange membrane, which also incorporates terms for the valences, mobilities and transference numbers of ions i and j.

The potentiometric selectivity coefficient, $K_{i,j}^{(pot)}$, is given by

$$K_{i,j}^{(pot)} = \frac{u_i}{u_j} K_{ij} \tag{7}$$

where u is the mobility of the ionic form in the membrane and $K_{i,j}$ the equilibrium constant of the reaction

$$[I]^{z_i} \text{ (membrane)} + [J]^{z_i} \text{ (aq)} = [J]^{z_i} \text{ (membrane)} + [I]^{z_i} \text{ (aq)} \tag{8}$$

C. Reference Electrodes

As with pH measurements using the glass electrode, ion activity measurements require the ISE to be coupled with an external reference electrode. The composition of the reference solution in the latter must, however, be chosen such that it does not affect the concentration of the specific ion in the sample or introduce significant concentrations of interfering ions. It is also essential that the liquid junction potential (LJP) between the reference electrode and the calibrating and test solutions of the specific ion be invariant. The LJP depends on the nature of the junction and the interdiffusion of the external and internal solutions. Principally, three devices are popular for reducing LJP. The first is a filling solution in a *single junction electrode* (Fig. 4a) in which cation and anion have the same mobility, e.g., a cation-anion pair chosen from Cs^+ (67.5), Tl^+ (65.3), K^+ (64.2), NH_4^+ (64.3), $\frac{1}{2}Pb^{2+}$ (61.0), and $\frac{1}{2}SO_4^{2-}$ (68.5), Br^- (67.3), I^- (66.3), ClO_4^- (64.0), $\frac{1}{2}C_2O_4^{2-}$ (63.0), NO_3^- (61.6), $\frac{1}{2}CO_3^{2-}$ (60.0), where the values in parentheses represent ionic mobility or conductance. The second is *a double junction electrode* (Fig. 4b) in which the composition and concentration of the filling solution in the outer chamber is chosen such that the LJP with the test solution (Table 3) is minimal. The third involves addition of an *inert* electrolyte to the solutions in both electrode compartments in direct contact (single junction), or through a salt bridge of the inert electrolyte at the same concentrations (double junction); this effectively serves to decrease the relative concentration of the potential determining ion, and also to maintain its activity coefficient invariant. A reference electrode must be stable, have a high exchange current to maintain its constant potential under high current demand, and be reproducible so that the temperature and concentration dependence of its EMF is not subject to hysteresis.

ISEs also require an internal reference electrode, in which the reference solution has a composition unrelated to the requirements of the test and standard solutions but is primarily concerned with the ion-selectivity mechanism of the ISE membrane.

Most of the reference electrodes in current practice are anion-responsive. The few cation-responsive electrodes in use are for special applications. The external reference electrode of the saturated KCl-calomel variety, because of its fixed potential, is used extensively but the Ag/AgCl/saturated KCl electrode has been adopted increasingly for general use principally because the $Hg/Hg_2Cl_2/KCl$ electrode is a toxic hazard if damaged. A saturated KCl solution deposits crystals that clog the electrode junction when the temperature fluctuates, hence undersaturated solutions, 3.5 or 4.0 M KCl, are used increasingly. The insoluble chlorides, especially AgCl, are more soluble in concentrated KCl solutions, which results in drifts of the fixed potential of the reference electrode. In some applications, where continuous monitoring of solutions containing dilute electrolyte solutions is required,

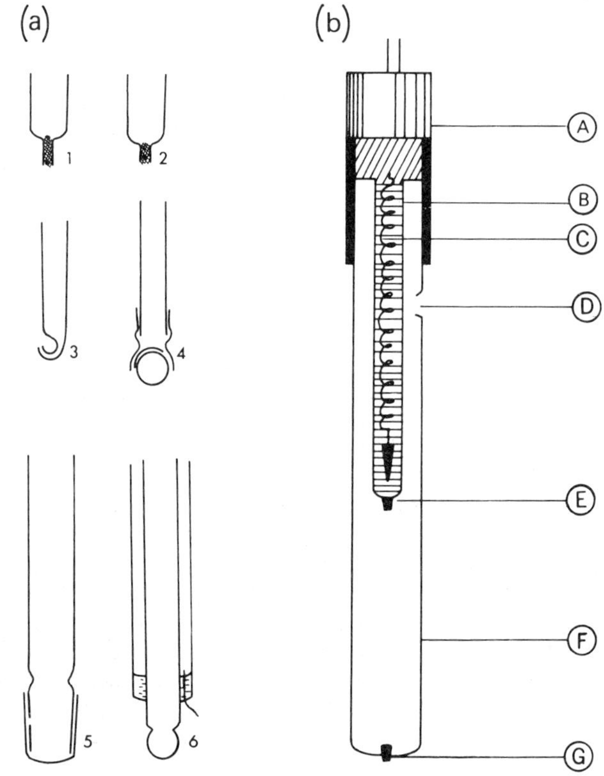

Figure 4 Types of liquid junctions in reference electrodes: (a) *Single junction* with restrained flow: (1) palladium annulus; (2) ceramic plug; (3) junction to overcome density gradient; (4) cellophane membrane; (5) ground glass sleeve; (6) fiber wick in combination glass electrode. (From Ref. 8.) (b) *Double junction*: (A) cap with air vent; (B) removable reference element; (C) reference element chamber containing concentrated KCl solution; (D) filling hole closed with teat during use; (E) ceramic diaphragm; (F) chamber containing bridging electrolyte; (G) ceramic diaphragm. (Courtesy of Corning Medical Ltd., Halstead, Essex, England.)

the use of dilute KCl, KNO_3, K_2SO_4, NH_4NO_3, $(NH_4)_2SO_4$, and NH_4F solutions is sometimes preferred for specific applications (Table 3). The anion chosen is tailored to the specific ion under test; the simplest and most obvious example is the use of NO_3^- or SO_4^{2-} filling solutions in chloride ion assay.

Table 3 Filling Solutions Recommended for Single (S) and Double (D) Junction Reference Electrode Half-Cells, Used with Various ISEs

Specific ion	Type of junction	Filling solution	Source[a]
Ba^{2+}	S	Equitransferent	O
	D	0.1 M $NaNO_3$	C
Ca^{2+}	S	Equitransferent	O
	S	4 M KCl (Ag)	O
	D	1 M KNO_3	O
	D	1 M KCl+sucrose	O
	D	0.1 M KNO_3	C
Cu^{2+}	S	Equitransferent	O
	D	1 M KNO_3	O,C
K^+	S	CCl_3COOLi	O
	S	0.06 M NaCl	O
	D	0.1 M Et_4NCl	C
Na^+	S	CCl_3COOLi	O
	D	1 M KNO_3	O
	D	0.1 M $CaCl_2$	C
Br^-	S	Equitransferent	O
	D	1 M KNO_3	O,C
Cl^-	S	Equitransferent	O
	D	1 M KNO_3	O,C
	D	1 M KCl+sucrose	O
CN^-	S	Equitransferent	O
	D	1 M KNO_3	O,C
F^-	S	Equitransferent	O
	D	1 M KNO_3	C
BF_4^-	D	0.1 M NH_4F	O
I^-	S	Equitransferent	O
	D	1 M KNO_3	C
NO_3^-	D	0.1 M KCl	O
	D	0.02 or 0.04 M $(NH_4)_2SO_4$	O
	D	0.1 M NaF	C
S^{2-}	S	1 M KNO_3 or $NaNO_3$	O,C
	S	4 M KCl (Ag)	O

[a]O: *Analytical Methods Guide*, Orion Research Inc., East Grinstead, West Sussex, England.
C: N. Walton, Corning Medical Ltd., Halstead, Essex, England.

Other variations are necessitated by the use of extracting solutions in analytical practice, or test solutions containing appreciable concentrations of other chemicals, for recovering the specific ion from an adsorbed phase. Direct ISE measurements on the extract necessitate the use of the same concentration of the extracting agent, or foreign chemical compound, in the salt bridge of a double junction electrode. Similarly, with test solutions containing nonaqueous solvents, the salt bridge has to include a solvent, usually ethanol, which is miscible with water and the nonaqueous solvent.

D. Calibration Methods

The method adopted for calibrating an ISE/reference electrode combination depends on (1) the versatility of the meter used [9]; (2) the magnitude of the variations in the ionic strength, or the concentrations of foreign chemical compounds, or the pH, of the test solutions; (3) the deviation of the ISE from the theoretical Nernstian response because of temperature changes, interfering ions, etc.

The most direct procedure is the calibration curve method relating the measured EMF to the concentration of the specific ion, taking steps normally necessary to account for item (2) above, or the activity of the ion if its activity coefficient can be calculated in the standard and test solutions. The method is accurate but laborious. For a small number of observations, greater precision is possible by using two standard solutions of concentrations, first above and then below that of the test solution, whose concentration can then be interpolated, independent of deviations from Nernstian slope.

With modern meters, the more usual practice is to use two solutions, whose specific ion concentration (or activity) ratio is known accurately, to calibrate the ISE/reference electrode/meter system. This concentration ratio is conventionally 10 and the range spans that in which the concentrations of the test solutions occur. If the ISE does not have a true Nernstian EMF response because of factors stated in item (3) above, such meters incorporate a scale-expansion facility that modifies the experimental slope to the theoretical one, and also a temperature compensating adjustment.

For solutions in which variations in ionic strength make accurate measurements difficult, or where hydrogen ion activity interferes with the specific ion measurement, or where the specific ion needs to be converted to an ion pair or ion complex, the use of ionic strength adjusters, pH buffers, and releasing agents added to standard and test solutions alike is recommended. Such combination buffer solutions are known as *total ionic strength adjusters* (ISA or TISAB). The composition of such "cocktails" can be tailored to the specific needs of a particular experimental technique [10]. TISAB contains 114 ml glacial acetic acid, 116 g NaCl, and 0.6 g Na citrate per liter H_2O titrated to pH 5.0-5.5 with 5 M NaOH. The addition of 1,2-cyclohexylenediaminetetraacetic acid to complex Al in solution is also recommended (Orion Research Inc. Applications Guide No. 5).

The most precise and rapid way of preparing calibration standards is the series addition method, in which the standard (in the TISAB or ISA appropriate to the application) at the high-concentration end of the range in which the test solutions lie, is successively diluted with the appropriate TISAB or ISA to give a full range of standards.

E. Methods of Analysis

The direct methods of measurement involve the use of a calibration curve, or the differential standards procedure for small numbers of samples in which two standards of concentrations close to the test solution are used; both have been indicated in the previous section. The precision with which the concentrations (or activities) of the specific ion is assayed depends on (1) the sensitivity of the meter (i.e., its response in millivolts to increments in concentration), its stability, and its precision; (2) the steps taken to ensure that the ionic strength, pH, concentrations of interfering ions, and presence of other compounds are normalized in calibration standards and test solutions; (3) temperature control; and (4) the extent to which measurements on standards and test solutions are replicated.

Several indirect methods are in general use, for which greater precision than the direct methods is claimed by their originators. However, these are only suitable for rapid routine processing of large numbers of samples, assuming preprogrammed microprocessor-operated meters are available [9] or simple computer programs are used. These methods are summarized below.

Standard Solution Addition to Test Solution

A volume v_i of the specific ion i at concentration c_i, to be determined, is spiked with a known increment of the ion (volume v_s of a standard solution of concentration c_s where $c_s \sim 100c_i$). The change in the measured EMF (positive for a cation and negative for an anion) is given by Thomas [11]:

$$E = E_{(i,s)} - E_i$$

$$= S \log \frac{v_i c_i + v_s c_s}{c_i (v_i + v_s)} \tag{9}$$

which follows from

$$E_{(i,s)} = E_0 + S \log \left(x_{(i,s)} f_{(i,s)} \frac{v_i c_i + v_s c_s}{v_i + v_s} \right) \tag{10}$$

(where S is the calibration slope and $x_{(i,s)}$ and $f_{(i,s)}$ are the fractions of uncomplexed ions and the ion activity coefficient in the spike solution). Equation (10) is obtained by assuming that x and f do not change on spiking, a simplification that follows if the conditions specified in item (2) are fulfilled. This simplified equation can be transformed to calculate c_i thus:

$$c_i = \frac{c_s}{10^{\Delta E/S}(1 + v_i/v_s) - v_i/v_s} \tag{11}$$

If the spike volume v_s is <1% of the test solution volume, this simplifies to

$$c_i = \frac{c_s v_s}{(10^{\Delta E/S} - 1)\, v_i} \tag{12}$$

Equation (10) can be further extended by measuring $E_{(i,s)}$ for two successive spikes, and by iterative calculation of c_i and S starting from $c_i = 0$, the true convergent value of c_i is obtained.

Test Solution Addition to Standard Solution

Here, a volume v_i of the test solution is added to a known volume v_s of the standard solution under the same equilibrium conditions as for the case above except that $c_s < c_i$; then

$$c_i = c_s \left[10^{\Delta E/S}\left(1 + \frac{v_s}{v_i}\right) - \frac{v_s}{v_i}\right] \tag{13}$$

Subtraction of Specific Ion from Test Solution

This is achieved by adding a known amount of a complexing or precipitating agent such that a stoichiometric amount of the specific ion is removed. The EMF of this solution is then given by

$$E_{(i,c)} = \text{constant} + S \log\left(x_{(i,c)} f_{(i,c)} \frac{v_i c_i - v_{(i,c)} c_c}{v_i + v_{(i,c)}}\right) \tag{14}$$

where the subscript i,c refers to the test and complexing solutions mixture and c to the complexing solution.

Titration Methods

In these, a standard titrant is added in small increments to the test solution such that the specific ion is precipitated or chelated and the EMF is measured with an ISE/reference cell combination. Alternatively, an excess of the standard titrant is added and this excess determined by back-titration with a standard solution of the specific ion. The ISE used could be sensitive to the specific ion or to the reacting ion in the standard titrant. Potentiometric titrations are considered to be more precise than the direct methods described above, but require more time for experimentation and data processing to obtain the desired result than even the indirect methods. Their essential principle is the measurement of the change of EMF with titrant addition, which is maximum at the equivalence point, is considerably more than the response slope of the ISE (Secs. II.B and D), and is affected little by liquid junction potentials and activity coefficients. Sharper breaks in the measured EMF at the end point, greater stoichiometry of the titration reaction, and a higher reaction velocity all lead to greater precision.

Potentiometric titration curves (EMF versus titrant volume) are sigmoidal (Fig. 5a), from which the equivalence point is most conveniently obtained graphically by plotting the first and second derivatives (Fig. 5b and c) of the sigmoidal relationship. Figure 5d is another variant – the reciprocal of the first derivative versus the titrant volume, in which the two arms intersect the volume axis at the equivalence point.

An even more laborious and more precise procedure is to determine the equivalence point potential by the direct method first, then add the titrant until this EMF is attained, thus giving the exact titration value of the amount of the specific ion in the test solution.

The Gran Plot

In very dilute solutions of the specific ion, the inflexion point in the sigmoidal curve is ill-defined and difficult to evaluate (Fig. 6), when the EMF values measured on adding increments of the titrant to the test solution are plotted against titrant volume, or indeed by any of the indirect methods described above. For acid-base titrations, Sorensen [13] proposed that the antilogarithm of the pH should be plotted against titrant volume, whose final value substantially exceeds the equivalence point. Gran [14] extended this to precipitation, complexation, and oxidation-reduction reactions, incorporating ions of different valences and also including a correction factor for increasing volume of the mixture when more titrant was added. This converts a curvilinear antilog plot to a linear form, making possible a precise extrapolation to the equivalence point (Fig. 7).

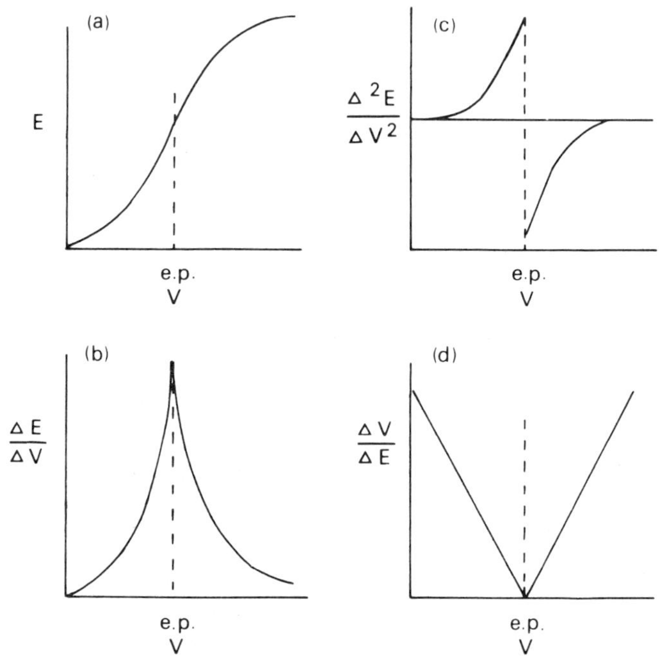

Figure 5 Titration curves: (a) EMF versus titration volume; (b) first derivative curve; (c) second derivative curve; (d) extrapolated $\Delta V/\Delta E$:v curve. e.p. = equivalence point. (From Ref. 12.)

In infinitely dilute solutions (where the activity coefficient f_i of ion i = 1) or in solutions where f_i remains unchanged on adding titrant, and in the absence of complexation in the test solution of foreign ions (i.e., x = 1), the theoretical basis of the Gran plot can be seen by rewriting Eq. (10) in the antilogarithmic form thus:

$$10^{(E_{i,s} - E_0)/S} = \frac{v_i c_i + v_s c_s}{v_i + v_s} \tag{15}$$

In practice, the appropriate volume correction factor can be tabulated for application to the antilogarithm of the measured EMF values of the control, or blank, and the test solutions.

A simpler expedient is to plot the measured EMF values on an antilogarithmic scale, as the ordinates, against the added titrant volumes on a linear scale, as the abscissas (Fig. 8). The volume correction is made by giving a finite slope to the abscissa lines (above the x-axis) corresponding to the percentage increase in the initial

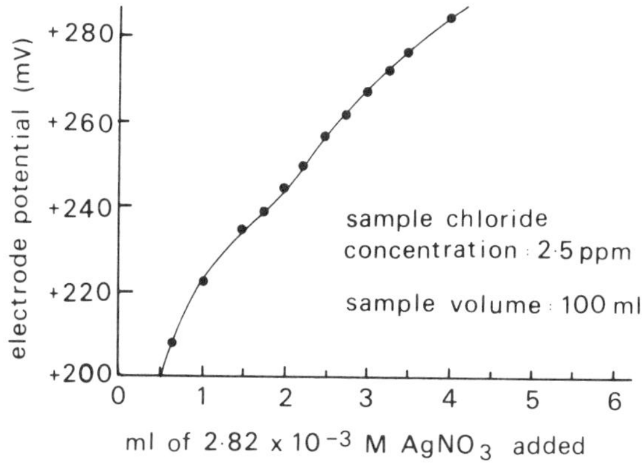

Figure 6 Conventional plot of titration data for low chloride concentrations. (From *Orion Analytical Methods Guide*, 9th edition, Orion Research Inc., East Grinstead, West Sussex, England, December 1978, p. 34.)

solution volume v_i when a predetermined excess volume of titrant, $(v_s)_f$, has been added finally, i.e., $100(v_s)_f/v_i$. The EMF at the origin is that measured on the control or blank before adding the titrant, and forms the base line to which the EMF values of the titrated test solution are extrapolated. The scale on the antilog axis is increasingly positive for cation assay and negative for anion assay. The titrant volume obtained by extrapolating to the base line the

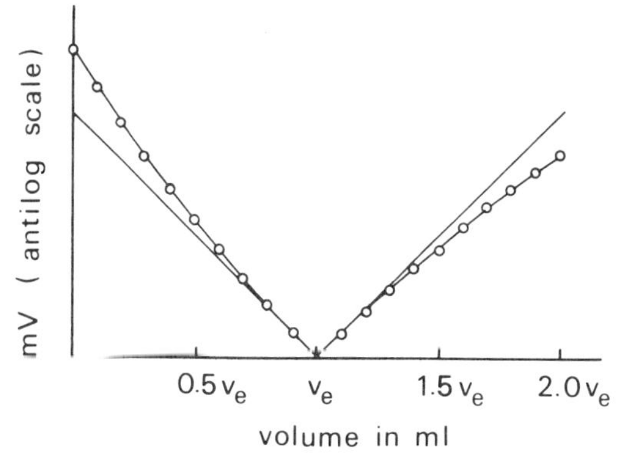

Figure 7 Titration of 40 ml 0.025 M HCl with 0.1 M NaOH; — o — o — o, Sorenson plot; ——, Gran plot. (From Ref. 14.)

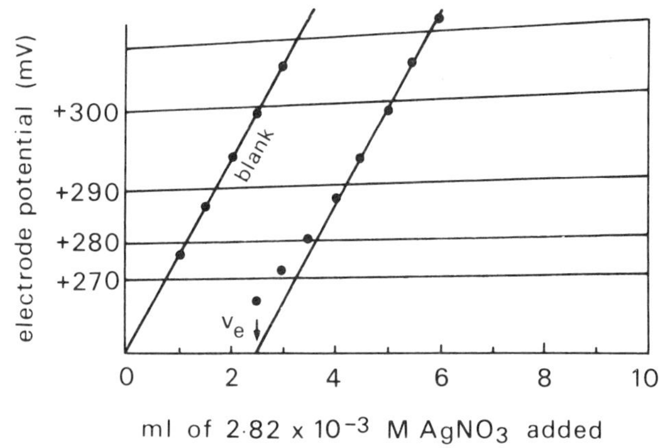

Figure 8 Gran plot of titration data for low chloride concentrations (from Fig. 6) using an AgS electrode. (From *Orion Analytical Methods Guide*, 9th edition, Orion Research Inc., East Grinstead, West Sussex, England, December 1978, p. 34.)

linear part of the EMF:titrant volume curve at high titrant volumes furthest away from the equivalence point, gives the equivalence point in volume units v_e of added titrant.

Differential Potentiometry

The principle is illustrated in the determination of chloride ions with a cell of the type

Ag, AgCl / A standard Cl^- solution // ion-selective membrane // B test Cl^- solution / AgCl, Ag

in which the reference electrodes can be replaced by matched chloride ISEs [15]. The cell is calibrated by measuring the EMF values of a series of standard solutions of the specific ions put in the half-cell B to produce a calibration curve, from which the concentration of the test solution can be interpolated.

III. COMMONLY AVAILABLE ELECTRODES

Earlier (Sec. II.A and Table 1) a summary list was given of the types of ISEs that are feasible in principle. All of these have been prepared and tested in research laboratories, and most are commercially available and have acceptably high sensitivity, reproducibility, and ruggedness. Their basic construction and performance characteristics are described in this section. Data sheets supplied by the manufacturers at the time of writing and compilation have been used to prepare the appropriate tables and diagrams. The information given thus is intended as a guide to potential users in soil and plant science so that they can select the ISE that best fits their environmental conditions. However, the reader is strongly advised to obtain from the manufacturer the up-to-date characteristics of the ISE selected before planning an experiment and buying the ISE.

A. Solid-State and Liquid Membranes

Solid-state membranes consist of (1) crystalline pure or mixed halides of silver and copper, or a halide-sulfide mixture, for Cu^{2+}, Br^-, Cl^-, I^-, CN^-, and S^{2-}; (2) lanthanum fluoride for F^-; (3) glass for Na^+ and K^+; (4) a silver phosphate-chalcogenide for phosphate; (5) a mixture of silver and copper sulfate and sulfide for SO_4^{2-}. Details of types (1)-(3) are given in Tables 4-8 and are briefly discussed in the succeeding sections. The phosphate and sulfate electrodes (types 4 and 5) have not as yet been produced commercially but are of sufficient importance to soil and plant science to merit special mention here.

The phosphate-sensitive electrode [16] gives a linear response in the concentration range 10^{-8}-10^{-3} M with a sensitivity of 14-15 mV per decade change in PO_4^{3-} concentration instead of the theoretical 19.38 mV at 20°C. The response time of the electrode is <30 sec and the EMF stability is better than 0.5 mV day^{-1}. The SO_4^{2-} and NO_3^- ions do not interfere, but K_{ij} for the chloride ion is 10^2 (see Sec. III.C). The limiting pH at which the electrode will function thus is given by

$$pH = 11.566 - \frac{1}{3}\, p(PO_4) \tag{16}$$

because of the complementary hydrolytic relationship between Ag_3PO_4 and AgOH.

The sulfate-sensitive electrode [17] of composition $PbSO_4$-PbS-Ag_2S-Cu_2S, has a linear response between 10^{-4} and 10^{-1} M, with a slope "only slightly less than the theoretical" at 18-20°C which remains unchanged over 2 months. Preliminary measurements with less satisfactory homogeneous $BaSO_4$ electrodes suggest that Cl^- and NO_3^- ions interfere only slightly but CO_3^{2-} and phosphate ions do so seriously. However, K_{ij} values for the heterogeneous composition electrodes are not given, nor is the effect of pH on electrode performance.

Most commonly available ISEs are of the *liquid membrane* type (Table 4) of various constructions (Sec. III.B) and basic principles (Sec. II.A and B) involving charged sites (i.e., ionizing organic salts), or neutral complexing carriers dissolved in a water-immiscible solvent, or impregnated in solid solution with an inert carrier.

Table 4 Commonly Available Ion-Selective Electrodes

Ion		Type of electrode membrane	Manufacturers[a]	Applications references			
Direct	Indirect			Soil pastes and suspensions	Soil extracts	Plant extracts	Others
Cations							
Ammonium		Liquid (in polymer)	P	37	-	-	-
Barium	Sulfate	Liquid	C, O	-	-	-	-
Calcium		Liquid (in polymer)	C, O, P	21	34, 35	78	-
Calcium + magnesium		Liquid (in polymer)	C, O	21, 36, 79	-	-	-
Copper		Solid state (crystal)	C, O, P	-	-	-	66
Potassium		Solid state (glass)	B, E	26,27,29,30,52	-	-	-
		Liquid (in polymer)	C, O, P	-	33	-	-

Sodium		Solid state (glass)	B, C, E	23-25,28,31,32	18-20, 22	-	77
		Solid state (crystal)	O	-	-	-	-
		Liquid (in polymer)	B	-	-	-	-
Anions							
Bromide		Solid state (crystal)	C, O, P	-	61, 62	61	-
Chloride		Solid state (crystal)	C, E, O, P	60	56-59	86	-
Cyanide		Solid state (crystal)	C, O, P	-	-	-	-
Fluoride	Aluminum	Solid state (crystal)	C, O, P	-	63, 64	64,83,84	85
Fluoro-borate	Boron	Liquid (free flowing)	O	-	-	-	-
Iodide		Solid state (crystal)	C, O, P	-	-	-	-
Nitrate		Liquid (in polymer)	C, E, O, P	38,48,50,52-54	38-47	46,48,49, 80-82	-
Sulfide		Solid state (crystal)	C, O, P,	-	65	-	-

[a]B, Beckman; C, Corning; E, EIL; O, Orion; P, Philips.

Table 5 Cation-Selective Electrodes: Performance Characteristics

Electrode	Manufacturer (see Table 4)	Membrane composition	Range (M)[a]
NH_4^+	P	–	10^{-6}-1
Ba^{2+}	C	Antarox	5×10^{-5}-1
	O	–	10^{-6}-1
Ca^{2+}	C	$Ca(Me_4 \cdot Bu \cdot Ph \cdot PO_4)$	5×10^{-6}-1
	O	–	5×10^{-7}-1
	P	–	10^{-5}-1
(Ca + Mg) and water hardness	C	$Ca(Me_4 \cdot Bu \cdot Ph \cdot PO_4)$	$>10^{-4}$
	O	–	6×10^{-6}-1
Cu^{2+}	C	$Cu_xS + Ag_2S$	$>10^{-6}$
	O	–	10^{-8}-satd.
	P	–	10^{-7}-1
K^+	B	Glass	10^{-6}-1
	C	Valinomycin	5×10^{-5}-1
	E[c]	–	–
	O	–	10^{-6}-1
	P	–	10^{-6}-1
Na^+	B	Glass	10^{-6}-1
	C	Glass	10^{-6}-1
	E	Glass	5×10^{-9}-10^{-1}
	O	Glass	10^{-8}-1
	P	–	10^{-6}-1

[a]The lower limit is taken to indicate also the detection limit.
[b]Usually at specific ion concentrations $\geq 10^{-3}$ M; larger times for more dilute solutions.
[c]Performance characteristics not supplied by manufacturers.

Sensitivity (mV per decade molarity change)	Response time (sec)[b]	Operational life (months)	pH range	Temperature range (°C)
56 ± 3	<30	6	4-10	0-50
–	<90	6	2-11	–
23 ± 2	–	6	3-11	0-50
–	<90	6	4-9	–
24	–	–	2.5-11	0-40
27 ± 2	<30	6	3-10	0-50
–	<90	6	4.5-8	0-50
24	–	–	–	0-50
–	<90	3	3-7	0-80
26	–	–	–	0-80
27 ± 2	<20	–	1-14	0-50
–	–	–	–	–
–	<90	6	4-9	–
–	–	–	–	–
54	–	–	2-12	0-40
56 ± 3	<30	6	2-12	0-50
–	–	–	–	–
–	–	>12	–	–
–	15-90	–	>p(Na)+3	0-80
55	<90	>12	>p(Na)+3	0-80
55 ± 3	<30	6	2-12	0-50

Table 6 Cation-Selective Electrodes: Interferences by Ions Commonly Found in Soil Water

Electrode	Manufacturer (see Table 4)	H	NH_4^+	Na^+	K^+
NH_4^+	P	3×10^{-2}	1	2×10^{-3}	2×10^{-1}
Ba^{2+}	C	–	–	4×10^{-4}	9×10^{-3}
	O	6×10^{-2}	4×10^{-3}	3×10^{-3}	2×10^{-2}
Ca^{2+}	C	–	–	1.5×10^{-4}	$<10^{-6}$
	O[a]	–	–	3×10^{-1}	–
	P	–	10^{-5}	3×10^{-4}	2×10^{-4}
(Ca + Mg) and water hardness	C	–	–	2×10^{-2}	–
	O(10^{-3} M)[a]	–	–	3×10^{-2}	–
Cu^{2+}	C	–	–	–	–
	O[a]	–	–	–	–
	P	–	–	–	–
K^+	B[b]				
	C	–	3×10^{-1}	2.6×10^{-3}	1.0
	E[b]				
	O(10^{-3} M)[a]	10^{-2}	6×10^{-3}	2.0	1.0
	P	10^{-5}	10^{-2}	3×10^{-6}	–
Na^+	B[b]				
	C	(pH≤pNa+3)	2×10^{-2}	1.0	3×10^{-2}
	E[b]				
	O[b]				
	P	5×10^{-1}	2×10^{-1}	1.0	5×10^{-1}

[a]For Orion ISEs: Maximum molarity of interfering ion giving 10% error at specific ion molarity given in brackets.
[b]K_{ij} values not supplied by manufacturers.

Selectivity Ratio $K_{ij} = a_i/a_j$						
Mg^{2+}	Ca^{2+}	Zn^{2+}	Cu^{2+}	Pb^{2+}	Fe^{2+}	Others
10^{-5}	3×10^{-5}	–	–	–	–	Ba^{2+}: 3×10^{-5}
–	2.5×10^{-3}	–	–	2.4	–	–
8×10^{-6}	2×10^{-4}	–	–	–	–	–
2.5×10^{-4}	1.0	1.0	–	0.1	0.8	–
1.0	–	1.0	4×10^{-2}	1×10^{-5}	2×10^{-2}	Ba^{2+}: 3×10^{-2}
4×10^{-4}	1.0	–	–	–	–	Ba^{2+}: 4×10^{-4}
–	–	–	–	–	3×10^{-3}	–
–	–	3×10^{-5}	3×10^{-5}	–	6×10^{-5}	Ba^{2+}: 6×10^{-4}
–	1×10^{-4}	–	1.0	$5 \times 10^{+3}$	–	Mn^{2+}: 6.4×10^{-5}
–	–	–	–	–	–	$S^{2-} \leq 10^{-7}$M; high Cl^-
–	–	–	–	–	–	High Cl^-
1.9×10^{-3}	2.5×10^{-3}	–	–	–	–	–
–	–	–	–	–	–	–
6×10^{-6}	3×10^{-5}	–	–	–	–	Ba^{2+}: 4×10^{-4}
–	–	–	–	–	–	–
8×10^{-4}	2×10^{-3}	–	–	–	–	Ba^{2+}: 5×10^{-3}

Table 7 Anion-Selective Electrodes: Performance Characteristics

Electrode	Manufacturer (see Table 4)	Membrane composition	Range (M)[a]	Sensitivity (mV per decade molarity change)	Response time (sec)	Operational life (months)	pH range	Temperature range (°C)
Br^-	B	$AgBr/Ag_2S$	$>10^{-7}$	–	–	–	0-14	–
	C	$AgBr/Ag_2S$	5×10^{-6}-1	–	<90	<36	2-12	–
	O	–	5×10^{-6}-1	57	–	–	2-12	0-80
	P	–	10^{-6}-1	56 ± 3	<30	12	1-11	0-50
Cl^-	B	$AgCl/Ag_2S$	$>5 \times 10^{-5}$	–	–	–	0-14	–
	C	$AgCl/Ag_2S$	$>5 \times 10^{-5}$	–	<90	<36	3-10	–
	E	–	10^{-6}-1	–	40	–	1-12	0-100
	O	–	$>8 \times 10^{6}$(liquid); $>10^{-5}$(solid)	57	–	–	2-11	0-80
	P	–	10^{-5}-10^{-1}	56 ± 3	<60	12	1-10	0-50
CN^-	C	AgI	$>10^{-6}$	–	<90	<36	–	–
	O	–	10^{-6}-10^{-2}	54	–	–	11-13	0-80
	P	–	10^{-7}-10^{-1}	56 ± 3	<30	12	1-12	0-50

F^-	B	–	$>10^{-6}$	–	–	–	4-13	–
	C	LaF_3	$>5 \times 10^{-7}$	–	<90	<36	5-8	–
	O	–	10^{-6}-satd.	56	–	–	5-8	0-80
	P	–	10^{-6}-10^{-1}	56 ± 3	<60	12	4-8	0-50
BF_6^-	O	–	3×10^{-6}-satd.	56	–	–	3-10	0-50
I^-	B	AgI	$>10^{-8}$	–	–	–	0-14	–
	C	AgI	$>10^{-5}$	–	<90	<36	3-12	–
	O	–	7×10^{-6}-1	55	–	–	2.5-11	0-40
	P	–	10^{-7}-1	56 ± 3	<20	12	1-12	0-50
NO_3^-	C	$(C_8H_{17})_4NNO_3$	$>10^{-5}$	–	<90	6	4-11	5-40
	E	–	10^{-5}-10^{-1}	–	–	<2	–	5-40
	O	–	6×10^{-6}-1	55	–	–	3-10	0-50
	P	–	10^{-5}-1	56 ± 3	<30	6	1-9	0-50
S^{2-}	B	Ag_2S	$>10^{-7}$	–	–	–	>12	–
	C	Ag_2S	$>10^{-5}$	–	<90	<36	13-14	–
	O	–	10^{-7}-1	28	–	–	13-14	0-80
	P	–	10^{-6}-10^{-1}	27 ± 2	<15	12	1-14	0-50

[a]The lower limit is taken to indicate also the detection limit.

Table 8 Anion-Selective Electrodes: Interference by Ions Commonly Found in Soil Water

Electrode	Manufacturer (see Table 4)	Selectivity ratio K_{ij}				
		OH^-	F^-	Cl^-	Br^-	I^-
Br^-	C	3×10^{-5}	–	2.5×10^{-3}	1.0	$5 \times 10^{+3}$
	O[b]	–	–	–	–	(2×10^{-4})
	P	1×10^{-3}	–	6×10^{-3}	–	$2.0 \times 10^{+1}$
Cl^-	C	1.3×10^{-2}	–	1.0	$3 \times 10^{+2}$	$2 \times 10^{+6}$
	E	–	–	1.0	c	c
	O[a]	(10^{-4})	(7×10^{-4})	1.0	(4×10^{-5})	(8×10^{-6})
	P	2.4×10^{-2}	–	1.0	1.2	$8.7 \times 10^{+1}$
CN^-	C	10^{-8}	–	10^{-6}	2×10^{-4}	V. large
	O[b]	–	–	(10^{+6})	$(<5 \times 10^{+3})$	(10^{-1})
	P	–	–	2×10^{-5}	2×10^{-4}	3
F^-	C	10^{-1}	1.0	–	–	–
	O[b]	(10^{-1})	1.0	–	–	–
	P	10^{-1}	–	–	–	–
BF_6^-	O[a]	(2×10^{-1})	(2×10^{-1})	(2×10^{-1})	(2×10^{-1})	–
I^-	C	10^{-8}	–	10^{-6}	2×10^{-4}	1.0
	O[a]	–	–	–	–	1.0
	P	–	–	6.6×10^{-6}	6.5×10^{-5}	1.0
NO_3^-	C	–	10^{-6}	9.4×10^{-3}	5×10^{-2}	4.1
	E	–	–	4×10^{-3}	Large	Large
	O[a]	–	–	(4×10^{-2})	(8×10^{-4})	(6×10^{-6})
	P	10^{-3}	5×10^{-3}	10^{-2}	3×10^{-1}	10
S^{2-}	C	–	–	–	–	–
	O[a]	–	–	–	–	–
	P	–	–	–	–	–

[a] Orion ISEs only: Value given is the maximum molarity m_j of the interfering ion that gives a 10% error in the measured concentration of the specific ion $(m_i)_{apparent}$ when its true concentration is 10^{-3} M.

[b] Orion ISEs only: Value given is the ratio m_j/m_i.

[c] K_{ij} values not supplied by manufacturer, but solubilities of Ag salts of these ions are less than that of AgCl, so the Cl^- electrode response is affected by the presence of these ions.

$= a_i/a_j$

CN^-	SO_4^{2-}	S^{2-}	PO_4^{3-}	NO_2^-	Others
$1.2 \times 10^{+4}$	–	V. large	–	–	NH_4^+: 3×10^{-1}
–	–	($<10^{-7}$)	–	–	–
$2.5 \times 10^{+1}$	–	–	–	–	CO_3^{2-}: 2.3×10^{-3}
$5 \times 10^{+6}$	–	V. large	–	–	NH_4^+: 8.3
c	–	c	–	–	–
–	(3×10^{-5})	–	–	–	(NO_3^-: 3×10^{-5}), (HCO_3^-;Acetate$^-$: 4×10^{-4})
$4 \times 10^{+2}$	–	–	–	–	CO_3^{2-}: 3×10^{-3}
1.0	–	V. large	–	–	–
1.0	–	($<10^{+7}$)	–	–	–
1.0	–	–	–	–	CO_3^{2-}: 3.6×10^{-4}
–	–	–	–	–	–
–	–	–	–	–	–
–	–	–	–	–	–
–	(2×10^{-1})	–	–	–	(NO_3^-: 2×10^{-2})
2.5	–	V. large	3×10^{-10}	–	AsO_4^{3-}: 3×10^{-10}
–	–	($<10^{-7}$)	–	–	–
3.4×10^{-1}	–	–	–	–	–
–	3.5×10^{-3}	–	10^{-6}	3×10^{-2}	–
–	–	–		4×10^{-2}	HCO_3^-;Acetate$^-$: 3×10^{-4}
(2×10^{-4})	–	(HS^-: 4×10^{-3})		(8×10^{-4})	HCO_3^-;Acetate$^-$: 3×10^{-2}
–	–	–	–	7×10^{-2}	HCO_3: 10
–	–	1.0	–	–	Hg^{2+}: V. large
–	–	1.0	–	–	Hg^{2+}: $<10^{-7}$
–	–	1.0	–	–	–

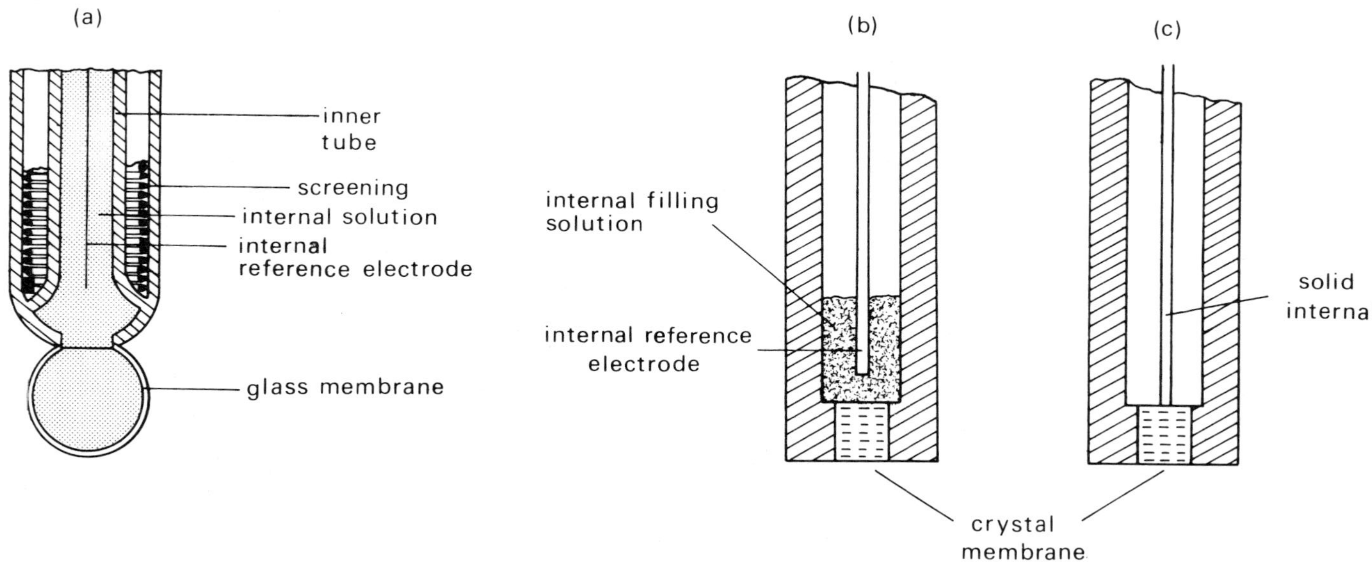

Figure 9 Solid state membranes: (a) glass; (b) crystalline, with internal reference solution; (c) crystalline, with internal solid connection. (From Ref. 87.)

B. Constructional Details

The three basic types of *solid-state membranes* are shown in Fig. 9. The glass electrode (Fig. 9a) is essentially the same as the pH electrode except that the membrane is less robust and of a composition that is sensitive to the specific ion. The membrane in the two crystal electrodes (Fig. 9b and c) has the same construction in that the small crystals of the active solid are pelleted into a disk to form a continuous ionic matrix in an inert cementing material of fine porosity. Electrical continuity with the external lead is through an internal reference electrode and a suitable filling solution (Fig. 9b), as with the glass electrode, or by cementing the internal conducting wire (mostly Ag or Pt) directly to the crystalline pellet (Fig. 9c). The filling solution can be replenished in a fully or partly demountable container (for easy cleaning), or be in a sealed unit, the latter being considered by the authors to be much more satisfactory than the former.

Liquid state membranes also occur in three forms, all with an internal reference electrode and filling solution. In the earlier, and now obsolescent form, the ion-specific liquid is restrained within the electrode by an inert porous membrane which can be replaced cheaply and quickly if contaminated or damaged. The small leakage rate (0.1-0.2 ml day^{-1}) can be greatly minimized by sealing the open end of the membrane with a thin impermeable and inert film when not in use. However, contamination of the internal electrode by the ion-specific liquid is a potential hazard to the long-term stability of the electrode. This and the possible leakage of the membrane liquid has led to the introduction of two alternative constructions. In one (Fig. 10), the membrane liquid is immobilized for up to 3 months in a plastic sponge providing a continuous pathway for charge transfer. The same effect is also achieved by incorporating films of the ion-specific liquid randomly in a thin plastic film which is inert to the charge transfer process but has sufficient chemical affinity for the membrane liquid to form a pseudo-solid solution which is physically homogeneous and stable (Fig. 11).

C. Performance Characteristics and Interferences

Table 4 summarizes ISEs that are currently available commercially and includes references to various types of applications in soil and plant science. Some of the more important of these applications are described in Sec. IV. Tables 5 to 8 provide a ready reference to the composition of the membranes, operational range of the specific ion concentration, sensitivity of response (i.e., millivolts per decade change in the specific ion concentration), response time, the pH and temperature ranges in which each electrode functions, and the interfering effects of foreign ions most likely to occur in the soil and plant environment. These data should be read in conjunction with Sec. II.C

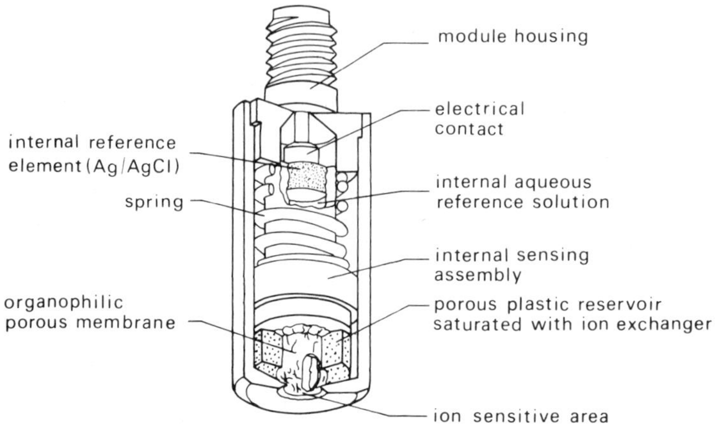

Figure 10 Free-flowing liquid membrane: liquid in porous plastic reservoir. [From *Orion Analytical Methods Guide*, data sheet on liquid membrane electrodes (electrode now obsolete). Orion Research Inc., Cambridge, Mass., 1978.]

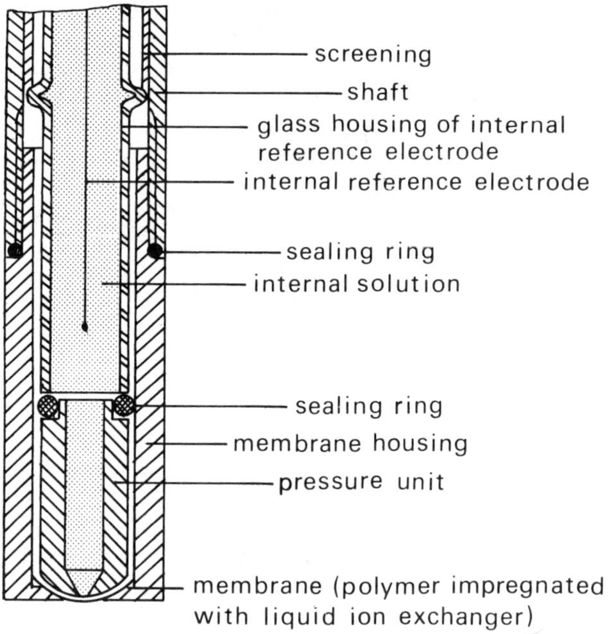

Figure 11 Liquid membrane immobilized in polymer. (From Philips Bulletin on Analytical Equipment, October, 1970, courtesy of Philips-Pye-Unicam Ltd., Cambridge, England.)

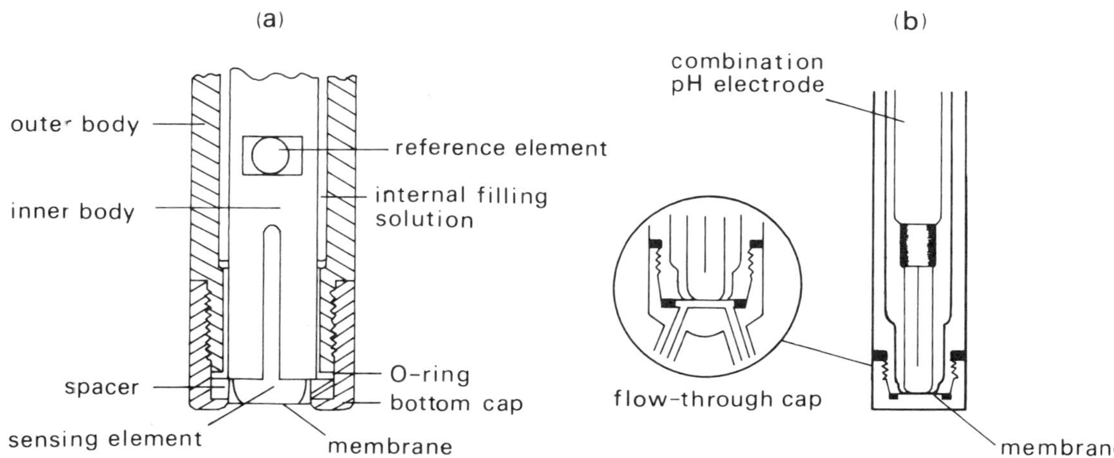

Figure 12 Gas sensing electrodes: (a) Ammonia electrode. (From *Orion Analytical Methods Guide*, Orion Research Inc., Cambridge, Mass., 1978.) (b) Sulfur dioxide probe. (Courtesy of Kent Industrial Measurements, Ltd., Chertsey, Surrey, England.) The sensing element is a pH glass electrode.

Table 9 Gas Sensing Probes for Dissolved Gases Incorporating a pH Glass Electrode

Gas	Manufacturer (see Table 4)	Detection limit (M)	Operational pH	Response time (s)	Interferences	Ref. (Soils)
NH_3	E	5×10^{-6}-5×10^{-2}	$\geqslant 11$	<60	Volatile amines	67-69
	O	10^{-6}-1	$\geqslant 11$	<60		
CO_2	O	10^{-4}-10^{-2}	5	<60	Volatile weak acids	
$NO + NO_2$	O	5×10^{-7}-10^{-2}	0-2	<60	Volatile weak acids	70
SO_2	E	5×10^{-5}-5×10^{-2}	1.7	<600	HF, HCl, Acetic acid	

to understand their full import. The reader's attention is drawn to the manner in which ion interference is expressed in Tables 6 and 8 for Orion ISEs (see footnote to the tables).

The lower limit of the operational concentration range in Tables 5 and 7 is taken as the detection limit of the specific ion, although the latter could well be a tenth of the lower limit under ideal conditions, e.g., good temperature stability, absence of interfering ions, and high stability of the measuring device. Response times are also adversely affected by these factors and could be as little as 1-10 sec under ideal conditions.

D. Gas Sensing Electrodes

Strictly, these are not sensitive to the specific ion, the sensing device being a pH glass electrode immersed in a film, directly connected to a small reservoir of a pH buffer solution within the electrode, between the glass membrane and an inert gas-permeable membrane (Fig. 12). The appropriate gas is generated from the specific ion solution by adjusting its pH (Table 9) in an enclosed microcell incorporating the gas-permeable membrane very close to the generating solution.

IV. APPLICATIONS IN SOIL SCIENCE AND RELATED DISCIPLINES

A. Introduction

Ion-selective electrodes offer a rapid method of determining the activities of a range of ions in soil extracts and in soil pastes and suspensions. In extracts they are a cheap alternative to spectrophotometric and colorimetric techniques, but the facility to measure the activity of an ion directly in a solution in equilibrium and in contact with the soil is unique. The extension of this to the determination of the activity of nutrients in the soil in situ is perhaps one of the ultimate goals in soil chemistry and plant nutrition studies.

Although ion selective electrodes have been in use in soil science for nearly 20 years, only rarely have they been established in routine analytical techniques. Their main limitation is imperfect selectivity for the measured ion, so that the presence of other ions (as is usually the case in soils) causes errors, although the selectivity of electrode membranes has been progressively improved. This imperfect selectivity has necessitated careful selection of extracting agents and of reference electrode salt bridges. With a calomel reference electrode, saturated KCl is usually used but is obviously not ideal when measuring a halide, ammonium, or nitrate ion and out of the question for K^+ and Cl^- determinations (see Sec. II.C).

That ISEs measure activities can be a great advantage, but often total concentrations are required. They can be calculated from activities but generally the problem is overcome by equalizing the ionic strength of the standards and the samples.

The following is a review of how ISEs have been employed in soil science.

B. Cation-Sensitive Glass Electrodes

Sodium

Bower [18] realized the value of glass sodium-selective electrodes in soil salinity investigations. He first used a Beckman sodium aluminosilicate glass electrode for determining sodium in soil water extracts. This glass, however, was equally sensitive to K^+ which consequently had to be precipitated with tetraphenyl boron. A later Beckman sodium electrode, made with lithium aluminosilicate glass, was much less sensitive to potassium. With this, Bower [19] measured Na^+ in soil water extracts mixed 1:1 with 1 M magnesium acetate to give a constant activity coefficient and to buffer the solution to pH >6, thus eliminating interference from H^+ ions. He also determined exchangeable sodium in the soil by measuring Na^+ in a 0.5 M $MgAc_2$ extract with the electrode, and similarly the cation exchange capacity (CEC) by saturating the exchange sites with sodium then displacing it with 0.5 M $MgAc_2$. He found good agreement between his electrode measurements and measurements made on the same extracts by flame photometry. This observation was confirmed by Fehrenbacher et al. [20], who obtained a correlation coefficient of 0.9982 between electrode and flame photometer measurements of sodium in 126 soil water samples obtained from field piezometers and percolation studies (Fig. 13). However, El-Swaify and Gazdar [21] found that the Beckman sodium electrode gave poor response slopes in 0.2, 0.5, and 1.0 M ammonium acetate and also in 0.1 and 0.25 M $MgAc_2$, although in 0.5 M $MgAc_2$ the response was close to the theoretical (see Sec. II.B). Using an electrode in a soil suspension or paste may produce errors caused by the surface charge on soil particles, e.g., the suspension effect with the sodium glass electrode [22].

Quite large errors may be involved in electrode measurements in soil pastes. Susini et al. [23] obtained 25-30% errors in sodium determinations in saturated soil pastes with a sodium glass electrode. They stated, though, that in salinity investigations a high degree of accuracy is not always necessary and the lack of it is compensated for by the speed of measurement. Nevertheless, over the wide concentration range which they measured (13-10,000 $\mu g\ g^{-1}$ Na), the correlations between the electrode measurements and measurements made on the supernatant solutions were very good ($r = 0.9986$). The regression coefficient ($b = 0.75$) was rather small, but this may have been in part due to the difference between activity measured by the electrode and concentration by the flame photometer.

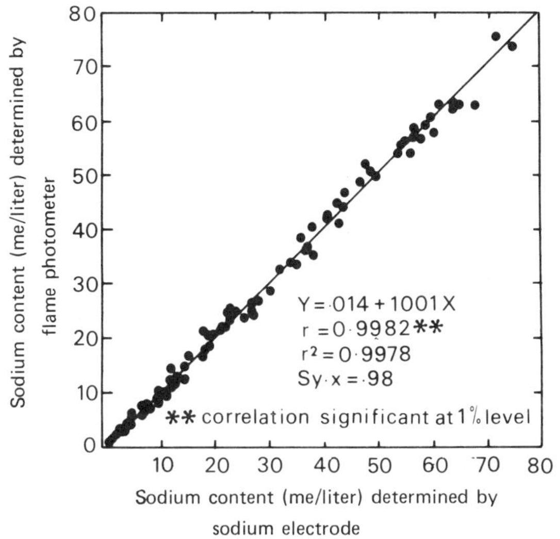

Figure 13 Relationship between the sodium contents of soil water as determined by the ISE and flame photometry. $(Na)_{photometry} = 0.014 + 1.001(Na)_{ISE}$; r = 0.9982**. [Reproduced from Ref. 20, Soil Sci. Soc. Am. Proc., *27*, March-April, No. 2, p. 153 (1963) by permission of the Soil Science Society of America.]

Orlov and Alzubaidi [24] measured sodium activities in 1:5 soil: water suspensions with a glass electrode to distinguish solonetz and saline soils. They found that these values correlated well with exchangeable sodium; solonetz soils had pNa values of 1.5 or less, and nonalkaline soils pNa values greater than 3.0. Alzubaidi [25] used the same method to measure the degree of dissociation of the exchangeable sodium in soils expressed as:

$$\text{Dissociation percentage} = \frac{a_{Na} \text{ in suspension}}{a_{Na} + \text{exch. Na}} \times 100$$

Using a technique described briefly in the next section on potassium, Krupskiy et al. [26-28] measured sodium ion activities in soil pastes with a glass electrode which could tolerate a 100:1 excess of Ca, 4:1 of K, and pH down to 1.6. They also made measurements in the field, and found that field soils at 15-20% moisture gave similar results to the same soils in the laboratory.

Potassium

Mortland [29] determined exchangeable potassium in 30 soils using the early Beckman cation selective electrode, with which Bower [18,19] had first measured sodium. This could only be achieved in soils containing little sodium and lithium because the glass membrane was equally sensitive to those ions. He eliminated H^+ interference by extracting the exchangeable K^+ with 0.5 M $MgAc_2$ in a 1:10 soil-solution ratio as Bower had done for Na^+, measuring K^+ in the suspension with the electrode and in the supernatant by flame photometry. The correlation between the two sets of values was satisfactory (r = 0.902). The standard curve for the electrode in KCl solutions, made up in 0.5 M $MgAc_2$, had a response slope of 55.3 mV per 10-fold change in a_{K^+} between 10^{-4} and 10^{-1} M. Mortland also compared the electrode measurements of exchangeable K with those made by flame photometry on M NH_4Ac extractions (r = 0.825). Surprisingly, a saturated KCl calomel reference electrode was used in these measurements.

Orlov and Alzubaidi [30] used a K^+ glass electrode to determine K^+ activities in soil suspensions (1:2, soil to water) and compared the results with flame photometer measurements made in the supernatant liquid. In forest soils the two determinations agreed well, but in a chestnut soil the electrode readings tended to be higher than those of the flame photometer, probably because of interference by sodium.

Very thorough investigations into the use of ion selective electrodes in soil pastes were made by Krupskiy et al. [26,27]. Measuring K^+ with a glass electrode, they found that the $p(K^+)$ value measured (= -log[K]) decreased linearly as the logarithm of the dilution factor of the paste increased. The regression coefficients were characteristic of individual soils. It was therefore necessary to select a standard moisture content in which to make measurements. These workers selected the lower liquid limit as determined by the Vasil'yev balance cone, at which the paste consistency is such that the cone sinks in it to a certain mark. Another interesting feature of their work was the use of lithium acetate in agar as a second salt bridge. They found lithium only interfered in the electrode measurement when the Li:K ion ratio was greater than 5:1. The acetate ion has a mobility similar to that of the Li^+ ion because of the considerable hydration capacity of Li^+.

The K^+ electrode was also sensitive to sodium and could not be used where the K:Na ion value was less than 1:1. Figure 14 shows that the electrode calibration curve was linear down to quite low K^+ concentrations, i.e., high $p(K^+)$ values. These critical $p(K^+)$ values increased with decreasing Na^+ concentrations. By fitting second-degree polynomials to a series of calibration curves, obtained in the presence of a range of sodium concentrations, Krupskiy et al. [26] were able to derive one equation to elucidate the effect on the EMF:K activity relationship at any given concentration of sodium.

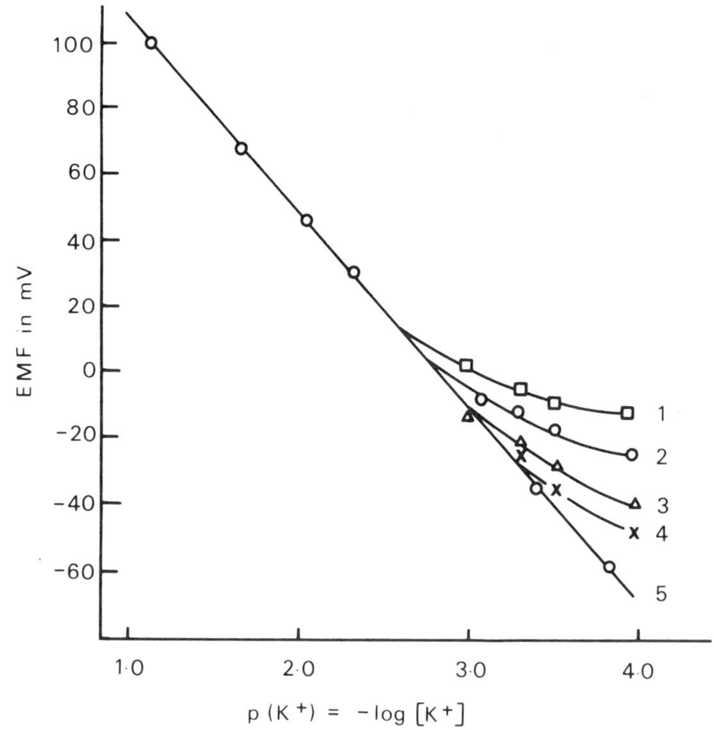

Figure 14 Dependence of the EMF of a potassium ISE on sodium activity in solution. p(Na) values: (1) 2.29; (2) 2.42; (3) 3.32; (4) 3.06; (5) no sodium. (From Ref. 26, reprinted with permission of Scripta Publishing Co., 7961 Eastern Ave., Silver Spring, Maryland 20910.)

Cation Exchange

The measurement of ion activities in suspensions lends itself to application in cation exchange studies. Kennedy and Brown [31] measured the rate of exchange of calcium and magnesium for sodium with a sodium electrode on sand-sized stream sediment and on kaolinite, montmorillonite, and illite. They added $CaCl_2$ solution to a suspension of clay or sand particles and measured the increase in Na^+ activity in the free solution with the electrode. The results were plotted on a chart recorder attached to the millivoltmeter. However, they found that a second method, in which the Na saturated clay or sand was added to a 0.085 M $MgCl_2$ solution, gave smoother curves. It proved possible to measure the rates of exchange up to 90% Na saturation. Beyond this point the rate of electrode response for the small change in activity was too slow.

Malcolm and Kennedy [32] also used this second method to measure rates of Na^+ and K^+ exchange with Mg^{2+}, Ca^{2+}, and Ba^{2+} on several standard clay minerals using Beckman electrodes for K^+ and Na^+. They added the K^+ or Na^+ saturated clay suspensions rapidly into 0.1 M Ca, Ba, or Mg chloride solutions and then monitored the release of Na^+ or K^+ into the solution with the electrode. However, all exchanges except K^+ for Ba^{2+} on vermiculite were as fast as, or faster than, the response rate of the electrode and hence could not be measured. Exchange on the illite and the kaolinite appeared to be even faster than the electrode response in the blank. The authors suggested this may be because suspensions of these two minerals were made more concentrated to compensate for their low exchange capacities and so were perhaps not well mixed. The electrode would then have measured artificially high local concentrations. They discounted any suspension effect because the percentage change in conductance in the liquid, produced by addition of the clay, was only of the order of 0.5%. The CECs of the clays, taken as the total K released into the solution after 15 min, agreed within 3% with values obtained using ^{85}Sr exchange.

C. Cation-Sensitive Liquid Membranes

Potassium, Calcium, and Magnesium

With the introduction of macrocyclic antibiotics as selective membranes, the problem of sodium interference in the measurement of K^+ was greatly reduced. Banin and Shaked [33] used a valinomycin-in-polymer K electrode with a calomel reference electrode and a lithium trichloroacetate salt bridge, to measure potassium activity in aqueous extracts of 16 soils, with soil:water ratios ranging from 1:10 to a saturated paste. The measurements were made in the filtered liquid. They obtained a high correlation between K concentrations measured by the ISE and the flame photometer but a low regression coefficient ($r = 0.92$, $b = 0.65$), i.e., 35% of the potassium in the extract was not measured by the electrode. Ionic strength effects were accounted for in the calculation of concentration from the activity measurements. The authors suggested that the potassium in the soil solution is bound by ligands and thus the concentration measured is determined by the dissociation constant of these ligands. However, we have often observed that submicron particles in filtrates from soil suspensions cause erroneously high flame photometer readings.

Calcium-selective liquid membrane electrodes have been employed to determine soil calcium by several workers. El Swaify and Gazdar [21] tested Orion Ca^{2+} and $Ca^{2+} + Mg^{2+}$ electrodes with extracting solutions for exchangeable cations. They found that the electrodes' response slopes generally decreased with increasing sodium or ammonium acetate concentrations up to 1 M and that only the calcium electrode in 0.2 or 0.5 M NaAc gave sufficient response to be of practical value in determining exchangeable calcium. The calcium electrode was sensitive to magnesium to such an extent that Mg could not be determined by the difference between the Ca^{2+} and $Ca^{2+} + Mg^{2+}$ electrode measurements.

Woolson et al. [34] compared electrode and atomic absorption determinations of calcium in 0.5 M NaAc extracts of soil diluted 10-fold. The correlation and regression coefficients for 12 soils were both 0.99. Similarly, Doiron and Chamberland [35] obtained, for 93 soils, good correlation between calcium concentrations measured by the Orion electrode and by atomic absorption in 0.5 M KCl soil extracts ($r = 0.91$). However, the regression coefficient was only 0.85 and they also suggested that the 15% Ca^{2+} not measured by the electrode might be complexed by ligands.

McLean and Snyder [36] investigated pH-dependent charges on illite and bentonite with the aid of ion selective electrodes. Activities of Ca^{2+} and Rb^{+} were measured in mixed suspensions of H^{+}, Ca^{2+}, and Rb^{+} saturated clay, Rb^{+} with a Beckman cation glass electrode, and Ca^{2+} with the Orion $Ca^{2+} + Mg^{2+}$ electrode. The Beckman glass electrode is sensitive to all monovalent cations, so a correction had to be made for hydrogen.

Ammonium

The Philips ammonium electrode has been used to measure ammonium activities in soil pastes [37] but is not ideally suited for use with soil suspensions because of the low NH_4^{+} concentrations normally present and because of its relatively high selectivity for potassium (see Sec. III.C).

D. Anion-Sensitive Liquid Membranes

Nitrate

The measurement of nitrate has probably been the most important application of ion selective electrodes in soil. Only rarely is nitrate adsorbed by the soil and so its total concentration in the soil is measurable in the soil solution. However, soils with some positively charged surfaces will adsorb nitrate; then a salt solution, usually a sulfate, is used to displace the nitrate. Except for chloride in saline soils, interfering ions are not usually present in sufficient concentration to cause serious errors in the electrode measurement of nitrate, hence its popularity in usage.

Several extracting solutions have been used to remove nitrate from the soil for measurement with the electrode. Myers and Paul [38] obtained 94% recovery of added nitrate, measured by the electrode in clear water extracts of soil (1:5 and 1:10 soil to water, shaken for 30 min). Electrometric determinations correlated well with measurements made on the same extracts by the phenoldisulfonic acid colorimetric method (r = 0.991 for 75 soils) and the relationship between the two determinations was almost 1:1 (b = 0.98). Extractants containing $CuSO_4$ (with Ag_2SO_4 added to precipitate chloride) and saturated $CaSO_4$ were also useful. Other workers used 0.01 M $CuSO_4$ [39], 0.005 M $CuSO_4 + Ag_2SO_4$ [40], saturated $Ca(OH)_2$ [41], 0.25 M K_2SO_4, saturated $CaSO_4$ [42], and $KAl(SO_4)_2$ [43] as extractants for nitrate.

Soil:solution ratios varied from 1:5 to 1:20. Too great an excess of solution could give too small a concentration of nitrate. Shaking periods used were generally between 15 min and an hour. These methods mainly correlated well with determinations by other recognized techniques, including phenoldisulfonic acid colorimetry and steam distillation.

Black and Waring [44] extracted NO_3^- from an oxisol as efficiently with 0.05 M K_2SO_4 as with 4 M KCl (oxisols have positively charged sites that adsorb anions). Measurements made with the Orion electrode in the supernatant 0.05 M K_2SO_4 extract gave a correlation coefficient of 0.99 with the steam distillation determinations made on the same solutions and the regression coefficient was 0.97 (Fig. 15). Except in such soils, the extracting solutions serve mainly to keep a uniform

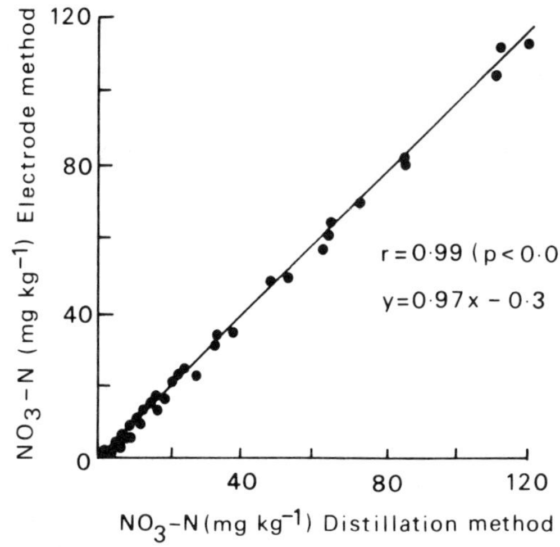

Figure 15 Nitrate levels in field samples determined by the nitrate ISE and by a distillation-colorimetry method. [From Ref. 44, A. S. Black and S. A. Waring, Nitrate determination in an Oxisol using K_2SO_4 extraction and the nitrate-specific ion electrode, *Plant and Soil, 49*(1): 211 (1978), reproduced by permission of Martinus Nijhoff Publishers BV, The Hague, The Netherlands.]

ionic strength. They may also prevent dispersion of the clay particles which can cause errors in the measurement [42].

The most abundant interfering ion in these soil extracts is generally chloride. As already mentioned, Ag_2SO_4 has sometimes been used to precipitate the chloride. Raveh [45] added 0.1 M sodium citrate to aqueous soil extracts to overcome interferences (1 ml sodium citrate to 9 ml extract) and found it effective except where $Cl:NO_3 >$ 100:1.

Milham et al. [46] used a buffer of 0.01 M $Al_2(SO_4)_3$. 0.01 M Ag_2SO_4, 0.02 M boric acid, and 0.02 M sulfamic acid adjusted to pH 3 with sulfuric acid, to eliminate chloride, nitrite, and organic anion interferences in soil extracts. This was added 1:1 to the aqueous soil extracts prior to nitrate determination with the electrodes. Francis and Malone [47] added sulfanilamide and 0.01 M H_2SO_4 to soil extracts to complex nitrite as a diazonium salt. The H_2SO_4 aids the reaction, destroys HCO_3^-, and maintains uniform ionic strength.

It is also possible to measure nitrate with an electrode directly in soil suspensions and pastes, and in moist soil in situ in the field. Bremner et al. [48] developed a technique for determining nitrate in 1:2 soil-water suspensions. The mixture was stirred and left for an hour, then the measurement was made while the suspension was magnetically stirred. They obtained almost perfect agreement between results from this method, from electrode measurements made in the supernatant solution, and from the steam distillation method. They also achieved 100% recovery of added nitrate on eight contrasting soils.

Øien and Selmer-Olsen [39] recovered 98% of added nitrate in 2:5 soil: water pastes but found that increasing the ratio decreased the recovery. Onken and Sunderman [49] obtained good recoveries in 513 soils in 1:5 soil-water suspensions with only 15 min shaking.

Bound [50] measured nitrate with a polymer membrane electrode in soil pastes made up in saturated $CaSO_4$. He found a better correlation between results obtained with an ISE and analyses done on a Technicon AutoAnalyzer when the ISE was coupled to an Ag-AgCl reference electrode with a salt bridge immobilized in PVC, than to a saturated calomel reference electrode. With an immobilized salt bridge there is no liquid junction potential set up [51].

Measurements of soil nitrate in situ in the field were made on the Broadbalk Classical Experiment at Rothamsted by Nair and Talibudeen [52]. Permanent holes, lined with plastic tubes, were maintained in the soil at different depths, throughout the summer months, under winter wheat and in adjacent fallow soil. To make a measurement, the soil at the bottom of a hole was wetted with about 5 ml of deionized water then left to come to equilibrium for about 10 min. A Corning ISE, with a micro-calomel reference electrode, was inserted and the reading on an Orion Ionalyzer taken after 60 sec. Sometimes it was difficult to wet the soil when it was dry and cracked, so the method

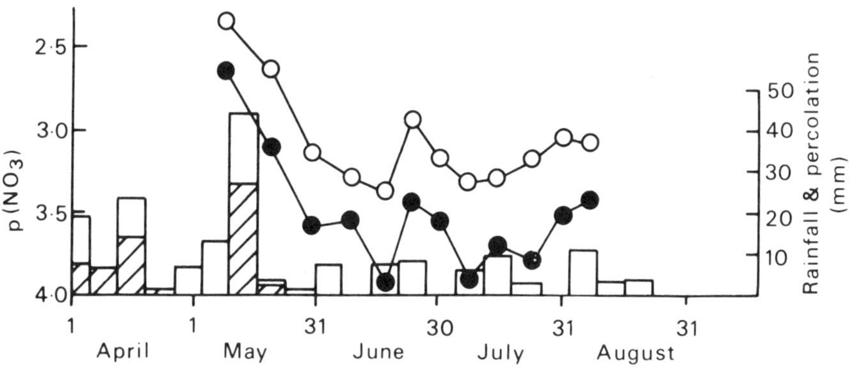

Figure 16 Change of $p(NO_3) = -\log[NO_3]$ in fallow and cropped soil at 15 cm depth with time on the Broadbalk Classical Experiment, Rothamsted, Plot 07, and histograms of weekly rainfall and percolation through 0-50 cm soil (shaded), for April-August, 1175. o — o — o: fallow o — o — o: cropped [From Ref. 53, M. B. Page and O. Talibudeen, Nitrate concentrations under winter wheat and in fallow soil during summer at Rothamsted, *Plant and Soil, 47*(3):527, (1977), published by Martinus Nijhoff Publishers BV, The Hague, The Netherlands.]

was later modified [53]; the soil was stirred to a paste immediately after the addition of the water and the measurement made 2 or 3 min later. The soil was then taken to be approximately at water holding capacity. Both methods proved to be reproducible. Typical results are shown in Fig. 16, showing the effect of the crop on nitrate levels on a plot given mineral nitrogen fertilizer. Using this method the weekly uptake of NO_3^- -N by a winter wheat crop was estimated and compared with weekly uptakes as measured by regular sampling and analysis of the crop [54]. The correlation between the two methods was 0.77 ($p < 0.01$).

It seems quite likely that ion-selective electrodes will be increasingly used in monitoring nutrient ion concentrations in soil-free composts used in glasshouse crop production. In the relatively small volumes of composts used per plant, ion concentrations and salinity levels are critical. Preliminary investigations have been made on the use of electrodes in determining NO_3^-, K^+, and NH_4^+ concentrations in water extracts (with added ionic strength adjuster; see Sec. II.D) of sedge peat compost [55]. High recoveries of added nitrate have been achieved (90-110%) but only 50-60% of K^+ and NH_4^+, probably because of the cation exchange properties of the peat.

E. Crystalline Solid-State Membranes

Chloride

The halide electrode most important to soil science is that which is sensitive to chloride. In saline soils, the natural chloride concentration is important, because of its potential toxicity to plants; and in artificial systems created for cation exchange studies, chloride is often the only anion present. There is, therefore, much scope for the use of chloride-sensitive electrodes in soil science, although up to now reports of their use are few.

Hipp and Langdale [56] compared Cl^- determinations in water extracts of soil by titration, both with a silver/silver chloride electrode and a chloride ion selective electrode as end point detectors, and by direct measurement with the chloride-selective electrode. The titrations gave identical results, but those from direct measurements were slightly higher. The regression equations were:

$$y_1 = 1.055x - 0.085$$

$$y_2 = 1.046x + 2.58$$

where

x = results obtained by titration using the Ag-AgCl electrode
y_1 = results obtained by titration
y_2 = results by direct measurement, using the Cl^- electrode

Selmer-Olsen and Øien [57] recovered all the chloride added to a soil by shaking it with 0.5 M NH_4NO_3 (1:5, soil:solution) and measuring Cl^- in the filtered acidified extract with an Orion electrode. They found no significant difference between measurements made with the electrode and by colorimetry.

The chloride electrode can be used in a rapid method for the determination of the cation exchange capacity (CEC) of soils. The soil is washed with ethanol to remove excess salts, then leached successively with 1 M NH_4Cl and 0.04 M NH_4Cl, and finally with 0.5 M Na_2SO_4. The Na_2SO_4 leachate is analyzed for NH_4^+ and Cl^- and the CEC calculated as the difference, in meq, between the two. Smart et al. [58] followed this procedure, analyzing the leachate for chloride by titration and with an Orion solid state electrode. Calibrating the electrode with standard ammonium chloride solutions showed good agreement between the two methods, although with NaCl standards the comparison was poor. This was because NH_4Cl accounts for any complexing between NH_3 and the silver of the electrode. Greenhill and Peverill [59] used basically the same method, but with a mixed displacing solution of 1.5 M KNO_3 and 0.25 M $Ca(NO_3)_2$ and an ammonia probe to measure the ammonium.

Krupskiy et al. [60] made measurements of Cl^- in soil pastes with a solid-state electrode. The pastes were prepared with the Vasil'yev cone as described above. The reference electrode was a calomel with a KNO_3-agar salt bridge. The disadvantage of this bridge is the slow electrode response it creates, 10-15 min in this case. The same electrode was also used to make measurements in situ in the field, with soil moisture contents of 13-16%, but without the salt bridge.

Although the *bromide* ion is not of great importance in soil science, some workers have used electrodes to measure it. Abdalla and Lear [61] recovered an average of 97% of bromide added to soil by extraction with water (1:2, soil:water) and measurement in 0.01 M $NaNO_3$ with a solid state bromide electrode. Saffigna et al. [62] determined bromide in lysimeter leachates and in 0.05 M K_2SO_4 soil extracts.

Fluoride

Fluoride is considered as a pollutant on agricultural land, hence increasing importance is attached to its measurement. It was determined in soils, rocks, and plants with an ISE after it had been rendered water soluble by fusion of the sample with an alkali [63,64]. After fusion, McQuaker and Gurney [64] dissolved the sample in HCl, bringing the solution to pH 8. Aluminum, iron, calcium, and magnesium and also silicates were then filtered off as insoluble compounds to prevent complexation of the fluoride ion. The ISE measurement was made in a TISAB buffer (see Sec. II.D) at pH 5.2. A slightly acid pH was chosen because hydroxyl ions interfere. The citrate ion in TISAB also helps prevent Al-F complexes forming; Crenshaw and Ward [63] increased the usual citrate concentration to enhance this effect, but without success. The EMF measurements were made in the buffered sample solutions by known addition techniques (see Sec. II.E). McQuaker and Gurney [64] recovered between 95 and 100% of fluoride added to grass, hemlock, and soil samples by this method.

Sulfide

Sulfides form under anaerobic conditions in soil, and free hydrogen sulfide, toxic to plants even at low concentrations, may accumulate. Allam et al. [65] determined H_2S concentrations in submerged rice soils by measuring S^{2-} concentrations, with a silver sulfide membrane electrode, and pH. Hydrogen sulfide concentrations were calculated using the equilibrium constant (K) for the reaction

$$H_2S \rightleftharpoons 2H^+ + S^{2-}$$

Thus,

$$K = \frac{(a_{(H^+)})^2 \, a(S^{2-})}{a_{(H_2S)}} = 10^{-20.9}$$

Electrode potential measurements were made in situ in the field and in samples taken back to the laboratory using a silver/silver chloride reference with a potassium nitrate salt bridge.

The sulfide ISE was calibrated in serially diluted sodium sulfide solutions (made with 1 M sodium hydroxide and purged with nitrogen) down to a sulfide concentration of 10^{-6} M. To obtain a calibration curve for sulfide concentrations below this value, the electrode potential was measured in a solution of constant *total* sulfide concentration at different pH values. In such a solution the *free* sulfide ion and hydrogen ion concentrations are inversely proportional [Eq. (18)]. From this and an electrode potential/pH response curve, the sulfide ion concentrations for different pH values were calculated and a calibration curve for sulfide ion concentrations down to 10^{-20} M plotted.

Copper

Copper is complexed by soil organic matter. Bloomfield and Sanders [66] measured Cu^{2+} concentrations, with an Orion ISE, in solutions 10^{-4} M with respect to Cu^{2+} and 0.1 M with respect to KNO_3 (for constant ionic strength) in the presence of various amounts of organic matter. The latter consisted of the dialysate and colloidal fractions of extracts of composted lucerne, soil, and peat. The dialysate complexed more Cu and hence reduced the solution concentration of Cu^{2+} more than did the colloidal fraction. The degree of complexing decreased with pH.

F. Gas Sensing Electrodes

Ammonia

The ammonia gas probes have found two particular niches in soil science and agronomy: (1) in the determination of the cation exchange capacity of soils and clays and (2) ammonia determination in Kjeldahl digests for total N. An example of the use of an ammonia probe in soil CEC determinations has already been mentioned [59]. A simpler method, that only involves one measurement, has been devised for clays [67]. Salt-free ammonium-saturated clay (50-100 mg) is suspended in magnetically stirred ammonia-free water. 0.5 ml of 10 M NaOH solution is added to convert the adsorbed ammonium to ammonia, which is then determined directly with the probe (see Sec. III.D). This method gave satisfactory results for kaolinites, illites, and montmorillonites.

Ammonia formed in Kjeldahl digests is generally determined either by an automated colorimetric technique or by steam distillation and titration. The former method requires expensive equipment and the latter is slow and tedious. The probe offers a cheap and rapid alternative to these. Bremner and Tabatabai [68] described a method in which a sample containing approximately 1 mg of N was digested and the final digest made up to 100 ml. A 10 ml sample of this was mixed with 10 ml of 2.5 M NaOH and 80 ml of water. This solution was magnetically stirred and the NH_3 probe inserted to make the measurement.

Exchangeable and water soluble ammonium in soil can also be determined with the ammonia probe [69]. Concentrations in filtered KCl and water extracts measured with the probe correlated almost perfectly with those measured by steam distillation, giving a regression coefficient of 1.

Nitrogen Oxides

The nitrogen oxides gas sensing probe can be used to determine nitrite in acidified solutions. Tabatabai [70] tested it in standard nitrite solutions made up with reagents commonly used as extractants for soil nitrate and nitrite (2 M KCl, 1 M KCl, saturated $CaSO_4$, 0.1 M LiCl, 0.15% $CaCl_2$, 500 μg ml^{-1} P as $Ca(H_2PO_4)_2 \cdot H_2O$). One part of a $H_2SO_4 + Na_2SO_4$ solution (190 g anhydrous Na_2SO_4 + 53 ml concentrated H_2SO_4 per liter) was added to 10 parts of the test solution to give a final pH of 1.2 for nitrite measurement with the probe. Of the extractants, only 2 M KCl caused significant error, all the others gave results within 2% of the known nitrite concentration.

Nitrite measurements made in actual soil extracts and water samples with the probe and by the colorimetric method of Barnes and Folkard [71] correlated perfectly (r = 1, b = 1). Also, more than 99% of nitrite added to soil extracts was measurable with the probe. Ammonium, Na^+, K^+, Ag^+, Ca^{2+}, Mg^{2+}, Cu^{2+}, Hg^{2+}, nitrate, chloride, sulfate, and phosphate did not cause any interference in the measurements when present in the test solutions at 0.1 M concentration. Of these, Cu^{2+} and Hg^{2+} are particularly important because they are often added to soil extracts to prevent microbial activity and they do cause errors in the colorimetric methods.

G. Continuous Flow Measurements

Because of the rapid response time of most ISEs, they can be used in continuous flow systems, either to measure ion concentrations in a series of solutions passed through sequentially or to monitor the concentration of a flowing solution. Greenhill and Peverill [59] used the flow cell shown in Fig. 17 for the measurement of NH_3 and Cl^- concen-

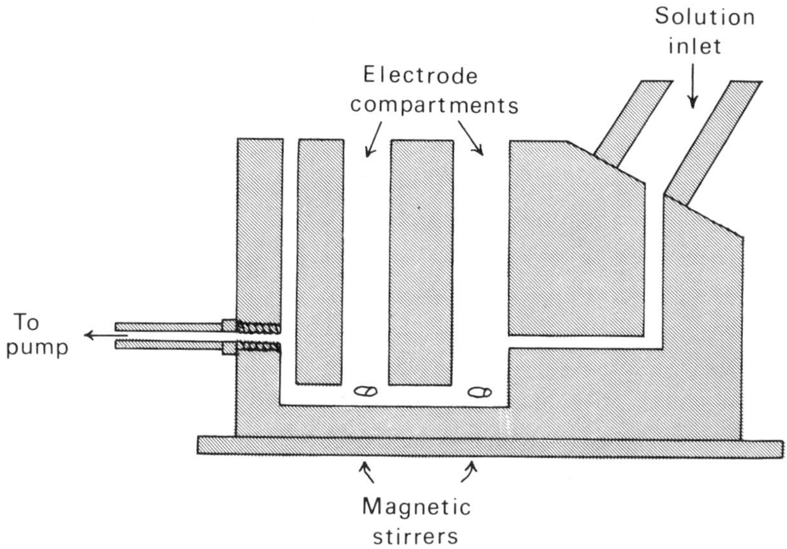

Figure 17 Flow cell for measuring ammonia and chloride ions. (From Ref. 59.)

trations in their CEC determinations. They found that it was not possible to make both measurements concurrently as hydroxyl ions, from NaOH added to the samples for NH_3 determination, interfere with the chloride determinations. Determining one ion at a time, the flow cell gave results comparable to those made individually.

Mansell and Elzeftawy [72] designed a flow cell incorporating a combination chloride and reference electrode to monitor chloride percolating through a soil column (Fig. 18). Testing it with chloride solutions passed through a column of glass beads, they demonstrated that the response time of the electrode was sufficiently small to enable it to monitor solutions flowing at 35 ml hr^{-1}.

In a study of plant growth and nutrient uptake in flowing culture solutions, Clement et al. [73] used nitrate and potassium ISEs to monitor and control concentrations of these ions. In this system, output from the electrodes activates an automatic titrator which adds solutions to the system to bring the ion concentration up to a preset level. The electrodes operate in a flow cell with a reservoir which is filled at fixed intervals from a bypass pipe via solenoid valves (Fig. 19). One particularly interesting feature in the electrode system is the use of a servo-system incorporating a potentiometer to correct the potential between the ion selective and reference electrodes in a stan-

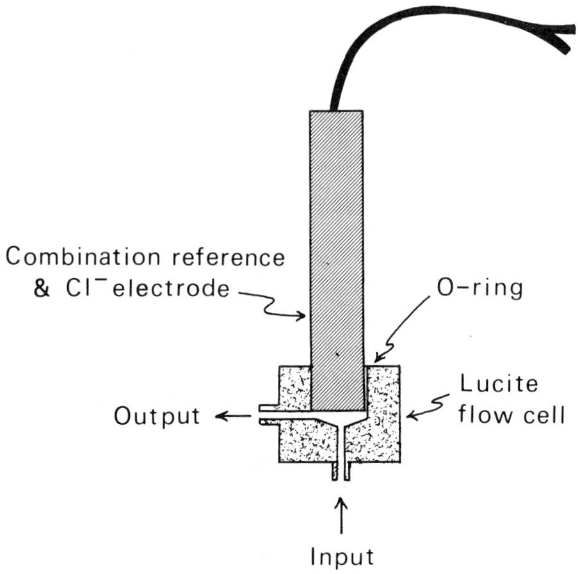

Figure 18 Diagram showing a combination reference and chloride ion electrode in a lucite flow cell. [From Ref. 72, *Soil Sci. Soc. Am. Proc.*, *27*, March-April, No. 2, p. 378 (1972) by permission of the Soil Science Society of America.]

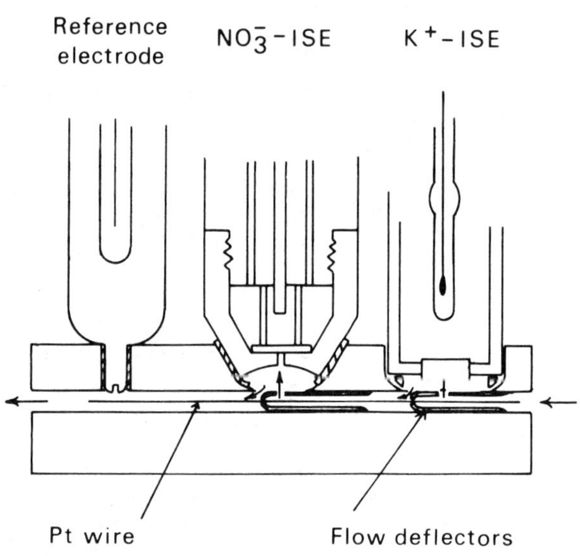

Figure 19 Details of a flow cell with two ion-selective electrodes. (From Ref. 73.)

dard solution to a preset value on the autotitrator controller. This enables the output of an electrode pair at a standard ion concentration to be maintained for several weeks within ±0.75 mV.

H. Plant Analysis

This book is concerned with soil analysis but, because in many laboratories soil and plant analyses go on side by side, it is appropriate to mention briefly the use of ISEs in the analysis of plant material. The subject has been adequately reviewed by Moody and Thomas [74].

Plant materials are processed into three forms for such measurements:

1. Extracts from dried material, made directly or after ignition
2. Filtered extracts of fresh material
3. Macerates and juices of fresh material

Generally, problems concerned with electrode measurements in extracts prepared from dried plant material differ little from those in soil extracts; indeed the same extraction techniques are often used [45,46,61,64]. However, the greatest value of ISEs in plant analysis lies in the possibility of rapid measurements of nutrient or pollutant ions in fresh material. The technique has been used to determine levels of potentially toxic ions in foodstuffs (e.g., nitrate, cyanide, bromide [74]). No great success seems to have been achieved in using ISEs in the diagnosis of plant nutritional problems. In concentrated aqueous macerates of wheat leaves, a Corning nitrate electrode did not give reproducible results, apparently because of poisoning of the liquid ion exchanger [75]. However, Moody and Thomas [74] found it possible in preliminary experiments to measure chloride with a combination electrode directly on the freshly cut surfaces of oranges, apples, bananas, potatoes, carrots, and cucumbers.

For nitrate ion assay in plant material, principally fresh or dried leaves, the most common extracting agents are cold or boiling water and dilute aluminum sulfate solutions, the latter because the NO_3^-/SO_4^{2-} free ion ratio in plant material is usually small. Also, the sulfate ion does not seriously interfere. By contrast, the chloride ion frequently occurs in appreciable concentrations in plant material and interferes more than the sulfate ion with nitrate ISE measurements. In such conditions, the known additions method (Sec. II.E) has to be used to evaluate the correct nitrate concentration. This is doubly useful in decreasing Cl^- interference by increasing the NO_3^-/Cl^- ion ratio, and in giving a reliable measure of the extent of interference. Similar direct analysis of such aqueous extracts has been reported for Cl^- ions, and should also be feasible for Na^+, K^+, Mg^{2+}, and Ca^{2+} in dilute acid extracts. For measurements of the fluoride, cyanide, and sulfur contents of plant material, however, plant material has to be digested such that the respective ionic forms are formed in the final extract by the hydrolysis, or oxidation, of the very stable covalent organic compounds of fluorine, cyanogen, or sulfur that can occur in plant material.

I. Microelectrodes in Plant Studies

In view of their rapid response times, ion selectivity, sensitivity, and portability, the logical extension of ISEs in plant studies is to the measurement of ion fluxes in various parts of the plant during growth. This is of special significance to research in the soil-plant environment if such measurements can be related to concomitant changes in ion concentrations in the root zone of the plant. The development in the last 20 years of microelectrodes with ion-sensitive tips, varying in length from 100 μm initially to a few micrometers at the present time, makes it feasible to measure ion concentrations in plant cells. No such measurements have been reported hitherto (although animal cells have been examined thus), neither have such electrodes been produced commercially, but their methods of manufacture are reported in detail in a recent book [76], from which this brief account is abstracted.

Some examples of solid and liquid membrane microelectrodes are shown in Fig. 20, which are principally used for measuring H^+, Na^+, K^+, Ca^{2+}, and Cl^- ion concentrations. Figure 20a and b illustrates the design of the ion-sensitive glass electrode in which the membrane tip protrudes from a protective and insulating sheath of lead glass, the protrusion being overlong originally, as shown. In modern versions, this protrusion is about 10 μm long. The two glasses are joined together by a glass-to-glass seal (type a) or by dental wax (type b). For better protection of the ion-sensitive membrane, a recessed-tip electrode was later invented (type c) in which the lead glass sheath, drawn to an open jet point with an outer diameter of about 1 μm, encloses the ion-selective membrane. Such electrodes have been used for H^+, Na^+, K^+, and Cl^- assay, the Cl^--sensitive element being an Ag/AgCl wire sealed by wax to the insulating glass.

The liquid ion selective microelectrode is illustrated in Fig. 20d, which is self-explanatory. The tip diameter is of the order of 1 μm and is drawn by a mechanical device from borosilicate tubing. Hitherto, calcium and chloride ion assays by such electrodes have been reported, but liquid membranes for Cl^- are not ion-selective enough for use in soil-plant research. For in vivo plant studies, the ion-selective liquid membranes, reported in Sec. III, presage in the near future the universal use of such microelectrodes for measuring also NH_4^+ and NO_3^- concentrations, especially if enterprising manufacturers take up the challenge of marketing them.

An account of the applications of ion selective microelectrodes to biomedical research [88] contains some additional details of technique that may interest and help potential users in the examination of plant cells.

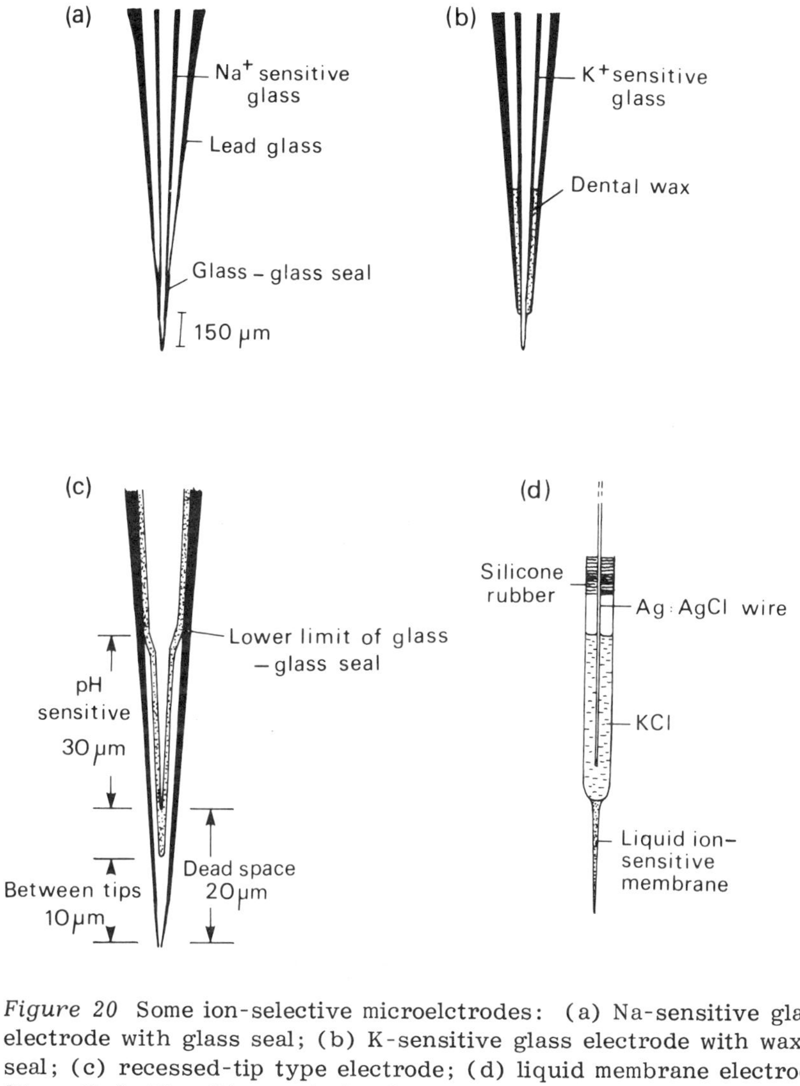

Figure 20 Some ion-selective microelctrodes: (a) Na-sensitive glass electrode with glass seal; (b) K-sensitive glass electrode with wax seal; (c) recessed-tip type electrode; (d) liquid membrane electrode. [From Ref. 76, with permission from R. C. Thomas, *Ion-Sensitive Intracellular Microelectrodes*, 1978. Copyright by Academic Press, Inc. (London) Ltd.]

REFERENCES

1. R. P. Buck, in *Ion Selective Electrodes* (E. Pungor and I. Buzas, Eds.), Elsevier, Oxford, 1978, p. 21.
2. G. Eisenman, in *Ion Selective Electrodes* (R. A. Durst, Ed.), National Bureau of Standards, Washington, D.C., Special Publ. 314, 1969, p. 1.
3. J. P. Sandblom, G. Eisenman, and J. L. Walker, J. Phys. Chem., *71*:3862 (1967).
4. W. E. Morf, E. Lindner, and W. Simon, Anal. Chem., *47*:1596 (1975).
5. W. E. Morf, G. Kahr, and W. Simon, Anal. Chem., *46*:1538 (1974).
6. R. P. Buck, Anal. Chem., *40*:1432 (1968).
7. W. E. Morf and W. Simon, in *Ion Selective Electrodes* (E. Pungor and I. Buzas, Eds.), Elsevier, Oxford, 1978, p. 149.
8. A. K. Covington, in *Ion Selective Electrodes* (R. A. Durst, Ed.), National Bureau of Standards, Washington, D.C., Special Publ. 314, 1969, p. 107.
9. G. J. Moody and J. D. R. Thomas, Lab. Practice, *28*:125 (1979).
10. M. S. Frant and J. W. Ross, Anal. Chem. *40*:1169 (1968).
11. J. D. R. Thomas, in *Ion Selective Electrodes* (E. Pungor and I. Buzas, Eds.), Elsevier, Oxford, 1978, p. 175.
12. R. A. Durst, in *Ion Selective Electrodes* (R. A. Durst, Ed.), National Bureau of Standards, Washington, D.C., Special Publ. 314, 1969, p. 375.
13. P. Sorensen, Kem. Maanedsbl., *32*:73 (1951).
14. G. Gran, Analyst (Lond.), *77*:661 (1952).
15. T. M. Florence, J. Electroanal. Chem., *31*:77 (1971).
16. J. Tacussel and J. J. Fombon, in *Ion Selective Electrodes* (E. Pungor and I. Buzas, Eds.), Elsevier, Oxford, 1978, p. 567.
17. W. Misniakiewicz and K. Raszka, ibid. p. 467.
18. C. A. Bower, Soil Sci. Soc. Am. Proc., *23*:29 (1959).
19. C. A. Bower, Trans. 7th Int. Congr. Soil Sci., *2*:16 (1960).
20. J. B. Fehrenbacher, L. P. Wilding, and A. H. Beavers, Soil Sci. Soc. Am. Proc., *27*:152 (1963).
21. S. A. El Swaify and M. N. Gazdar, Soil Sci. Soc. Am. Proc., *33*:665 (1969).
22. C. A. Bower, Soil Sci. Soc. Am. Proc., *25*:18 (1961).
23. J. Susini, M. Rouault, and A. Kerkeb, Cah. ORSTROM Pedologie, *10*:309 (1972).
24. D. C. Orlov and A. Alzubaidi, Agrokhimiya, No. 2, 135 (1965).
25. A. Alzubaidi, Iraq J. Agric. Sci., *4*:13 (1969).
26. N. K. Krupskiy, L. A. Rappoport, and A. M. Aleksandrova, Sov. Soil Sci., *6*:111 (1974).

27. N. K. Krupskiy, A. M. Aleksandrova, and L. A. Rappoport, Sov. Soil Sci., *6*:241 (1974).
28. N. K. Krupskiy, A. M. Aleksandrova, L. A. Rappoport, and Yu. I. Lapkina, Agrokhimiya, No. 7, 152 (1974).
29. M. M. Mortland, Quart. Bull. Michigan Agric. Exp. Sta., *43*:491 (1961).
30. D. C. Orlov and A. Alzubaidi, Agrokhimiya, No. 7, 137 (1965).
31. V. C. Kennedy and T. E. Brown, Clays and Clay Minerals, *13*:351 (1964).
32. R. L. Malcolm and V. C. Kennedy, Soil Sci. Soc. Am. Proc., *33*:247 (1969).
33. A. Banin and D. Shaked, Agrochimica, *15*:238 (1971).
34. E. A. Woolson, J. H. Axley, and P. C. Kearney, Soil Sci., *109*:279 (1970).
35. E. B. Doiron and E. Chamberland, Commun. Soil Sci. Plant Anal., *4*:205 (1973).
36. E. O. McLean and G. H. Snyder, Soil Sci. Soc. Am. Proc., *33*:388 (1969).
37. D. Hornby and M. E. Brown, Plant Soil, *48*:455 (1977).
38. R. J. K. Myers and E. A. Paul, Can. J. Soil Sci., *48*:369 (1968).
39. A. Øien and A. R. Selmer-Olsen, Analyst (Lond.), *94*:888 (1969).
40. M. Willems, Tidsskrift for Planteavl., *79*:495 (1975).
41. M. K. Mahendrappa, Soil Sci., *108*:132 (1969).
42. A. R. Mack and R. B. Sanderson, Can. J. Soil Sci., *51*:95 (1971).
43. A. Cottenie and G. Velghe, Meded. Fakulteit Landbouwwetenschappen, Rijksuniversiteit Gent, Belgium, *38*:560 (1973).
44. A. S. Black and S. A. Waring, Plant Soil, *49*:207 (1978).
45. A. Raveh, Soil Sci., *116*:388 (1973).
46. P. J. Milham, A. S. Award, R. E. Paul, and J. H. Bull, Analyst (Lond.), *95*:751 (1970).
47. C. W. Francis and C. D. Malone, Soil Sci. Soc. Am. Proc., *39*:150 (1975).
48. J. M. Bremner, L. G. Bundy, and A. S. Agarwal, Anal. Lett., *1*:837 (1968).
49. A. B. Onken and H. D. Sunderman, Commun. Soil Sci. Plant Anal., *1*:155 (1970).
50. G. P. Bound, J. Sci. Fd. Agric., *28*:501 (1977).
51. G. P. Bound and B. Fleet, J. Sci. Fd. Agric., *28*:431 (1977).
52. P. K. R. Nair and O. Talibudeen, J. Agric. Sci., Camb., *81*:327 (1973).
53. M. B. Page and O. Talibudeen, Plant Soil, *47*:527 (1977).
54. M. B. Page, J. L. Smalley, and O. Talibudeen, Plant Soil, *49*:149 (1978).
55. C. Turner and W. Carlisle, Private communication, 1979.

56. B. W. Hipp and G. W. Langdale, Commun. Soil Sci. Plant Anal., *2*:237 (1971).
57. A. R. Selmer-Olsen and A. Øien, Analyst (Lond.), *98*:412 (1973).
58. R. St. C. Smart, A. D. Thomas, and D. P. Drover, Commun. Soil Sci. Plant Anal., *5*:1 (1974).
59. N. B. Greenhill and K. I. Peverill, Commun. Soil Sci. Plant Anal., *8*:579 (1977).
60. N. K. Krupskiy, A. M. Aleksandrova, and Z. L. Boriskova, Sov. Soil Sci., *5*:749 (1973).
61. N. A. Abdalla and B. Lear, Commun. Soil Sci. Plant Anal., *6*:489 (1975).
62. P. G. Saffigna, D. R. Keeney, and L. L. Hendrickson, Commun. Soil Sci. Plant Anal., *7*:691 (1976).
63. G. L. Crenshaw and F. N. Ward, Bull. U. S. Geol. Surv., No. 1408, 77 (1975).
64. N. R. McQuaker and M. Gurney, Anal. Chem., *49*:53 (1977).
65. A. I. Allam, G. Pitts, and J. P. Hollis, Soil Sci., *114*:456 (1972).
66. C. Bloomfield and J. R. Sanders, J. Soil Sci., *28*:435 (1977).
67. E. Busenberg and C. V. Clemency, Clays and Clay Minerals, *21*:213 (1973).
68. J. M. Bremner and M. A. Tabatabai, Commun. Soil Sci. Plant Anal., *3*:159 (1972).
69. W. L. Banwart, M. A. Tabatabai, and J. M. Bremner, Commun. Soil Sci. Plant Anal., *3*:449 (1972).
70. M. A. Tabatabai, Commun. Soil Sci. Plant Anal., *5*:569 (1974).
71. H. Barnes and A. R. Folkard, Analyst (Lond.), *76*:599 (1951).
72. R. S. Mansell and A. Elzeftawy, Soil Sci. Soc. Am. Proc., *36*:378 (1972).
73. C. R. Clement, M. J. Hopper, R. J. Canaway, and L. H. P. Jones, J. Exp. Bot., *84*:81 (1974).
74. G. J. Moody and J. D. R. Thomas, J. Sci. Fd. Agric., *27*:43 (1976).
75. M. B. Page and O. Talibudeen, unpublished.
76. R. C. Thomas, *Ion-Sensitive Intracellular Electrodes*, Academic Press, London, 1978.
77. J. H. Puffer and R. S. Cohen, Chem. Geol., *15*:217 (1975).
78. D. T. Cooke, Commun. Soil Sci. Plant Anal., *6*:501 (1975).
79. K. L. Chang, J. C. Hung, and D. H. Prager, Microchem. J., *18*:256 (1973).
80. J. L. Paul and R. M. Carlson, J. Agric. Fd. Chem., *93*:729 (1968).
81. A. S. Baker and R. Smith, J. Agric. Fd. Chem., *17*:1284 (1969).
82. A. W. M. Sweetsur and A. G. Wilson, Analyst (Lond.), *100*:485 (1975).
83. R. C. Baker, Anal. Chem., *44*:1326 (1972).
84. P. M. McElfresh, J. Agric. Fd. Chem., *26*:276 (1978).

85. J. E. Harwood, Water Research, *3*:273 (1969).
86. R. L. LaCroix, D. R. Kerrey, and L. M. Walsh, Commun. Soil Sci. Plant Anal., *1*:1 (1970).
87. T. S. Light, in *Ion Selective Electrodes* (R. A. Durst, Ed.), National Bureau of Standards, Washington, D.C., Special Publ. 314, 1969, p. 349.
88. R. N. Khuri, ibid., p. 287.

3

Continuous-Flow and Discrete Analysis

KEITH A. SMITH and ALBERT SCOTT The Edinburgh School of Agriculture, Edinburgh, Scotland

I. INTRODUCTION

Routine chemical analysis involves a number of operations, such as pipetting, diluting, mixing, filtering, and transference of solutions to measuring instruments, which are repetitive and lend themselves to automation. Where instrumental methods of measurement are involved, the recording of results may also be automated by the use of chart recorders or data recording systems. The trend toward the elimination of manual operations, particularly where large numbers of samples are involved, has grown very rapidly in the last two decades. A major contribution to this phenomenon was the development of *continuous-flow analysis* (CFA) by Skeggs [1], which was subsequently further developed by the Technicon Instruments Corporation. More recently, a variant of the technique known as *flow injection analysis* (FIA) has been developed, and another system, known as *discrete*, or *batch*, analysis, has provided an alternative approach to automation. Both continuous-flow and discrete analysis have their origins in clinical chemistry, but the former system has been extensively applied in soil and plant analysis and now discrete systems are also finding similar applications.

In this chapter, the principles underlying the systems of continuous-flow and discrete analysis are described, and the applications of these techniques to analysis of soils and other environmental materials are reviewed.

II. PRINCIPLES

A. Continuous-Flow Analysis

The basic system is that devised for colorimetric analysis, in which solutions of sample and reagents are made to flow through plastic tubes by the action of a peristaltic pump. Interaction between successive samples is minimized by segmentation of the liquid stream by air bubbles. The relative quantities of reagents and sample are controlled by the diameters of the tubes. Streams are combined via T or Y pieces, mixing occurs during passage through coils of tubing, usually made of glass, and the color reaction takes place. The liquid then passes through the flow cell of a spectrophotometer, where the intensity of the color is measured and recorded on a chart. Comparable flow systems may be used for analyses involving flame photometry, atomic absorption spectrometry, or ion-selective electrodes as the method of measurement.

Dispersion and Interaction between Samples

The flow of a segmented liquid stream through a tube which is wetted by the liquid is illustrated in Fig. 1. The thin film of liquid, thickness

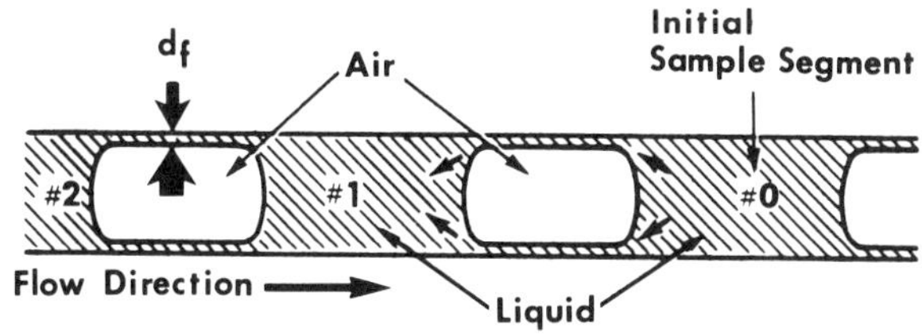

Figure 1 Segmented flow through an open tube. (From Ref. 2, reprinted with permission from L. R. Snyder and H. J. Adler, Dispersion in segmented flow through glass tubing in continuous-flow analysis: the ideal model, *Anal. Chem. 48*:1017. Copyright 1976 American Chemical Society.)

d_f, remaining on the tube wall from the passage of segment No. 0 will be overtaken by and mixed into segment No. 1, thus solute present in the former segment will become transferred to the latter. The process is repeated between segments 1 and 2, 2 and 3, and so on over many following segments. Thus the initially sharp transition in concentration (Fig. 2a), between a series of sample segments containing the analyte and preceding and following series containing a wash solution at zero concentration becomes transformed into a more gradual one (Fig. 2b). The relationship between concentrations in successive segments, and sample-sample interactions, have been investigated experimentally, mainly by Thiers and his co-workers. Mathematical models have been developed and used as a basis for computer programs to correct analytical data for interaction effects [2-7].

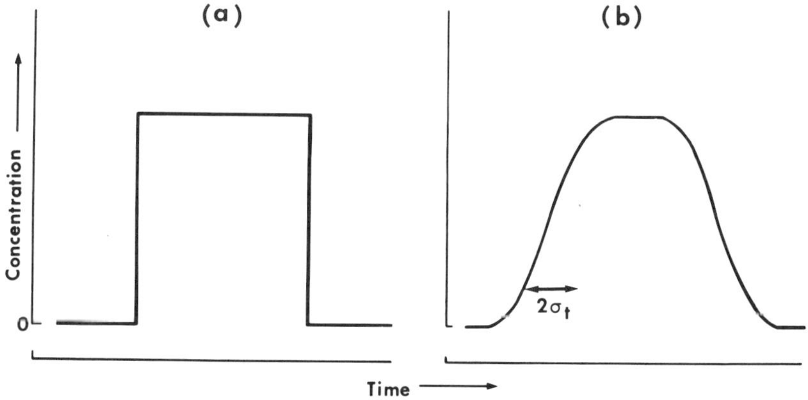

Figure 2 Sample concentration profiles in segmented continuous-flow analysis, (a) at injection, (b) at detector. (After Snyder, Ref. 8.)

The dispersion process limits the spacing of samples and the resulting throughput rate. In Fig. 2b the increasing and decreasing parts of the curve are integrated Gaussian curves [2,8], and the extent of sample dispersion is given by σ_t, which is the standard deviation of the Gaussian curve, and is measured in seconds. The throughput rate f depends on the spacing of sample bands in units of σ_t. Snyder [8,9] has derived a value of $8\sigma_t$ for curves with flat plateaux as shown in Fig. 2b. Thus, for the latter, the throughput rate f (in samples per hour) is given by

$$f = \frac{3600}{8\sigma_t} = \frac{450}{\sigma_t} \tag{1}$$

According to the theory worked out by Snyder and Adler [2,10] and Snyder [11], the variance σ^2 of the dispersion occurring during segmented flow through a length of tubing is given by

$$\sigma^2 = \left[\frac{(\pi^2/72) d_t^{\,4} u^{5/3} \eta^{2/3}}{\gamma^{2/3} V_s D_m'} + 1 \right] \frac{0.5\pi L d_t^{\,2} (u\eta/\gamma)^{2/3}}{V_s} \tag{2}$$

where d_t is the internal diameter and L the length of the tube, u is the linear velocity of the liquid of viscosity η and surface tension γ, V_s is the liquid-segment volume, and D_m' is a mass transfer coefficient that varies with the nature of the sample (solute) molecules and the viscosity of the liquid. For other liquids and/or temperatures D_m' is related to the value of D_m for water at 25°C ($D_{w;25}$) and to liquid viscosity by the relationship

$$D_m' = D_{w,25} \left(\frac{\eta}{0.0089} \right)^{-5/3} \tag{3}$$

The dispersion σ in Eq. (2) can be expressed in terms of σ_t [Eq. (1)]:

$$\sigma_t = \frac{\sigma}{n} \tag{4}$$

where n is the air-bubble rate. A number of other variables in Eq. (2) can be replaced by easily measured parameters such as the flowrate F_1. The dependence of sample dispersion on F_1 and n is shown in Fig. 3. Equation (2) has been tested over a wide range of values of the various parameters, and general agreement within ±10% has been found [8]. However, continuous flow systems are generally run at lower throughput rates than those indicated by Eq. (1) and (2). This is because of the additional contribution to dispersion resulting either from de-bubbling or, for the latest generation systems in which bubbles pass through the detector, from the requirement that each liquid segment be large enough to more than fill the detector flow cell [8].

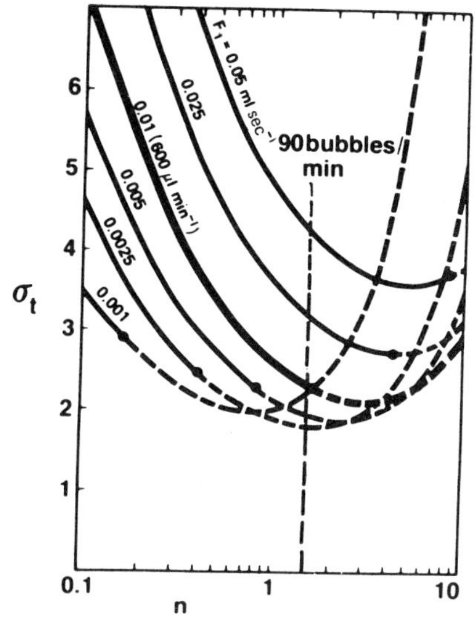

Figure 3 Dispersion as a function of F_1 and n (bubbles per second). (From Ref. 8.)

Mixing of Sample and Reagents

Mixing is required whenever a reagent or a diluent is added, and may be repeated several times in the course of one analysis. Complete mixing must occur in each liquid segment; otherwise an irregular detector output will result. Mixing is brought about by the natural fluid motions that occur within a short, moving segment of liquid bounded at each end by a gas-liquid interface. These motions, known as *bolus* flow (Fig. 4), result in rapid longitudinal mixing along the segment and in slow radial mixing [10]. In straight lengths of tubing, radial mass transfer occurs mainly by molecular diffusion, which is slow in liquids, but this can be overcome by inducing convective mixing simply by substituting a helical coil for the straight tube. Both the physical properties of the liquid and tube geometry are important in determining the rate of mixing. Viscosity, density and flow rate all affect the process, with viscosity the most significant factor. Tube internal diameter, helix coil diameter, and segment length are also important [12].

Flow Stability

To achieve a stable detector output, the proportions of sample to reagents must be constant for all segments derived from the same sample, which requires constant flow rates. However, pulsations do occur. A particular cause, in the case of Technicon peristaltic pumps, is roller

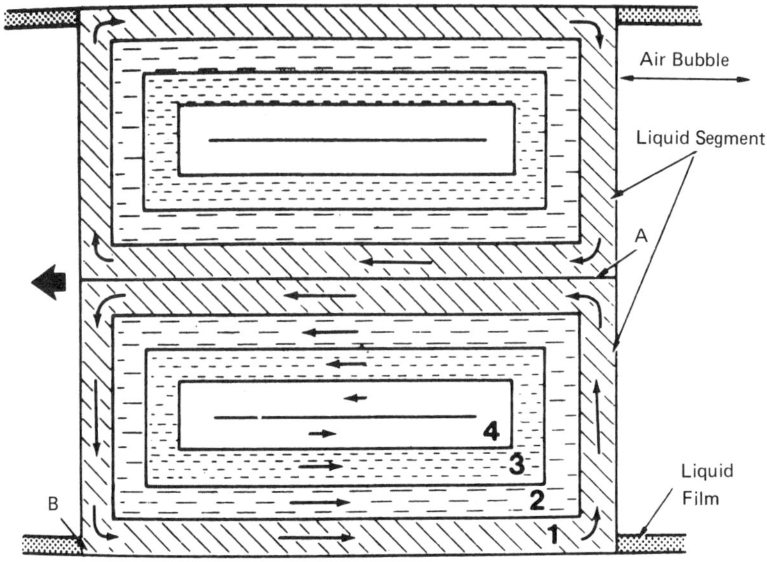

Figure 4 Diagrammatic representation of the flow patterns within a moving liquid segment (bolus flow). Dye added at A or B is immediately dispersed into a figure-of-eight pattern (region 1), then moves more slowly into adjacent regions 2, 3, 4. (From Ref. 10, reprinted with permission from L. R. Snyder and H. J. Adler, Dispersion in segmented flow through glass tubing in continuous-flow analysis: the non-ideal model, *Anal. Chem.* *48*:1022. Copyright 1976 American Chemical Society.)

lift-off from the platen, which affects some segments and not others. A way of minimizing this effect is to synchronize the pulsations of all sample and reagent streams by injecting the air bubbles in phase with roller lift-off [12].

Flow Injection Analysis

In the mid-1970s Růžička et al. [13,14] introduced a new concept of continuous flow analysis, known as flow injection analysis (FIA). In contrast with the conventional system involving a segmented sample stream, the flow injection method is based on the rapid injection of the sample solution into a carrier stream of reagent. The injected sample forms a well-defined zone which is transported by the carrier stream toward the detector. Mixing of sample and reagent occurs as a result of turbulence, which is induced by fast flow rates and small tube diameters.

The sample is injected as a "plug" into the reagent stream, but due to dispersion the characteristic detector response is a bell-shaped

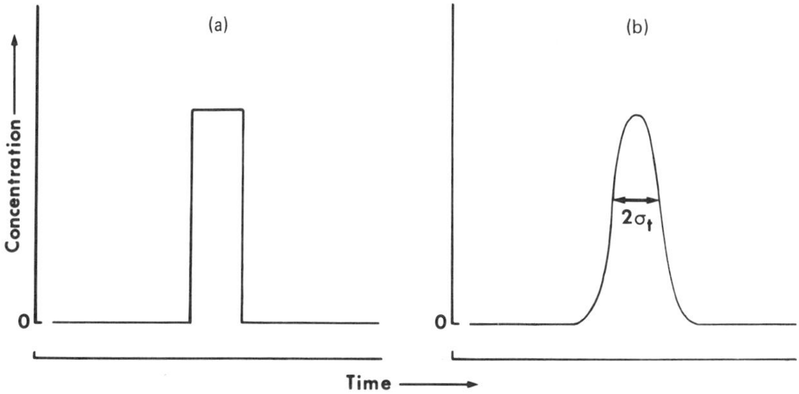

Figure 5 Sample concentration profiles in flow injection analysis, (a) at injection, (b) at detector. (After Snyder, Ref. 8.)

(Gaussian) peak (Fig. 5). The throughput rate depends on the spacing of sample bands in units of σ_t and, by analogy with the related problem in chromatography of the resolution between adjacent bands, $4\sigma_t$ is an appropriate spacing for this Gaussian distribution [8]. This is half the spacing for the band shape characteristic of segmented flow systems (see "Dispersion and Interaction between Samples" in Sec. II.A).

Although FIA systems are physically simpler than their CFA counterparts, the prediction of σ_t values is less well developed, at least for flow in coiled tubes. Existing theory indicates, however, that values for FIA are generally much higher than those for CFA, for systems of similar dimensions. This reflects the marked ability of air segmentation to break up laminar flow and decrease sample dispersion [8]. The absence of air in FIA, however, allows the degree of dispersion to be manipulated more readily, and it is claimed [15] that the control of dispersion may well be the most important aspect of the technique. Tubing of very small diameters (down to 0.25 mm) can be used in FIA, which significantly reduces the dispersion, but practical difficulties arise when such miniaturization is attempted with air-segmented systems [8].

An important factor in the achievement of satisfactory throughput is the time taken for chemical reactions to take place. In FIA, provided that the time taken for each sample-reagent mixture to pass through the system to the detector remains constant, and provided that this time is long enough to produce an adequate amount of colored or other detectable product, there is no need for chemical equilibrium to be reached in order to obtain valid analytical results [15].

The relative merits of segmented continuous-flow and flow injection analysis have been summarized by Snyder [8]:

FIA has advantages over CFA (as now practised) for the rapid analysis of samples that require short incubation times (e.g., of the order of a few seconds). But this limits FIA applications to very simple situations, and to dedicated as opposed to flexible or multichannel analyzers. Many automated assays require longer incubation times, and CFA then offers higher possible throughput rates. In addition, automated chemical analysis derives much of its appeal from its ability to mimic other laboratory manipulations: dialysis, evaporation-to-dryness, solvent extraction, etc. For these situations, the increased dwell time t within the system clearly gives the advantage to CFA over FIA. It thus appears that there is a place for both FIA and CFA in automated chemical analysis. The choice of one technique over the other should be made on the basis of the various compromises involved.

B. Discrete Analysis

"Discrete" or "batch" analyzers are designed to simulate the operations commonly used in manual procedures. The advent of microprocessors has made possible very sophisticated systems, which can be programmed to suit the needs of each type of analysis required.

In contrast with continuous-flow analysis, in which each sample in turn traverses a common pathway of tubes to a detector, in discrete analysis each sample is contained in its own particular test tube or reaction vessel until it reaches the detector. In a typical integrated system, disposable plastic cups containing the samples are loaded, together with reaction tubes, into racks on a conveyor. A predetermined aliquot of sample is transferred by means of a pneumatic or motor-driven syringe into the tube. One or more reagents are dispersed in a similar manner, and the mixture is stirred mechanically. The time interval between additions may be programmed to allow the necessary reactions to take place. During this sequence of operations the tubes are immersed in a constant temperature water bath to standardize reaction conditions, if necessary. The solution is finally pumped into the flow cell of a spectrophotometer and then to waste.

In many applications of the discrete analysis principle, only one or two of these operations may be employed. For example, automatic dispensing and dilution of samples may be all that is required prior to determination of metal ions by flame photometry or atomic absorption spectrometry. For pH determinations the automated steps could include the addition of water or calcium chloride solution to the soil, stirring, and insertion of the pH electrode into the suspension. In general, the large integrated analyzers are intended mainly for clinical chemical applications, although they can be readily adapted to a variety of other uses, whereas the simpler dispensers, diluters, and samplers are the instruments which have become more readily adopted for agricultural and environmental analysis.

III. PRACTICAL SYSTEMS – CONTINUOUS FLOW

A. General Aspects

For some years after the first publication of the continuous-flow method [1] the only commercially available equipment was the AutoAnalyzer produced by the Technicon Instruments Corporation, Tarrytown, New York. This equipment, in its newer versions, still occupies a major part of the market, but similar systems are now available from manufacturers in several countries.

A continuous-flow system consists of a number of modules linked together in series, the basic ones being a sampler, peristaltic pump, mixing coil, colorimeter, and recorder. Heating baths and units for Kjeldahl digestion, distillation, dialysis, and filtration may also be included, and a flame photometer may be used for the detection of cations such as K^+ and Na^+. Because of the modular nature of the system, it has been common practice in many analytical laboratories to assemble "hybrids" involving apparatus originating from several manufacturers or including items made in laboratory workshops, rather than to purchase a complete assembly from a single source. An example of such a hybrid system is shown in Fig. 6.

B. Pumps and Pump Tubes

Pumps

The flow of samples and reagents through a continuous-flow system is brought about by the action of a peristaltic pump. Constriction of a flexible tube, followed by displacement of the constriction along the tube, induces flow of the fluid within it. In the Technicon AutoAnalyzer system the pump takes the form of a series of chain-driven rollers which press the tubes against a flat metal platen (see Fig. 6), and other manufacturers have produced pumps which operate in a similar way, even though they may differ somewhat in roller and platen geometry. Pumps such as the Watson-Marlow Delta series (Watson-Marlow Ltd, Falmouth, Cornwall, England) operate on a different principle, whereby the tube is wrapped around a three-sided rotor with concave faces, the corners of which are machined to small radii. The tube becomes kinked at three points, and as the rotor rotates the kinks travel with the radii, thus causing a displacement of the contents of the tube.

Some types of pump operate at a single speed, in which case flow rates are determined only by tube sizes, but a wide range of speed settings is available on other models. Reference has already been made to the need for a constant flow rate, to achieve a stable detector output. This requirement is generally more important than the absolute value of the flow rate achieved.

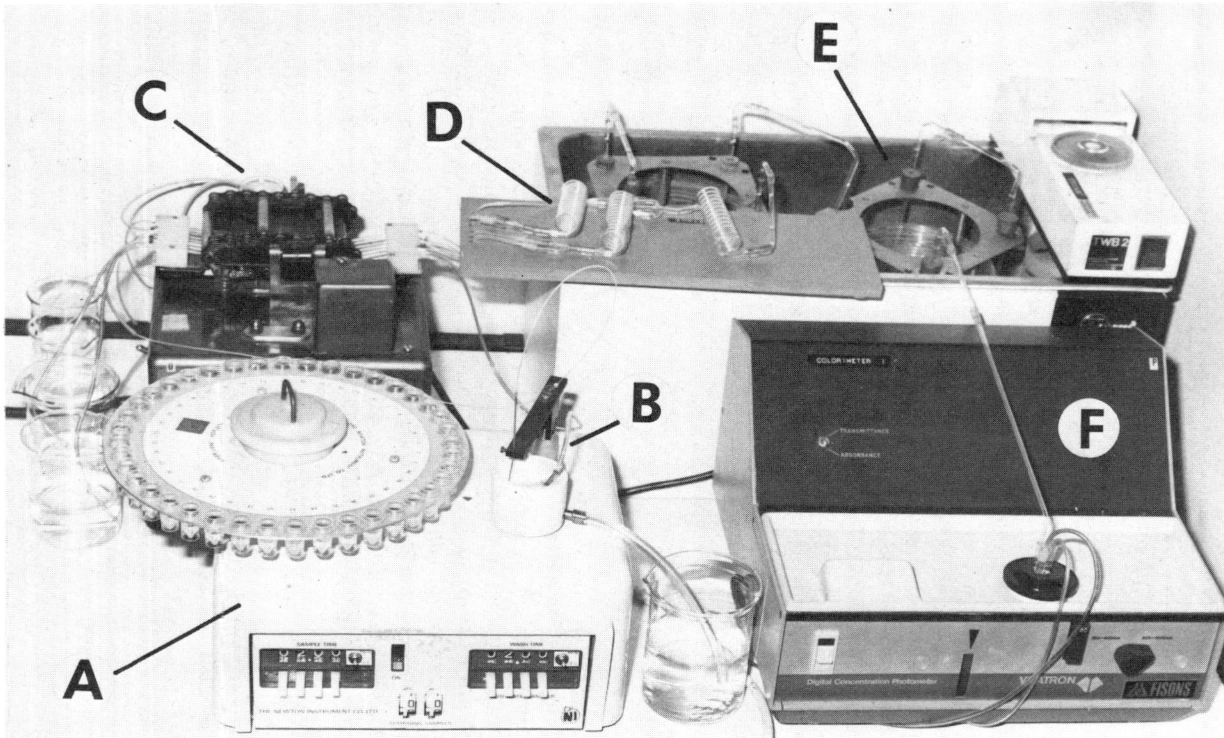

Figure 6 Typical laboratory-assembled continuous-flow system. A, sampler; B, sampling probe and wash reservoir; C, peristaltic pump; D, platter with mixing coils; E, water bath containing delay coils; F, colorimeter.

Table 1 Recommended Tubing for Specific Chemicals

Tubing	Chemical
Clear standard (polyvinyl chloride)	Acetaldehyde (dilute) Acids, mineral, dilute (to 50%) (not HF) Aqueous solutions Formaldehyde Glycerine Glycol Sodium hydroxide Water
Silicone	Acetic acid 95%, water 5% Acetone Alcohols (low molecular weight) Dioxane Pyridine
Acidflex (fluoro-elastomer)	Acids, concentrated (>50%) Benzene Chloroform Ethylene dichloride Styrene Toluene Trichloroacetic acid Xylene
Solvaflex	Carbon tetrachloride Hexane MIBK – ethanol (50:50) Methanol, ethanol, 2-propanol

Source: Ref. 16.

Pump Tubes

A number of different materials are available for use as pump tubes, depending on the chemicals passing through them. Those recommended for use in the Technicon AutoAnalyzer II system are shown in Table 1.

If the compatibility of tubing for a given solvent is not known, it may be tested as follows: Immerse a small piece of tubing in the solvent for about 30 min. If there is no sign of swelling or softening proceed with a dynamic test, by fitting a pump tube of the same material and pumping the solvent through it for 20-30 min. Then, measure the flow rate by pumping into a graduated cylinder for 5-10 min. Pump with the tube for an hour or more and again check the flow rate. If the two values are similar, rapid degradation of the tube has not occurred [16].

Table 2 Internal Diameters and Color Codes of Tubes Used in Continuous-Flow Analysis

Tube number		Internal diameter		
Acidflex, Solvaflex	Silicone	in.	mm	Color
-01	-	0.005	0.13	Orange & black
-02	-	0.0075	0.19	Orange & red
-03	-	0.010	0.25	Orange & blue
-04	-	0.015	0.38	Orange & green
-05	-	0.20	0.51	Orange & yellow
-06	-07	0.025	0.64	Orange & white
-07	-08	0.030	0.76	Black
-08	-09	0.035	0.89	Orange
-09	-10	0.040	1.02	White
-10	-11	0.045	1.14	Red
-11	-12	0.051	1.30	Gray
-12	-13	0.056	1.42	Yellow
-	-01	0.058	1.47	Clear
-19	-14	0.060	1.52	Yellow & blue
-13	-15	0.065	1.65	Blue
-14	-16	0.073	1.85	Green
-15	-17	0.081	2.06	Purple
-16	-18	0.090	2.29	Purple & black
-17	-19	0.100	2.54	Purple & orange
-18	-20	0.110	2.79	Purple & white

Source: Ref. 16.

Generally, if a pump tube is incompatible with a given liquid, there will be visible signs of deterioration during the above test. The tube can swell, crack, or snake, or can become cloudy or brittle, or turbidity may develop in the liquid as constituents leach from the tube [16]. If a pump tube cannot be found to pump a given solvent, it may be necessary to change to another solvent or to adopt a different chemical reaction as the basis of the desired determination.

If neither alternative is practical, the displacement bottle technique can be used to introduce the solvent. A liquid which *can* be pumped satisfactorily, and which is immiscible with the solvent and either much heavier or much lighter than it, is pumped into a stoppered flask containing the solvent, thus displacing it through an outlet tube at a flow rate equal to the pumping rate of the other liquid.

Temperature affects chemical resistance, and the recommendations in Table 1 apply to room temperatures only. By diluting a solvent with a relatively inert diluent, it is possible to decrease the chemical effect on the tube [16]. Tubes may have a life of up to 500 hr, but this will be affected by many factors and cannot be specified very precisely.

At a given rate of rotation of the pump, flow rate is determined by tube internal diameter, and for accurate proportioning of samples and reagents the dimensions must be accurately known. Standard pump tubing is made from extrusions of polyvinyl chloride manufactured to very fine tolerance limits and also treated to relieve mechanical stress. Two grades of this tubing are currently available: the normal grade, with internal diameters varying from 0.13 mm (0.005 in.) to 2.79 mm (0.110 in.) (Table 2), which is usually quoted with approximate flow rates, and the more expensive flow-rated grade in which the tubes have precise flow rates varying from 0.015 to 3.90 ml min^{-1}.

All manifold tubes have two colored identification collars exactly 6 in. (152 mm) apart to hold the tube in position when pumping. The color of the collars indicates the internal diameter, a standard color coding system having been universally adopted (Table 2).

C. Transmission Tubing and Flow System Accessories

Tubing is required for interconnection of modules and for use as sleeving to connect glass tubing to glass tubing or plastic nipples. Similar considerations of solvent tolerance to those for pump tubes obviously apply, and when, for instance, acid-tolerant tubing is being used to pump a reagent, sleeving and transmission tubing downstream in the system must be made of the same material or of glass. It is recommended that glass transmission tubing should be used where possible, to promote a smooth flow, and that the internal diameter of tubing, coils, and fittings should be the same, or at least within ±0.2 mm ([16].

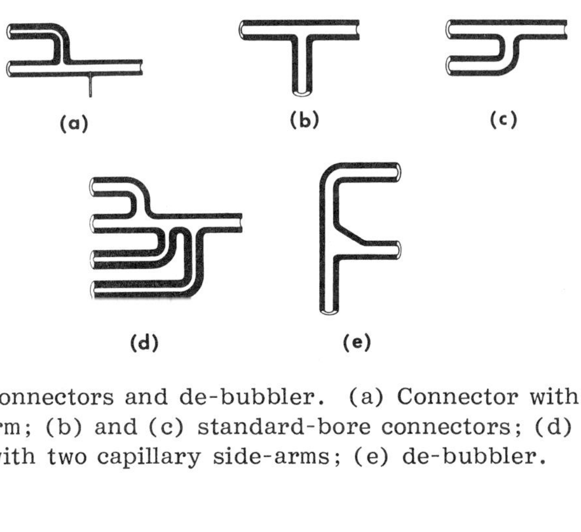

Figure 7 Connectors and de-bubbler. (a) Connector with metal capillary side-arm; (b) and (c) standard-bore connectors; (d) multiple connector with two capillary side-arms; (e) de-bubbler.

Plastic transmission tubing or sleeving should not be used for pump tubes because of the greater dimensional tolerances and lack of stress relief.

Connectors

Combination of the sample stream with one or more reagents is achieved with connectors ("cactus fittings") of the type shown in Fig. 7. The connector may have either a glass or platinum capillary side arm (Fig. 7a), the choice being dependent on the flow of liquid through it, i.e., platinum (Fig. 7a) for less than 0.25 ml min^{-1} and glass between 0.25 and 1.2 ml min^{-1}. For sample flow rates greater than 1.2 ml min^{-1} a standard bore glass connector (Fig. 7b and c) has to be used.

Air injection into the liquid stream is usually accomplished with an injection connector with a glass capillary side arm. However, if large flow rates are involved, this will result in the entry of a large number of small bubbles which will tend to aggregate in the mixing coils and upset the regularity of the segmented flow. In these conditions a connector with a wide-bore side arm is used instead. The larger the air bubbles are, the larger the cross section of mixing and delay coils downstream can be before the segmentation breaks down.

In first-generation Technicon AutoAnalyzers, the air tube size was calculated by summing all reagent tube volumes and dividing by four. Second-generation instruments, however, normally use smaller flow rates and tubing of smaller diameter, and air is steadily injected into the system by means of an *air bar* which compresses a short piece of silicone tubing attached to the air inlet line. The air bar moves up and down either because of the mechanical action of the peristaltic pump or under separate electronic control, so allowing a constant size of bubble to be injected into the liquid stream.

Mixing and Delay Coils

When any two streams are combined they are mixed by passage through a coil. Commercially available coils are usually made of glass, except where alkaline solutions are being pumped at elevated temperatures, for which Teflon or Kel-F coils are available. The glass coils are available in a number of standard sizes, the most commonly used being "single" (14-turn) and "double" (28-turn) helixes with a helical radius of 8 mm, corresponding to a length of tubing of approximately 0.7 and 1.4 m, respectively (Fig. 8a and b).

Single and dual inlet versions are available, the latter (Fig. 8c) avoiding the need for a separate Y-connector to be attached to the inlet end. For the Technicon SMA system, one coil consists of two helical sections side by side, to make the assembly more compact (Fig. 8d). Jacketed coils are used where cooling is required (Fig. 8e).

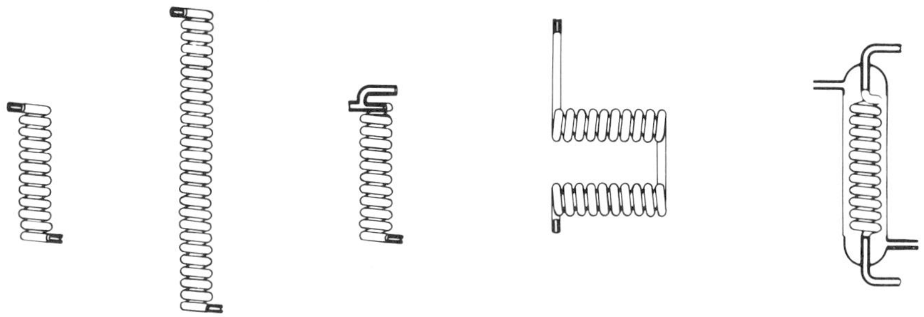

(a) (b) (c) (d) (e)

Figure 8 Mixing and cooling coils.

To facilitate immersion in controlled-temperature baths, delay (reaction) coils are commonly constructed with vertical inlet and outlet tubes and with the axis of the helix also vertical. Delay coils are available with helical radii of either 80 mm or slightly less, so that where two are required in a given system one can be fitted inside the other and so save space in the bath. The length of tubing in delay coils may be as much as 12 m, and at typical flow rates, solutions may have a total residence time within the coil of 10 to 40 min. The inlet to the coil should always be at the bottom, so that the liquid flows upward against a positive pressure. This is necessary to maintain the stability of the bubble pattern in the tube.

Cheap but satisfactory flow systems have been constructed for some applications in which plastic tubing has been used throughout, the coils being constructed by winding the tubing in a helix round some suitable cylinder such as a wooden dowel or glass test tube. If polyethylene tubing is used it may be molded by immersion in boiling water, and the cylinder dispensed with [17,18]. Glass helixes of any desired dimensions can also be contructed in the laboratory by a competent glassworker. In general, glass tubing is preferable to plastic because it is easier to keep clean.

In multichannel analyzer systems it may be necessary to adjust the times of arrival of different streams at the detectors. This is accomplished by using so-called phasing coils, which are merely small delay coils of known internal volume.

Solvent extraction, or liquid-liquid extraction as it is now more generally termed, may be carried out in continuous-flow systems. When the species of interest is readily extractable into the solvent, a series of mixing coils may be used to effect the extraction. In instances where extraction is more difficult this approach is impracticable because of the large number of coils required [19]. More efficient extractors containing glass beads or rings have the disadvantage of large dead volumes.

Much better mixing of the solvent and aqueous solution can be achieved by vibrating the coils on an ordinary laboratory shaker, as was demonstrated by Carter and Nickless [19], when extracting copper diethyldithiocarbamate into carbon tetrachloride. Using four coils in series, the extraction efficiency increased from 28% for stationary coils to 100% when vibrated at frequencies in excess of 40 Hz, without inducing significant carryover between successive samples.

Standard separators are available for separating the liquid phases after extraction; the design depends on whether the nonaqueous phase is heavier or lighter than water. When the complex extracted is colored it may be measured directly by passing the nonaqueous phase through the colorimeter flow cell.

Manifolds

An arrangement of tubes, coils, and connectors designed for one particular analysis is known as a manifold. This can be mounted on a platter around the pump which can be easily removed and replaced by another one when the apparatus is required for a different analysis. The removable platter may include the pump tubes, as in, for example, the Technicon AutoAnalyzer I system or, more simply, consist of the mixing coil assembly only (see Fig. 6). Other manifold systems can be in the form of an "analytical cartridge" (Fig. 9) either supplied complete by a manufacturer or made up in the laboratory. Such a cartridge contains all the necessary coils, connectors, and other accessories for a specific analysis, including, if necessary, a miniaturized heating bath.

When designing an analytical manifold it must be recognized that the size and type of tubes and connectors to be used will be governed by the volumes of liquid flowing through the system. Pumping large volumes through fine-bore tubing will create surging and back-pressure, which may eventually lead to a leak at one of the connections, which should be kept to a minimum. Several examples of manifold layouts are shown diagrammatically in Sec. V.

D. Samplers

After air segmentation of the liquid stream, the sample solution is added by means of an automatic sampler. This device normally consists of a motor-driven turntable carrying a number of sample cups, typically of 2 or 4 ml capacity, into each of which in turn a sampling probe dips and aspirates solution into the continuous-flow system. The probe then aspirates a wash solution from a reservoir before the sample turntable moves the next cup into position, so that successive samples are separated in the flow system by wash solution.

The sample probe, which is usually made from stainless steel, is connected to the sample tube with a short length of polyethylene

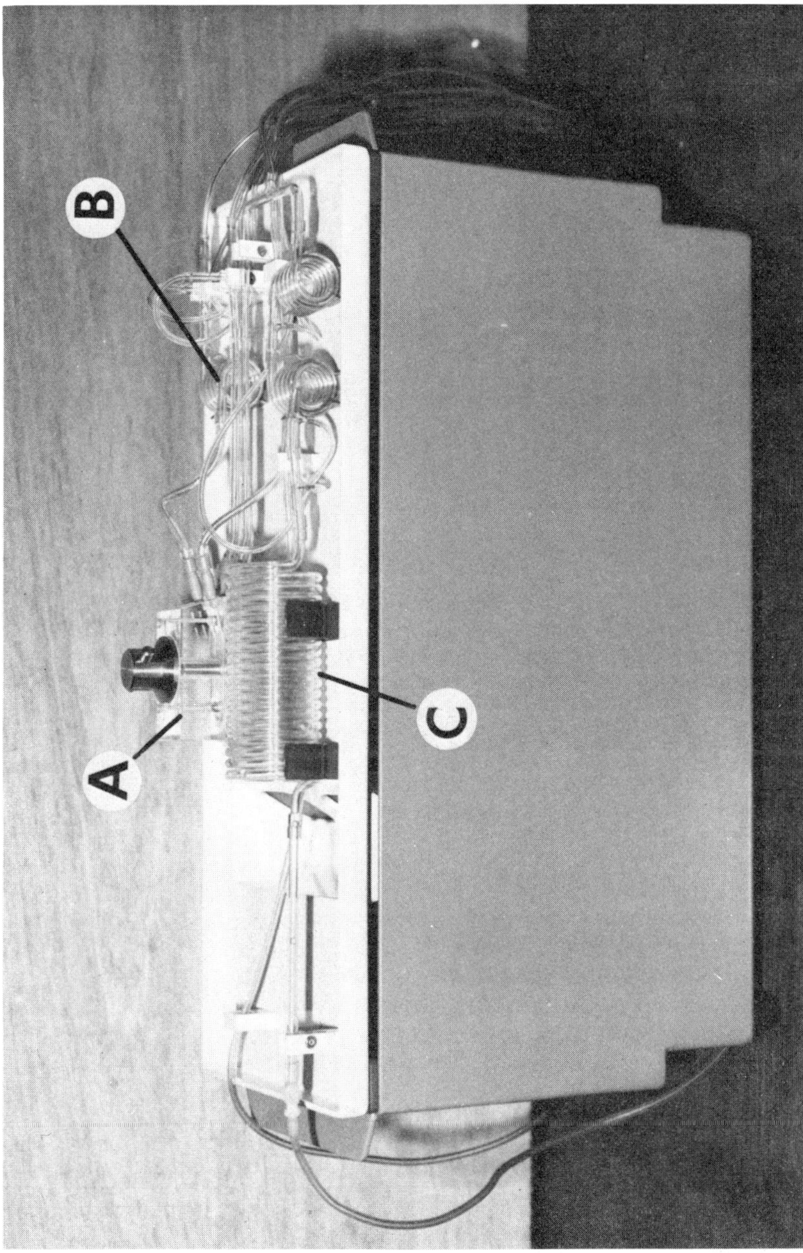

Figure 9 Analytical "cartridge". A, dialyzer block; B, mixing coils; C, delay coil.

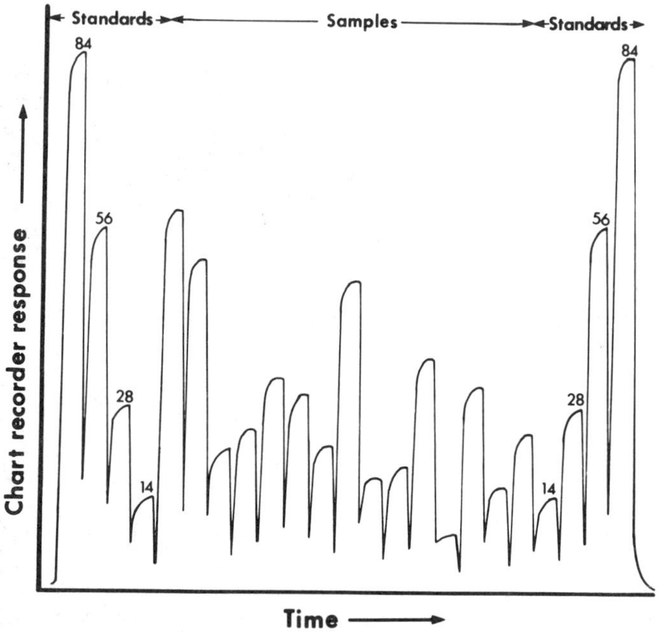

Figure 10 Recorder trace of adequately resolved peaks in segmented continuous-flow analysis (of ammonium in soil extracts). Standards are 14, 28, 56, and 84 μg N ml^{-1}. Analysis rate 60 samples per hour; sample:wash ratio 5:1. (From Ref. 20.)

tubing. This is used in preference to other materials because of its nonwetting properties, and should be the same diameter as the sample tube or smaller.

Both sampling and wash times are controlled by a cam or an electronic timer, and are adjusted in practice according to the requirements of the particular chemical reactions being employed. For optimum results, the minimum sampling time is that taken for the recorder trace of the most concentrated standard to reach a plateau and remain there for 5 sec. The time taken for the recorder to return to the baseline gives the maximum wash time. For most analyses it is not necessary to have complete resolution of the individual peaks, and the wash time can be progressively reduced to the point where significant interference between a sample of high concentration and a following one of low concentration is just avoided. A typical recorder trace of adequately resolved peaks is illustrated in Fig. 10.

A detailed discussion of sampler mechanisms, and the errors and inaccuracies which can arise in practice, can be found in the book by Furman [21].

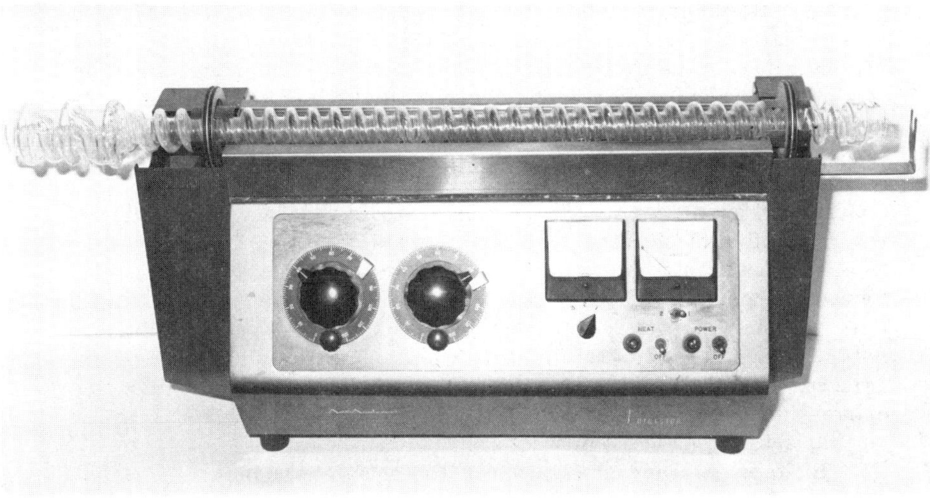

Figure 11 Technicon digestor (inlet and outlet connections not shown).

E. Other Modules

Digestor

The automation of Kjeldahl nitrogen determinations is normally restricted to the determination of ammonium, after digestion of the sample by conventional manual procedures, but in some circumstances it may be extended to include the digestion stage and also the distillation of ammonia where required (see below).

The Technicon Digestor (Fig. 11) is designed to bring about a rapid digestion of materials which can be relatively easily dissolved. It consists of a 0.9 m glass tube with a helical groove running along its length which is rotated at a few revolutions per minute and heated to 300-400°C. Sample solution and a digestion mixture made up of sulfuric, perchloric, and selenous acids are introduced at one end of the tube, and flow in a thin film along the groove as the tube rotates. The digestion proceeds rapidly during the passage along the tube, and conditions are adjusted so that it is complete before the mixture reaches the end. The time taken is normally a few minutes, which is much less than the time for the conventional manual digestion. At the outlet end the digested sample is diluted, part of it is aspirated for further stages of the continuous-flow analysis, and the excess is aspirated to waste.

The nitrogen (and phosphorus) contents of fertilizers and many plant materials have been determined in this way, as they pass readily into solution in sulfuric acid or sulfuric acid/hydrogen peroxide mixtures and may therefore be transferred to the sampler in this form

[22-25]. However, a detailed study by Davidson et al. [26] showed that results for the nitrogen content even of easily decomposed biological materials tended to be slightly low when the Technicon digestor was used, compared with those obtained by conventional Kjeldahl digestion, and recoveries of nitrogen from pure compounds could be as low as 70%. They concluded that where 1% accuracy was required the automatic digestor method was not sufficiently reliable, unless the characteristics of the sample material were well defined and suitable reference standards were available. In spite of this, in situations where an occasional value which was 5% low could be tolerated, the increased speed of analysis using automated digestion was worthwhile. Far greater problems arise with soils because of the more resistant nature of much of the organic matter and the insolubility of the mineral fraction, and the technique has been generally rejected in favor of conventional digestion followed by continuous-flow analysis of the resulting solution of ammonium ions.

Even where easy dissolution of the sample is achievable, many workers prefer off-line digestion methods, because of the hazards associated with pumping concentrated sulfuric acid and the problems of removing obnoxious acid vapors. Modifications to the Technicon Digestor to deal with corrosion and vapor problems have been described by Bosset and Jenni [27] but there is no obvious way of permanently avoiding the rupture of a pump tube and consequent spillage of the highly corrosive digestion acids.

Distillation Apparatus

It is possible to incorporate a distillation stage into continuous-flow analytical procedures. This is valuable for the determination of ammonia in digests of soil samples which contain sediments, colored substances, or high concentrations of heavy metals which could cause interference with the colorimetric determination [28,29].

In the method described by Keay and Menagé [28], the sample, together with 10% sodium hydroxide solution, is pumped through a coil in a bath at 116°C. As the mixture emerges from the coil a stream of air is injected, which blows the mixture into a series of splash-heads. The trapped liquid is pumped to waste, while the air stream containing steam and ammonia passes through the splash-heads and merges with a stream of 0.1 M hydrochloric acid which absorbs the ammonia. Because of the corrosive effect of hot alkali on Pyrex glass, a steel coil was used by Keay and Menagé; Skjemstad and Reeve [30] went further by also using a Teflon still.

As an alternative to distillation, ammonia may be separated from Kjeldahl digests by stripping from solution with a stream of air, then trapping in dilute acid prior to measurement by an ion-selective electrode [31] or a colorimetric method [32].

Heating Bath

Many reactions used for colorimetric determinations take an unacceptably long time to proceed to completion, but this is unnecessary, provided that sufficient color has developed to give adequate sensitivity and provided that all samples and standards have had an identical reaction time and temperature. Heating baths play a vital part in autoanalytical procedures by maintaining uniform reaction conditions.

Baths may be filled with either water or mineral oil for temperatures up to 80 or 90°C. Above that, mineral oil or another suitable liquid such as glycerol [28] must be used. Uniformity of temperature for all samples requires rapid circulation of the liquid in the bath, and a sensitive temperature control device; ordinary thermostats may result in sufficient variation in temperature to produce a clearly visible cycling of reaction conditions as shown by the output signal from the detector. This problem is discussed in detail by Furman [21].

Dialyzer

The use of dialysis to separate ions or small organic molecules from macromolecular species which might interfere with the analysis is widespread in clinical applications of continuous-flow systems. The usefulness of the technique in soil analysis relates particularly to situations where extracts contain strongly colored humic substances or suspended matter which interfere in colorimetric [33-35] or nephelometric [36] determinations. Also, dialysis affords a convenient method of separating a desired species in reactions where colloids or precipitates are formed, and can be used as an alternative to a dilution/recycling stage when sample concentrations are too high for satisfactory colorimetric determination.

A dialyzer is usually constructed from acrylic sheet, and typically comprises two circular plates containing matching spiral grooves which may be up to 2 m in length. A membrane, usually made of cellulose acetate, is sandwiched between the plates, which are then clamped together. The assembly is immersed in a constant-temperature bath, in a horizontal position. The segmented stream containing the sample is pumped through the groove over the membrane, while a segmented receiver stream (of similar composition, apart from the analyte) flows at a similar rate below. The substance to be determined passes through the membrane and reaches a concentration in the receiver stream proportional to the concentration in the original sample. The rate of dialysis depends on the nature of the membrane and the temperature; the final concentration in the receiver stream will depend also on the length of membrane. For some analyses, very much shorter path lengths are adequate and simple dialyzers are now available made from thick acrylic blocks containing grooves 7.5-60 cm in length (see Fig. 9). These are not immersed in a bath; where required, elevated temperatures are achieved by preheating the donor solution in a coil and relying on the excellent thermal insulation properties of the acrylic blocks [37].

Reductor

A common method for determining low concentrations of nitrate in water and soil extracts involves its reduction to nitrite, which is then determined colorimetrically. A method using copperized cadmium for this reduction [38] was modified for continuous-flow analysis by Armstrong et al. [39] and first applied to soil extracts by Henriksen and Selmer-Olsen [33]. The reductor is easily constructed in the laboratory. It consists of a 140 mm long, 4 mm i.d. vertical glass tube packed with 40-60 mesh (250-420 μm) cadmium which has been stirred briefly in copper sulfate solution. The solution enters the bottom via a 2 mm i.d. glass tube after passing through a de-bubbler. Gas bubbles may form in the column, particularly after prolonged use, and when this occurs they can be removed by locating the de-bubbler after the reductor column [33] (see Fig. 18, p. 155). Air is then re-injected into the stream in the normal controlled manner to create a segmented sample once again.

F. Measuring Instruments

The liquid stream emerging from the manifold must enter some suitable measuring instrument to allow the concentration of the species of interest to be determined. A single instrument may be used where only one species is being determined, but where several measurements are required on a single sample, more than one may be involved. Multiple instruments are usually operated in parallel, following the splitting of the sample stream.

The most common instrument is the colorimeter, or spectrophotometer, used to measure the products of color-producing reactions. A variety of other instruments may also be used, e.g., flame photometers, atomic absorption spectrometers, ion-selective electrodes, nephelometers, and radioactivity detectors.

Colorimeter and Spectrophotometer

The colorimeter measures the transmission of light of a particular wavelength through a solution of a colored species which absorbs strongly at that wavelength. The relationship between transmission and concentration is given by the Beer-Lambert law:

$$I = I_0 e^{-kcd} \tag{5}$$

where I and I_0 are the intensities of the light transmitted and the incident light, respectively; c is the concentration of the species (moles per liter) and k is its *molar extinction coefficient* for the particular wavelength; and d is the path length of the light through the solution.

Rearranging Eq. (5) and taking logarithms,

$$-kcd = 2.303 \log \frac{I}{I_0} \tag{6}$$

Often it is preferable to express the ratio in terms of the *absorbance*, A (optical density), because this quantity is directly proportional to concentration:

$$A = \log \frac{I_0}{I} = Kc \tag{7}$$

The Beer-Lambert law is an idealized relationship applying to monochromatic radiation. In actual practice, using filters, prisms, or gratings, it is impossible to obtain perfect monochromatic radiation. However, the narrower the band which can be obtained, the closer experimental data conform to the law [40]. Departures from a linear relationship may occur in systems involving ionization or dissociation equilibria which may be affected by such factors as pH and dilution, or where high concentrations cause changes in the refractive index of the solution [40,41].

A colorimeter was used as the basic detection system of the first AutoAnalyzers. In this device, the output of a single tungsten-filament lamp is divided and collimated to provide two light beams, each of which passes through a filter of the desired wavelength and is focused on a separate photocell. The output voltage from one photocell is used to measure the optical transmission of sample solutions and standards, and the output from the other is used as a reference. The ratio of the sample voltage to the reference voltage is measured in a null balance system in a chart recorder.

Linearizing devices which convert the logarithmic concentration-transmission relationship to a linear one are commonly used, to obtain output data in a more convenient form.

Frequently, nowadays, more sophisticated spectrophotometers are adapted for use with continuous-flow systems. These instruments work on the same basic principle as the simple colorimeter, but instead of using filters they are adjustable to any desired wavelength by means of grating or prism monochromators. They are also likely to have superior sensitivity and better linearity of response, and may be equipped with digital readout as well as analog recording facilities. Colorimeters designed specifically for continuous-flow analysis have also undergone considerable development. Phototubes have replaced the photocells, and multichannel devices incorporating quartz-fiber optics are now available, in which up to six or eight determinations can be carried out in parallel. For full details of the features currently available, recent manufacturers' literature should be consulted.

Flow Cells. Flow cells may be tubular or square in cross section, with an internal diameter of 1.5-3.5 mm and a path length of 10-50 mm. Second-generation flow cells have a much reduced internal volume compared with the first AutoAnalyzer devices. A typical size is 1.5 mm diameter × 15 mm path length, which corresponds to an unsegmented volume of 26 μl. Smaller flow rates are required to wash out these small cells, e.g., 1 ml min^{-1} for an analysis rate of 60 samples per hour, compared with 3 to 4 ml min^{-1} for older types of cell [37].

Many types are equipped with either an external or an internal de-bubbling facility (the former is illustrated in Fig. 12) because the passage of air bubbles through the cell results in large and rapid changes in the optical transmission, and thus in large oscillations of the recorder. Circuits have been devised to isolate the recorder during the passage of a bubble through the cell and thus make de-bubbling unnecessary [42], and this approach has been further developed and adopted in recent commercial instruments.

Turbidimetric Methods. Colorimeters or spectrophotometers may be used for turbidimetric determinations such as sulfate as barium sulfate [43-45]. Barium chloride solution containing a suspension-stabilizing agent such as gelatine or polyvinyl alcohol is added to the sample, and the absorbance caused by the barium sulfate precipitate is measured in the flow cell. The technique requires precautions to be taken to minimize adsorption of barium sulfate on the flow cell wall, which causes recorder baseline drift. The usual method is to wash with dilute HCl containing a wetting agent [45] or an EDTA buffer solution [43].

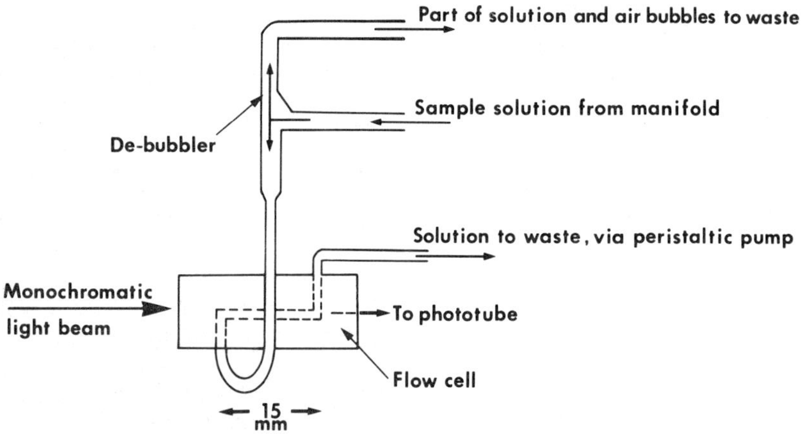

Figure 12 Flow cell with external de-bubbler.

Flame Photometer and Atomic Absorption Spectrometer

The incorporation of flame photometers into automatic analysis systems has been widespread for a number of years – principally for the determination of sodium and potassium. The use of the atomic absorption spectrometer has been less common, but this instrument is particularly relevant to routine advisory soil analysis because it affords the most satisfactory way of determining magnesium in soil extracts. Full details of these instruments are given in Chap. 1 and only the aspects relevant to continuous-flow analysis are dealt with here.

Modern techniques used in the production of hollow-cathode lamps have now practically eliminated one of the inherent difficulties in automating atomic absorption analysis, i.e., the drifting or decaying of the radiant energy source. These improved lamps, together with the use of double-beam instrumentation, have resulted in stable energy outputs being maintained over long periods.

Both types of instrument measure simple elemental species, thus the role of the continuous-flow system here is in automating not chemical reactions but sampling, dilution, and transfer to the aspirator-burner assembly. A major advantage of the continuous-flow system is that it is capable of making large order dilutions with a high degree of accuracy and consistency.

The only modifications to standard procedures for flame photometric and atomic absorption measurements usually involved when methods are automated are in the introduction of the sample into the aspirator-burner assembly. The air-segmented stream must be debubbled before being aspirated, to maintain a stable output, and a number of systems devised for this purpose have been reviewed by Furman [21].

Nephelometer

The determination of sulfate may be carried out by nephelometric (light scattering) methods instead of the turbidimetric method referred to above [36,46,47]. A nephelometer is essentially a modified fluorescence spectrophotometer (or *fluorimeter*) in which a square glass or quartz flow cell replaces the normal tubular one. A light beam which falls on the flowing stream is scattered by the suspended particles of barium sulfate, and some of the scattered light passes through the optical system to the detector. The special flow cell is necessary to reduce the background scattering to an acceptable level [21].

Nephelometric sulfate determinations have been found to be critically dependent on the crystalline form and size distribution of the light-scattering particles [48]. These factors are affected by temperature and standing time before measurement, so the standardization of reaction conditions is important. This is precisely the type of situation in which a fully automated method of analysis has an advantage over manual methods, quite apart from any considerations of speed or convenience.

Ion-Selective Electrodes

Ion-selective electrodes or probes (see Chap. 2) are finding increasing application in continuous-flow systems. A higher precision is normally attainable with continuous-flow than with manual methods, because of the greater standardization of the conditions in which the sample is presented to the electrode [49].

Ions such as Na^+, K^+, Ca^{2+}, F^-, Cl^-, Br^-, NO_3^-, NH_4^+ (or ammonia), and boron (as fluoroborate) can be routinely determined by ion-selective electrodes in digests or extracts of soils, fertilizers, and plant materials. Adaptation of such determinations to continuous-flow systems (either segmented-flow or flow injection variants) have been reported for ammonium in Kjeldahl digests [50,51] nitrate [52,53], and potassium [53], and it is likely that a number of other flow-through electrode systems (see Refs. 21 and 49) will find a similar application in agricultural and environmental analysis.

To incorporate the ion-selective electrode into a continuous-flow system, it is normally fitted into a flow-through cap or cell (Fig. 13a). Where a separate reference electrode is required, this can be located in a reservoir downstream from the indicator electrode (Fig. 13b), so that the cell dead volume can be kept as small as 5-10 μl [52]. It is sometimes necessary to de-bubble a segmented stream before entering the detector cell, to avoid an erratic response. However, the ammonia probe has no external reference electrode, and the partial pressure of ammonia in the air bubbles is in equilibrium with the solution, so a steady response can be achieved without de-bubbling [49].

G. Optimization of Analytical Conditions

Sensitivity

To improve the sensitivity of an analytical system a number of conditions may be altered [16]:

1. Sample size can be increased by increasing the diameter of the sample pump tube, and reagent concentration can be increased to counteract the resulting dilution of the final reaction mixture.
2. The path length of the flow cell may be increased, because according to the Beer-Lambert law the absorbance is directly proportional to path length.
3. Where a reaction is dependent on elevated temperatures or takes a considerable time, the temperature of the heating bath can be raised, a larger volume heating coil can be used, or a time delay coil added to allow the reaction to progress further.
4. Where dialysis is involved, the path length can be increased, the temperature increased, or a more permeable membrane substituted for the one normally used.

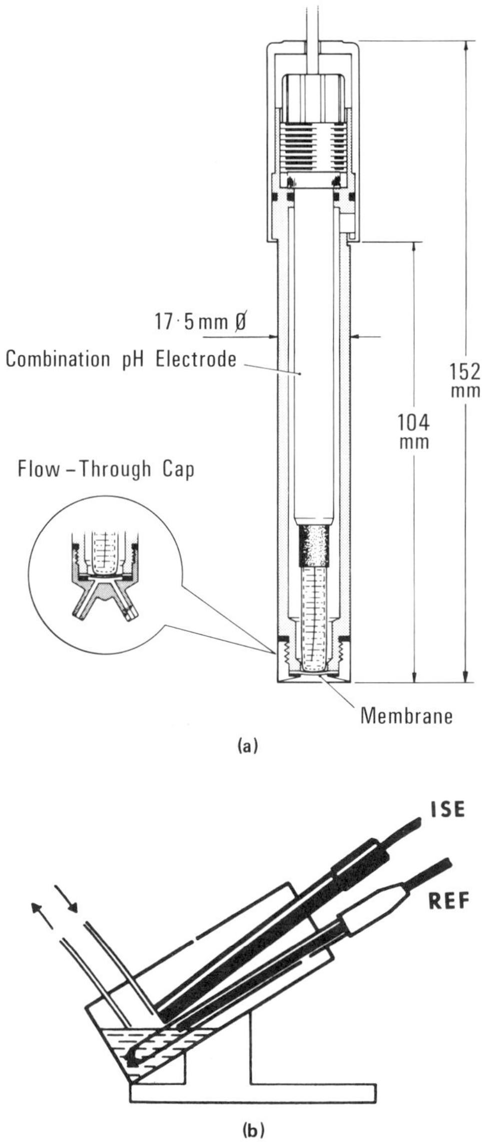

Figure 13 (a) Combination electrode fitted with flow-through cap for use in continuous-flow analysis. (Adapted from Ref. 51, courtesy of P. L. Bailey, Kent Industrial Measurements Ltd., Chertsey, Surrey, England.) (b) Flow cell accommodating ion-selective electrode (ISE) and a reference electrode (REF) in reservoir downstream. (From Ref. 52.)

A method of determining the optimum reagent concentration has been described [16]. In this procedure, a "midrange" standard is aspirated continuously into the analytical system. Two solutions of the reagent are made up: one (A) at 5-10 times the concentration recommended for the method, and another (B) one-tenth the concentration of A. Solution B is added to a stoppered flask on a magnetic stirrer and connected to the pump, and as B is pumped out of the flask, A is syphoned in at the same rate. A mark is made on the recorder paper at the time pumping begins, and another mark is made when an increase in absorbance begins. The interval indicates the delay time of the system. After a further period of time the recorder trace should form a plateau, indicating that the optimum reagent concentration has been reached. This concentration can be calculated from the formula

$$C_t = C_A(1 - e^{-Rt/V}) \tag{8}$$

where C_A is the concentration of solution A, R is the pumping flow rate (milliliters per minute), V is the initial volume of B (milliliters), and t is the time taken for the response to reach a plateau.

Accuracy and Precision

Absorbance. For greatest accuracy, it is desirable that absorbance should be about 0.4 unit because this is near the point at which photometric error is minimized [16]. This is obviously much easier to accomplish for samples which are of similar concentration than for those which vary widely, and repeated analyses after dilution may be desirable for samples giving rise to very high absorbance. Departures from the Beer-Lambert law occur for various reasons (see above, Sec. III.F) and this also usually puts an upper limit on the range of concentrations which may be determined accurately. For example, NH_4^+-N can only be determined in the range of 0-2.5 μg N ml^{-1} by the nitroprusside-catalyzed indophenol reaction [20,54] and the molybdate-antimony-ascorbic acid method for phosphate [55] has an upper limit of 1.2 μg P ml^{-1} [35].

Drift and Noise. Calibration drift may occur due to changes in baseline or method sensitivity, or both [56]. Causes may include changes in ambient, module or reagent temperatures, stray light and manifold stretching [57], coating of the flow cell, reagent degradation or degassing [16], or instrumental drift. Attempts should be made to eliminate the cause of drift, for example, where coating occurs wetting agents may be employed to overcome the problem. Where drift cannot be eradicated, midrange standards and water blanks should be inserted at intervals to monitor it.

If recorder tracings are noisy, all fittings and connections should be checked to ensure that joints have been properly made. Uniformity of air-bubble injection should also be checked, transmission tubing made as short as is practicable, and wetting agents used where appropriate [16].

Sample and Wash Times. Reference has already been made above (Sec. III.D) to the need for sampling times to be long enough to allow a response plateau to be sustained for at least 5 sec. If a digital printer is to be used, then a few more seconds should be added. Interaction between samples may be checked experimentally by analyzing several low and high concentration samples alternately, followed by several high concentration ones in succession. If the second low sample produces a higher peak than the first, or if the second successive high sample is higher than the first, the wash time should be extended until all peaks from replicate samples agree within the limits required [16].

H. Modifications for Flow Injection Analysis

Systems for flow injection analysis (FIA) have generally been developed in the laboratory, using conventional continuous-flow apparatus together with one or two items which are either specially made or "borrowed" from other branches of analytical science. However, commercially produced systems are now becoming available.

Flow is induced in the normal way by a peristaltic pump, but sample and reagent are brought together by a method which is fundamentally different from that employed in conventional segmented-flow systems. The sample is introduced into a moving stream of reagent, either by manual injection from a syringe through a septum, as in gas chromatography (Fig. 14a), or by means of a switching valve (Fig. 14b). Nonuniformity in the speed of manual injection results in poor reproducibility, and much better precision is obtainable by use of a valve [58]. The use of an eight-port valve such as that illustrated in Fig. 14b makes it possible to fill one loop while the contents of the other are being injected and vice versa. The filling of the loop may be accomplished either by using one channel of the peristaltic pump to aspirate solution from a sampler [58] or by manual injection from a syringe [59].

The frequency of sampling is much higher in FIA, resulting in a sample throughput of 75-200 samples per hour (Fig. 15), which is several times the rate typically achieved with segmented-flow methods. Sample carryover is overcome by increased pumping rates and decreased diameters of tubing, which result in turbulent rather than laminar flow. The increased pumping rate also results in an increase in reagent consumption, but this is canceled out by the increased frequency of sampling. Manifolds are normally made from polyethylene or tygon (vinyl) tubing with internal diameters varying from 0.38 mm [61] to 1.0 mm [62].

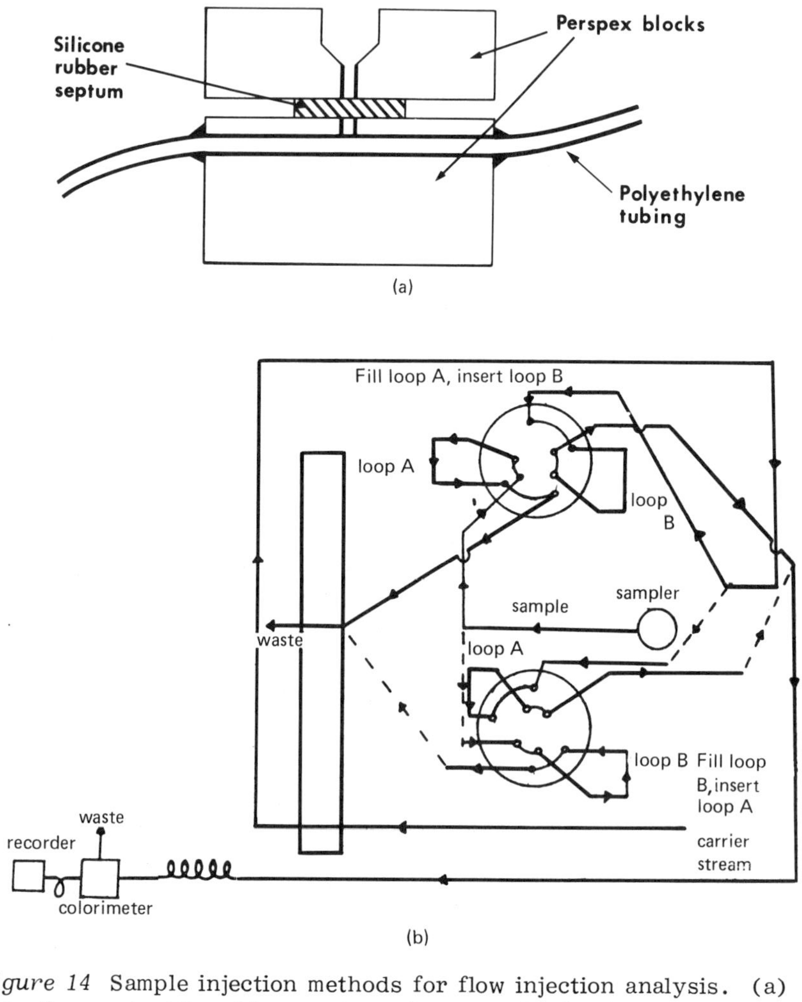

Figure 14 Sample injection methods for flow injection analysis. (a) Injection port with rubber septum, for manual injection by syringe. (From Ref. 13.) (b) Eight-port sampling and switching valve used in conjunction with sampler and peristaltic pump. Diagram shows the two alternative positions of the valve. (From Ref. 58.)

Line length is defined as the portion of the tubing within which sample and reagents are mixed, the color developed and the absorbance measured. The distance between the point of sample injection and the flow cell should be kept as short as possible, yet should be long enough to allow sufficient time for color development [58]. A good illustration of this principle is afforded by the work of Růžička and Stewart [14], who studied the rate of color development in the molyb-

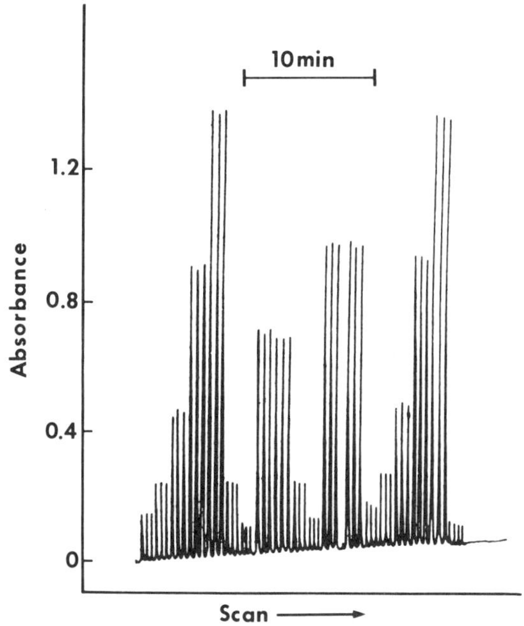

Figure 15 Recorder trace from flow injection analysis of ammonia in soil extracts. From left to right are shown a series of ammonia standard solutions (0.5, 1.0, 2.0, 4.0, and 6.0 μg NH_4^+-N ml^{-1}), followed by eight sample solutions, a second set of standards, and finally a blank, all solutions being injected in triplicate. (From Ref. 60.)

denum blue method for phosphate. They found an optimum line length of 2.5-3.5 m; below 2 m the sensitivity was decreased and the errors greatly increased. A further point to note about line length is that it should be coordinated with injection timing so that a sample is never injected when the maximum concentration of the previous sample is passing through the flow cell. This is because the transient increase in pressure in the system results in increased peak heights [58].

IV. PRACTICAL SYSTEMS – DISCRETE ANALYSIS

A. General Aspects

Major improvements have occurred in recent years in discrete or batch analyzers. Instruments designed to carry out a number of routine functions, such as pipetting, diluting, adding reagents, stirring, and transferring solutions to a flow cell or flame photometer are available from a number of manufacturers in several different countries, and have also been purpose-built in laboratory workshops and described in the scientific literature. The functions of single pieces of apparatus vary much more widely than in the case of continuous-flow systems, which are essentially modular in construction. It is convenient to consider separately here those multistage analyzers which carry out a complete sequence of operations from those which are intended to perform one or two functions only.

B. Automation of Individual Processes

Sample Dispensers and Diluters

The simplest form of sampler is essentially the same as that used for continuous-flow analysis, and may be used to automate the supply of samples to a dispenser or to instruments such as colorimeters, flame photometers, and atomic absorption spectrometers. Sample cups or tubes are mounted on a turntable or a conveyor and presented sequentially to a sampling probe. The sampling time can be varied as desired and a wash sequence introduced between samples. In instruments such as the Hook and Tucker A 40 Sampler, used in the authors' institute, a mixer unit is available to mix sample and reagent thoroughly, prior to transfer to the analyzer. When used with colorimeters which do not have their own aspiration facilities, suitable accessories for filling flow cells can be attached.

Where the concentration of the original sample solution is too high for direct analysis, an automatic diluter unit may be used. In a typical system, a preset volume of sample is withdrawn from a cup on one conveyor and, along with the desired quantity of diluent, transferred to a cup on a second conveyor where mixing takes place if required.

Very high rates of operation are possible with automated discrete systems involving only one or two processes. For example, the Hook and Tucker T 40 Sampler can dilute and feed samples to a colorimeter flow cell, flame photometer, or atomic absorption spectrometer at a rate of 360 samples per hour, compared with rates up to 60 samples per hour achievable with typical segmented continuous-flow systems.

It has already been pointed out (Sec. III) that continuous-flow systems can be used for precise, large-factor dilutions, and that this is invaluable for photometric determinations. Discrete samplers and diluters have at least as high precision, and their greater sample throughputs and relative simplicity have led to them being preferred for those applications in an increasing number of laboratories.

Systems for Determination of pH

The determination of soil pH is one of the most common tasks in soil analytical laboratories, and it is the type of repetitive operation which is admirably suitable for automation. In spite of this, no commercial instrument is available, but several laboratory-made systems have been described.

In Baker's system [63], 500 polystyrene containers are mounted in rows of 10 on aluminum plates, which are joined to form an endless conveyor belt. A fixed volume of each soil for analysis is introduced into each of the containers in the loading area, together with a small plastic-covered bar magnet. As each row of containers moves forward stepwise, a fixed volume of distilled water is automatically added to each, and the suspensions are mixed by a row of magnetic stirrers under the belt, stirring continually for 12 min. Ten pH electrodes mounted on a Perspex plate are automatically lowered into the suspensions. Each electrode is switched in sequence to a pH meter which is interfaced to a digital voltmeter and printer. After the readings have been made the containers are inverted and washed with distilled water, and the bar magnets are collected in an aluminum mesh tray and recovered for reuse. The system has been used to analyze samples at the rate of 20,000 per year.

Goodman [64] developed a system to cope with batches of 6-60 samples which used motor-driven glass paddles for stirring and a single electrode. The apparatus could be fitted not only with a pH electrode but also an ion-selective electrode, and has been applied to nitrate analysis. The paddles were necessary to break up the lumps of fresh soil that were necessarily used when nitrate was to be measured.

Grigg et al. [65] have devised yet another approach to pH measurement, based on commercial instruments. An LKB Radirac fraction collector rotator is fitted with a 60-position aluminum turntable carrying PVC sample cups. To 12 ml soil samples, 30 ml of water is added from a dispenser, 10 cups at a time. The cups are stored at constant temperature overnight and then loaded onto the turntable, together with cups of buffer solutions. Stirrers and a combination glass-calomel electrode, attached to a single beam, are lowered into each cup in turn by means of a motor-driven cam. After 40 sec, a series of 12 measurements is made at 1 sec intervals. A programmable calculator rejects the highest and lowest value, works out the mean of the other 10, and adjusts the result for any drift in the pH meter during a run of samples. This result is then printed out. After the measurements, the electrode is withdrawn and automatically washed by a water jet.

C. Integrated Systems

Some examples of commercially available discrete analyzer systems are given in Table 3. Practical details differ between one instrument and another, but many basic principles are common. The Pye Unicam AURA system, which has a number of features fairly typical of currently available equipment, is described below as an illustration.

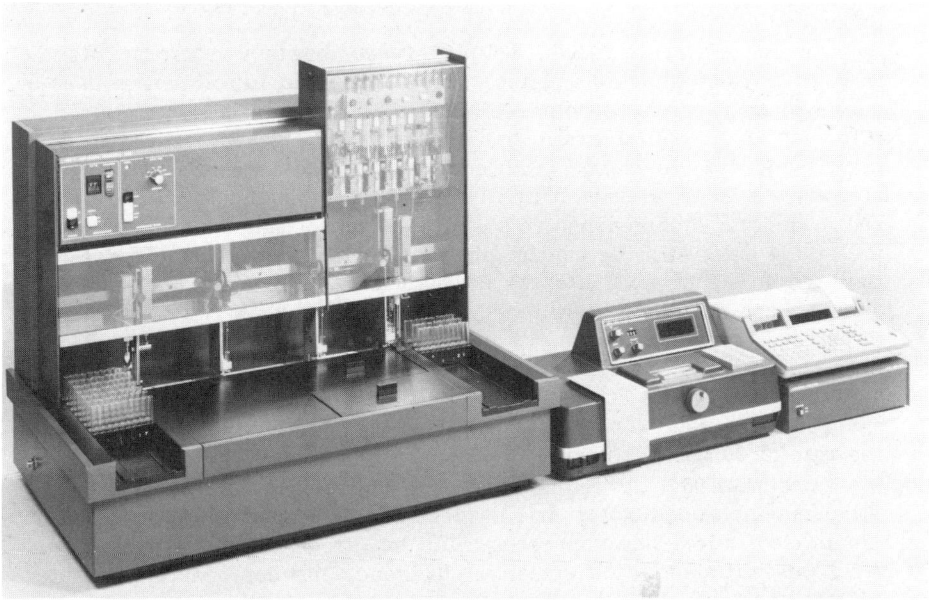

(a)

(b)

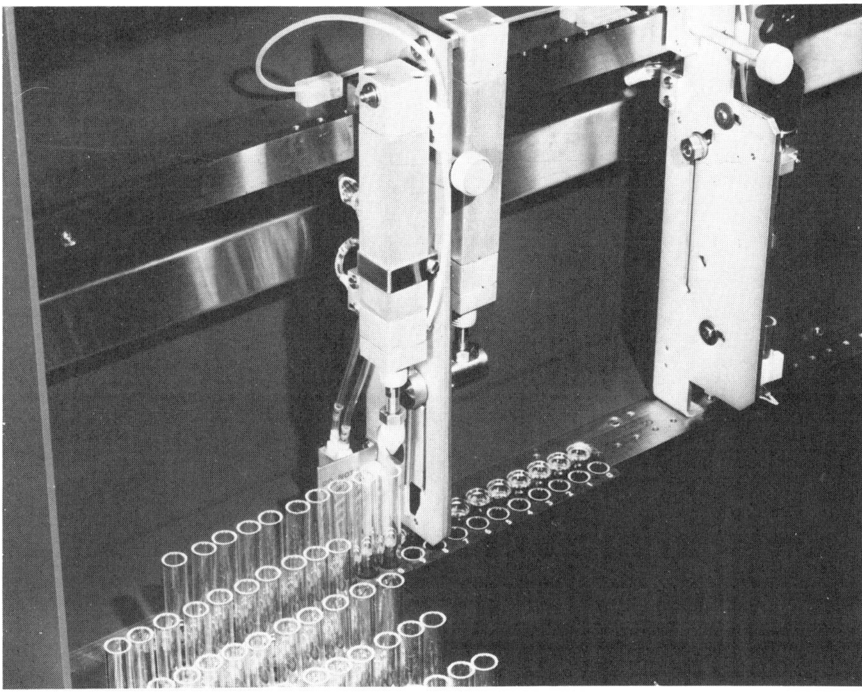

(c)

Figure 16 The Pye-Unicam Aura Automatic Chemistry System. (a) The AC 1 automatic chemistry unit with spectrophotometer and printer; (b) syringe bank; (c) diluter. (Courtesy of Pye Unicam Ltd, Cambridge, England.)

The AURA system (Fig. 16) combines the AC 1 automatic chemistry unit with a spectrophotometer and printer, and is capable of up to 240 colorimetric analyses per hour. Transfers of liquids are brought about by pneumatically operated syringes. The sampling syringe (normal size 120 μl) is connected to another syringe which dispenses diluent. Up to three more syringes can be fitted to dispense reagents, and two more control the filling and emptying of the flow cell. The speed at which the syringes are operated is individually adjustable, to cope with liquids of varying viscosities. Apart from the sampling syringe, which is glass, disposable plastic syringes are used. They are less prone than glass ones to jamming or breakage, can be cheaply replaced, and function with satisfactory precision [66]. In some other analyzers, diaphragm valves are used instead of syringes for bringing about the transfer of liquids.

Up to 10 racks, each containing 10 sample cups and 10 test tubes can be placed on the loader of the AC 1 unit. The movement of racks

Table 3 Some Examples of Systems for Discrete Analysis

Manufacturer	System	Maximum number of samples per batch	Detector system
Gilford (U.S.)	System 3500	112	Colorimetric
Hook and Tucker (U.K.)	A40/T40/K40 samplers + diluters and dispensers	40	Not supplied. Any suitable colorimeter, flame photometer or atomic absorption spectrometer can be added
Instrument Specialties Co (U.S.)	Model 1560	210	Colorimetric
LKB (Sweden)	Ultrolab	100	Colorimetric
Pye Unicam (U.K.)	AURA	100	Colorimetric
Saitron (Italy)	6059 and 9119	30	Colorimetric
	9601	80	Colorimetric

to the processing stations is brought about by pneumatic activators. At the processing stations, sample dispensing, dilution, reagent addition, and stirring can all be automatically carried out. The sample is then finally transferred, via a probe, to the spectrophotometer cell. The transfer tube and cell are flushed through with most of the reaction mixture prior to measurement, to ensure an adequate wash and thus to minimize carryover. A volume of about 1.5 ml is required in the test tube for this purpose.

The AC 1 unit was designed to interface with several Pye Unicam UV or visible spectrophotometers, and could readily be attached to equipment originating from other manufacturers. Combined with the SP-6 visible spectrophotometer, it has been used for a number of water analyses, including ammonia, phosphate, chloride, nitrate, and nitrite, which are of relevance to soil scientists and environmental analysts as well as water authorities. The system has been found to cost less and to have a better analytical output than a three-channel continuous flow system operating under optimum conditions [67].

Multichannel discrete analyzers have so far been restricted to applications in clinical analysis, but there seems to be no reason why, in principle, these instruments should not also be used for the routine analysis of natural waters, soil extracts, and plant materials.

V. APPLICATIONS

A. Nitrogen

Total Nitrogen and Ammonium

Wet chemical methods for the determination of total nitrogen in soils are all ultimately based on the conversion of the nitrogen to ammonium ion and the determination of that ammonium in solution either by a colorimetric method or by an ion-selective electrode. The methods generally involve an off-line Kjeldahl digestion, after which aliquots of digest solutions are taken for automated continuous-flow or discrete analysis. The Technicon digestor has been used as an integral part of continuous-flow systems for the analysis of fertilizers [24,25] and plant tissue [22]; the manifold for the latter application is shown in Fig. 17. However, the presence of insoluble mineral matter renders soil samples unsuitable for analysis by this method (see Sec. III).

A typical application of continuous-flow methods to soil analysis is that of Schuman et al. [68]. They digested soil and plant samples in 75 ml tubes in a Tecator block digestor and determined the ammonium contents of the digests, using a Technicon AutoAnalyzer, by the "indophenol method," in which the ammonium present produces a blue color with alkaline phenol and sodium hypochlorite. Sodium potassium tartrate was added to prevent precipitation of hydroxides in the alkaline solution [34].

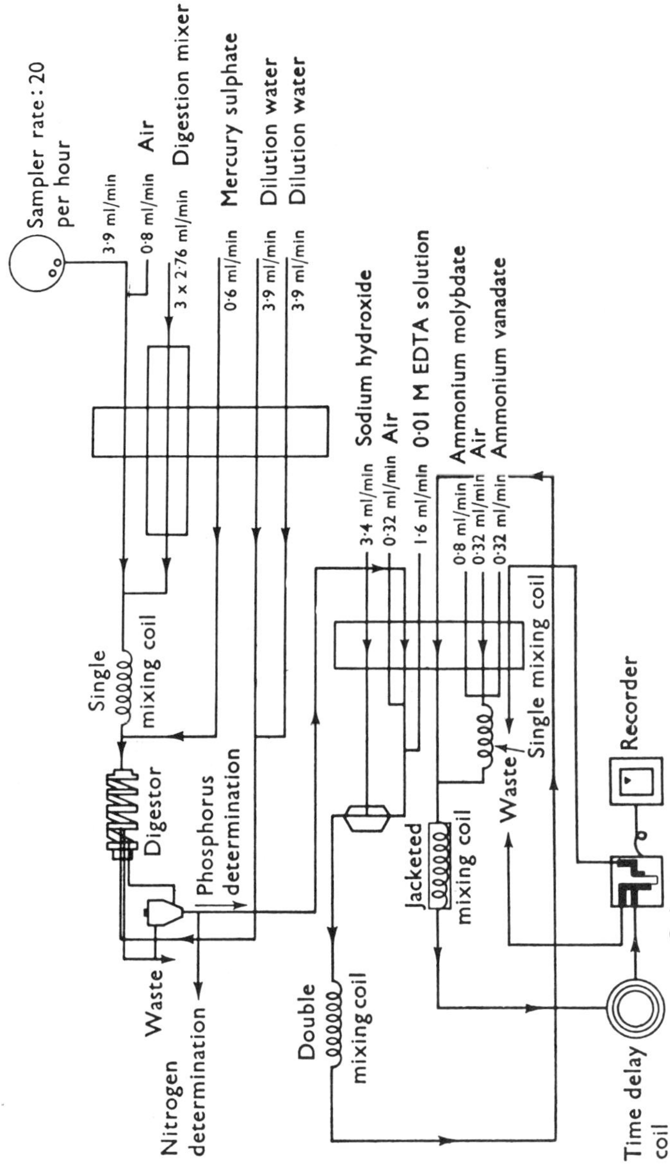

Figure 17 Manifold for continuous-flow analysis of nitrogen and phosphorus in plant material, including automatic digestion. (From Ref. 22.)

Table 4 Regression Analysis of AutoAnalyzer and Distillation Values for Cation Exchange Capacity (CEC) and Total Nitrogen

	CEC (me %)[a]	Total nitrogen (%)
Range	0-30	0-0.5
Regression[b]	y = 0.9932x + 0.0947	y = 0.9907x + 0.0005
r	0.9987***	0.9991***

[a] 1 me. % - 1 mmol per charge on free ion per 100 g oven dry soil.

[b] x = AutoAnalyzer values, y = distillation values.

Source: Ref. 20.

The intensity of the indophenol blue color is strongly dependent on pH, and enhanced by iron, chromium, and manganese (II) ions [69]. Conflicting evidence of such enhancement resulting from digests of plant material has been reported [70,71] but it is likely that such effects will be more acute with soil samples. Accordingly, Keay and Menage [28] and Crowther and Evans [72] have introduced an automated distillation step into the continuous-flow procedure. Skjemstad and Reeve [30] found that the glass still used by Keay and Menage was subject to alkali attack and replaced it by a Teflon cylinder with stainless steel tubing on the inlet side. Using their modified distillation procedure they compared the results obtained with the Honda [73] rapid digestion technique (using a phosphoric-sulfuric acid mixture) with the conventional technique of Bremner [74]. Only very small differences between the two digestion methods were found, which were regarded as being too small (0.001% N) to be of any real concern in routine analysis. The automated Honda digestion/distillation method was also demonstrated to give high precision when the same batch of soils was analyzed on consecutive days. It may be readily adapted to include nitrite and nitrate by using the reduction procedure described by Bremner and Shaw [75].

The quality of agreement between continuous-flow procedures and the manual Kjeldahl distillation and titration method was explored by Searle [20]. Table 4 shows his results for 50 soils, both for total nitrogen and for the measurement of cation exchange capacity, in which the soils were first saturated with ammonium ions and then leached with sodium chloride. The values obtained by both methods were very highly correlated ($p < 0.001$) and the regression coefficients were very close to unity, indicating no significant bias between the methods. In Searle's work, modifications of the indophenol method were employed in which interfering metals were complexed with sodium citrate-tartrate reagent [76]. Sodium salicylate and sodium dichloro-s-triazine-2,4,6-trione (NaDTT) were used instead of the traditional alkaline phenol and hypochlorite reagents, and nitroprusside was used to catalyze the reaction [54]. Sample analysis rates up to 100 hr^{-1} were achieved.

Many colorimetric determinations of ions in soil extracts may be affected by interference by colored humic substances or suspended matter. Selmer-Olsen [34] overcame the problem when determining ammonium in 2 M potassium chloride extracts, by incorporating a dialyzer into the manifold. He obtained slightly higher values with the distillation method [74] than by continuous-flow analysis, which he attributed to interference from labile organonitrogen compounds destroyed by the steam distillation with magnesium oxide and determined as ammonium.

A flow-injection method for ammonium in soil extracts and natural waters has been described by Krug et al. [60]. Nessler's reagent (mercury (II) chloride and potassium iodide in alkaline solution) were used, and the colored turbidity produced was measured at 410 nm. This method was found suitable for extracts containing 0.5-6.0 μg NH_4^+-N ml^{-1}, provided that no colloidal material was present, and an analysis rate of 120 samples per hour was achieved.

Nitrite and Nitrate

Most continuous-flow methods for the determination of low concentrations of nitrate in aqueous solution are based on its reduction to nitrite, with subsequent colorimetric determination of the nitrite formed plus any nitrite originally present in the sample. The latter may be separately determined by leaving out the reduction stage, and the nitrate obtained by difference. Henriksen and Selmer-Olsen [33] applied this approach to 2 M potassium chloride extracts of soil as well as to natural water samples. Copperized cadmium was used as reducing agent and the nitrite was determined by diazotizing sulfanilamide and coupling with N-1-naphthylethylenediamine to form an azo-dye, which was measured at 520 nm. When colored or turbid samples were being analyzed, a dialyzer was incorporated in the manifold (Fig. 18). Best [77], using hydrazine as the reducing agent for nitrate, found very good agreement with results obtained by a distillation method [78].

The same azo-dye colorimetric procedure as used in these segmented-flow methods has been adapted to flow-injection analysis [79] and to discrete analysis, with a throughput of 240 samples per hour [67]. Hansen et al. [52] determined nitrate in soil extracts and fertilizer solutions by employing the flow-injection principle together with a nitrate electrode in a flow-through cell. At a sampling rate of 90 samples per hour the typical coefficient of variation was 1-2%. However, results by this procedure were consistently lower than those obtained by spectrophotometry.

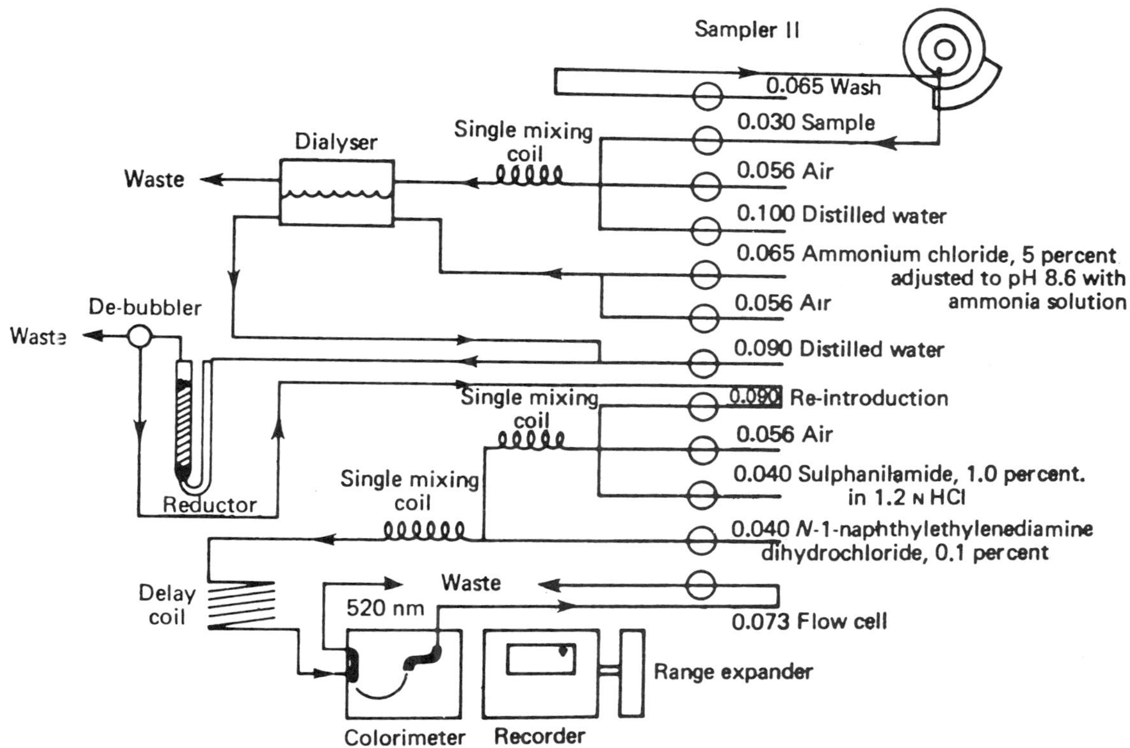

Figure 18 Manifold for continuous-flow analysis of nitrate and nitrite in soil extracts. (From Ref. 33.)

Table 5 Applications of Automated Methods to the Determination of Phosphorus in Soil Extracts

Extractant	References
0.001 M H_2SO_4/0.002 M $(NH_4)_2SO_4$, pH 3 (Truog)	35,41
0.5 M $NaHCO_3$/NaOH, pH 8.5 (Olsen)	35,80,81,86
0.3 M NH_4F/0.025 M HCl (Bray and Kurtz No. 1)	35,87,93
0.3 M NH_4F/0.1 M HCl (Bray and Kurtz No. 2)	35,85
0.2 M Ca-lactate, pH 3.8 (Egner-Riehm)	82
0.1 M NH_4-lactate/0.4 M acetic acid (Egner-Riehm-Domingo)	82,94
0.5 M NH_4-acetate/0.5 M acetic acid, pH 4.8 (modified Morgan)	83
0.4 M acetic acid, pH 2.6	41
0.01 M $CaCl_2$	81
0.01 M H_2SO_4	88,89
0.05 M HCl/0.012 M H_2SO_4	93
0.5 M Na_2SO_4, after sorption on an ion exchange resin	95
Chang and Jackson's solutions (1 M NH_4Cl, 0.5 M NH_4F, 0.1 M NaOH, 0.3 M Na-citrate/$Na_2S_2O_4$)	35

In his purpose-built discrete system for soil pH and nitrate determinations, Goodman [64] first used an Orion 92-07 nitrate ion-selective electrode but replaced this with a Corning liquid-junction electrode which gave more satisfactory results. The system had been applied to the analysis of several thousand samples and proved very reliable in routine operation.

B. Phosphorus

A wide variety of extractants has been used in soil fertility studies for the determination of labile phosphorus, and others have been used in the vain hope of extracting only the phosphorus present in the soil in particular chemical combinations. Automated methods of analysis (both continuous-flow and discrete) have been developed which are applicable to the majority of these extractant solutions (Table 5).

Colorimetric determination of phosphorus by reduction of a molybdophosphate complex to "molybdenum blue," first developed for manual methods, has been widely employed also in the automated versions, although modifications have been required. The use of tin (II) chloride as reducing agent causes difficulties, arising from precipitation and surface coating of the flow cell, and it has been frequently replaced by ascorbic acid. Color development is slow at moderate temperatures, with the latter reagent, and heating bath temperatures of 90-95°C have been used [80-82]. However, the reaction can be sufficiently accelerated by the addition of potassium antimony tartrate [55] to allow color development to take place at room temperature, using two large delay coils [81].

The ascorbic acid method is less sensitive than reduction with tin (II) chloride, and this can cause problems at very low phosphate concentrations. Therefore, for routine soil analysis in the authors' institute, a reducing agent consisting of tin (II) chloride and hydrazine is used, and the apparatus is periodically cleaned out with acid. This procedure was first adopted for continuous-flow analysis, but has also proved very satisfactory for discrete analysis. A change to the latter system was introduced because of the greater flexibility, economy of reagent consumption, and increased sample throughput rate – about six times greater than with continuous flow [83]. Hydrazine [41] and hydroxylammonium chloride [84] have been demonstrated to reduce interference from iron (III) by reduction to iron (II). Another interference arises from the presence of organic matter. Grigg [35] found that the high organic matter content of New Zealand soils resulted in a precipitate being formed on acidification of the extract, so the organic matter had to be removed by dialysis.

In addition to being applied to the extractants listed in Table 5, continuous-flow methods are also suitable for solutions resulting from wet digestion procedures, using perchloric, nitric, and sulfuric acids, or sulfuric acid and hydrogen peroxide [41,85] or from sodium carbonate fusions [85].

Carbon dioxide is liberated when acidic ammonium reagent is added to alkaline bicarbonate (Olsen) extracts of soil. Zandstra [86] found that, as a result, the flow rate increased during the passage of a sample and decreased during the wash cycle, and Allen et al. [41] have suggested that the Olsen method is not suitable for continuous-flow analysis. However, the problem seems to have been overcome satisfactorily by passing the liquid stream through a de-bubbler to remove CO_2, after acidification, either before the introduction to the color-reaction manifold [35,80,81] or at the molybdate addition stage [86].

Orr [87], Burrows and Meyer [88], and Burrows [89] have described discrete methods for phosphorus in soil extracts, based partly or wholly on LKB equipment (LKB Instruments, Bromma, Sweden). In Orr's method, ammonium molybdate and Elon reagent (sodium metabisulfite, p-methylaminophenol sulfate, and sulfonated oleic acid) were added manually, and 15 min allowed for color development prior to automatic introduction to the colorimeter. A single operator could accomplish 360 determinations per hour with this system. Burrows [89] used an LKB Ultrolab system to determine phosphorus as the 12-molybdoantimonylphosphate complex, at a rate of 200 samples per 80 min compared with the 8 hr previously required.

Another colorimetric method for phosphorus is the molybdovanadophosphate yellow method [90], adapted for continuous-flow analysis by Varley [91], Basson et al. [22], and Steckel and Flannery [92] (see Fig. 17, Sec. V.A). This method is less sensitive but more stable than the molybdenum blue method, and has therefore been mainly applied to plant analysis [22,83,91], although Flannery and Markus [93] have analyzed Bray No. 1 and HCl-H_2SO_4 extracts of soil in this way.

C. Potassium, Calcium, Magnesium, and Sodium

In routine advisory soil analysis, some or all of the elements potassium, calcium, and magnesium, and occasionally sodium, have been determined by automated methods. Often, these determinations have been made, using continuous-flow methods, in parallel with the colorimetric determination of phosphorus (and sometimes ammonium or nitrate) and on the same soil extract. Zsoldos [82] has described such a system, fed by a 200-position sampler, in which potassium and sodium were determined by flame photometry. Interference from calcium in the determination of potassium was avoided, even in extracts of calcareous soils, by using a low flame temperature (propane-butane/compressed air flame). Lithium was used as an internal standard.

Flannery and Markus [93] used independent flow systems when analyzing phosphorus, magnesium, and potassium plus calcium in soil extracts. These workers adopted a modified lake procedure for magnesium, in which $Mg(OH)_2$ was precipitated in alkaline solution and Magnesium Blue dye was adsorbed on the precipitate in the presence of a detergent (Brij 35) and suspending agent (polyvinyl alcohol). Potassium and calcium were measured simultaneously by flame photometry.

Clinton [96] has developed a four-channel photometric system for soil leachates and plant digests in which calcium, sodium, and potassium are determined by flame emission and magnesium by atomic absorption, at a rate of 200 samples per hour. A recent development by staff of the Perkin-Elmer Corporation [97] indicates the sophistication of photometric measurement now possible as part of an automated analysis system. They programmed a Perkin-Elmer Model 5000 atomic absorption and flame emission spectrometer equipped with an automatic sampler to determine Ca, Mg, K, and Na in 1 M ammonium acetate extracts of soil without any dilution. A 50 mm burner head was used which could be rotated through 90°, to vary the sensitivity, and all the parameters necessary for each determination were inserted on magnetic cards.

D. Aluminum, Silicon, and Iron

Lancaster and Balasubramaniam [98] developed a continuous-flow method for aluminum in Malayan soils and plant digests, using the complexing dye sodium alizarin-3-sulfonate (alizarin red S). This reagent gave a very reproducible and linear response up to 15 μg Al ml^{-1}; the interfering effect of iron was removed with thioglycollic acid. A similar method has been routinely applied at the U.K. Hill Farming Research Organization to the determination of soluble aluminum in acidic hill soils (Sheppard, personal communication). Allen et al. [41] recommend for aluminum a method using the complex formed with eriochrome cyanine R, and for silicon the reduction of molybdosilicic acid to a "molybdenum blue" complex, i.e., a reaction analogous to that commonly applied to the determination of phosphorus. In the method for silicon, phosphorus interference is prevented by pH control and the presence of oxalic acid. Up to 40 samples per hour in a concentration range 0-5 μg ml^{-1} can be determined by this procedure.

Discrete analysis has been applied to extractable aluminum in soils by Burrows and Meyer [88], using LKB Ultrolab equipment. Aluminum was extracted with 0.2 M ammonium chloride [99] and determined colorimetrically as a pyrocatechol violet complex, which was linear over the range 0-0.8 μg Al ml^{-1} (corresponding to the range 0-200 μg Al g^{-1} soil).

Iron is a major component of most soils, and total iron or the fractions which are extractable with reagents such as oxalic acid or pyrophosphate are frequently measured in pedological investigations. For the determination of plant-available iron, 2.5% acetic acid is commonly used. Allen et al. [41] have adapted the manual colorimetric method for iron using sulfonated 4,7-diphenyl-1,10-phenanthroline [100] for use with the AutoAnalyzer and applied it to 3% oxalic acid and 2.5% acetic acid extracts. A 50-fold dilution of the former is recommended for most soils. The same method has also been used to determine exchangeable iron extracted with 1 M potassium chloride (Sheppard, personal communication).

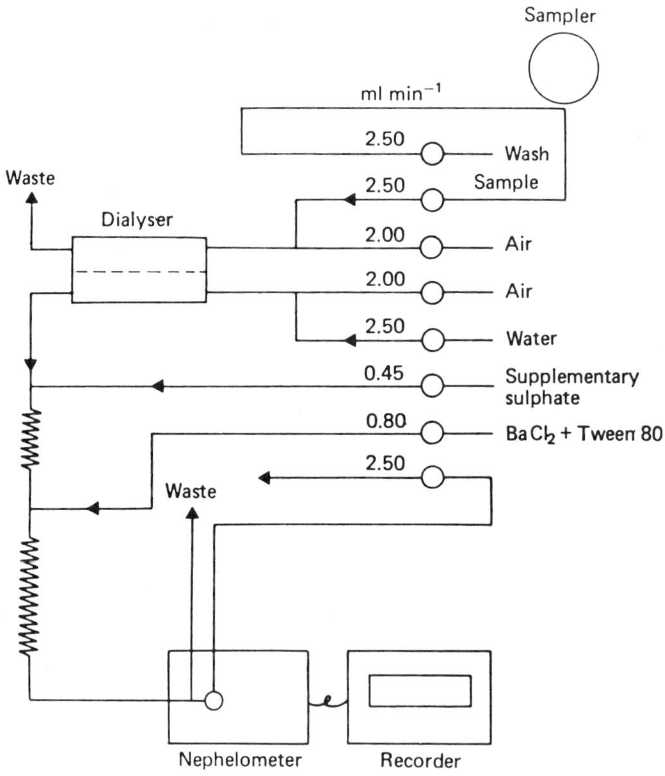

Figure 19 Manifold for continuous-flow analysis of sulfate in water samples and soil extracts containing large amounts of humic compounds. (From Ref. 36.)

E. Sulfur

Published methods for sulfur (as sulfate) in soil extracts and natural water samples have involved turbidimetric, nephelometric, and colorimetric methods of measurement. Sinclair [44] developed a barium sulfate turbidimetric procedure suitable for the determination of sulfate in 0.01 M calcium phosphate extracts of soils. The limit of detection was 0.5 μg S g^{-1}. Good agreement was obtained with the reduction/distillation method [101], provided that the soil extracts were decolorized by shaking with charcoal prior to automatic analysis. Mulcahy [45] also used turbidimetry for total sulfur in soils, following nitric/perchloric acid digestion. He found that, to obtain reliable results, it was necessary to carry out a prolonged digestion at 190°C (to decompose methionine) and to precipitate interfering ions with sodium hydroxide. The solution was dialyzed before the formation of barium sulfate for turbidimetry.

Bettany and Halstead [47] used Technicon AutoAnalyzer equipment together with a modified Turner Model 111 Fluorometer for the nephelometric determination of sulfate in extracts of sulfur-deficient Saskatchewan soils (<1 μg SO_4^{2-} -S ml^{-1} aqueous extract). Interference from organic matter and common soil cations was prevented by clarifying the soil extract with sodium peroxide. Ogner and Haugen [36] also used nephelometry for rather higher sulfate concentrations (1-100 μg SO_4^{2-} -S ml^{-1}) in water samples and soil extracts containing large amounts of humic compounds. The colored organic matter was removed effectively by dialysis, using a membrane of pore size 20-35 nm (the flow system is shown in Fig. 19). Better results were obtained with the nephelometric technique than with a turbidimetric method. Aluminum at concentrations lower than 15 μg ml^{-1} showed no significant interference in sulfate determinations, but at 30 μg ml^{-1} the results were 3% lower than for samples without aluminum present.

Keay et al. [102] determined sulfate in soil extracts and a wide range of other biological materials by reduction to hydrogen sulfide, distillation, and formation of colloidal bismuth sulfide which was measured in a colorimeter at 410 nm. A methylthymol blue color reaction [103,104] can also be used for sulfate, but is unsuitable for soil solutions and natural waters containing colored organic matter. Cronan [105] overcame this problem by decolorizing the solutions prior to continuous-flow analysis by irradiation with UV light.

F. Trace Elements

Boron

Basson et al. [106] developed a continuous-flow colorimetric method for boron in plant tissue, using azomethine-H as the color reagent. This has been adapted with only slight modification to the analysis of hot water extracts of soils [41]. Boron has also been determined in soil extracts, plants, and irrigation waters by a manual spectrofluorimetric method using carminic acid [107]. This procedure has been adapted to continuous-flow analysis for boron in plant ash by Ogner [108], and this automated procedure should be readily applicable to soil extracts also. Yet another automated method for boron in plant ash was based on the color reaction with quinalizarin [109].

Copper and Molybdenum

Copper in plant materials has been analyzed, simultaneously with iron, by an automated method at the rate of 50 samples per hour [110]. The determination of molybdenum in plants has also been automated [111] but no reports have been traced of the adaptation of these methods to soil extracts.

Chlorine, Bromine, and Iodine

Chlorine (as chloride) can be determined by a reaction in which chloride releases thiocyanate ions from undissociated mercuric thiocyanate; these ions then react with iron (III) to form a red color [41]. A method for bromide in natural waters has recently been developed, based on the catalytic effect of Br^- on the oxidation of iodine to iodate by potassium permanganate in sulfuric acid solution [112]. Twenty samples per hour can be analyzed, down to concentrations of 0.01 $\mu g\ Br^- \ ml^{-1}$. Iodine in plant tissue and sodium hydroxide extracts of soil has been determined by an AutoAnalyzer procedure based on the catalytic action of iodine on the oxidation of arsenic (III) by cerium (IV) [113].

Pollutant Elements

Vijan et al. [114] have determined arsenic in digests of soils and vegetation by reducing it to arsenic gas with sodium borohydride, separating the gas with a stream of argon and passing it through a quartz cell in the path of an atomic absorption spectrometer equipped with an arsenic hollow-cathode lamp. Forty samples per day could be analyzed by this method, with a detection limit of 0.15 $\mu g\ As\ g^{-1}$. Coefficients of variation were <4% and 7% for vegetation and soil samples, respectively. The same technique has been applied to the determination of As, Sb, and Se in natural waters [115].

The determination of mercury in soils and sediments has also been automated, using a somewhat similar procedure involving oxidation with persulfate followed by reduction to elemental mercury and detection by the cold-vapor method [116,117].

Recently, the application of flow injection analysis to the determination of chromium (VI) in natural waters and soil extracts has been reported by Jorgensen and Regitano [118]. The complex with 1,5-diphenylcarbazide was used, being highly selective for chromium if the pH was kept at 1-2. The method was applicable to a concentration range of 0.1-20 $\mu g\ Cr\ ml^{-1}$, with a coefficient of variation of about 1.3% and a rate of analysis of at least 70 samples per hour. Satisfactory agreement was found with results obtained by an atomic absorption procedure.

G. pH

Reference has already been made (Sec. IV.B) to several systems purpose-built for the automated measurement of soil pH. Discussion here is confined to the comparisons made with manual methods, which show the automated methods to give very satisfactory results.

Baker [63] compared results for 200 soil samples (136 temperate and 64 tropical) with pH ranging from 4.6 to 8.2, organic matter ranging from <1% to nearly 70%, and including the whole textural range. The following regression equation was obtained:

$$pH_{automated} = 1.0064\ (\pm 0.0191)\ pH_{manual} - 0.1460$$

and the correlation coefficient was 0.966. Goodman [64] found that the standard deviation for an individual measurement with his apparatus was 0.044 pH unit, compared with 0.039 for a manual measurement. The two sets of results were highly correlated ($r = 0.999$), and the regression gave a slope of 0.966.

Grigg et al. [65] found a significant but small bias between automated and manual methods of 0.038 pH unit, which they showed to be due to the slow attainment of equilibrium. They concluded that this difference was of little practical importance, that 95% of the pH values determined lay within 0.05 unit of the true value, and that the precision of the automated method was thus of the same order as that to which the results were reported (0.1 pH unit).

H. Organic Herbicides

Instrumental methods for the analysis of organic herbicides and pesticides in soil have mainly involved the use of gas and liquid chromatography (see Chap. 11), but Guth and Voss [119] have reported the use of a continuous-flow analyzer for this purpose. They determined eight phenylurea, N-phenyl carbamate, and aniline herbicide residues which had been extracted from soil and separated by TLC by a colorimetric method involving diazotization and coupling with N-(1-naphthyl)-ethylenediamine. The method had a limit of detection of 0.05-0.06 $\mu g\ g^{-1}$ and an analysis rate of 20 samples per hour.

An AutoAnalyzer system has also been used for automated extraction and clean-up of atrazine residues in soil prior to gas chromatography [120,121]. Soil samples (40 g) were dispensed by a Solid-prep sampler and extracted with CH_3CN/water, and the atrazine extracted into hexane-ether. The method included a novel application of the Technicon Digestor, as an evaporator to concentrate the CH_3CN extract.

VI. CONCLUSION

Automated methods for wet chemical analysis have now been widely adopted in agricultural and environmental laboratories. The most widespread application has been in the routine analysis of soil extracts and plant materials in connection with advisory and extension services. The first phase of the switch to automated methods concentrated on the use of segmented continuous-flow systems, but more recently the greater flexibility, efficiency and speed offered by discrete analysis has led to its increasing popularity. This trend seems likely to continue. The other departure from segmented-flow methods, i.e., flow injection analysis, will doubtless find new applications in the future, but in terms of speed and economy of reagent use it is unlikely to prove a match for discrete methods.

The development of automated chemical procedures has, until recently, outstripped the development of methods for handling the data produced. Just as the widening of one city street in isolation inevitably results in shifting the traffic jam to a new location, rather than eliminating it, so the automation of the chemical procedures has resulted in the analyst calculating peak heights from reams of chart recorder paper. This had led to the use of devices such as peak height readers and digitizers, as an interim solution, but the newer measuring instruments are equipped with digital display and printout systems which overcome the problem in a more direct way. There is much scope for further development of data output and data transfer systems which are flexible enough to allow either the acquisition of the result directly from the measuring instrument or the automatic transfer of data to a computer, and which also allow the analyst to continue to monitor the performance of his analytical system in as direct a way as hitherto. Such developments are likely in the near future. It is to be hoped that they will not only be available as part of entirely new instrument systems but also as relatively cheap units to add to existing instruments which have years of useful life in them.

ACKNOWLEDGMENT

The authors wish to thank Dr. P. Crooks, Head of the Central Analytical Laboratory of the Edinburgh School of Agriculture, for his many useful comments and suggestions during the preparation of this chapter.

REFERENCES

1. L. T. Skeggs, Am. J. Clin. Pathol., *28*:311 (1957).
2. L. R. Snyder and H. J. Adler, Anal. Chem., *48*:1017 (1976).
3. R. E. Thiers and K. M. Oglesby, Clin. Chem., *10*:246 (1964).
4. R. E. Thiers, R. R. Cole, and W. J. Kirsch, Clin. Chem. *13*:451 (1967).
5. R. E. Thiers, J. Meyn, and R. F. Wilderman, Clin. Chem., *16*:832 (1970).
6. M. A. Evenson, G. P. Hicks, and R. E. Thiers, Clin. Chem., *16*:606 (1970).
7. R. E. Thiers, A. H. Reed, and K. Delander, Clin. Chem., *17*:42 (1971).
8. L. R. Snyder, Anal. Chim. Acta, *114*:3 (1980).

9. L. R. Snyder, in *Advances in Automated Analysis, 1976*, Technicon International Congress, Mediad Press, Tarrytown, N.Y., 1977, p. 76.
10. L. R. Snyder and H. J. Adler, Anal. Chem., *48*:1022 (1976).
11. L. R. Snyder, J. Chromatogr., *125*:287 (1976).
12. L. R. Snyder, J. Levine, R. Stoy, and A. Conetta, Anal. Chem., *48*:942A (1976).
13. J. Růžička and E. H. Hansen, Anal. Chim. Acta, *78*:145 (1975).
14. J. Růžička and J. W. B. Stewart, Anal. Chim. Acta, *79*:79 (1975).
15. J. Růžička and E. H. Hansen, Anal. Chim. Acta, *114*:19 (1980).
16. Technicon Instruments Corporation, Tarrytown, New York, Manual TN1-0170-01 (1972).
17. E. Amador, Clin. Chem., *18*:164 (1972).
18. W. E. Neeley, Clin. Chem., *18*:509 (1972).
19. J. M. Carter and G. Nickless, Analyst (Lond.), *95*:148 (1970).
20. P. L. Searle, N. Z. J. Agric. Res., *18*:183 (1975).
21. W. B. Furman, *Continuous-Flow Analysis*, Marcel Dekker, New York, 1976.
22. W. D. Basson, D. A. Stanton, and R. G. Bohmer, Analyst (Lond.), *93*:166 (1968).
23. H. G. Glento and C. E. Daugherty, in *Advances in Automated Analysis*, vol. II, Technicon International Congress, 1970, Thurman Associates, Miami, 1971, p. 75.
24. C. W. Gehrke, L. L. Wall, J. S. Killingley, and K. Inada, J. Assoc. Off. Anal. Chem., *54*:651 (1971).
25. C. W. Gehrke, J. S. Killingley, and L. L. Wall, J. Assoc. Off. Anal. Chem., *55*:467 (1972).
26. J. Davidson, J. Mathieson, and A. W. Boyne, Analyst (Lond.), *95*:181 (1970).
27. J. O. Bosset and U. Jenni, Lab. Practice, *22*:578 (1973).
28. J. Keay and P. M. A. Menagé, Analyst (Lond.), *94*:895 (1969).
29. R. S. Hanawalt and J. E. Steckel, in *Automation in Analytical Chemistry*, vol. 1, Technicon Symposia 1966, Mediad Inc., White Plains, N.Y., 1967, p. 133.
30. J. O. Skjemstad and R. Reeve, Commun. Soil Sci. Plant Anal., *7*:229 (1976).
31. R. B. Roy, J. Jansen, and M. Sahn, J. Assoc. Off. Anal. Chem., *63*:931 (1980).
32. P. D. Robertson, U.K. Agricultural Research Council Letcombe Laboratory Annual Report for 1974, HMSO, London 1975, p. 73.
33. A. Henriksen and A. R. Selmer-Olsen, Analyst (Lond.), *95*:514 (1970).
34. A. R. Selmer-Olsen, Analyst (Lond.), *96*:565 (1971).
35. J. L. Grigg, Commun. Soil Sci. Plant Anal., *6*:95 (1975).

36. G. Ogner and A. Haugen, Analyst (Lond.), *102*:453 (1977).
37. W. J. Smythe, M. H. Shamos, S. Morgenstern, and L. T. Skeggs, in *Automation in Analytical Chemistry*, vol. 1, Technicon Symposia 1967, Mediad Inc., White Plains, N.Y., 1968, p. 105.
38. E. D. Wood, F. A. J. Armstrong, and F. A. Richards, J. Mar. Biol. Assoc. U.K., *47*:23 (1967).
39. F. A. J. Armstrong, C. R. Stearns, and J. D. H. Strickland, Deep Sea Res., *14*:381 (1967).
40. A. B. Prince, in *Methods of Soil Analysis* (C. A. Black, Ed.), Am. Soc. Agron., Madison, Wis., 1965, p. 866.
41. S. E. Allen, H. M. Grimshaw, J. A. Parkinson, and C. Quarmby, *Chemical Analysis of Ecological Materials*, Blackwells, Oxford, 1974.
42. R. L. Habig, B. W. Schlein, L. Walters, and R. E. Thiers, Clin. Chem., *15*:1045 (1969).
43. W. D. Basson and R. G. Böhmer, Analyst (Lond.), *97*:266 (1972).
44. A. G. Sinclair, N. Z. J. Agric. Res., *16*:287 (1973).
45. O. R. Mulcahy, Lab. Practice, *26*:679 (1977).
46. G. Toennies and B. Bakay, Anal. Chem., *25*:160 (1961).
47. J. R. Bettany and E. H. Halstead, Can. J. Soil Sci., *52*:127 (1972).
48. E. Canals, A. Charra, and G. Riety, Glasn. Hem. Drust. Beograd, *12*:1055 (1945) (cited in Ref. 43).
49. P. L. Bailey, *Analysis with Ion-Selective Electrodes*, Heyden, London, 1978.
50. G. K. Buckee, J. Inst. Brewing, *80*:291 (1974).
51. P. L. Bailey and M. Riley, Analyst (Lond.), *100*:145 (1975).
52. E. H. Hansen, A. K. Ghose, and J. Růžička, Analyst, (Lond.), *102*:705 (1977).
53. E. H. Hansen, F. J. Krug, A. K. Ghose, and J. Růžička, Analyst (Lond.), *102*:714 (1977).
54. F. Holz and H. Kremers, Landw. Forsch., *26*:177 (1971).
55. J. Murphy and J. P. Riley, Anal. Chim. Acta, *27*:31 (1962).
56. A. Bennet, D. Gartelmann, J. I. Mason, and J. A. Owen, Clin. Chim. Acta, *29*:161 (1970).
57. J. A. Nisbet and E. Simpson, Clin. Chim. Acta, *39*:339 (1972).
58. W. D. Basson, Lab. Practice, *26*:541 (1977).
59. J. W. B. Stewart, J. Růžička, H. Bergamin Filho, and E. A. G. Zagatto, Anal. Chim. Acta, *81*:371 (1976).
60. F. J. Krug, J. Růžička, and E. H. Hansen, Analyst (Lond.), *104*:47 (1979).
61. W. D. Basson and J. F. van Staden, Lab. Practice, *27*:863 (1978).
62. J. Růžička, J. W. B. Stewart, and E. A. G. Zagatto, Anal. Chim. Acta, *81*:387 (1976).

63. K. F. Baker, Analyst (Lond.), *95*:885 (1970).
64. D. Goodman, Analyst (Lond.), *101*:943 (1976).
65. J. L. Grigg, H. J. Flewitt, G. A. Baird, R. B. Jordan, and K. V. Vo, Analyst (Lond.), *105*:1 (1980).
66. J. E. Steward, *The Pye Unicam AC 1 Automatic Chemistry Unit*, Pye Unicam Ltd. Tech. Report, Cambridge, England, 1975.
67. C. T. Cottrell, *Automatic Chemistry Systems for Water Analysis*, Pye Unicam Ltd. Tech. Report, Cambridge, England, 1978.
68. G. E. Schuman, M. A. Stanley, and D. Knudsen, Soil Sci. Soc. Am. Proc., *37*:480 (1973).
69. J. A. Russell, J. Biol. Chem., *156*:457 (1944).
70. G. Suzzarini, in *Automation in Analytical Chemistry*, vol. II, Technicon Symposia 1966, Mediad Inc., White Plains, N.Y., 1967, p. 53.
71. C. H. Williams and J. R. Twine, CSIRO Div. Plant Industry Tech. Paper No. 24, Melbourne, 1967.
72. J. Crowther and J. Evans, Analyst (Lond.), *105*:841 (1980).
73. C. Honda, J. Sci. Soil Manure, Tokyo, *33*:195 (1962).
74. J. M Bremner, in *Methods of Soil Analysis* (C. A. Black, Ed.), Am. Soc. Agron., Madison, Wis., 1965, p. 1149.
75. J. M. Bremner and K. Shaw, J. Agric. Sci. Camb., *51*:22 (1958).
76. B. Dabin, Cahiers ORSTOM, Serie Pedologie, *3*:338 (1965) (cited in Ref. 20).
77. E. K. Best, Queensland J. Agric. Anim. Sci., *33*:161 (1976).
78. J. M. Bremner and D. R. Keeney, Soil Sci. Soc. Am. Proc., *30*:577 (1966).
79. M. F. Gine, H. Bergamin Filho, E. A. G. Zagatto, and B. F. Reis, Anal. Chim. Acta, *114*:191 (1980).
80. J. D. Colwell, Chem. Ind., 893 (1965).
81. P. D. Salt, Chem. Ind., *18*:584 (1968).
82. L. Zsoldos, Hung. Sci. Instr., *39*:19 (1977).
83. P. Crooks and A. Scott, unpublished work.
84. S. McLeod and A. R. P. Clarke, Analyst (Lond.), *103*:238 (1978).
85. S. K. Ng, J. Sci. Fd. Agric., *21*:275 (1970).
86. H. G. Zandstra, Can. J. Soil Sci., *48*:219 (1968).
87. C. H. Orr, Commun. Soil Sci. Plant Anal., *2*:85 (1971).
88. J. R. Burrows and J. H. Meyer, Proc. S. Afr. Sugar Technol. Assoc., *50*:114 (1976).
89. J. R. Burrows, Proc. S. Afr. Sugar Technol. Assoc., *52*:195 (1978).
90. R. E. Kitson and M. G. Mellon, Ind. Eng. Chem., Anal. Ed., *16*:379 (1944).
91. J. A. Varley, Analyst (Lond.), *91*:116 (1966).

92. J. E. Steckel and R. L. Flannery, in *Instrumental Methods for Analysis of Soils and Plant Tissues* (L. M. Walsh, Ed.), Am. Soc. Agron., Madison, Wis., 1971.
93. R. L. Flannery and D. K. Markus, J. Assoc. Off. Anal. Chem., *63*:779 (1980).
94. K. Darab and E. Akos, Agrokem Talajtan (Hungary), *21*:115 (1972); Chem. Abs., *77*:150847 (1972).
95. J. Hislop and I. J. Cooke, Soil Sci., *105*:8 (1968).
96. O. E. Clinton, Chem. N. Z. *43*:143 (1979).
97. M. Cooksey and W. B. Barnett, At. Absorpt. Newsl., *18*:1 (1979).
98. L. A. Lancaster and R. Balasubramaniam, J. Sci. Fd. Agric., *25*:381 (1974).
99. N. G. Reeve and M. E. Sumner, Soil Sci. Soc. Am. Proc., *34*:263 (1970).
100. J. P. Riley and H. P. Williams, Mikrochim. Acta, 804 (1959).
101. C. M. Johnson and H. Nishita, Anal. Chem., *24*:736 (1952).
102. J. Keay, P. M. A. Menagé, and G. A. Dean, Analyst (Lond.), *97*:897 (1972).
103. Technicon Instruments Corporation, Tarrytown, New York, Method No. 226-72W, 1969.
104. M. R. McSwain, R. J. Watrous, and J. E. Douglass, Anal. Chem., *46*:1329 (1974).
105. C. S. Cronan, Anal. Chem., *51*:1333 (1979).
106. W. D. Basson, R. G. Böhmer, and D. A. Stanton, Analyst (Lond.), *94*:1135 (1969).
107. R. Gabriëls and W. Van Keirsbulck, Lab. Practice, *26*:620 (1977).
108. G. Ogner, Analyst (Lond.), *105*:916 (1980).
109. A. L. Willis, Commun. Soil. Sci. Plant Anal., *1*:205 (1970).
110. J. V. Purves, Commun. Soil Sci. Plant. Anal., *5*:261 (1974).
111. E. G. Bradfield and J. F. Stickland, Analyst (Lond.), *100*:1 (1975).
112. G. S. Pyen, M. J. Fishman, and A. G. Hedley, Analyst (Lond.), *105*:657 (1980).
113. H. van Vliet, W. D. Basson, and R. G. Böhmer, Analyst (Lond.), *100*:405 (1975).
114. P. N. Vijan, A. C. Rayner, D. Sturgis, and G. R. Wood, Anal. Chim. Acta, *82*:329 (1976).
115. P. D. Goulden and P. Brooksbank, Anal. Chem., *38*:1431 (1974).
116. H. Agemian and A. S. Y. Chau, Analyst (Lond.), *101*:91 (1976).
117. A. M. Jirka and M. J. Carter, Anal. Chem., *50*:91 (1978).
118. S. S. Jørgensen and M. A. B. Regitano, Analyst (Lond.), *105*:292 (1980).

119. J. A. Guth and G. Voss, Weed Res., *11*:111 (1971).
120. D. E. Ott, G. Formica, G. F. Liebig, Jr., D. O. Eberle, and F. A. Gunter, J. Assoc. Off. Anal. Chem., *54*:1388 (1971).
121. N. D. Hörmann, G. Formica, F. Ramsteiner, and D. O. Eberle, J. Assoc. Off. Anal. Chem., *55*:1031 (1972).

4

Automated Instruments for Determination of Total Carbon, Nitrogen, and Sulfur in Soils by Combustion Techniques

M. ALI TABATABAI and JOHN M. BREMNER Iowa State University, Ames, Iowa

I. INTRODUCTION

Dry combustion techniques have not been used extensively for determination of total carbon, total nitrogen, or total sulfur in soils because until recently the techniques available were complicated and time-consuming compared with wet oxidation methods. These defects of dry combustion methods of analysis have been eliminated by the development of automated techniques that are very simple and rapid compared with their predecessors. The purpose of this chapter is to describe the automated combustion instruments that have been evaluated for the determination of total C, total N, and total S in soils and to review current information concerning the use of these instruments for soil analysis. Several automated instruments besides those discussed have been used to analyze soils, but no work to evaluate these instruments for soil analysis has been reported. Most of these instruments were designed for analysis of pure compounds or homogeneous materials having high C, N, or S contents, and it seems unlikely that they will prove satisfactory for analysis of soils or other heterogeneous materials with low contents of these elements.

II. CARBON ANALYZERS

The only automated combustion instrument thus far evaluated for total C analysis of soils is the Leco 70-Second Carbon Analyzer developed by the Laboratory Equipment Corporation, St. Joseph, Mich. This instrument was originally designed for total C analysis of iron and steel, but its speed and simplicity have led to its use for analysis of nonferrous as well as ferrous materials, and Tabatabai and Bremner [1] have shown that it is very satisfactory for the determination of total carbon in soils.

A. Principles

The Leco 70-Second Carbon Analyzer (Fig. 1) consists of an O_2-purification train, a high-frequency induction furnace, and a carbon dioxide analyzer that utilizes the difference in thermal conductivity between CO_2 and O_2. In total C analysis by this instrument, the sample is treated in a ceramic crucible with combustion accelerators (usually iron plus tin and tin-coated copper) and heated to a high temperature (>1650°C) in a stream of purified O_2, the high temperature being generated by using an induction furnace to induce an electrical field in the accelerator-treated sample. The gases evolved are passed through a dust trap (to remove metal oxides), a trap containing activated MnO_2 (to remove S oxides, N oxides, and halogen gases), a heated catalyst (hopcalite) tube (to convert CO to CO_2), and a trap containing anhydrous $Mg(ClO_4)_2$ (to remove water vapor) (see Fig. 1). The purified

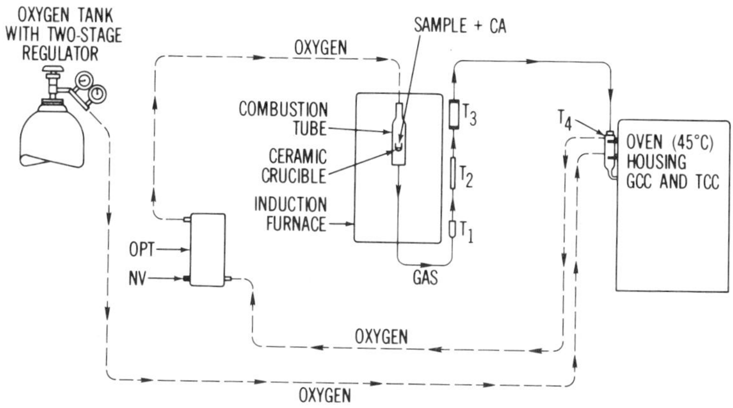

Figure 1 Schematic diagram of Leco 70-Second Carbon Analyzer. CA, combustion accelerators; T_1, dust trap; T_2, trap containing activated MnO_2; T_3, trap containing heated catalyst (hopcalite); T_4, trap containing anhydrous $Mg(ClO_4)_2$; OPT, O_2-purification train; NV, needle valve; GCC, gas collection cylinder; TCC, thermal conductivity cell.

CO_2-O_2 mixture thus obtained is collected in a cylinder maintained at a constant temperature (45°C) and is analyzed for CO_2 by measuring its thermal conductivity with a thermistor-type thermal conductivity cell. The gas collection cylinder (Fig. 2) is maintained at 45°C to eliminate temperature variations that would cause an imbalance in the thermal conductivity cell. This cell consists of a pair of matched thermistors in two arms of a Wheatstone bridge. The reference thermistor is kept in an environment of constant gas pressure, gas flow, and temperature, and the measuring thermistor is maintained in the same environment, except that the gas composition varies according to the CO_2 content of the sample. The sensitivity of the analysis for CO_2 is governed by the difference in thermal conductivity between O_2 and CO_2. The thermal conductivities of O_2 and CO_2 are 24 and 14 mJ m^{-1} sec^{-1} $°C^{-1}$ (57 and 33 μcal cm^{-1} sec^{-1} $°C^{-1}$), respectively.

The Leco 70-Second Carbon Analyzer is designed so that it functions automatically once the accelerator-treated sample is placed in the induction furnace and the cycle switch is pressed to start the analysis. The sequence of events initiated when this switch is pressed is as follows: (1) The O_2 flow is turned on and solid state timers are activated in sequence through the contacts of a "holding" relay tube. (2) The piston in the gas collection cylinder drops, thereby purging gas from the cylinder. (3) O_2 is swept through the system and the gas collection cylinder for a period of 20 sec (the induction furnace is off during this time). (4) The induction furnace is turned on (this is indicated by a red light) to initiate a 10 sec pre-burn cycle (O_2 is

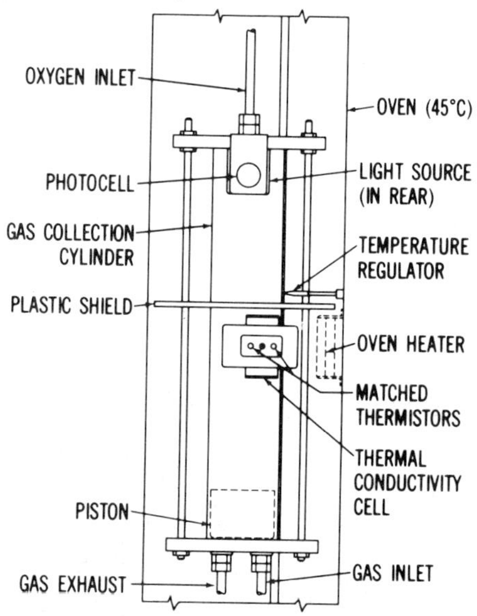

Figure 2 Details of gas collection cylinder in Leco 70-Second Carbon Analyzer.

supplied to the combustion tube during this time, but the flow of O_2 to the cylinder is stopped). (5) At the end of the 10 sec pre-burn, a solenoid valve is opened to admit the gas from the combustion tube to the bottom of the cylinder. As the gas flows into the cylinder, the piston is forced upward until it interrupts the light source of the photocell (Fig. 2). This causes a "holding" relay tube to turn off the induction furnace and close the gas inlet at the bottom of the cylinder. (6) After a 20 sec delay, a clamp controlling a voltmeter is released to allow it to respond to the measuring thermistor in the conductivity cell. (7) At the end of a 10 sec "read" period, the output of the thermal conductivity cell is amplified by a DC amplifier, and the output of the amplifier is recorded automatically as milligrams of carbon on a digital meter. The response of the digital meter is linear between 0 and 8 mg C.

B. Calibration

Before use, the thermal conductivity cell is balanced so that, with pure O_2 in the gas collection cylinder, it gives zero output on the digital meter. With the cell thus balanced, its output as measured on the digital meter is proportional to the amount of CO_2 in the cylinder. Before use, the analyzer is calibrated with carbon standards. The metallic-carbon standards supplied by the Laboratory Equipment Corporation are convenient for this purpose.

C. Commercial Instruments

The Laboratory Equipment Corporation supplies three automated carbon analyzers besides the Leco 70-Second instrument, namely, the Leco IR-12 Carbon Determinator, the Leco DC-12 Duo-Carb Analyzer, and the Leco CS-46 Simultaneous Carbon/Sulfur Determinator. These instruments have not been evaluated for soil analysis, but they clearly deserve evaluation. The DC-12 Duo-Carb instrument permits determination of both total C and organic C, and the CS-46 instrument allows determination of both total C and total S.

In total C analysis by the Leco IR-12 instrument, the sample is mixed with a metal combustion accelerator (usually Cu or Fe) in a ceramic crucible and is heated in an induction furnace in an atmosphere of O_2 to oxidize C to CO and CO_2. The gas mixture thus obtained is dried, passed through a heated Pt catalyst to convert CO to CO_2, and collected in a cell for CO_2 analysis by an infrared detector. This instrument has been used successfully for determination of total C in complex materials such as automotive catalyst support formulations [2], which contain about 97% of inorganic material, such as clay, talc, and alumina, and about 3% of organic extrusion aids, such as methyl cellulose and diglycol stearate.

In total C analysis by the Leco DC-12 Duo-Carb Analyzer, the sample is mixed with vanadium pentoxide and heated to about 1000°C in a stream of O_2 in an induction furnace. The CO_2 thus produced is collected in a cylinder and measured by a thermal conductivity detector. To determine organic C, a second sample is combusted at about 600°C, and the CO_2 thus produced is determined as in the procedure for determination of total C. Inorganic C in the sample is calculated as the difference between the results of the analyses for total C and organic C. This instrument was designed for rapid determination of total C and organic C in shales, soils, and limestones, but it has not been evaluated for soil analysis.

In total C analysis by the Leco CS-46 instrument, the sample is mixed with a combustion accelerator and combusted in O_2 in an induction furnace. The CO and CO_2 thus produced are measured individually by separate infrared detectors, and the outputs of these detectors are added electronically to obtain total C. The SO_2 produced by combustion is measured by a third infrared detector. This instrument differs from previous carbon analyzers in that a recirculating O_2 system replaces the conventional flow-through system, solid state infrared detectors replace the conventional Luft cell for CO_2 analysis, and direct measurement of CO replaces catalytic conversion of CO_2. It permits simultaneous determination of total C and total S within 30 sec and clearly deserves evaluation for routine determination of total C and S in soils.

Another automated combustion instrument now available for simultaneous determination of total C and total S is the I.R.-Matic "C-S" VK-111 AS Analyzer supplied by Kukusai Electric Co., Japan. In determination of total C and total S by this instrument, the sample is mixed in a ceramic crucible with combustion accelerators (iron powder and tungsten chips) and combusted in a high-frequency induction furnace using O_2 as the combustion and carrier gas and N_2 as purge gas. The CO_2 and SO_2 thus produced are measured by Luft-type nondispersive infrared detectors. The instrument is supplied with an automatic balance and accessories such as linearizers, integrators, blank adjusters, single-point calibrating adjusters, and weight compensators, and it permits determination of both total C and total S in less than 60 sec. It has given satisfactory results when used for analysis of standard rocks, NBS minerals, and geological exploration reference samples and it seems promising for determination of C and S in soil samples [3].

Stewart et al. [4] demonstrated that the Coleman Model 29 Nitrogen Analyzer (an automated Dumas combustion instrument described in Sec. III) can be adapted for determination of total C in soils.

Dugan [5] has described an automatic combustion instrument developed by Hercules, Inc., Wilmington, Del. for simultaneous determination of C, H, N, and S. The procedure in analysis by this instrument involves an uncatalyzed, flash-combustion of the sample in a quartz tube in an O_2-He atmosphere. The CO_2, H_2O, N_2, and SO_2 formed by this combustion are separated and determined by a gas chromatographic procedure involving use of a thermal conductivity detector. This instrument has not been evaluated for determination of C, N, or S in soils, but it seems unlikely to prove useful for soil analysis.

Another automated combustion instrument that involves flash combustion in a He-O_2 atmosphere is the Model 1106 Elemental Analyzer developed by Carlo Erba Strumentazione, Milan, Italy for determination of C, H, N, O, and S. For C, H, and N analysis, the gases produced by flash combustion are passed over Cr_2O_3 to ensure quantitative combustion, and the excess O_2 is removed by Cu at 650°C before separation and determination of N_2, CO_2, and H_2O by gas chromatography. For S analysis, the gases produced by flash combustion are passed over WO_3 to ensure quantitative conversion of S to SO_2, which is determined by gas chromatography.

III. NITROGEN ANALYZERS

The only automated combustion instruments thus far evaluated for total N analysis of soils are the Model 29 and Model 29A Nitrogen Analyzers developed by Coleman Instruments, Inc., Maywood, Ill. (now Coleman

Table 1 Coleman Nitrogen Analyzers (Manufacturer's Specifications)

Specification	Analyzer model	
	29	29A
Sample size, mg	2-50	25-500
Maximal amount of N in sample, mg	4	40
Number of analyses per hour	4-5	3-4
Capacity of nitrometer syringe, ml	5	50
Reproducibility of nitrometer syringe, μl	±2	±10
Operating cycle, min	8	12
Readout	Digital counter	Motor-driven digital counter

Instruments Division of the Perkin-Elmer Corporation, Oak Brook, Ill.). These analyzers are automated Dumas combustion instruments (see Table 1 for specifications).

A. Principles

A schematic diagram of the Coleman Model 29 Nitrogen Analyzer is shown in Fig. 3. In total N analysis with this instrument, the sample is mixed with combustion catalyst (usually CuO) and heated to above 900°C in a quartz combustion tube filled with catalyst. This heating to pyrolyze the sample is carried out in an atmosphere of high-purity CO_2, and the gases evolved are swept by purified CO_2 through a post-heater tube containing Cu in the first half of its length and CuO in its terminal section. This tube is heated by an auxiliary furnace (usually to 600-700°C). Its function is to reduce N oxides to N_2, oxidize any H_2 to H_2O, and convert CO to CO_2. The gas stream from the post-heater tube passes into a glass nitrometer containing concentrated alkali (KOH solution), where it is released from a nozzle under mercury as a stream of small bubbles. These bubbles are brought into contact with the alkali by a magnetic stirrer, which ensures complete absorption of CO_2, and the N_2 collects in the body of the nitrometer, displacing alkali into a reservoir. The N_2 in the nitrometer is then measured volumetrically by drawing it into a precisely calibrated stainless steel syringe. The plunger of this syringe is driven by a micrometer screw linked to a precision digital counter. Manual adjustment of this counter provides readout of N_2 volume directly in microliters. The Model 29 Analyzer has a syringe capacity of 5 ml, which permits consecutive analyses of microsamples without resetting the N_2 measurement system. The Model 29A Analyzer has a syringe capacity of 50 ml. Its syringe plunger is driven to approximate adjustment by a reversible electric motor, final adjustment being made manually.

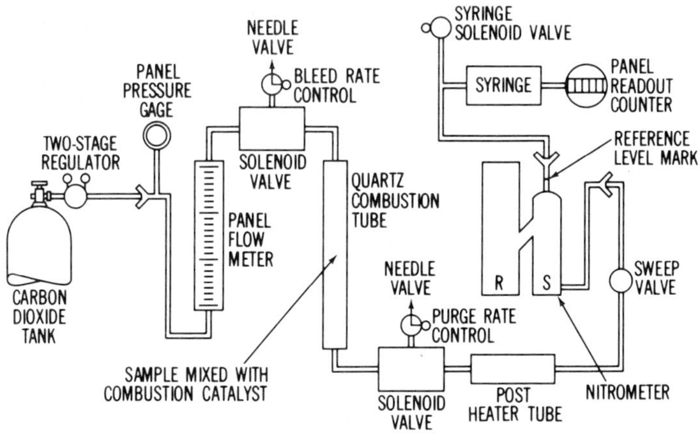

Figure 3 Schematic diagram of Coleman Model 29 Nitrogen Analyzer. Diagram does not show furnace used to heat combustion tube. Nitrometer has reservoir (R) and magnetic stirrer (S) and contains concentrated KOH solution.

B. Calculation of Results

The percentage of N in the sample analyzed is calculated from: R_1 (initial counter reading), R_2 (final counter reading), P (barometric pressure), t_1 (initial syringe temperature), and t_2 (final syringe temperature). The steps in the calculation are as follows:

1. Record the observed volume of N_2, V_o:

$$V_o = R_2 - R_1$$

where V_o is the observed N_2 volume (in microliters) and R_1 and R_2 are as defined above.

2. Determine the corrected N_2 volume, V_c:

$$V_c = V_o - (V_b + V_t)$$

where

V_b = volume of blank (in μl)
V_t = volume correction for temperature (in μl)
= $C_f\,(t_2 - t_1)$

where C_f is a correction factor based on the final counter reading (this factor is readily available from tables supplied by the manufacturer) and t_2 and t_1 are in degrees Kelvin.

3. Determine the corrected barometric pressure P_c:

$$P_c = P_o - (P_b + P_v)$$

where P_o is the observed barometric pressure (in torr), P_b is the barometric temperature correction, and P_v is a correction for the vapor pressure of the KOH solution in the nitrometer (P_b and P_v are readily available from tables supplied by the manufacturer).

An empirical approximation of $P_b + P_v = 11.0$ is sufficiently accurate for P_o between 740 and 780 torr (99-104 kPa) and syringe temperatures between 298 and 305 K.

4. Calculate the percentage of N in the sample from the following equation:

$$\%N = \frac{P_c}{T} \times \frac{V_c}{W} \times 0.04493$$

where T is the final syringe temperature in degrees Kelvin, W is the sample weight in milligrams, and P_c and V_c are as defined above.

C. Commercial Instruments

The Laboratory Equipment Corporation has recently developed three automated nitrogen analyzers (the Leco UO-14SP Nitrogen Determinator, the Leco TN-15 Nitrogen Determinator, and the Leco TC-36 Nitrogen/Oxygen Determinator) in which total N is converted to N_2 by fusing the sample in a graphite crucible at a high temperature in an inert (He) atmosphere. These instruments were developed for rapid determination of total N in ferrous materials, but it seems likely that they could be adapted for routine determination of total N in soils.

In total N analysis by the Leco UO-14SP instrument, the sample is mixed with CuO-Pt catalyst, encased in tinfoil, and fused (ca. 2600°C) in a graphite crucible in an impulse furnace. The gases evolved are passed over CuO at 450°C (to convert N oxides to N_2), through ascarite (to remove CO_2), and through $Mg(ClO_4)_2$ and P_2O_5 (to remove H_2O). The remaining gases are passed through a molecular sieve column maintained at 10 to 11°C. This column separates N_2 from other gases and permits quantitative determination of N_2 by a thermal conductivity detector. The analysis time is approximately 7 min. There seems very little doubt that this instrument will prove satisfactory for the determination of total N in soils, because Wong and Kemp [6] found that it gave satisfactory results when used for total N analysis of 18 sediments containing from 0.02 to 2.3% N, and the problems in sediment analysis are similar to those encountered in soil analysis.

The procedure for total N analysis by the Leco TN-15 instrument is similar to that employed in the UO-14SP instrument, but the sample is fused in a He atmosphere in an electrode furnace at 3000°C, and the analysis time is only 80 sec.

The Leco TC-36 instrument permits simultaneous determination of nitrogen and oxygen. In analysis by this instrument, the sample is fused in a He atmosphere in an electric furnace at 2500°C, and a stream of He is used to carry the gases evolved over CuO (to oxidize CO to CO_2 and H_2 to H_2O) and through a column containing $Mg(ClO_4)_2$ (to remove H_2O). The N_2 and CO_2 in the stream of He emerging from this column are then separated and determined by a gas chromatographic procedure involving use of a thermal conductivity detector (oxygen in the sample is calculated from the CO_2 determined by this procedure). The analysis time is about 3.5 min.

Carlo Erba Strumentazione, Milan, Italy, has developed an automated combustion instrument (the Carlo Erba Model 1400 Nitrogen Analyzer) that merits evaluation for determination of total N in soils. In total N analysis by this instrument, the sample is oxidized in a furnace containing NiO and O_2 maintained at 1000°C. The gases thus produced are swept from the oxidation furnace by a stream of He and passed through a reduction furnace containing Cu at 750°C to reduce oxides of N to N_2. The N_2 thus obtained is then passed through traps to remove moisture, CO_2, and any inorganic acids in the gas stream and is determined by a gas chromatographic procedure using a thermal conductivity detector.

Antek Instruments, Inc., Houston, Texas, recently introduced automated N analyzers in which the sample is pyrolyzed in an O_2-Ar atmosphere to oxidize N to NO, which is determined by using a chemiluminescent detector to measure the light emitted by metastable NO_2 produced by mixing NO with O_3. These instruments have been used to determine N in petroleum fractions [7] and to measure gaseous loss of N from plants during transpiration [8], but they have not been evaluated for total N analysis of soils.

IV. SULFUR ANALYZERS

The only automated combustion instrument thus far evaluated for total sulfur analysis of soils is the Leco Sulfur Analyzer supplied by the Laboratory Equipment Corporation, St. Joseph, Mich. Like the Leco Carbon Analyzer discussed in Sec. II, this instrument was originally designed for analysis of iron and steel.

A. Principles

The Leco Sulfur Analyzer consists of an O_2-purification train, a high-frequency induction furnace fitted with a timer to select the time of induction, and an automatic sulfur dioxide (SO_2) titrator (Fig. 4). In total S analysis by this instrument, the sample is treated in a ceramic crucible with combustion accelerators (usually iron plus tin, copper, or tin-coated copper), and the crucible is covered with a porous ceramic cover and heated to a high temperature (ca. 1600°C) in a stream of purified O_2, the high temperature being generated by using an induction furnace to induce an electrical field in the acceler-

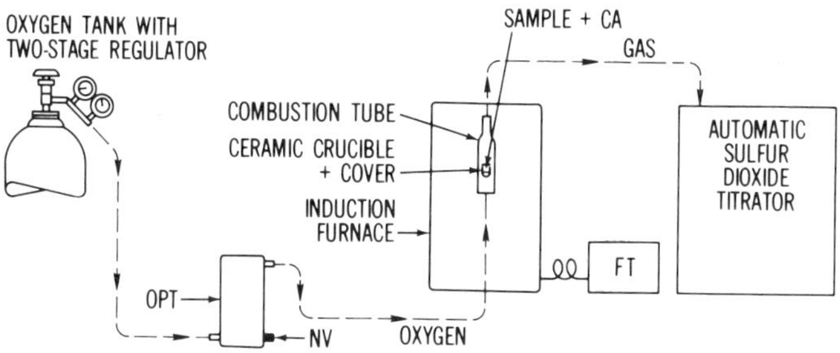

Figure 4 Schematic diagram of Leco Sulfur Analyzer. CA, combustion accelerators; OPT, O_2-purification train; NV, needle valve; FT, furnace timer.

ator-treated sample. The SO_2 liberated by combustion of the sample is collected in a dilute solution of hydrochloric acid containing KI, starch, and a trace amount of KIO_3 and is determined by titration of this solution with standard KIO_3 solution. This titration is performed automatically through use of a photocell to maintain the initial optical density of the blue color formed by the reaction of starch with the iodine formed by the reaction of KI and the KIO_3 in the presence of HCl. The reactions that occur during this titration are indicated by the following equations:

$$5KI + KIO_3 + 6HCl = 3I_2 + 6KCl + 3H_2O$$

$$SO_2 + I_2 + 2H_2O = H_2SO_4 + 2HI$$

The Leco Sulfur Analyzer functions automatically once the accelerator-treated sample is placed in the induction furnace and the timer switch is pressed to initiate the analysis.

B. Calibration

Before use, the automatic sulfur dioxide titrator is calibrated by analyzing sulfur standards. The metallic-sulfur standards supplied by the Laboratory Equipment Corporation are convenient for this purpose.

C. Commercial Instruments

The Laboratory Equipment Corporation has recently developed two automated sulfur analyzers that have not thus far been evaluated for soil analysis, but clearly deserve evaluation. One is the Leco CS-46 Simultaneous Carbon/Sulfur Determinator described in Sec. II.C. The other is the Leco SC-32 Sulfur Analyzer. In total S analysis by

Table 2 Precision of Results Obtained in Total Carbon Analysis of Soils by Leco 70-Second Carbon Analyzer and by Wet Combustion Method of Allison

Soil	Method of analysis[a]	Number of analyses	Total C content (%)		
			Range	Mean	SD[b]
Buckner	L	8	0.29-0.33	0.32	0.02
	A	5	0.28-0.34	0.31	0.02
Ida	L	8	1.06-1.13	1.10	0.02
	A	6	1.10-1.18	1.15	0.03
Hayden	L	8	3.02-3.11	3.06	0.02
	A	6	2.97-3.11	3.07	0.02
Oxbow[c]	L	5	3.30-3.38	3.36	0.03
	A	5	3.34-3.46	3.38	0.03
Glencoe	L	8	5.76-5.84	5.80	0.13
	A	6	5.65-5.82	5.67	0.13
Peat	L	5	33.9-34.5	34.2	0.25
	A	5	32.5-34.5	33.5	0.90

[a]L, Leco C Analyzer; A, Allison method [9].
[b]SD, standard deviation.
[c]26.5% $CaCO_3$.
Source: Ref. 1.

this instrument, the sample is combusted at about 1350°C in a stream of O_2, and the SO_2 thus produced is measured by a solid state infrared detector. The analysis time is only 2 min.

V. APPLICATIONS

A. Carbon Analyzers

Table 2 and Fig. 5 show results obtained by Tabatabai and Bremner [1] in work to evaluate the Leco 70-Second Carbon Analyzer for total C analysis of soils. The 46 soils used in this work were selected so that they differed markedly in organic-matter content (0.3-34% organic C), texture (2-94% sand, 1-43% clay), and carbonate content (0-26.5% $CaCO_3$), and the total C values obtained in analysis of these soils by the Leco 70-Second Carbon Analyzer were compared with those obtained by the wet combustion method of Allison [9], which has gained acceptance as a reliable method of determining total C in soils. This comparison showed that the total C values obtained by the two methods

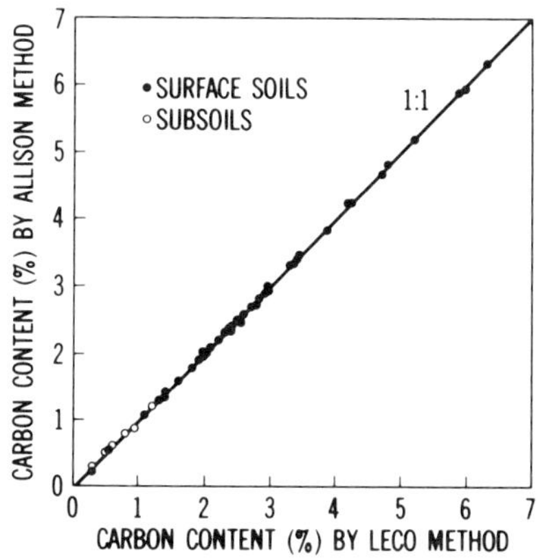

Figure 5 Results of total carbon analysis of 46 soils by Leco 70-Second Carbon Analyzer versus results by method of Allison. [Reprinted from Ref. 1, Soil Sci. Soc. Am. Proc., *34*:607-610 (1970) by permission of the publisher, Soil Science Society of America.]

agreed closely (Fig. 5) and that the Leco method of analysis was at least as precise as Allison's method (Table 2). Other studies reported by Tabatabai and Bremner [1] showed that the Leco 70-Second Carbon Analyzer gave quantitative recovery of carbonate-C and that the results of soil analysis by this instrument were not affected by the presence of substantial amounts of chloride. A study of the effect of sample mesh-size on total C analysis of mineral soils by the Leco Carbon Analyzer showed that the precision of the results obtained with this instrument usually increased with decrease in soil mesh-size but that, with most soils studied, the results with <40 mesh (420 μm) soil were almost as precise as those obtained with <100 mesh (150 μm) or <300 mesh (50 μm) soil (Table 3). Soil samples containing 2-10 mg of C (usually 0.2-0.3 g of mineral soil) were found satisfactory for total C analysis by the Leco Analyzer. With several soils, incomplete combustion was observed when the amount of soil taken for analysis exceeded 0.5 g or contained more than 12 mg of C.

Total C analysis of soils with the Leco 70-Second Carbon Analyzer is very simple and rapid, and a single operator can easily perform more than 150 analyses in a normal working day (including weighing of soil samples). Well over 250 analyses of preweighed samples can be performed within 8 hr.

B. Nitrogen Analyzers

Stewart et al. [10] analyzed 47 soils containing 0.02-0.47% N by the Coleman Model 29 Nitrogen Analyzer and compared the results with those obtained by a Kjeldahl method commonly used for total N analysis

Table 3 Effect of Sample Mesh-Size on Total Carbon Analysis of Soils by Leco 70-Second Carbon Analyzer

Soil	Max. particle size		Total C content by Leco analyzer (%)[a]		
	mesh	μm	Range	Mean	SD
Lindley	10	2000	1.16-1.41	1.29	0.11
	40	420	1.31-1.38	1.33	0.03
	100	150	1.35-1.38	1.36	0.01
Sharpsburg	10	2000	2.30-2.45	2.36	0.06
	40	420	2.28-2.34	2.31	0.02
	100	150	2.28-2.32	2.29	0.02
	300	50	2.26-2.29	2.28	0.01
Grundy	10	2000	2.60-2.69	2.63	0.04
	40	420	2.63-2.75	2.68	0.04
	100	150	2.62-2.69	2.66	0.03
	300	50	2.60-2.68	2.64	0.03
Glencoe	10	2000	5.65-6.05	5.81	0.15
	40	420	5.68-5.86	5.78	0.07
	100	150	5.77-5.84	5.80	0.03

[a]Five analyses. SD, standard deviation.
Source: Ref. 1.

Table 4 Comparison of Results Obtained in Total Nitrogen Analysis of 15 Soils by Coleman Model 29 Nitrogen Analyzer and by Kjeldahl Method

	Total N content (%)	
Soil number	Coleman analyzer	Kjeldahl method
1	0.060	0.061
2	0.102	0.103
3	0.103	0.110
4	0.154	0.155
5	0.141	0.145
6	0.168	0.167
7	0.189	0.191
8	0.194	0.196
9	0.215	0.217
10	0.299	0.300
11	0.330	0.339
12	0.349	0.358
13	0.481	0.481
14	1.178	1.202
15	1.313	1.316

Source: Ref. 4.

of soils. They found that the results by these two methods agreed closely for soils containing less than about 0.3% N, but that the Model 29 Analyzer gave 5-20% higher values than the Kjeldahl method for soils containing substantial amounts of organic matter. This confirmed previous work [11,12] showing that Dumas and Kjeldahl procedures gave almost identical results with mineral soils, but that the Dumas method gave significantly higher values than the Kjeldahl method with organic soils. Stewart et al. [10] showed that the higher values obtained by the Dumas procedure with organic soils were caused by incomplete combustion, which resulted in methane (CH_4) being formed and measured with the N_2 produced by combustion. They later found [4] that this problem could be eliminated by use of CuO-Pt catalyst instead of CuO in the post-heater tube of the Model 29 Analyzer and showed that, when the Model 29 Analyzer was so modified, the total N values obtained with both mineral and organic soils by this instrument agreed closely with those obtained by Kjeldahl analysis (Table 4).

The Coleman Model 29 Analyzer was not designed for analysis of soils or other heterogeneous materials having low N contents, and its use for total N analysis of such materials is complicated by sampling problems imposed by sample-size limitations [4]. To overcome this

Table 5 Comparison of Results Obtained in Total Nitrogen Analysis of 12 Soils by Coleman Model 29A Nitrogen Analyzer and by Kjeldahl Methods

		Total N content (%)		
Soil	Sample depth (cm)	Coleman analyzer	Kjeldahl method	HF-Kjeldahl method
Buckner	0-15	0.028	0.028	0.028
Yolo	38-58	0.086	0.083	0.092
Promise	75-112	0.065	0.055	0.068
Clarence	38-58	0.092	0.092	0.093
Ida	0-15	0.108	0.108	0.109
Monona	0-15	0.169	0.169	0.170
Marshall	0-15	0.212	0.211	0.213
Galva	0-15	0.281	0.280	0.281
Clyde	0-15	0.411	0.410	0.412
Glencoe	0-15	0.887	0.890	0.889
Howard peat	0-15	1.64	1.64	1.63
Colo peat	0-15	2.34	2.34	2.33

Source: Ref. 13.

difficulty and eliminate the need for fine grinding of samples and use of a microbalance in analysis of heterogeneous materials having low N contents, the Coleman Company developed the Model 29A Nitrogen Analyzer, which permits analysis of relatively large samples of heterogeneous materials (see Table 1).

Keeney and Bremner [13] evaluated the Coleman Model 29A Nitrogen Analyzer for total N analysis of soils. They found that a technique involving combustion of soil with CuO-Pt catalyst at 940°C for 12.6 min in a tube filled with this catalyst and use of a post-heater tube containing Cu and CuO-Pt catalyst maintained at 700°C gave higher results than other techniques studied and that the total N values obtained by this technique with surface samples of mineral and organic soils agreed closely with those obtained by a Kjeldahl method previously evaluated for soil analysis (Table 5). This combustion technique differed from that used by Stewart et al. [4] for soil analysis by the Model 29 Analyzer mainly in that the time of combustion was 12.6 instead of 4.5 min and the soil sample was mixed with CuO-Pt catalyst instead of CuO and was pyrolyzed in a tube filled with this catalyst.

Stewart et al. [10] found that the coefficient of variation of the results obtained by the technique they originally adopted for total N analysis of soils by the Coleman Model 29 Analyzer was about 2% for soils containing less than 0.1% N and about 1% for soils containing more than 0.1% N. They subsequently found [4] that, when this combustion technique was modified to eliminate interference by CH_4, the average relative standard deviation for the mean of the results obtained with mineral and organic soils was 1.61%, which was slightly higher than the corresponding deviation for the mean of the results obtained by Kjeldahl analysis (1.42%). Keeney and Bremner [13] found that the precision of the technique they adopted for total N analysis of soils by the Coleman Model 29A Analyzer was slightly lower than that of the Kjeldahl method. For example, the coefficient of variation of the results they obtained in six analyses of a Marshall soil (0.212% N) was 1.5% for the Model 29A Analyzer and 0.9% for the Kjeldahl method.

Stewart et al. [4] used 30-500 mg of soil (0.2-0.3 mg of N) for determination of total soil N by the Coleman Model 29 Analyzer and emphasized the need for fine grinding of the soil sample to eliminate sampling error. Keeney and Bremner [13] used 0.2-1.0 g of soil (2.8-4.7 mg of N) for total N analysis by the Coleman Model 29A Analyzer but found that this instrument permitted analysis of up to 3 g of mineral soil and 1 g of organic soil and that the sample analyzed need not be finely ground (satisfactory results were obtained with <40-mesh (420 μm) soil). They also found that total N analysis of soil by the Model 29A Analyzer was not affected if the sample analyzed contained a substantial amount of water. In contrast, Stewart et al. [10] obtained evidence that a sweeping action caused by vaporization of water during combustion could vitiate total N analysis of undried soil samples by the Model 29 Analyzer.

Stewart and Porter [14] showed that indigenous fixed ammonium N in profile samples of two clay soils was fully recovered by the Kjeldahl method only when these samples were treated with HF before analysis. Stewart et al. [10] found that their procedure for total N analysis of soils using the Coleman Model 29 Analyzer did not recover any more or less of the fixed ammonium N in these profile samples than did the Kjeldahl method. Keeney and Bremner [13] found that the technique they adopted for total-N analysis of soils by the Model 29A Analyzer gave higher values than the Kjeldahl method with two subsoils showing a pronounced HF effect in Kjeldahl analysis, but did not give quantitative recovery of the fixed ammonium N in these subsoils (Table 5).

Stewart et al. [4] obtained 94.5-100.0% recovery of nitrate N in analysis of soil by the Coleman Model 29 Analyzer, and Keeney and Bremner [13] obtained 96.0-97.6% recovery of both nitrate N and nitrite N in analysis of soil by the Coleman Model 29A Analyzer. This suggests that soils contain materials that promote recovery of nitrate N and nitrite N in Dumas analysis because difficulties have been encountered in total N analysis of nitrates and nitrites by Dumas procedures [15-17].

The possibility that Coleman Nitrogen Analyzers or other automated Dumas combustion instruments can be adapted successfully for

Table 6 Comparison of Results Obtained in Total Sulfur Analysis of Soils by Leco Sulfur Analyzer and by Other Methods

	Total S content (μg S g^{-1} soil)			
		Other methods[a]		
Soil	Leco S analyzer	A	T	S
Buckner	60	56	56	55
Tama	257	246	248	238
Marshall	330	310	315	294
Hayden	354	334	333	324
Webster	405	410	411	392
Primghar	405	418	418	392
Marcus	472	438	441	440
Mean	326	316	317	305

[a]A, acid oxidation method of Arkley [27]; T, alkaline oxidation method of Tabatabai and Bremner [28]; S, dry ashing method of Steinbergs et al. [29].
Source: Ref. 26.

^{15}N analysis in tracer studies of N transformations in soils deserves attention because the Dumas combustion technique converts labeled N directly to N_2 and is simple and rapid compared with the three-stage technique currently used to convert labeled N to N_2 for mass spectrometer assay of ^{15}N (see Ref. 18). Attempts to adapt the Coleman Model 29 Nitrogen Analyzer for ^{15}N analysis have been reported by Barsdate and Dugdale [19] and by Desaty et al. [20].

Several studies have indicated that Coleman Nitrogen Analyzers can be adapted successfully for total N analysis of plant materials [4, 21-25].

In summary, work to evaluate Coleman Nitrogen Analyzers for determination of total N in soils indicates that these instruments can be used satisfactorily for total N analysis of most soils but do not give quantitative recovery of fixed ammonium N in some soils. These analyzers have the advantage that they require very little bench space compared with the equipment needed for total N analysis by the Kjeldahl method. However, they have gained little acceptance because, besides requiring considerable skill and experience for satisfactory use, they permit only 30-40 analyses per day and do not compare favorably with semiautomated Kjeldahl methods for routine total N analysis of soils.

Table 7 Precision of Results Obtained in Total Sulfur Analysis of Soils by Leco Sulfur Analyzer and by Other Methods

		Total S content (μg S g^{-1} soil)[b]		
Soil	Method[a]	Range	Mean	SE
Tama	L	224-272	252	5.0
	A	236-254	244	1.8
	T	241-252	247	1.3
	S	227-247	237	2.6
Webster	L	368-445	398	8.7
	A	406-424	411	1.6
	T	406-420	410	1.3
	S	369-406	391	3.5
Marcus	L	394-506	470	10.0
	A	430-447	438	1.6
	T	434-446	441	1.1
	S	399-446	432	7.0

[a]L, Leco S Analyzer; A, acid oxidation method of Arkley [27]; T, alkaline oxidation method of Tabatabai and Bremner [28]; S, dry ashing method of Steinbergs et al. [29].
[b]Ten analyses. SE, standard error.
Source: Ref. 26.

C. Sulfur Analyzers

Tabatabai and Bremner [26] attempted to adapt the Leco Sulfur Analyzer for determination of total S in soils but were unable to obtain accurate and precise results with this instrument. The procedure giving results in closest agreement with those obtained by three procedures used as reference methods involved use of methionine to calibrate the burette of the automatic SO_2 titrator and combustion of <100 mesh (150 μm) soil (usually about 0.25 g) for 7 min using iron and tin-coated copper as combustion accelerators. However, the results obtained by this procedure were less precise and usually higher than those obtained by the methods used for comparison (Tables 6 and 7). The combustion technique evaluated by Tabatabai and Bremner was adopted after systematic study of factors affecting analysis of soils by the Leco Sulfur Analyzer. The factors studied included amount and mesh-size of soil sample, amount and type of combustion accelerator (iron, copper, tin, tin-coated copper), method of treating soil sample with accelerator, type of combustion crucible, time of combustion, and method of calibrating the burette of the automatic SO_2 titrator. The results showed that <100 mesh (150 μm) soil gave more reproducible results than did <60 mesh (250 μm) soil and that a combustion time of 7 min gave higher sulfur values than did a combustion time of 4 min. They also showed that none of the ceramic crucibles tested gave satisfactorily low and precise blank values and that the results of soil analysis by the Leco Sulfur Analyzer depended upon the type of S standard used to calibrate the burette of the automatic SO_2 titrator. The total S values obtained when inorganic S standards were used to calibrate the burette of the titrator were consistently lower than those obtained when calibration was performed with organic S standards.

The Leco method of S analysis is based on the assumption that the combustion technique in this method converts S quantitatively to SO_2. Since the procedure used to determine SO_2 in this method does not recover S evolved as sulfur trioxide (SO_3), Tabatabai and Bremner [26] studied the possibility that their failure to obtain accurate and precise results in total S analysis of soils by the Leco Sulfur Analyzer might be due to formation of SO_3 as well as SO_2 during combustion of soils by the Leco technique. To check this possibility, they used 1.0 M NaOH instead of dilute HCl to absorb the S-containing gases formed by combustion of soil samples in the Leco instrument and analyzed this solution for S by the Johnson-Nishita procedure [30] (SO_2 and SO_3 are rapidly absorbed by 1.0 M NaOH with formation of sodium sulfite and sodium sulfate, and both sulfite-S and sulfate-S are determined quantitatively by the Johnson-Nishita procedure). They found, however, that this modification did not improve the accuracy of precision of the Leco method.

Searle [31] described a method for total S analysis of soils in which the soil sample is heated with molybdenum trioxide, chromium trioxide, and iron powder in a Leco high-frequency induction furnace, and the S-containing gases evolved are absorbed in 1 M NaOH, which is subsequently analyzed for S by the Johnson-Nishita procedure [30]. He evidently found that it was necessary to use catalytic oxidants, such as molybdenum and chromium trioxides, to obtain complete combustion of soil samples with the Leco induction furnace. However, his results showed that, when catalytic oxidants were used, the recovery of soil S was far from quantitative if it was determined by absorption and titration of the SO_2 evolved as in the normal Leco technique. Searle concluded that catalytic oxidants promote formation of SO_3 during combustion of soils by the Leco technique and cited work by Green [32] indicating that catalytic oxidants promote formation of SO_3 during S analysis of cast iron by an induction-furnace technique. But he did not report results of S analysis of soils by the normal Leco procedure or give any basis for his conclusion that use of catalytic oxidants is essential for complete combustion of soil samples with the Leco induction furnace. Moreover, examination of his data shows that the modified Leco method he proposed for total S analysis of soils had low precision and gave results that did not agree closely with those

Table 8 Sulfur Analyses of Some Alberta Soil Profile Samples

Soil	Depth of sample (cm)	Sulfur content (μg S g^{-1} soil)[a]			
		T	H	C	H + C
1	0-10	343	338	231	569
	10-18	139	.143	73	216
	18-28	87	70	21	91
	28-43	96	89	32	121
2	0-12	869	465	484	949
	12-25	148	253	20	273
	25-37	78	145	14	159
	37-51	183	310	-	-
3	0-3	1040	590	643	1233
	3-18	396	189	144	333
	18-38	204	129	68	207
	38-88	72	76	21	97

[a]T, total S (Leco S Analyzer); H, HI-reducible S; C, carbon-bonded S.
Source: Ref. 34.

obtained by an unpublished acid oxidation procedure used for comparison.

Tiedemann and Anderson [33] attempted to adapt the Leco Sulfur Analyzer for total S analysis of both soil and plant material. Initially they employed a Leco method recommended for total S analysis of hydrocarbons, but they found that this method was unsatisfactory for soil or plant analysis and recommended a procedure involving four modifications of this method, including modifications to eliminate interference by "N and Cl from the sample." However, the procedure recommended gave only 80-90% recovery of inorganic S (potassium sulfate) or organic S [p-(methylamino) phenol sulfate] added to soil or plant material, and it was not evaluated by comparing its results with those obtained by accepted methods of determining total S in soils or plants.

Lowe [34] used the Leco Sulfur Analyzer in work to characterize the S in some Canadian soil profiles but did not fully describe the procedure he used for total S analysis of soils with this instrument. He noted, however, that the procedure did not have good precision, and his results showed that it gave very poor recovery of total S in the soil samples analyzed. This is evident from Table 8, which shows that the total S values he obtained by the Leco Analyzer with several profile samples were lower than the values obtained for HI-reducible S and that, with many samples, they were much lower than the sum of the values obtained by analysis for HI-reducible S and carbon-bonded S.

Kaplan et al. [35] found that the Leco Sulfur Analyzer did not give precise results when used for total S analysis of marine sediments and that the total S values obtained by this instrument with surface samples of these sediments were much lower than those obtained by a wet combustion method of determining total S. They attributed the low results with surface samples to incomplete combustion of organic matter and reported that combustion of these samples produced smoke that interfered with the operation of the automatic photoelectric titrator used for determination of SO_2 in the Leco method of analysis. Tabatabai and Bremner [26] did not experience this smoke problem in analysis of soils.

Smith [36] recently described a method for total S analysis of soils and plant material using the Leco Sulfur Analyzer, but he reported no evidence that this method was satisfactory for soil analysis.

VI. SUMMARY

Several instruments permitting rapid determination of total C, total N, or total S by automated combustion methods have been evaluated for soil analysis. The principles of these methods are discussed, and studies to evaluate these instruments for soil analysis are reviewed.

Work to evaluate the Leco 70-Second Carbon Analyzer has shown that this instrument permits rapid, accurate, and precise determination of total C in soils and is very satisfactory for routine soil analysis. The Coleman Model 29 and Model 29A Nitrogen Analyzers have been adapted successfully for total N analysis of soils, but they are not simple and rapid enough for routine soil analysis. Attempts to adapt the Leco Sulfur Analyzer for accurate and precise determination of total S in soils have not thus far been successful, but this instrument may prove useful for S analysis in some types of soil investigations. Recently developed automated combustion instruments that merit evaluation for soil analysis include the Leco range of instruments (the DC-12 Duo-Carb Analyzer, the IR-12 Carbon Determinator, the CS-46 Simultaneous Carbon/Sulfur Determinator, the SC-32 Sulfur Analyzer, the UO-14SP Nitrogen Determinator, the TC-36 Nitrogen Analyzer, and the TN-15 Nitrogen Determinator) and the Carlo Erba Model 1400 Automatic Nitrogen Analyzer.

REFERENCES

1. M. A. Tabatabai and J. M. Bremner, Soil Sci. Soc. Am. Proc., *34*:608 (1970).
2. B. A. Swinehart, Anal. Chim. Acta, *91*:417 (1977).

3. S. Terashima, Anal. Chim. Acta, *101*:25 (1978).
4. B. A. Stewart, L. K. Porter, and W. E. Beard, Soil Sci. Soc. Am. Proc., *28*:366 (1964).
5. G. Dugan, Anal. Lett., *10*:639 (1977).
6. H. K. T. Wong and A. L. W. Kemp, Soil Sci., *124*:1 (1977).
7. H. V. Drushel, Anal. Chem., *49*:932 (1977).
8. C. A. Stutte and R. T. Weiland, Crop Sci., *18*:887 (1978).
9. L. E. Allison, Soil Sci. Soc. Am. Proc., *24*:36 (1960).
10. B. A. Stewart, L. K. Porter, and F. E. Clark, Soil Sci. Soc. Am. Proc., *27*:377 (1963).
11. A. W. J. Dyck and R. R. McKibbin, Can. J. Res., *13B*:264 (1935).
12. J. M. Bremner and K. Shaw, J. Agric. Sci. Camb., *51*:22 (1958).
13. D. R. Keeney and J. M. Bremner, Soil Sci., *104*:358 (1967).
14. B. A. Stewart and L. K. Porter, Soil Sci. Soc. Am. Proc., *27*:41 (1963).
15. G. M. Gustin and C. L. Ogg, in *Treatise on Analytical Chemistry*, Part II, vol. 2 (I. M. Kolthoff and P. J. Elving, Eds.), Interscience, New York, 1965, pp. 405-498.
16. D. F. Ketchum, Anal. Chem., *36*:957 (1964).
17. A. Steyermark, *Quantitative Organic Microanalysis*, Academic Press, New York, 1961, p. 152.
18. J. M. Bremner, in *Methods of Soil Analysis*, Part 2 (C. A. Black, Ed.), Am. Soc. Agron., Madison, Wis., 1965, pp. 1256-1286.
19. R. J. Barsdate and R. C. Dugdale, Anal. Biochem., *13*:1 (1965).
20. D. Desaty, R. McGrath, and L. C. Vining, Anal. Biochem., *29*:22 (1969).
21. M. E. Ebeling, J. Assoc. Off. Agric. Chem., *50*:38 (1967).
22. M. E. Ebeling, J. Assoc. Off. Agric. Chem., *51*:766 (1968).
23. S. N. Farmer, C. J. Howarth, and L. B. Hughes, Chem. Ind. (Lond.), *No. 4*:154 (1967).
24. F. G. Hamlyn and J. K. R. Gasser, Chem. Ind. (Lond.), *No. 35*:1142 (1970).
25. G. F. Morris, R. B. Carson, D. A. Shearer, and W. T. Jopkiewiez, J. Assoc. Off. Agric. Chem., *51*:216 (1968).
26. M. A. Tabatabai and J. M. Bremner, Soil Sci. Soc. Am. Proc., *34*:417 (1970).
27. T. A. Arkley, Ph.D. Dissertation, University of California, Berkeley, 1961.
28. M. A. Tabatabai and J. M. Bremner, Soil Sci. Soc. Am. Proc., *34*:62 (1970).
29. A. Steinbergs, O. Iismaa, J. R. Freney, and N. J. Barrow, Anal. Chim. Acta, *27*:158 (1962).
30. C. M. Johnson and H. Nishita, Anal. Chem., *24*:736 (1952).

31. P. L. Searle, Analyst (Lond.), *93*:540 (1968).
32. H. Green, Metallurgia, *60*:229 (1959).
33. A. R. Tiedemann and T. D. Anderson, Plant Soil, *35*:197 (1971).
34. L. E. Lowe, Can. J. Soil Sci., *49*:375 (1969).
35. I. R. Kaplan, K. O. Emery, and S. C. Rittenberg, Geochim. Cosmochim. Acta, *27*:297 (1963).
36. G. R. Smith, Anal. Lett., *13*:465 (1980).

5

X-Ray Fluorescence Analysis

CAROLYN WILKINS* Rothamsted Experimental Station, Harpenden, Hertfordshire, England

*Former affiliation.

I. INTRODUCTION

X-Ray fluorescence spectrometry (XRFS) is a method of elemental analysis which assesses the presence and concentration of various elements by measurement of secondary X radiation from the sample which has been excited by an X-ray source. The method is rapid, does not destroy the sample, and, with automatic machines, is relatively foolproof for routine operation.

In recent years, greatly increased interest in problems of environmental pollution and trace element deficiencies and toxicities in agriculture has meant that data are required on elements which were not, in the past, investigated. These elements are often tedious or very difficult to determine at the concentrations required by other methods, such as atomic absorption or emission spectrometry and classical "wet" chemistry. In many cases XRFS offers a rapid and accurate alternative. Elements such as Br, Se, U, and S can easily be determined by XRFS while presenting many problems with other techniques. XRFS is, however, only useful when determining elements of atomic number greater than 8.

In the past few years, much progress has been made in the field of energy-dispersive X-ray spectrometry, which has several advantages, as well as some drawbacks, over the previously more widely used wavelength-dispersive spectrometry. This chapter will, however, deal mainly with wavelength-dispersive spectrometry, which has been applied to many problems in soil analysis, while discussing the advantages and disadvantages of the energy-dispersive systems that may be used increasingly for soil analysis in the next few years. Other forms of X-ray spectrometry, such as electron spectroscopic methods, are not yet widely used and will not be discussed here but have been briefly reviewed by Jenkins [1].

In this chapter a brief description is given of the theory of X-ray spectrometry and the component parts of the spectrometer, to enable a new operator to understand the spectrometer, to be aware of the pitfalls, and to be able to assess the possibilities of the various techniques. Methods of sample preparation and other practical considerations are discussed, and applications to the quantitative analysis of different types of samples are reviewed.

II. X-RAY PRODUCTION AND CHARACTERISTICS

Before describing the design and working of an X-ray spectrometer it is necessary to discuss briefly the behavior and nature of X rays. Accounts of the physics of X rays and information related to XRFS can be found in several books, including Liebhafsky et al. [2], Bertin [3], Jenkins and de Vries [4], and Jenkins [1], the last mentioned being a useful general book on many aspects of XRFS. Up-to-date reviews of all aspects of XRFS can be found biennially in *Analytical Chemistry* (e.g., [5,6]).

A. Production of X Rays

Characteristic X radiation is produced when excited electrons of an atom return from outer orbitals to inner levels and the atom returns to the ground state. When an atom is bombarded by particles of sufficient energy, the electrons filling the subshells are excited and may fill outer orbitals. On returning to the original atomic configuration, the atom will emit energy, in the form of X radiation, with wavelengths depending on the excited and the final orbital in which the electron rests. Many such transitions take place in one atom and a number of characteristic wavelengths can be detected when any element has been excited. Fortunately, the X-ray spectrum of a particular element is simple and the main analytical lines used in XRFS rarely overlap. The commonly used analytical lines have wavelengths in the region 0.02-2 nm (0.2-20 Å) although X rays can range in wavelength from 0.01 to 20 nm, bounded on one side by the vacuum ultraviolet region and on the other side by γ rays.

The wavelengths of radiation emitted from any one transition can be found from the following equation:

$$E = h\nu = \frac{hC}{\lambda} \tag{1}$$

where E is the difference in energy of the electron in its final and excited states, h is Planck's constant, ν is the frequency of the radiation, C is the velocity of light, and λ is the wavelength of the radiation.

From (1), relationship (2) can be derived:

$$\lambda = \frac{1.24}{E} \tag{2}$$

where λ is in nanometers and E is in kilo-electron volts (keV).

Continuous radiation is produced when high-speed electrons enter matter and decelerate, and in X-ray fluorescence spectrometry will be produced in the X-ray tube. Secondary excitation, e.g., by X rays, will not produce continuous radiation.

B. Types of Radiation

The two types of radiation, continuous and characteristic, are illustrated in Fig. 1. It can be seen that the former has a continuous range of wavelengths. The intensity of this radiation emitted by an X-ray tube varies with tube current, voltage, and atomic number of the material of the anode. Continuous radiation usually provides the main part of the radiation for specimen excitation.

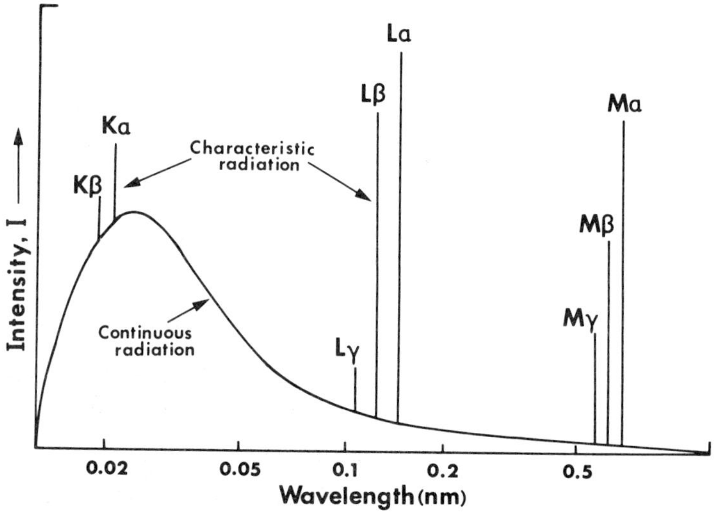

Figure 1 X-Ray spectrum from cellulose sample irradiated by tungsten tube, showing continuous and characteristic radiation.

Characteristic radiation is unique for every element and the spectrum is governed by the electron configuration within the atom. If the reader is not familiar with the structure of the atom and the nomenclature for the shells around the nucleus (K, L, M, etc.), Bertin [3] has an easily followed summary of the subject. The characteristic X-ray spectrum of an element is formed by X-ray photons emitted as the atom returns from an excited state to the ground state. An electron falling into a vacancy in the K shell will give K series lines, i.e., $K\alpha_1$, $K\alpha_2$, $K\beta$, etc., as well as a series of transitions in outer shells. All lines in one series result from electron transitions from different levels to one shell and the number of transitions is limited by various rules applying to quantum numbers (see Bertin [3]). For general analytical work $K\alpha$, $K\beta$, $L\alpha$, and $L\beta$ spectral lines are most important. For the same spectral line, wavelength decreases as atomic number increases (Moseley's law):

$$\frac{1}{\lambda} = K_1(Z - K_2)^2 \tag{3}$$

where λ is the wavelength, K_1 and K_2 are constants, and Z is the atomic number.

To eject an electron from the inner shells of an atom, the incident electron must have sufficient energy. The energy needed is called the excitation potential and is measured in electron volts.

C. X-Ray Characteristics of Importance in XRFS

When X rays interact with matter, they are both absorbed and scattered. Relative to absorption, scatter is small or very small except for elements of low atomic number at short wavelengths.

Absorption

As X rays pass through matter, radiation is absorbed and the intensity of the emergent beam is given by

$$I = I_0 e^{-\mu x} \tag{4}$$

where I, I_0 are the radiation intensities of a monochromatic beam of X rays as they enter (I_0) and leave (I) an absorber of thickness x with a linear absorption coefficient μ. I is always less than I_0 and the difference is mainly due to photoelectric absorption caused by bombardment of the sample and ejection of electrons by photons. The mass absorption coefficient of a sample is the most useful measure of absorption for practical purposes and the term is derived from Eq. (4) as follows:

$$I = I_0 e^{-(\mu/\rho)\rho x} \tag{5}$$

where μ/ρ is the mass absorption coefficient in square centimeters per gram, x is the thickness, and ρ the density of the sample.

μ/ρ is a function of the wavelength of the incident radiation and the atomic number of the absorber. If the absorber contains more than one element, μ/ρ is equal to the sum of the weight fractions multiplied by their individual mass absorption coefficients, i.e.,

$$\frac{\mu}{\rho} = W_1\left(\frac{\mu_1}{\rho_1}\right) + W_2\left(\frac{\mu_2}{\rho_2}\right) + \cdots \tag{6}$$

where W_1, W_2 are weight fractions.

Tables of mass absorption values are available [2] and examination of these will show a number of sudden drops in μ/ρ with increasing wavelengths (Fig. 2). These coincide with the K, three L, five M, etc., absorption edges which equate to the binding energies of electrons at these levels. Incident radiation of shorter wavelength than an absorption edge can excite electrons from the same level. In practical terms, if the mass absorption coefficient of a sample is determined either side of the absorption edge due to a major component, very different values will be obtained and, conversely, two samples containing high and low concentrations of a particular element will have very different mass absorption coefficients at some wavelengths.

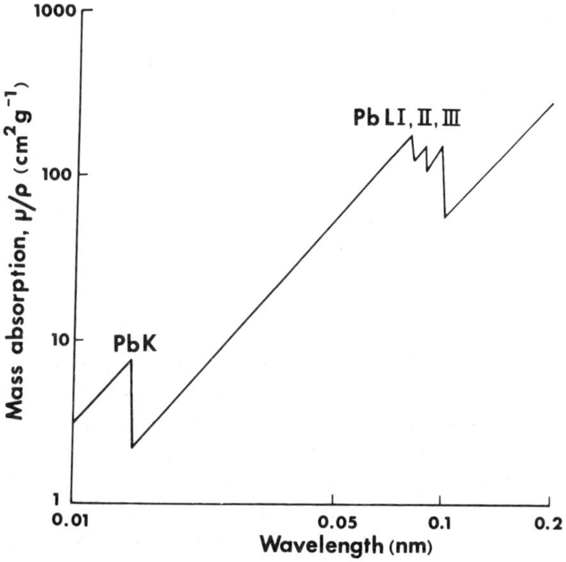

Figure 2 Variation of mass absorption with wavelength for lead.

Scattered Radiation

There are two forms of scattered radiation, known as coherent and incoherent scatter. Coherent scatter occurs when a photon hits an electron and is deflected without losing energy. Incoherent, or Compton, scatter is caused by a photon hitting an electron with loss of energy (see also Chaps. 6 and 7). In an X-ray spectrum, this Compton scatter is shown as a diffuse peak of wavelength slightly greater than lines caused by the tube target (coherently scattered radiation). The difference in wavelength (λ) varies only with the angle between scattered and unscattered X rays, and for most commercial spectrometers, $\Delta\lambda \cong 0.0024$ nm [3]. The ratio of intensity between the coherent and Compton scatter is related to the atomic number of samples and the wavelength of the radiation. Compton scatter can be seen very clearly when looking at samples of low Z and also when using an X-ray tube with high energy radiation. For example, more Compton scatter would be apparent when irradiating an organic soil than a mineral soil and would also be greater for Mo Kα radiation (0.071 nm) than Cr Kα radiation (0.229 nm) (see Sec. II.D). Care must be taken to account for any Compton scattered radiation when interpreting X-ray spectra, especially if the wavelength coincides with that of some element present in the sample.

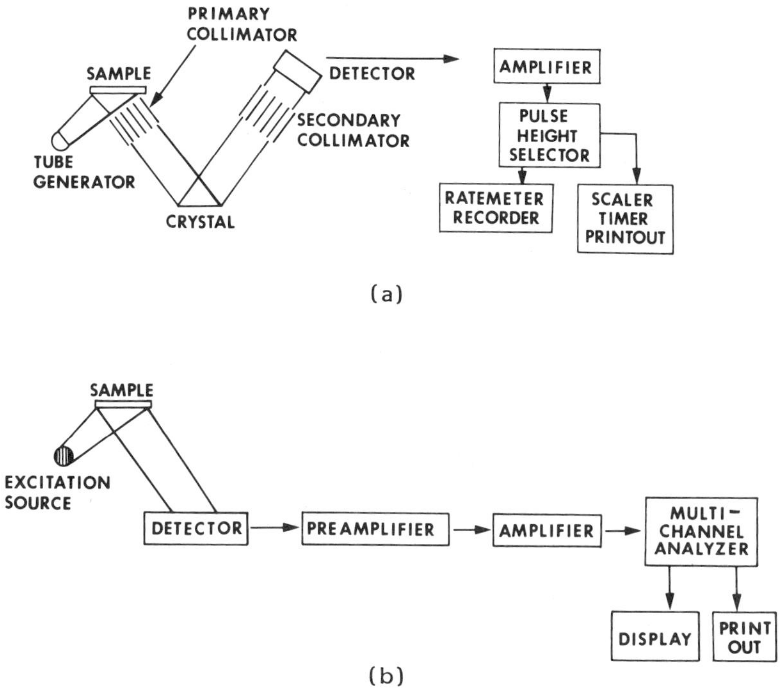

Figure 3 Diagrammatic representations of (a) wavelength-dispersive spectrometer, (b) energy-dispersive spectrometer.

Diffraction

When X rays come into contact with a crystal lattice, they are diffracted according to Bragg's law:

$$n\lambda = 2d \sin \theta \tag{7}$$

where n is the line order (e.g., 1, 2, 3), d is the distance between diffracting planes of the crystal, and 2θ is the angle through which the X rays are diffracted.

This equation can be used in XRFS to identify the elements present in analytical samples.

D. Instrumentation

Figure 3 shows generalized diagrams of the components of the two main types of X-ray spectrometer. The bulk of the discussion on instrumentation, however, will refer to the wavelength-dispersive spectrometer. Most other sections of this chapter are equally applicable to both types of spectrometer.

Excitation of Sample

The excitation source of an X-ray fluorescence spectrometer must be stable and have sufficient output to excite the sample. An X-ray tube (generally the Coolidge type) is most commonly used in XRFS, especially in wavelength dispersive spectrometers, and has a high stability, although excitation is only possible if $Z \geqslant 9$ owing to low production of X rays with long wavelengths and absorption of X rays by the tube window.

X-Ray tubes can be made with targets composed of various elements, commonly Cr, Ag, Au, Mo, W, and Rh. Choice of tube is governed by the wavelength of the element of interest, with regard to both the excitation of this wavelength and the likelihood of interference by characteristic tube radiation. The Cr tube is most suitable for light elements, as a large proportion of long wavelength X rays are produced. Jenkins [1] has given a description of Coolidge and other X-ray tubes. X-Ray tubes without windows are being marketed and can be useful for the measurement of very light elements.

Radioisotopes are often used as sources of excitation in energy dispersive X-ray spectrometers. However, the low photon yield of these sources makes the output too low to be of use in a wavelength dispersive spectrometer.

Crystal Dispersion

Radiation which is emitted by an excited sample is collimated, usually by a set of Soller slits, and is then diffracted by a crystal of known 2d spacing, d being the interplanar spacing between diffracting planes of the crystal. The choice of crystals is large (Table 1) and some crystals can be cut in different ways to give more than one 2d spacing.

The crystal sledge inside the goniometer chamber of the spectrometer is linked to the detectors, and as the goniometer is turned the crystal travels half the angular distance of the detector (see Fig. 3). The goniometer is rarely used outside a range for 2θ of 0 to 147°. Because of this, and because a crystal cannot diffract wavelengths greater than its own 2d spacing, all elements cannot be determined using one crystal. Tables of wavelengths are available [7,8] and include information on energies, wavelengths, and 2θ for various elements and crystals.

Crystals can be curved or flat, depending on the geometry of the spectrometer. A curved crystal can give a strongly focused beam of radiation and, therefore, greater intensity than a flat crystal. As the optimum curvature of the crystal alters with wavelength, curved crystals are often used in fixed-channel spectrometers where channels are

Table 1 Common[a] Diffraction Crystals Most Efficient for Soil Analysis

Substance	2d (nm)	Useful range (nm)	Reflection	Comments
Lithium fluoride				
LiF 220	0.2848	0.025-0.272	High	Similar range to LiF 200 used when greater resolution is needed.
LiF 200	0.4028	0.035-0.384	Very high	Useful for Ca and heavier elements.
Germanium[b]				
Ge	111			
	0.6532	0.057-0.623	High	Mainly used for S, P, and Cl.
Pentaerythritol[b]				
PET 002	0.8742	0.076-0.834	High	Temperature affected 2d spacing. Mainly used for K, Al, and Si.
Acid phthalates[b]				
KAP 001	2.664	0.22-2.5	Medium	All unstable in acid environments.
RbAP 001	2.612		High	TlAP is most efficient, mainly used
TlAP 001	2.575		High	for F, Na, and Mg.

[a]Less common but potentially very useful crystals include: Graphite 002 (2d=0.6715 nm), for determination of P, S, Cl, and K, very expensive. Lead stearate (2d=10.04 nm), for determination of very light elements (Mg-B).

[b]Crystal is sometimes mounted 30° advanced in crystal sledge, therefore 2θ angle set on goniometer is 30° less than actual 2θ.

set up for many analyses at one particular wavelength, while flat crystals are used in spectrometers for more varied purposes.

Detection of X Rays

After the radiation has been diffracted it is again collimated and is detected, usually by a scintillation or gas flow proportional counter. The photons which register in the counter are turned into electrical pulses which can then be measured.

Gas Flow Proportional Counter. Photons enter the counter via a thin window generally made of Mylar, polypropylene, or polycarbonate (usually 1-6 μm) and react with the inert gas which flows through the counter (generally argon plus 10% methane) to form ion pairs of one positive ion and one electron. As the electrons move toward the wire anode which runs through the center of the counter, they ionize argon atoms, causing a large gain in the number of electrons which reach the anode. The mean number of ion pairs produced, and the size of the electronic output pulse, is proportional to the energy of the radiation. The "dead time" is the recovery time after the counter has received a pulse and it cannot function until the effect has dispersed (1-2 μsec).

A peak resulting from characteristic radiation of the flow counter gas will occur if the energy of photons entering the detector is greater than the absorption edge energy of the counter gas. This is called an escape peak and has energy slightly less than that of the main peak being analyzed. This peak can usually be removed by pulse height selection (see below) but causes interference in some analyses.

Scintillation Counter. Photons are converted into light pulses when they fall on a phosphor (generally a sodium iodide crystal). These light pulses are converted into electrical energy by a photomultiplier. The "dead time" is slightly shorter in the scintillation counter than the gas flow proportional counter, but the resolving power with regard to photon energy is less. It should be noted, however, that if a flow proportional counter wire is allowed to become very pitted or dirty, resolution will deteriorate.

The most useful wavelength range for the scintillation counter is 0.02-0.2 nm, whereas the flow proportional counter becomes sensitive at about 0.15 nm and is useful until about 2 nm if a very thin (1 μm) window is fitted.

Semiconductor Detectors. These are used in energy-dispersive spectrometers where the detector has to resolve all the pulses of different energies. Technology in this field is advancing rapidly and information on this type of detector and the accompanying preamplifier can be found in the journal *X-Ray Spectrometry*. Descriptions of semiconductor detectors can be found in Bertin [3] and Jenkins [1]. More detailed discussion of radiation detector systems can be found in Chap. 6.

Pulse Height Selection. In the wavelength-dispersive spectrometer, pulses arising from radiations which are diffracted at the same angles but which have different voltages can be sorted by pulse height selection. This can arise when a first order line for one element has the same diffraction angle as a higher order line for a different element or, as mentioned above, if any escape peak of different energy overlaps the peak for the element to be analyzed. A lower level voltage barrier is set and a "window" of an acceptable voltage allows some pulses to pass while blocking those of very high or very low voltage. The multichannel analyzer of an energy-dispersive spectrometer can be thought of as a large number of individual pulse height selectors.

After pulse height selection, the pulses are counted by scaler and timer and can be recorded on a chart for visual appraisal or can be printed. In a wavelength-dispersive spectrometer, the scaler and timer readouts near and at the peak of sample and standard will give information with which to assess the concentration of a particular element. In some cases, dead time correction and correction for interference by other elements must be applied. Information from the multichannel analyzer of an energy-dispersive system must be subjected to peak stripping, live time corrections, escape peak interference corrections, and others before interpreting. This is usually done by computer. Reed and Ware [9] discuss procedures for energy-dispersive analysis.

III. CHOICE OF OPERATING CONDITIONS FOR WAVELENGTH-DISPERSIVE SPECTROMETER

When choosing optimum operating conditions for various parts of the wavelength-dispersive spectrometer, many factors must be considered, such as type of sample, element to be determined, elements likely to interfere, concentration of the element, and availability of equipment.

It is assumed, when discussing operating conditions, that the sample is solid, of a size suitable to be held in a standard sample holder, and infinitely thick (i.e., a sample of sufficient thickness that an increase does not result in an increase in intensity).

A. Generator Setting and X-Ray Tube

Tube voltage should be chosen to maximize the difference between peak and background relative count rates, and the current (which affects both peak and background equally) should be chosen so that the spectrometer runs at slightly less than the maximum quoted power rating.

Although some new spectrometers are not designed to enable easy changing of the X-ray tube for different analyses, if selection is possible it is governed by three main factors:

1. Whether tube lines may interfere with analyte lines. (It may be possible to remove such interference by placing a filter which will absorb this radiation over the tube window.)
2. The wavelength of the analyte line and the thickness of the tube window. (A tube with a thin window, e.g., a Cr tube, transmits greater amounts of long wavelength radiation which more efficiently excites elements of low atomic number.)
3. The energy of the characteristic tube radiation with regard to optimum excitation of the sample. (The optimum is radiation of a wavelength slightly shorter than the absorption edge of the analyte element.)

The higher the atomic number of the X-ray tube target, the more energy is produced by the continuous radiation.

B. Primary Collimator

A fine collimator will resolve adjoining peaks better than a coarse collimator, but intensity will be lower. If high resolution is needed the fine collimator should be used, but for analysis of elements with low atomic numbers, where peaks of neighboring elements are widely separated, the coarse collimator is often adequate. However, when analyzing for Mg with a TlAP crystal, the use of the fine collimator is recommended, due to overlap of the Mg $K\alpha$ line by the Ca $K\alpha$ (111) line in calcareous samples [10].

C. Analyzing Crystal

It is not practical to use one analyzing crystal to diffract the entire range of wavelengths necessary in XRFS, therefore a selection of crystals is needed (Table 1). For the determination of any element, the crystal with the most appropriate 2d spacing, greatest reflectivity, and greatest angular resolution should usually be chosen. However, the crystal which gives greatest resolution rarely gives the greatest intensity and choice depends on individual analyses. For example, when analyzing soil samples for Zr, Y, Sr, Rb, and Br, interfering $K\beta$ lines (i.e., Sr $K\beta$ with Zr $K\alpha$, etc.) are minimized by using a lithium fluoride crystal in which the 220 lattice planes are parallel to the reflecting surface (LiF 220). If greater intensity is needed, the LiF 200 crystal is used and correction factors are applied to reduce interference. (Two lines which cannot be resolved are Pb $L\alpha$ and As $K\alpha$, so a correction for Pb must be made if determining As using As $K\alpha$, and for Pb determinations, the Pb $L\beta$ line should be used.)

Care should be taken when handling crystals and when using them in the spectrometer. Although many crystals are fairly stable and robust, the following need special attention:

1. Pentaerythritol (PET). This crystal is commonly used for K, Si, and Al analysis but has a large coefficient of thermal expansion. Therefore, the 2d spacing, and hence the goniometer reading for the peak of a particular element, varies with temperature. Jenkins and de Vries [4] have plotted temperature sensitivities of various analyzing crystals. It is recommended that the peak position angle be checked frequently during analysis.
2. Thallium acid phthalate (TlAP). This crystal is used for Mg analysis. It is attacked by airborne acid vapors, and should, therefore, be stored in a dessicator when not in use.

Reflectivity can decrease as crystals age, so old crystals are often not as efficient as new crystals.

D. Detector

Most wavelength dispersive spectrometers have a choice of two detectors, the scintillation counter and the flow proportional counter, which can be used separately or, in some spectrometers, in tandem.

The scintillation counter is most efficient for the detection of short wavelengths and can be used up to about 0.3 nm, and the flow proportional counter is more efficient for longer wavelengths (above 0.15 nm). A thin window, often of aluminum-coated Mylar (the aluminum coating on the inside of the counter enables a symmetrical field to be maintained around the counter wire) is fitted to the entry port of the flow proportional counter, and although the 6 μm window is popular and robust, much greater intensity for the analysis of low atomic number elements can be obtained by using a thinner window of 1 or 2 μm which does not absorb so much long wavelength radiation.

E. Path

The path of the X rays in the spectrometer is usually through air or vacuum. X Rays are absorbed by air, the extent of absorption increasing with wavelength. At wavelengths of greater than 0.15 nm a vacuum path is more efficient and becomes increasingly necessary for longer wavelengths. It is, however, not possible to analyze liquids in a conventional sample cell if the system is under vacuum; to overcome this problem either a special sample cell is used which will withstand a vacuum [11], or the system can be flushed with helium which will not absorb long wavelength radiation to the same extent as air [1].

If the flow proportional counter is being used and the sample is being analyzed in air, a slow leakage of gas takes place through the flow counter window into the goniometer chamber. X Rays are absorbed to a greater extent by the counter gas than by air, causing a

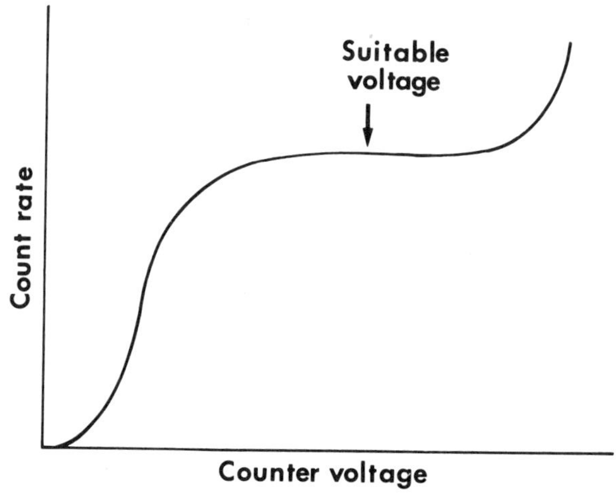

Figure 4 Plot of X-ray count rate against counter voltage, showing suitable voltage for analysis.

decrease in intensity. To overcome this effect, the gas should be removed by evacuation at intervals.

F. Electronic Parameters

Optimum counter voltage (EHT) will vary with attenuation. The attenuation level should be chosen to be just high enough to eliminate most of the amplifier noise. Any higher setting (lower amplifier gain) necessitates an increase in EHT, which in turn leads to drift in pulse height and broadening of the pulse amplitude distribution. Pulses which are of too large or too small voltage are rejected by upper and lower level discriminators, respectively. Only those falling in the "window" between these levels are recorded.

Figure 4 shows a plot of count rate against EHT when no discrimination is in operation. The voltage range of the plateau varies with wavelength and attenuation, and an EHT setting should be chosen which is fairly near the middle of the plateau. Figure 5 shows a plot of count rate against increasing lower discriminator level using a very narrow window and constant EHT. For analytical applications the lower discriminator level should be set at A and the upper at B, giving a window of B-A volts and allowing all the counts falling in the peak to be recorded.

Choice of count rates and counting times is discussed in Secs. V and VI.

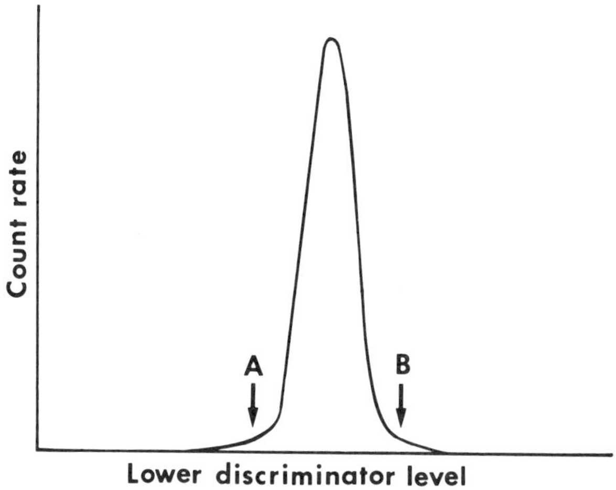

Figure 5 Variation in count rate with lower discriminator level, using a narrow window (0.05 V). A is a suitable lower level, and the interval A-B a suitable window, for analysis.

IV. SAMPLE PREPARATION

There are many different ways of preparing a sample for XRFS analysis and the choice depends on several factors, including desired accuracy and precision of results, amount and nature of the sample, and concentration of the element of interest.

For qualitative analysis, very little sample preparation is usually necessary. For example, a rock sample can be placed in a sample holder, dry soil can be poured into a sample cup which has Mylar or polycarbonate at the base, or a powder can be spread on Scotch tape and fixed into a sample holder. Sample preparation for quantitative analysis, however, must be very much more carefully done to ensure homogeneity of the sample, sufficiently fine particle size, a flat surface of the sample, and other factors to be discussed below.

A. General Procedures

When choosing methods of sample preparation for XRFS, it is necessary to consider sampling in general. It is difficult to obtain a sample of soil representative of a particular area, and intrafield variation of various soil properties is often as great as interfield variation. However, when a sample of soil enters the laboratory, care can then be taken to ensure that the specimen taken for analysis is as representative and homogeneous as possible. First, if the sample is to be dried, it must be realized that some elements may volatilize at fairly low temperatures, e.g., S, Se, Hg [12,13]. Also, elements will be more easily lost from peaty soils or herbage samples than from mineral samples, in which they have more stable forms. The following scheme of sample preparation should ensure that a fairly representative specimen is analyzed:

1. Air-dry soil, and dry plant material at less than 100°C.
2. Select a random subsample of the whole, if necessary by quartering or riffling.
3. Grind the subsample in a dry state.

For most chemical analyses, dry grinding of the sample is necessary and this should be done by a method which will not contaminate the sample with the elements of interest. To check for contamination, duplicate samples can be ground by separate methods, e.g., using apparatus made of different materials, and analytical results compared. Information on the materials used in grinding equipment can be obtained from the manufacturers. When determining some metals in soils after sieving there is a possibility of contamination by solder from brass sieves. For minor element determinations, if sieving is necessary, nylon sieves should be used.

B. Sample Preparation for XRFS

Undiluted Samples

In soils, for the analysis of elements of atomic number greater than 23, or for the analysis of some of the very light elements (e.g., F, Na, Cl, S), a convenient method of sample preparation is to grind the dry soil finely (until it feels like flour when rubbed between the fingers; approximately 50 μm diameter) and then to press the soil into an aluminum cap or a boric acid backing of an appropriate size to fit in the sample holder. A detailed account of sample grinding for XRFS has been written by Jenkins and de Vries [4], and dies for pressing disks can be obtained commercially or constructed from the drawings of Norrish and Chappell [14]. Samples and standards for comparison must be pressed at the same pressure; 150 MPa (1.5 t cm^{-2}) is usually used in the author's laboratory but some samples, e.g., quartzose soils, are still friable at this pressure and it may be necessary to increase the pressure on the sample or add a binding agent. Useful binding agents for soils are polyvinyl alcohol, urea, and cellulose, added in the minimum quantities possible (thus reducing dilution of the sample and matrix changes). A sample of about 2 g will make an adequately thick disk for analysis of most soils using K spectra (up to Z = 40) but for herbage samples, and possibly for very peaty soils, more sample may be needed to make an infinitely thick disk.

Fused Samples

Fused samples are used in many laboratories when analyzing soils and other silicate-rich materials for some major elements (Mg, Al, Si, P, K, Ca, Ti, and Fe). Several methods of fusion are discussed in detail by Hutchinson [15], and choice of fusion mixture will depend on the chemical composition of the sample. Probably the most useful established method for major element determinations in soils is that of Norrish and Hutton [16]. In this method, the soil sample, usually ignited, is converted into a glass by fusing at 950° with a borate flux and oxidizer in the ratio 0.28 g soil:1.5 g flux:0.02 g $NaNO_3$. The borate flux is composed of lithium tetraborate (38.0 g), lithium carbonate (29.6 g), and lanthanum oxide (13.2 g), which are fused at 1000°C, quenched, and homogenized. Full details of the preparation of the fusion mixture and the glass disk are to be found in Norrish and Hutton [16], but the following points should be noted:

1. The fusion mixture is available commercially (Johnson Matthey Chemicals Ltd., Hatton Garden, London).
2. Glass disk preparation is made considerably easier if a mechanical plunger is used to press the disk.
3. The weight loss on ignition should be determined for each batch of fusion mixture by igniting a weighed amount, cooling and reweighing, using the same conditions as those used when fusing samples to make glass disks.
4. The recommended temperatures can be obtained easily with a compressed air/natural or coal gas burner, a gas/oxygen burner, or a propane/oxygen burner.
5. The asbestos parts of the apparatus can be replaced by silica or alumina equipment.
6. The graphite disks can be replaced by duralumin disks.

The pressing of the disks by this method takes a certain amount of skill but it is worth persevering. For a full explanation of experimental and theoretical considerations the reader should refer to Norrish and Hutton [16] and also to Harvey et al. [17], who suggested the modifications listed above to make sample preparation easier. Hutchinson [15] has given a very detailed and simple account of sample preparation for this method.

When performing major element analysis in some laboratories, it may be convenient to determine Na and Mg by other chemical methods, as Na determinations done by XRFS are not very sensitive and Mg determinations are open to errors caused, among other things, by background variations due to differences in composition in samples and standards. The Na and Mg determinations can be done by other methods on the same sample after XRFS analysis, if the $NaNO_3$ is omitted and the disk redissolved. (The determination of Mg is explained in detail by Norrish and Hutton [16] but it is likely to be the least accurate and precise of the elements determined.)

Another fusion method which has great potential for soil investigations has been described by Haukka and Thomas [18] and Thomas and Haukka [19]. A glass is made with a 2:1 mixture of sample and lithium metaborate which can be used to determine major and minor elements. Glasses are made in a similar way to that described by Norrish and Hutton [16] and matrix corrections are given. The fact that major and minor elements can be determined on the same sample, which has been diluted very little without the addition of a heavy absorber, is very useful, especially on automatic spectrometers and equipment with multichannel analyzers. However, matrix coefficients and standardization constants must be determined by the user.

Liquids

It is useful to be able to determine extractable concentrations of elements by XRFS, if the concentration is great enough, and the solution can be poured into a liquid sample holder to be analyzed directly. Solutions containing concentrations below the detection limit would have to be concentrated (see Sec. VII).

V. QUALITATIVE AND SEMIQUANTITATIVE ANALYSIS

X-Ray fluorescence spectrometry, especially energy-dispersive spectrometry, is extremely useful for identifying the elements contained in a sample. The wavelength-dispersive spectrometer, although spanning the same wavelength range as the energy-dispersive system but with greater sensitivity, has to be run using several changes of crystal, counter, and electronics, while the energy-dispersive spectrometer can display the whole spectrum at the same time, making a rapid assessment of major and moderate constituents very easy. For this purpose, an energy-dispersive system is often fitted to a scanning electron microscope and can then be used qualitatively to determine the constituents of individual grains in soils or soil fractions. Figure 6a shows the part of the spectrum from a soil sample obtained with a lithium fluoride crystal in a wavelength-dispersive spectrometer, and Fig. 6b shows the corresponding spectrum with an energy-dispersive system.

A. Operation of the Wavelength-Dispersive Spectrometer for Qualitative Analysis

The goniometer arm of the spectrometer, to which the detectors are coupled, is driven twice as fast as the diffraction crystal inside the crystal chamber. The goniometer can be driven automatically at

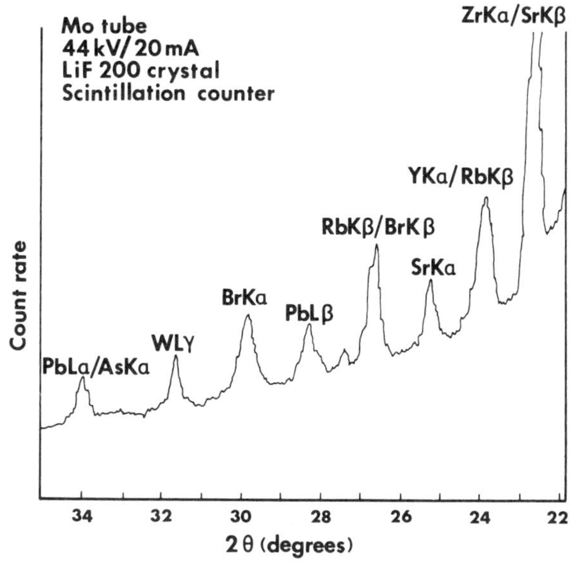

(a)

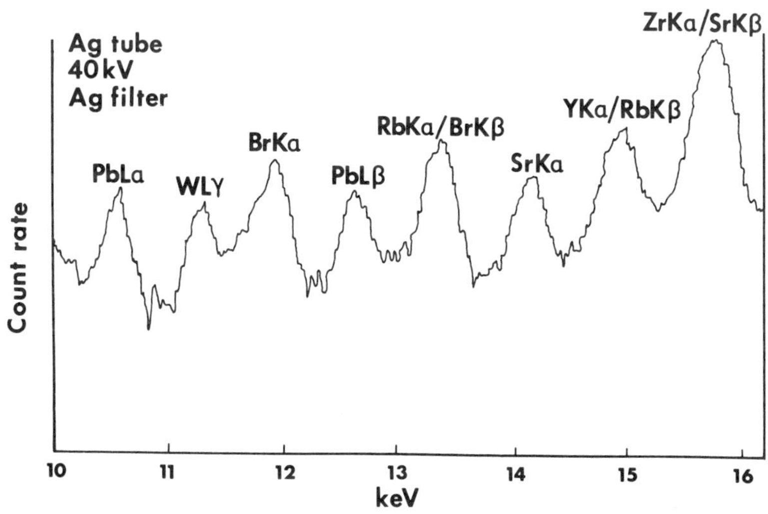

(b)

Figure 6 (a) Spectrum from soil sample, using wavelength-dispersive spectrometer. Soil concentrations ($\mu g\ g^{-1}$): Zr:403; Y:91; Sr:35; Rb:91; Pb:100; Br:89; As:4. The presence of W is due to contamination from the grinding barrel. (b) Spectrum from soil sample, using energy-dispersive spectrometer. Soil concentrations as shown for Fig. 6a.

various speeds, and the detector is maintained in the correct position to receive the diffracted X rays.

To identify the elements in a sample by XRFS, the procedure is as follows:

1. Present sample to spectrometer. For qualitative purposes, the sample can be stuck onto Scotch tape and placed in a liquid sample holder, absorbed onto a filter paper, or irradiated in solution.
2. Select generator current and voltage. Count rates of below 50,000 and above 100 counts per second should be aimed for. (N.B. Do not operate the generator above the approved safe power rating.)
3. Select primary collimator and crystal. For the wavelength range 0.035-0.38 nm, LiF 200 is the most useful crystal and combines very high reflectivity and resolution. For the lighter elements, PET gives adequate coverage.
4. Select counter. The scintillation counter is adequate for qualitative purposes for elements with an atomic number of above 20.
5. Select counter EHT. This can be done as described in Sec. III.F for an element in the range of the scan.
6. Select window. Use a very wide window or no window.
7. Record trace. A trace of the form of Fig. 6a will be recorded and the peaks can then be identified. The alignment of the chart with the 2θ angle marker pen makes identification of the 2θ angle straightforward.
8. Identify elements. Tables of X-ray wavelengths are available which give the 2θ angles for particular crystals and the main analytical lines can be picked out. It is easy to mistake a peak for one element when it is, in fact, a line resulting from the X-ray tube, including Compton scatter, contamination from the sample holder, or a less intense peak from a major constituent of the sample. When identifying peaks, make sure that the $K\beta$ peak is identified, as well as the $K\alpha$ peak (intensity is generally 5:1, $K\alpha$:$K\beta$) or the $L\alpha$, β, and γ peaks. It is useful to make a transparent overlay of tube and common elemental lines for easy identification in qualitative analysis.

B. Semiquantitative Analysis

If reasonably flat and homogenous samples are used, the X-ray spectrum described above can be used to measure intensities of different elements and, by comparison with standards of similar composition, a semiquantitative estimate of elemental concentrations can be made. For soil determinations, standards made up of Al_2O_3 or Al_2O_3 + SiO_2

(more friable) plus 5-10% Fe_2O_3 contain some of the main constituents of a natural soil and have a similar matrix to a mineral soil for the purpose of XRFS. Standards for plant analysis should be made from cellulose. Known additions of elements of interest enable the standards to be compared with samples.

This method of semiquantitative analysis was used by Williams [20] to obtain information on 12 elements (Mn, Fe, Ni, Cu, Zn, As, Br, Rb, Sr, Y, Zr, Pb) in a large number of soils. The peak height above background for a particular element was compared with that of standards, any interference from other peaks was subtracted, and a simple matrix correction was applied to allow for differences in composition between sample and standard. For the last 10 of the elements, the matrix correction was obtained very simply from measurement of the background between peaks, but iron and manganese needed a more complicated treatment depending on measured iron content.

When energy-dispersive XRFS is being used, energy tables can be used to identify peaks. However, care must be taken when interpreting the spectra, as the resolution is poorer than that of the wavelength-dispersive spectrometer and misleading peaks can arise, including escape peaks and $K\beta$ peaks.

VI. QUANTITATIVE ANALYSIS

In X-ray fluorescence spectrometry, the count rate obtained from a particular element when a specimen is irradiated is proportional to the concentration of that element. Thus, if the count rate for the element in a sample is compared with the count rate of a standard of known composition, an estimate of the amount of the element can be made. There are, however, factors caused by differences in composition in different specimens which will influence count rate and which must be corrected for.

If there are large amounts of the element of interest in the sample and the background near the peak position is low, the background can be ignored. In most minor element determinations, however, background count rates should be measured either side of the peak and as near to the peak as possible, without being influenced by radiation due to the element being determined or by any other peaks in the vicinity, and the background count rate at the peak position estimated. This can be done by measuring background and peak count rates on a specimen prepared from a pure chemical, e.g., silica, which does not contain the element to be determined. A peak-to-background ratio can be determined and used to correct peak count rates for background in samples.

Results obtained from comparing the count rates of samples with those of standards must be corrected for variations in mass absorption

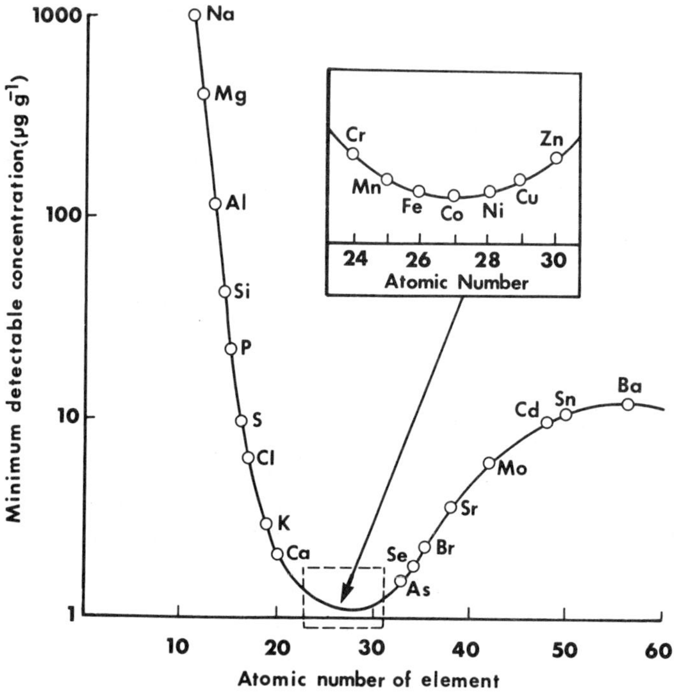

Figure 7 Approximate guide to minimum concentrations detectable by XRFS (After G. Brown, personal communication.)

coefficient between samples and standards. Methods of determining mass absorption coefficients are described in Sec. VI.B.

A. Sensitivity, Accuracy, and Precision

Limits of detection for elements being determined by XRFS vary with the matrix of the sample and with the equipment used. Figure 7 shows a rough guide to limits of detection on a wavelength-dispersive spectrometer. However, sensitivity, accuracy, and precision are based on careful measurement and correction procedures. The main effects which will influence analyses are absorption, particle size, and enhancement. The first two have a large effect in soil analysis but the enhancement effect is not usually so important. Enhancement will occur when photoelectric absorption of a characteristic line by a matrix element takes place. This is especially noticeable where radiation from a characteristic line is of slightly higher energy than the element being measured and has the effect of increasing the radiation emitted by this element. In soils, analysis of major elements may be affected by this phenomenon, but if fused samples are analyzed by the method of Norrish and Hutton [16], enhancement effects are accounted for in the computation of results. The determination of minor elements in soil Fe/Mn concretions can also be affected strongly by enhancement and absorption.

Bertin [3] lists sources of errors which are likely to mar the accuracy and precision of quantitative XRFS determination. They are dealt with, in turn, below.

Counting Errors

Counting statistics are dealt with in great detail by Jenkins and de Vries [4]. Many X-ray analysts find it convenient to set a fixed time for counting although this is not always as accurate as setting fixed counts. Counting errors become large at very high count rates due to dead time and pulse drift. At count rates of over 5000 sec^{-1} a dead time correction must be applied (see Chap. 6). Dead time can be calculated by several methods comparing ratios of high and low X-ray intensity [14,21,22]. Optimum distribution of counting time between peak and background varies with the difference in count rates of peak and background in the following way:

$$\frac{t_p}{t_b} = \left(\frac{C_p}{C_b}\right)^{1/2} \qquad (8)$$

where t_p and t_b are the counting times for peak and background, respectively, and C_p and C_b are the corresponding count rates.

Therefore, although the time should be divided equally between peak and background at very low concentrations, as concentration increases, so should the ratio $t_p:t_b$. The lower limit of detection Q, for a particular analysis is given by:

$$Q = \frac{3}{m}\left(\frac{C_b}{t_b}\right)^{1/2} \qquad (9)$$

where m is counts per second per unit of concentration (e.g., micrograms per gram or percent). However, the practical limit for a determination is higher than that evaluated statistically, and is discussed by Jenkins [1].

Instrumental Errors

These are generally caused by generator instability, pulse drift in counters, and crystal instability with temperature changes (mainly PET). Generators are usually stable in the short term provided that adequate warmup time is given (usually 1 hr) and the mains supply is reasonably stable. Changes in 2d spacing of temperature-sensitive crystals make it advisable to use a standard to every three or four samples and to check the 2θ angle frequently. Energy-dispersive systems have very stable electronics and the temperature of the detector is closely controlled.

Variations in Operating Conditions

Errors due to this cause have been drastically reduced by automatic and semi-automatic spectrometers. With manual spectrometers, care must be taken to reproduce operating conditions exactly when changing such parameters as 2θ angle, crystal, and EHT. Careful sample loading and labeling is, of course, essential.

Variations in Samples

Errors due to variations in samples are probably amongst the most important in quantitative analysis. The most important factors are homogeneity, particle size, flatness, and thickness.

Norrish and Chappell [14] outlined the theory involved in choice of particle size for XRFS and recommended grinding until there is no change in X-ray intensity. Care must be taken, however, that some crystals, e.g., mica, do not orientate themselves to form a film over the top of a pressed pellet of a micaceous sample; saline samples, also, may form a layer of salt on the surface. Berry et al. [23], Hunter and Rhodes [24], and Rhodes and Hunter [25] have all discussed particle size effects in detail.

If a sample is not infinitely thick with regard to X rays, there will be loss of intensity. The amount of sample necessary varies with wavelengths and samples for light element determinations can be thinner than samples for the heavier elements. The thickness necessary for a particular wavelength and type of sample can easily be found in practice by preparing samples of varying thickness until no increase in count rate is detected with increased thickness.

Line Interferences

These cause errors which can increase the intensity of the line being analyzed or of background determinations on either side of the peak. Choice of crystals can reduce interferences, but some peaks cannot be resolved, e.g., Pb Lα and As Kα. First order interferences, i.e., those where n = 1 in the Bragg equation (7), can cause large errors (e.g., Sr Kβ with Zr Kα; Rb Kβ with Y Kα); tube lines, and their Compton scatter, can interfere (e.g., Cr Kβ with Mn Kα, Mo Kα with Zn Kα); and high-order interferences ($n > 1$) can cause error although they can be reduced by pulse height discrimination. Escape peaks can also cause difficulties in some analyses.

Calibration Errors

Calibration errors can be kept to a minimum by the choice of carefully chosen and prepared standards. For soils, artificial standards can be made up in a "soil base" mixture of iron, aluminum, and silicon oxides or iron and aluminum oxides. Adequately analyzed natural standards are few for soils and it is common practice to use U.S. Geological Survey standard rocks [26]. It must be noted, however, that when using these rocks as standards the matrices vary greatly with resulting big differences in mass absorption coefficients. Recently, however, the Soil Research Institute, Ottawa, Ontario has started an analytical program between laboratories to assess the major and minor element concentrations of four soils. Although the amounts of some of the minor elements are very small, with little variation, these soils, when adequately analyzed, should provide useful additional standards.

B. Analytical Procedures

Quantitative analysis almost always involves more than a straightforward comparison between samples and standards. For samples all of similar composition, with similar mass absorption coefficients, this is possible but is unlikely in soil analysis. One of the few exceptions is the analysis of some of the very light elements in soils, e.g., F, Na, and S, where large differences are needed in major element concentrations to cause large differences in mass absorption. The Na peak, therefore, can be measured and intensity versus concentration can be read straight from a calibration. Sulfur, another fairly minor light element, can also be measured this way but Brown and Kanaris-Sotiriou [27] recommend a correction for mass absorption in highly organic soils.

Methods of correcting for different matrices are described below:

Addition of Internal Standard

The sample can be "spiked" with a known amount of element to be analyzed and the initial concentration of the element in the sample is derived by extrapolation. This can be a very accurate and precise way of determining the amount of a particular element in a sample, as matrix effects are avoided.

Dilution

Dilution of the sample is a very efficient way of decreasing matrix effects and also, for major element analysis, bringing the concentration down to a reasonable level so that dead time corrections, pulse drift, and other phenomena found at high count rates are reduced.

Table 2 Comparison of Mass Absorption Ratios Obtained by Quantitative and Semiquantitative Methods

	Mass absorption ratio[a] for Sr		Mass absorption ratio for Zn	
Sample	Quantitative[b]	Semi-quantitative[c]	Quantitative	Semi-quantitative
1	0.83	0.88	0.86	0.88
2	0.97	1.03	0.98	0.94
3	1.16	1.18	1.15	1.03
4	0.82	0.92	0.84	0.84
5	1.00	1.01	1.01	0.97

[a]Mass absorption coefficient expressed as ratio to that of artificial soil blank.
[b]Measured by attenuating X-ray beam [14].
[c]Calculated from background count rate.

Determination of Mass Absorption Coefficient

The mass absorption coefficient (MAC) can be measured directly for wavelengths shorter than around 0.3 nm by using a small, thin pellet of the sample to attenuate the X-ray beam (for very long X-rays, the beam becomes very greatly attenuated and it is difficult to make pellets thin enough to measure the MAC). The pellet can be placed in a holder between the scintillation counter and the diffraction crystal in a Philips PW 1540 spectrometer, but in several of the newer spectrometers it is difficult to position the pellet. Therefore, this method of MAC measurement is not suitable for all modern and especially the automatic spectrometers.

MAC is calculated from the equation

$$A = \frac{1}{\rho x} \ln \frac{I_0}{I} \tag{10}$$

where A is the MAC, ρx is in units of grams per square centimeter, and I_0 and I are the intensities of the unattenuated and attenuated X-ray beam, respectively.

The method and apparatus has been described in detail by Norrish and Chappell [14] and Hutchinson [15].

Indirect measurements of mass absorption can be achieved by several methods. A correction based on the intensity of background radiation near the peak is a convenient estimate of scattered radiation as this usually has to be measured if determining minor elements, [e.g., 20, 28]. Table 2 shows the relationship between background intensities and mass absorption coefficient, both expressed as ratios to an artificial soil base. This method is only reliable with samples and standards of fairly similar compositions, and will not be accurate if extrapolated past an absorption edge of an element of high and variable concentration (e.g., iron in soils).

The intensity of scattered tube radiation can also be measured to estimate matrix variations and is a good approximation if a coherently or incoherently scattered line can be found near the line being analyzed [29]. Norrish and Chappell [14] point out that there are limitations to these methods, including the following points, that background can be affected by the tail of the analyte line (this can be corrected for), by interference from other lines including satellite lines [1], and by crystal fluorescence. Measurements of Compton scattered radiation are less misleading in these respects.

If the major element composition of the sample is known, the mass absorption of the sample can be calculated from mass absorption tables using Eq. (6).

VII. EXAMPLES IN SOIL ANALYSIS

X-Ray fluorescence spectrometry has been applied to many aspects of soil science. Table 3 shows a summary of analytical conditions for common soil elements, gives examples of XRFS determination of these elements in soils, and also quotes the mean concentrations expected in soils [30] and the minimum detection limits for some elements [14]. The ease of sample preparation and the fact that the same sample can be used to determine many elements makes XRFS a useful tool for multi-element surveys for both major elements (e.g., [31-34]) and minor elements (e.g., [35-40]).

The determination of sulfur by XRFS is very rapid and sensitive and is of great value in soil science and plant analysis [41,42]. For example, XRFS has been used to determine S to locate acid sulfate soils [43], and S-deficient sites and S patterns in soils and herbage in a general survey [44]. The sulfur content can be determined on a pressed powder pellet and peak count rates of samples and standards can be compared directly. When analyzing highly organic samples, however, a matrix correction for organic carbon is necessary [27,45]. If the vacuum pump oil of the spectrometer contains S, the absorbent surface of the sample pellet may become contaminated. Kanaris-Sotiriou and Brown [46] recommended coating the sample with a polycarbonate film to cut down S absorption from contaminants. Sulfur has been determined by XRFS in soil extracts [47] and various soil fractions [48], and also in herbage by adding a variety of diluents to the sample to reduce matrix effects [49,50]. However, Bolton et al. [41] found that S could be determined rapidly on pressed powders and McLachan and Crawford [51] found the same with loose powders. When making standards for plant analysis, it must be realized that S can be lost from disks of cellulose containing elemental S [52].

Table 3 Conditions for XRF Analysis of Soils

Element	Mean soil concentration[30] (μg g^{-1})[a]	Lower limit of detection[14][b]	Radiation	Crystal	Line
F	200		Cr	TlAP	F Kα
Na	0.6%	15	Cr	TlAP	Na Kα
Mg	0.8% MgO	0.04% MgO	Cr	TlAP	Mg Kα
Al	13.4% Al_2O_3	0.016% Al_2O_3	Cr	PET	Al Kα
Si	70.5% SiO_2	0.025% SiO_2	Cr	PET	Si Kα
P	0.14% P_2O_5	0.008% P_2O_5	Cr	Ge 111	P Kα
S	700		Cr	Ge 111	S Kα
Cl	100		Cr	Ge 111	Cl Kα
K	1.7% K_2O	0.004% K_2O	Cr	PET	K Kα
Ca	1.92% CaO	0.002% CaO	Cr	LiF200	Ca Kα
Sc	7		Cr	LiF200	Sc Kα
Ti	0.83% TiO_2	0.003% TiO_2	Cr	LiF200	Ti Kα
V	100	1	Au	LiF220	V Kα
Cr	100	<1	Au	LiF200	Cr Kα
Mn	850	<1	Au/W	LiF200	Mn Kα
Fe	5.4% Fe_2O_3	<1	Au	LiF200	Fe Kα
Co	8	<1	Au	LiF200	Co Kα
Ni	40	<1	Au	LiF200	Ni Kα
Cu	20	<1	Au	LiF200	Cu Kα
Zn	50	<1	Au	LiF200	Zn Kα
Ga	<1-30	<1	Mo	LiF200	Ga Kα
Ge	1		Mo	LiF200	Ge Kα
As	6		Mo	LiF200	As Kα
Se	<1				
Br	5		Mo	LiF200	Br Kα
Rb	100	<1	Mo	LiF200	Rb Kα
Sr	300	<1	Mo	LiF200	Sr Kα
Y	50	<1	Mo	LiF200[c]	Y Kα
Zr	300	<1	Mo	LiF200[c]	Zr Kα
Mo	2				Mo Kα
Cd	0.06				
Sn	10				
I	5		Cr	LiF200	ILα

Operating conditions				
2θ	Counter[d]	Sample preparation	Interferences in soils	References
90.62	FPC	Pressed powder		
55.08	FPC	Pressed powder		
45.18	FPC	Glass disk	Ca	10
145.07	FPC	Glass disk		
109.21	FPC	Glass disk		
140.90	FPC	Glass disk		
110.68	FPC	Pressed powder		41-52
92.75	FPC	Pressed powder		70,71
50.69	FPC	Glass disk		
113.06	FPC	Glass disk		
97.65	FPC	Concentration necessary		72
86.12	FPC	Glass disk		55,56
123.16	FPC	Pressed powder	Ti	73
69.34	FPC	Powder/glass	V, Ba	
62.96	FPC	Powder/glass	Cr	53,74,75
57.51	FPC/SC	Powder/glass	Mn	53,73,76
52.79	FPC/SC		Fe	
48.66	FPC/SC	Pressed powder		54
45.02	FPC/SC	Pressed powder	Ni	53,75
41.79	FPC/SC	Pressed powder	Au(tube),Cu	36,53,77
38.91	SC	Pressed powder		
	SC	Concentration necessary		
33.99	SC	Pressed Powder	Pb	
		Concentration necessary		
29.96	SC	Pressed powder		78,79
26.91	SC	Pressed powder	Br	
25.14	SC	Pressed powder		35
23.79	SC	Pressed powder	Rb	57
22.55	SC	Pressed powder	Sr	34,57,58
	SC	Concentration necessary		
		Concentration necessary		
		Concentration necessary		
102.80	FPC	Concentration necessary		

Table 3 (Continued)

Element	Mean soil concen-tration[30] (μg g^{-1})[a]	Lower limit of detection[14][b]	Radiation	Crystal	Line
Ba	500	4	Au	LiF200	Ba Lβ
Hg	0.03-0.8				
Pb	10	1	Mo	LiF200	Pb Lβ
U	1	1	Mo	LiF220	U Lα

[a]Except where shown.
[b]Detection limits for minor elements in quartz matrix.
[c]Or LiF220.
[d]FPC: flow proportional counter; SC: scintillation counter.

Chlorine can also be determined by XRFS, but there are many forms of contamination when preparing samples and standards which may interfere with the analysis.

Many investigations of Mn, Co, Cu, Zn, and Mo have been done on soils, both to determine the total amounts of these important plant nutrients and to assess the extractable percentages. The determination of total Mn, Cu, and Zn in the range found in soils is straightforward by XRFS [53], although contamination in the Au tube anode by Ni, Cu, and Zn must be corrected for [54] and Au L Compton scatter overlaps the Zn Kα peak. The other two important nutrients mentioned here, Co and Mo, cannot be determined satisfactorily by XRFS at deficiency levels, interference by Fe Kβ masking the Co Kα peak and Mo often being below the detection limit.

Elements contained in resistate minerals, e.g., Zr and Ti, may often give information about the pedogenic history of a soil. These elements can both be easily determined by XRFS [55-58] although causing difficulties in some other methods of analysis.

The study of soil pollution has aroused much interest in the past few years and such elements as Cr, Ni, Zn, As, and Pb emitted during smelting are simply determined by XRFS. Unfortunately, three elements of importance, Cd, Sn, and Hg, are usually below the detection limit of XRFS in soils and preconcentration would be necessary before analysis. The quantities of polluting elements (except B) which often occur in flyash and sewage sludge are often within the range of detection of XRFS. Lead pollution of the atmosphere [59] and of urban

Operating conditions				
2θ	Counter[d]	Sample preparation	Interferences in soils	References
79.18	FPC	Pressed powder Concentration necessary	Ti	
28.23	SC	Pressed powder		59-61
37.30	SC	Pressed powder (Concentration)	RbKα	62

and rural soil and vegetation has been detected by XRFS determination [60,61].

Naturally large occurrences of many elements in soils are easily and rapidly detected by semiquantitative XRFS techniques and, in this way, unusually large amounts of uncommon elements have been discovered, e.g., [62].

It can be seen, from the brief discussion above, that XRFS can be used advantageously in many branches of soil science, often with very little sample preparation. In some cases, however, preconcentration is necessary, and the use of ion exchange concentration techniques would bring some of the elements for which XRFS is less sensitive into the range which can be detected (e.g., [62-67]). Coprecipitation, as a means of concentrating some minor elements onto filter paper to make a suitable sample for XRFS, has been described by Luke [68]. One of the special advantages of XRFS is the ability to scan a sample and make a qualitative or semiquantitative estimate of any unusual element, or any normal element which is present in particularly large or small concentration. In the nondispersive spectrometer, all the pulses are processed simultaneously, making multielement determinations very rapid.

Finally, the author would like to stress that X rays are dangerous and great care should be taken when dealing with X-ray equipment. The reader is recommended to read Jenkins and Haas [69] which outlines the hazards involved. However, providing precautions are taken, XRFS is a safe and reliable method for the determination of many elements in soils.

ACKNOWLEDGMENTS

The author would like to thank George Brown of Rothamsted Experimental Station for many helpful comments and Link Systems Ltd., High Wycombe, Buckinghamshire, England, for supplying information used in Fig. 6b.

REFERENCES

1. R. Jenkins, *An Introduction to X-Ray Spectrometry*, Heyden, London, 1974.
2. H. A. Liebhafsky, H. G. Pfeiffer, E. H. Winslow, and P. D. Zemany, *X-Ray Absorption and Emission in Analytical Chemistry*, Wiley, New York, 1960.
3. E. P. Bertin, *Principles and Practice of X-Ray Spectrometric Analysis*, Plenum Press, New York, 1970.
4. R. Jenkins and J. L. de Vries, *Practical X-Ray Spectrometry*, Macmillan, London, 1970.
5. L. S. Birks and J. V. Gilfrich, Anal. Chem., *48*:273R (1976).
6. G. L. Macdonald, Anal. Chem., *50*:135R (1978).
7. American Society for Testing and Materials, *X-Ray Emission Line Wavelength and Two Theta Tables*, ASTM Data Series DS 37, 1965.
8. M. A. Powers, *X-Ray Fluorescent Spectrometer Conversion Tables for Topaz, LiF, NaCl, EDDT and ADP Crystals*, Philips Electronic Instruments, New York, 1957.
9. S. J. B. Reed and N. G. Ware, X-Ray Spectr. *2*:69 (1973).
10. W. P. van der Burg, *The Analysis of Mg, Choosing the Crystal*, Philips Analytical Equipment Department Bulletin, Eindhoven, 1974.
11. D. J. Hughes and J. E. Davey, Analyst (Lond.), *95*:992 (1970).
12. J. M. McNeal and A. W. Rose, Geochim. Cosmochim. Acta, *38*:1759 (1974).
13. R. E. Collier and J. Parker-Sutton, J. Sci. Fd. Agric., *27*: 743 (1976).
14. K. Norrish and B. W. Chappell, in *Physical Methods in Determinative Mineralogy*, (J. Zussman, Ed.), Academic Press, London, 1977.
15. C. S. Hutchinson, *Laboratory Handbook of Petrographic Techniques*, Wiley, New York, 1974.
16. K. Norrish and J. T. Hutton, Geochim. Cosmochim. Acta, *33*: 431 (1969).
17. P. K. Harvey, D. M. Taylor, R. D. Hendry, and F. Bancroft, X-Ray Spectr. *22*:33 (1973).
18. M. T. Haukka and I. L. Thomas, X-Ray Spectr. *6*:204 (1977).
19. I. L. Thomas and M. T. Haukka, Chem. Geol., *21*:39 (1978).

20. C. Williams, J. Sci. Fd. Agric., *27*:561 (1976).
21. M. A. Short, Rev. Scient. Instrum., *31*:618 (1960).
22. D. J. Reed and A. A. Gillieson, Can. Spectr., *13*:67 (1968).
23. P. F. Berry, T. Furuta, and J. R. Rhodes, Advan. X-Ray Anal., *12*:612 (1969).
24. C. B. Hunter and J. R. Rhodes, X-Ray Spectr. *1*:107 (1972).
25. J. R. Rhodes and C. B. Hunter, X-Ray Spectr. *1*:113 (1972).
26. F. J. Flanagan, Geochim. Cosmochim. Acta, *37*:1189 (1973).
27. G. Brown and R. Kanaris-Sotiriou, Analyst (Lond.), *94*:782 (1969).
28. Z. H. Kalman and L. Heller, Anal. Chem., *34*:946 (1962).
29. C. W. Childs and R. J. Furkert, Geoderma, *11*:67 (1974).
30. H. J. M. Bowen, *Trace Elements in Biochemistry*, Academic Press, London, 1966.
31. S. H. Watts, Geochim. Cosmochim. Acta, *41*:1164 (1977).
32. T. Wakatsuki, H. Furukama, and K. Kyama, Geochim. Cosmochim. Acta, *41*:891 (1977).
33. C. Williams and J. H. Rayner, J. Soil Sci., *28*:180 (1977).
34. L. R. Drees and L. P. Wilding, Soil Sci. Soc. Am. Proc., *37*:82 (1973).
35. L. J. Evans and W. A. Adams, J. Soil Sci., *26*:319 (1975).
36. A. C. Oertel, J. Soil Sci., *12*:119 (1961).
37. R. J. Gilkes, G. Scholz, and G. M. Dimmock, J. Soil Sci., *24*:523 (1973).
38. C. W. Childs, Geoderma, *13*:141 (1975).
39. U. M. Cowgill, Dev. Appl. Spectrosc., *5*:3 (1966).
40. I. Bradley, C. C. Rudeforth, and C. Wilkins, J. Soil Sci., *29*:258 (1978).
41. J. Bolton, G. Brown, G. Pruden, and C. Williams, J. Sci. Fd. Agric., *24*:557 (1973).
42. M. A. Tabatabai and J. M. Bremner, Soil Sci. Soc. Am. Proc., *34*:417 (1970).
43. C. Bloomfield, J. K. Coulter, and R. Kanaris-Sotiriou, Trop. Agric., *45*:289 (1968).
44. C. Williams, J. Agric. Sci. Camb., *82*:189 (1974).
45. R. G. Darmody, D. S. Fanning, W. J. Drummond, and J. E. Foss, Soil Sci. Soc. Am. Proc., *41*:761 (1977).
46. R. Kanaris-Sotiriou and G. Brown, Analyst (Lond.), *94*:780 (1969).
47. S. Roberts and F. E. Kochler, Soil Sci., *106*:164 (1968).
48. R. G. McLaren and R. S. Swift, J. Soil Sci., *28*:445 (1977).
49. C. C. Evans, Analyst (Lond.), *95*:919 (1970).
50. N. Souty and R. Guennelon, Ann. Agron., *18*:653 (1967).
51. K. D. McLachan and M. J. Crawford, J. Sci. Fd. Agric., *21*:408 (1970).
52. R. T. King, X-Ray Spectr., *8*:9 (1979).

53. C. Wilkins, J. Agric. Sci., *92*:61 (1979).
54. V. S. Keramudas and D. S. Fanning, Soil Sci. Soc. Am. Proc., *40*:857 (1974).
55. D. C. Bain, J. Soil Sci., *27*:68 (1976).
56. A. E. Hubert and T. T. Chao, Anal. Chim. Acta, *92*:197 (1977).
57. E. Murad, J. Soil Sci., *29*:219 (1978).
58. D. S. Fanning and M. L. Jackson, Soil Sci., *103*:253 (1967).
59. P. A. Cawse, *A Survey of Atmospheric Trace Elements in the U.K. (1972-1973)*, U.K. Atomic Energy Authority, Harwell, 1974.
60. C. Williams, J. Agric. Sci. Camb., *82*:189 (1974).
61. C. Wilkins, Environ. Pollution, *15*:23 (1978).
62. C. Williams and G. Brown, Geoderma, *6*:223 (1971).
63. K. Gorindaraju, X-Ray Spectr., *2*:57 (1973).
64. W. J. Campbell, E. F. Spano, and T. E. Green, Anal. Chem., *38*:987 (1966).
65. G. R. Bradford, P. F. Pratt, F. L. Bair, and B. Goulben, Soil Sci., *100*:309 (1965).
66. C. W. Blount, R. E. Channel, and D. E. Leyden, Anal. Chim. Acta, *56*:456 (1971).
67. H. H. LeRiche, Geochim. Cosmochim. Acta, *32*:791 (1968).
68. C. L. Luke, Anal. Chim. Acta, *41*:237 (1968).
69. R. Jenkins and D. J. Haas, X-Ray Spectr., *2*:135 (1973).
70. B. P. Fabbi and L. F. Espos, Appl. Spectrosc., *26*:293 (1972).
71. N. B. Price, Geochim. Cosmochim. Acta, *41*:1769 (1977).
72. G. Nelson Eby, Anal. Chem., *44*:2137 (1972).
73. R. M. Taylor and J. B. Giles, J. Soil Sci., *21*:203 (1970).
74. L. A. Wolfe and H. Zeitlen, Anal. Chim. Acta, *51*:349 (1970).
75. J. H. Ellis, R. I. Barnhisel, and R. E. Phillips, Soil Sci. Soc. Am. Proc., *34*:866 (1970).
76. J. H. Ellis, R. E. Phillips, and R. I. Barnhisel, Soil Sci. Soc. Am. Proc., *34*:591 (1970).
77. H. Bergseth, Analyst (Lond.), *100*:96 (1975).
78. G. Brown and D. A. Jenkinson, Commun. Soil Sci. Plant Anal., *2*:45 (1971).
79. C. Wilkins, J. Agric. Sci. Camb., *90*:109 (1978).

6

General Radioisotope Techniques

KEITH A. SMITH and ROGER S. SWIFT* The Edinburgh School of Agriculture, Edinburgh, Scotland

---

**Present affiliation*: Lincoln College, Canterbury, New Zealand.

I. INTRODUCTION

Most substances may be measured either by instrumental methods, such as those described elsewhere in this book, or by conventional methods of analysis. Radioactive isotopes,* however, are universally measured by instrumental methods. Their essential characteristic, the emission of penetrating ionizing radiation, has been the basis of methods of measurement since the early part of this century, and a wide variety of instruments capable of detecting and recording this radiation has been developed.

The relative ease with which radioisotopes may be detected, and quantitatively measured, has led to their widespread use as tracers in physical and biological systems. In soil science the major application in this category has been to investigate the fate of labeled nutrient ions added to the soil, e.g., ^{32}P-labeled phosphate. The phenomenon of isotopic exchange, whereby ions present in the soil in exchangeable form can exchange with chemically identical but radioactively labeled ions in solution, has been widely exploited to demonstrate the dynamic nature of the exchange equilibrium and to determine the size of the labile pools of various nutrient ions. Measurements of this type can be made without destruction of the system, making in vivo applications possible. Another important application of radioisotopes has been elemental analysis of soil and plant materials, either by radio-activation techniques in which the activity is induced in the sample as a result of irradiation by neutrons or other nuclear particles, or by isotope dilution methods.

Over the last 30 years, much effort has been devoted to the determination, and the assessment of the biological significance, of radioisotopes distributed in the environment as a result of atmospheric explosions of nuclear weapons or releases from nuclear reactor installations and processing plants. Many of the techniques involved in tracer studies have also been employed in the investigation of these pollutant radioisotopes.

Consideration of all aspects of the use and measurement of radioisotopes in soil science and related fields of study would occupy a whole volume. Accordingly, the scope of this chapter is restricted

*This term will be employed throughout because of its general use among workers in soil science and related disciplines. Strictly speaking, we are dealing with "nuclides," which are defined by their atomic number (Z) and mass number (A). Two atoms are isotopes only if they have the same Z value but differ in A value.

to an examination of available methods for determining those radioisotopes used as tracers for biologically important elements in soils and soil-plant systems, radioisotope methods in elemental analysis, and the determination of the labile pools of nutrient ions in the soil and their interaction with soil materials.

An outline is given of the theory of radioactive decay, the properties of radioisotopes, and the principles of radiation detection, as a necessary basis for understanding and solving the practical problems involved in radioisotope methods. For more detailed treatments, and discussion of essential safety precautions associated with the handling of radioisotopes, a number of excellent books are available [1-10]; the only disadvantage of the older books is the outdated information on counting techniques. Another very useful source of information is the series of biennial reviews in *Analytical Chemistry* [11-15].

II. THEORY OF RADIOACTIVITY

A. Nature of Radioisotopes

Radioisotopes are characterized by having unstable nuclei which spontaneously disintegrate or *decay*, with the emission of charged particles or the capture of orbital electrons. This decay results in a change in the charge on the nucleus, giving a product of different atomic number, i.e., one which is an isotope of a different chemical element. The latter may be stable, or it, in turn, may be radioactive, in which case it will also decay. Ultimately a stable isotopic form will be achieved.

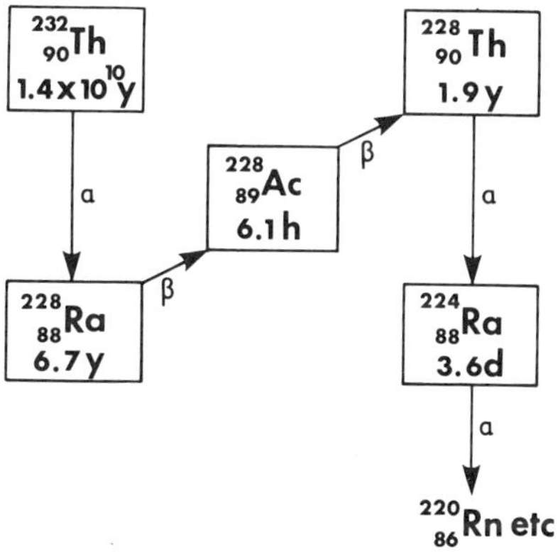

Figure 1 Changes in atomic mass and atomic number as a result of alpha and beta decay.

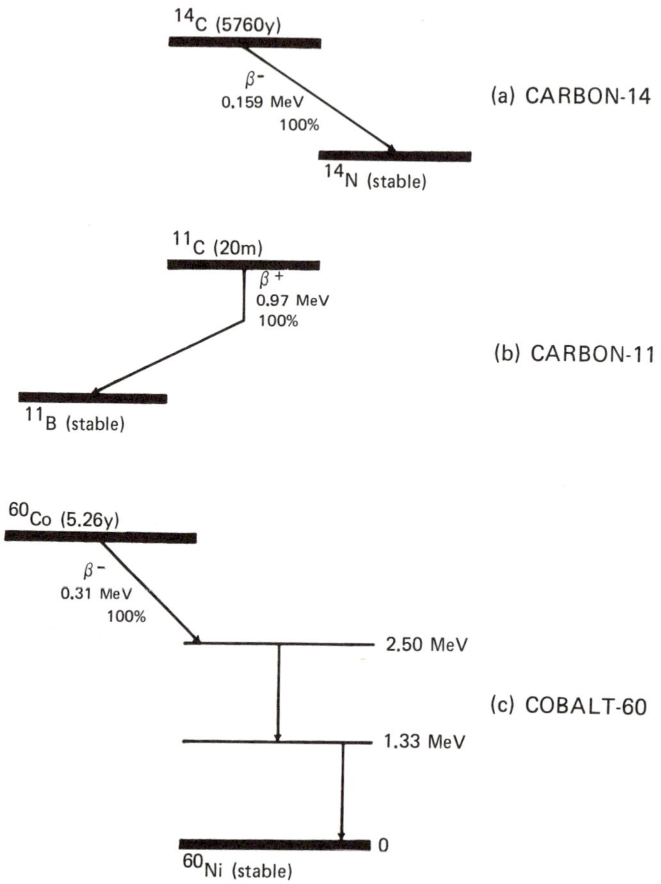

Figure 2 Beta-decay schemes (a) β^-, (b) β^+, (c) β^- accompanied by emission of gamma radiation (From Ref. 4.)

Some isotopes of those elements with an atomic number $\geqslant 82$, and a very small number of others, decay by the emission of an alpha (α) particle (a helium nucleus). This decreases atomic mass by 4 and atomic number by 2 (Fig. 1). Most other radioisotopes emit beta (β) radiation, i.e., electrons (β^-) or positrons (β^+), which increase and decrease atomic number, respectively, by 1, leaving mass unchanged (Fig. 2). If the daughter isotope produced by the decay is in an excited state, transition to the ground state occurs with the emission of the excess energy in the form of electromagnetic radiation, or gamma (γ) rays (Fig. 2c). Gamma emission leaves both atomic mass and atomic number unchanged. Another mode of decay involves the capture of an external electron by the nucleus. This results in the emission of X rays known as *Bremsstrahlung* or "braking radiation," with identical properties to γ radiation. Bremsstrahlung production also occurs when β particles lose kinetic energy by interaction with matter.

Decay schemes illustrating pure β^-, pure β^+, and β^- decay accompanied by γ radiation, respectively, are shown in Fig. 2. Many radioisotopes have much more complex decay schemes than that of ^{60}Co shown in Fig. 2c and decay frequently occurs by more than one process, e.g., by electron capture and positron emission. The relative proportions in which the two processes occur remain constant for any particular radioisotope. When a positron loses its energy by interaction with electrons it finally combines with an electron and the particles are annihilated, resulting in the liberation of two γ rays, each of which has an energy of 0.51 MeV. Thus a positron-emitting isotope always has this characteristic "annihilation radiation" associated with it, in addition to its β^+ activity.

B. Laws of Radioactive Decay

Radioactive decay processes are random, and the number of decay events which occur within a given period of time, i.e., the *activity*, A, is directly related to the number N of unstable atoms of the radioisotope present:

$$A = -\frac{dN}{dt} = \lambda N \tag{1}$$

where λ is the decay constant for the particular radioisotope in question.

Integration of Eq. (1) gives

$$N = N_0 e^{-\lambda t} \tag{2}$$

where N_0 is the number of radioactive atoms existing at time t = 0.

As $A = \lambda N$,

$$A = A_0 e^{-\lambda t} \tag{3}$$

Half-Life

This is the time taken for the activity to decrease to half the original value. Thus

$$A = \frac{A_0}{2} = A_0 e^{-\lambda t_{1/2}} \tag{4}$$

where $t_{1/2}$ is the half-life. Taking logarithms,

$$t_{1/2} = \frac{\ln 2}{\lambda} = \frac{0.693}{\lambda} \tag{5}$$

Table 1 Some Useful Tracer Radioisotopes

Radioisotope	Half-life[a]	Principal radiations (MeV)[a]		
		β^-	β^+	γ
^{3}H	12.3 years	0.02		
^{14}C	5700 years	0.2		
^{22}Na	2.6 years		0.5	1.275, 0.51[b]
^{24}Na	15.0 hours	1.4		2.75, 1.37
^{32}P	14.3 days	1.7		
^{33}P	25.3 days	0.2		
^{35}S	87.5 days	0.2		
^{36}Cl	3×10^5 years	0.7		
^{42}K	12.4 hours	3.5		1.525
^{45}Ca	163 days	0.3		
^{54}Mn	312 days			0.835
^{59}Fe	44.6 days	0.5, 1.6		1.10, 1.29
^{58}Co	70.8 days		0.5, 1.3	0.81, 0.51[b]

^{60}Co	5.3 years	0.3		1.33, 1.17
^{64}Cu	12.7 hours	0.6	0.7	0.51[b]
^{65}Zn	244 days		0.3	1.11, 0.51
^{75}Se	120 days			0.136, 0.265
^{86}Rb	18.7 days	1.8		1.08
^{85}Sr	64.9 days			0.51
^{89}Sr	50.5 days	1.5		
^{99}Mo	66 hours	1.2		0.74
^{131}I	8.0 days	0.6		0.36
^{134}Cs	2.1 years	0.7		0.605, 0.80
^{137}Cs	30.1 years	0.5, 1.7		0.67[c]
^{133}Ba	10.5 years			0.36
^{140}Ba	12.8 days	0.5, 1.0 (1.4[d])		0.54 (0.50, 1.6[d])

[a]Principal source of nuclear data: W. Seelman-Eggebert, G. Pfennig, and H. Münzel, *Chart of the Nuclides*, 4th ed., Gersbach, Munich, 1974.

[b]Annihilation radiation.

[c]From ^{137m}Ba daughter ($t_{1/2}$ = 2.6 min).

[d]From ^{140}La daughter ($t_{1/2}$ = 40 hr).

The observed activity of a radioactive source will be related to its absolute disintegration rate by a proportionality factor which will depend on the measuring instrument used and the particular conditions under which the measurement is made. This factor will remain constant if the conditions also remain constant, thus the observed activity of a source containing a single isotopic species will decrease exponentially wiith time, and, a plot of the logarithm of the activity against time will be linear. The half-life of a radioisotope can be determined from the slope of this plot; determination of half-life is the best criterion of radiochemical purity of a material.

If two or more unrelated isotopes are present in the same sample, the overall rate of decay is the composite of the individual exponential decay rates. Provided the half-lives of these isotopes are sufficiently different from each other, they may be determined by resolving the decay curve into its individual components.

Units

The basic S.I. unit of radioactivity, which has only very recently been adopted, is the becquerel (Bq); however, it is likely to take a considerable time for the familiar curie (Ci) to cease to be widely used.

$$1 \text{ Bq} = 1 \text{ disintegration per second} = 2.7 \times 10^{-11} \text{ Ci}$$

For most practical purposes radioisotopes are likely to be used in megabecquerel quantities or even higher, but sources which are actually measured ("counted") may have activities of the order of 0.1-100 Bq (i.e., approximately the picocurie to nanocurie range).

C. Production of Radioisotopes

Some radioisotopes occur in nature, for example ^{40}K, ^{14}C, and isotopes of elements with $Z \geqslant 82$ resulting from the decay of uranium and thorium. Others, including some now widely used as tracers, are produced by nuclear fission of uranium or plutonium. However, the great majority of useful tracers are produced artificially by the bombardment of target elements with α particles, protons, neutrons, or other particles to bring about the desired nuclear transformation [4]. For example, irradiation of phosphorus (which has only the single stable isotope ^{31}P) with thermal neutrons results in the capture of a neutron, and thus the formation of ^{32}P; excess energy is lost by the emission of a γ ray. This reaction is normally written in the form $^{31}P(n,\gamma)^{32}P$. The same product can also be produced from sulfur by the reaction $^{32}S(n,p)^{32}P$ if the neutrons are sufficiently energetic. The irradiation of elemental phosphorus transforms only a small fraction of the ^{31}P atoms to ^{32}P and thus the target material, after irradiation, contains both isotopes. The inactive ^{31}P is the isotopic carrier for the ^{32}P, so-called because it is normally present in sufficient quantity to be precipitated, and thus carry down the trace quantity of the radioactive isotope. The amount of radioactivity per unit mass of stable element (i.e., the *specific activity* of the isotope) will depend on the irradiation conditions. In contrast, the formation of ^{32}P from sulfur is an important example of the formation of a radioisotope in a *carrier-free* state, i.e., in a state in which the only atoms present, after chemical separation from the parent element, are radioactive ones of the new element.

Some of the radioisotopes which have been used in soil and plant studies are listed in Table 1. It will be seen that the half-lives vary from a few hours to 3×10^5 years (for ^{36}Cl). This greatly affects the usefulness of the various isotopes. Normally, experiments can only be carried out over a period of a few half-lives because of inadequate activity remaining for measurement, or because longer-lived activities present as impurities become increasingly significant. Even shorter-lived isotopes than those listed in Table 1 have been used in some studies, e.g., $^{13}N(t_{1/2} = 10\ \text{min})$ [16] and $^{15}O(t_{1/2} = 2\ \text{min})$ [17], but their usefulness is rather limited.

III. DETECTION AND COUNTING

A. General Aspects

The radiations emitted by radioisotopes when they decay cause ionization or molecular excitation in the substances through which they pass. It is these phenomena which are utilized in all the systems which have been devised to detect and count radioactive emissions. The ionization process can be detected electrically, as in ionization chambers, semiconductor detectors, and gas counters such as the familiar Geiger-Muller tubes, by the emission of light pulses, as in inorganic scintillation detectors, or by interaction with photographic emulsions; excitation of organic-crystal scintillators also results in emission of light.

In some laboratories a very narrow range of isotopes is used, leading to the use of a single specialist item of detection and counting equipment. For instance, many biological and medical laboratories use only the low-energy β emitters ^3H (tritium) and ^{14}C, and the counting is carried out almost exclusively by liquid scintillation counters. However, as work with soils involves the use of a wide range of isotopes (Table 1), emitting radiation of different types and energies, many forms of detector have been employed. An outline of the principal types is given in this section. More detailed information is available in a number of books [5,7,10,18-21] and the extensive literature cited therein.

Some of the detectors described here have existed in one form or another for many years. Although they are being superseded in some areas of research by newer devices, modern versions of traditional types of detector, e.g., Geiger-Muller tubes, are still commercially available. They are now normally combined with modern solid-state electronic circuitry, usually in modular form, e.g., the NIM series, and frequently form part of automated counting systems. Low-cost, manually operated instruments, however, continue to serve a useful function in soil science and related disciplines. Whatever the level of sophistication of the sample input or data output systems, satisfactory counting procedures depend on an appreciation of the characteristics of the particular radiation to be determined and selection of an appropriate detector.

Interactions of Radiation with Matter

Alpha particles are intensely ionizing, and lose their energy so rapidly by multiple collisions with bound electrons, that their tracks are only a few centimeters long in air, and they will not penetrate a thin sheet of plastic or paper. This has important implications for quantitative measurement of α activity. Unless the matrix material which contains the α-emitting isotope is spread in an extremely thin layer, self-absorption of the particles within the source may be severe. The ranges of the α particles from the isotopes of the Uranium and Thorium Series have been calculated to be 13-45 μm in soil and 20-70 μm in plant ash [22], and the ranges in precipitates of insoluble salts, following radiochemical separation, will be of the same order.

The path length of β particles will vary greatly, depending on their emission energy and the density of the medium through which they are traveling. A β particle with an energy of 1 MeV will have a range of approximately 3 m in air, 4 mm in water or biological tissue and somewhat less than this in solids. Absorption by air and by the "windows" of conventional gas ionization detectors (see Sec. III.B) always occurs, but it can be particularly important for α and low energy β particles, such as those emitted by ^{14}C. With such an isotope it is necessary to place the source very close to the detector and to make the window as thin as possible within practical limits. Self-absorption of β particles within the source is discussed in Sec. III.B.

It should be noted that the figures for β-particle energies listed in Table 1 are maximum values (E_{max}), and that particles are actually emitted with a range of energies, from very low values up to E_{max}. Thus β particles emitted from a radioisotope form a continuous spectrum of the type shown in Fig. 3. This can be important when considering the merits and disadvantages of β-counting systems. In some systems many of the lower energy particles are lost by absorption and scattering before they even reach the detector. The different

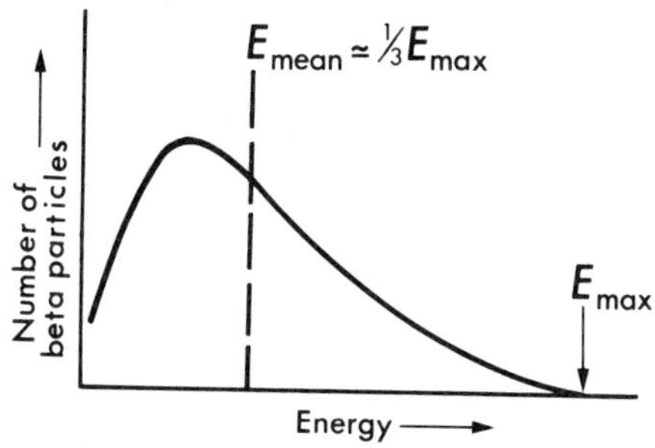

Figure 3 Beta spectrum. (From Ref. 7.)

β-energy profiles exhibited by different radioisotopes can be utilized to count two isotopes in a single sample, thus allowing double labeling experiments to be carried out [23].

Unlike α and β radiation, γ rays consist of electromagnetic radiation rather than charged particles. Consequently their interaction with matter is much reduced and their penetrating power is greatly increased. This greatly affects the methods used for satisfactory counting. While there is very little self-absorption in the source (which can therefore consist of up to several kilograms of soil or other material [24,25]), the detector must be large enough to absorb the energy of the γ ray, wholly or partially, in order to initiate some process which culminates in a detectable pulse.

Background and Shielding

The inevitable presence of cosmic radiation, and also natural radioactivity and fallout in the materials from which laboratories, and even counting equipment, are constructed, results in a background counting rate in the absence of a source. This background (particularly the cosmic ray element) is normally minimized by housing the detector in a 50 mm thick lead "castle."

In many tracer applications, source count-rates are sufficiently high for the background contribution to be insignificant, but in some circumstances it may be required to count very low activities. This requires special techniques to reduce the background, such as the use of anticoincidence circuitry. A typical anticoincidence system for β counting consists of a central detector surrounded by a guard detector or detectors. When radiation passes through the guard and the central detector (triggering both almost simultaneously) the pulse is not recorded. However, a particle emitted from a source into the central detector will not have sufficient energy to pass through it and into the guard detector, so only the former is triggered. The resulting pulse is recorded. Modern anticoincidence systems have backgrounds of less than 1 count per minute.

Other precautions include the use of old lead (in which the naturally occurring ^{210}Pb ($t_{1/2}$ = 22 years) has decayed away) for shielding [6,26], and materials of low potassium content in counter construction [26]. The performance of a counting system in detecting any particular source of radiation is normally assessed on the basis of a *figure of merit*, equal to S^2/B, where S and B are the sample and background counts, respectively.

Counting Efficiency and Standardization

Radiation is emitted from an active source in all directions; to maximize detection efficiency the sample source is usually placed as close to the detector as possible, so that almost all the radiation emitted within a solid angle of 2π steradians is intercepted. If the source is moved away from the detector, then the solid angle it subtends decreases and the count rate falls. Self-absorption of radiation by the source, and scattering, also have major effects on counting efficiency (see Sec. III.B).

For absolute standardization of β-emitting radioisotope preparations, 4π counters are used. As the name applies, the radiation emitted in all directions is counted. This can be achieved by evaporating a solution of a carrier-free, or virtually carrier-free, radioisotope onto a thin aluminum foil or plastic film. The foil is placed inside a gas-flow proportional counter (Sec. III.B) which has two anodes, one on each side of the foil, and particles emitted from the source in any direction are detected [27]. More sophisticated 4π β-γ coincidence techniques are widely employed for this purpose nowadays [28]. Most users are not equipped to undertake absolute standardization of the radioisotope preparations they wish to use. Standardized solutions are available from the major suppliers, and can be used to prepare sources for counter calibration. For many applications, measurements need only be made on a relative basis, and absolute calibration of counters is therefore not so important. It is usually necessary, however, to ensure that counter performance does not fluctuate over the period when experimental studies are being carried out, and it is desirable to check this by regular counting of standard sources.

Statistical Accuracy of Counting

Radioactive decay is a random process, thus the counting rate observed from a given source will vary from measurement to measurement. It is not possible to define a true counting rate, but only a mean rate about which the observed rate will vary according to a Poisson distribution. For a Poisson distribution, the standard deviation σ is given by

$$\sigma = (N)^{1/2} \tag{6}$$

where N is the number of events recorded in a given observation.

When low count rates have to be measured, fluctuations in the background rate become significant. It can be shown [2] that if the net count rate for a sample is R_S, with a standard deviation σ_S,

$$\sigma_s = \left(\frac{R_{(s+b)}}{t_{(s+b)}} + \frac{R_b}{t_b}\right)^{1/2} \tag{7}$$

where

$R_{(s+b)}$ = observed sample-plus-background count rate

$t_{(s+b)}$ = counting period for the sample

R_b = background count rate

t_b = background counting period

Thus for measurements of sample counting rates similar to the background, the error for the sample is much higher than that for the sample plus background. This factor must be taken into account when intending to count low-activity samples to a given statistical accuracy. Fuller treatments of the statistical theory are given in Ref. 2 and other standard radiochemistry texts.

B. Gas Ionization Detectors

In its simplest form, a gas ionization detector is a chamber containing a gas and two electrodes, across which an electric potential is maintained. When radiation passes through this chamber it causes the gas molecules within it to ionize. These ions then move toward the appropriate electrode and a pulse of current flows between the electrodes. Measurement of the total current or the number of pulses can be used to measure the amount of radiation reaching the chamber. This system is very flexible, and by varying the dimensions and materials of the gas chamber and its operating conditions, it can be used to detect α and β particles, γ rays, and neutrons. However, the counting efficiency for γ rays is very low, and depends on the emission of secondary electrons from a heavy metal absorber (see Sec. III.E).

The pulse height from the ionization chamber varies with the applied voltage as shown in Fig. 4, and this affects the way in which it can be used. At voltages less than V_1, any ions present within the chamber move relatively slowly toward the electrodes. Some of the ions recombine before reaching the electrodes, so that this region of the curve is unsuitable for counting purposes. Between V_1 and V_2, the ions travel sufficiently quickly to reach the electrode without recombination occurring, giving rise to a *plateau* (the saturation current). Under these conditions the chamber can be used as a counter and this region is known as the *ionization chamber region*.

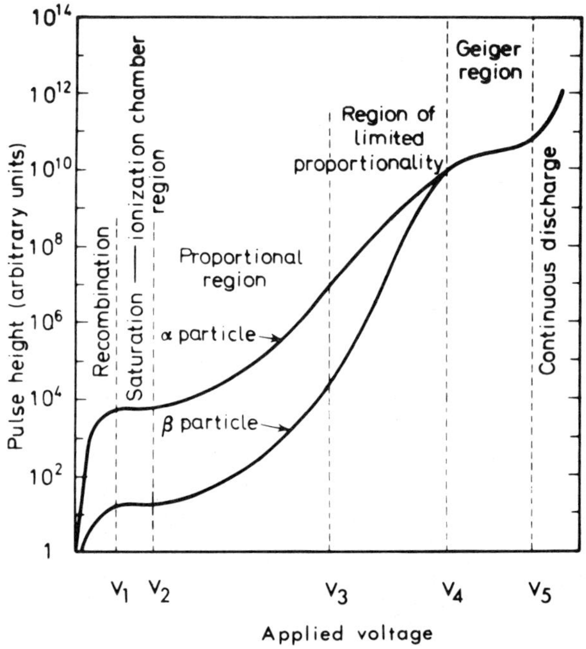

Figure 4 Relationship between pulse height and applied voltage for a gas ionization detector (From Ref. 3.)

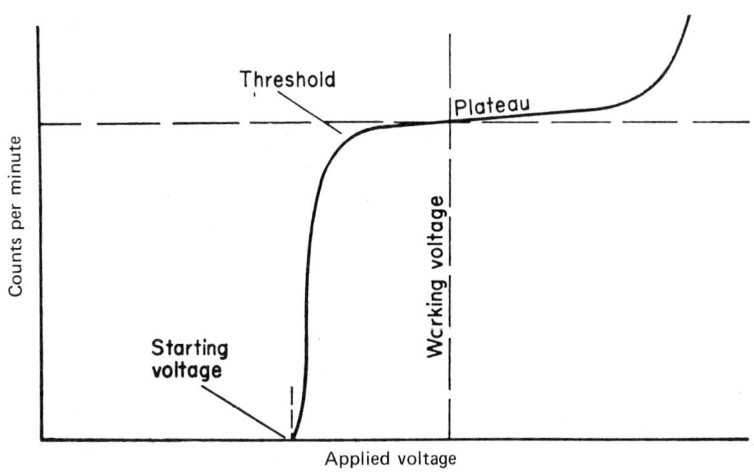

Figure 5 Relationship between counting rate and applied voltage for a G-M tube (From Ref. 7.)

Above V_2 (Fig. 4), the pulse size increases again with voltage. The ions are subject to greater acceleration, sufficient to cause the production of secondary ions by collision with further gas molecules. This behavior results in an avalanche of ions, and a corresponding increase in pulse size; this process is known as *gas amplification*. Furthermore, the size of the pulse obtained is proportional to the energy of the ionizing radiation, and the region between V_2 and V_3 is known as the *proportional region*. A detector operating under these conditions is known as a proportional counter, and can be used to differentiate between radiations of different energies.

Between V_4 and V_5, secondary ionization processes are so great that the pulse size always reaches its maximum value, and is independent of the energy of the incident radiation, although it does increase somewhat with voltage. This is known as the Geiger-Muller (G-M) region and is the basis of the familiar *Geiger counters*. Above V_5 the chamber begins to discharge continuously because new ions are continually being formed by the movements of the very highly accelerated charged particles within the chamber.

The relationship between counting rate and applied voltage for a G-M tube is shown in Fig. 5. For normal applications a tube is considered satisfactory if it has a plateau with a slope not greater than 2-2.5%/100 V [18,28].

Ionization Chamber. The ionization chamber finds its major use nowadays as a health physics dosemeter for measuring exposure to γ rays, X rays, or neutrons [28,29]. It can also be used for counting α particles [28,30], but for counting β particles, proportional and G-M counters are generally used.

Proportional Counter. The anode of a proportional counter usually consists of a loop of fine wire enclosed in a chamber which often contains argon or an argon/methane mixture. The gas can be sealed into the chamber at high or low pressures or can be allowed to flow slowly through the chamber under a slight positive pressure. This latter type is shown diagrammatically in Fig. 6a. The gas-flow system has the advantage of avoiding problems associated with deterioration of the gas, and because the pressure differential used is very low, the end-window can be very thin, thereby facilitating the penetration of low-energy β particles, such as those from ^{14}C, into the counter. The gas used must be purified before use to remove corrosive agents, dust particles, etc. A further modification for suitable radioisotopes, e.g., ^3H, in gaseous forms is to pass them through the chamber with the flow of counting gas.

Since the pulses produced by these counters are proportional to the energy of the incident radiation, it is possible, by use of suitable electronic equipment, to distinguish between different particle energies and carry out spectroscopic measurements, or count two isotopes separately when both are present, as in double labeling experiments, or simply to reject unwanted pulses.

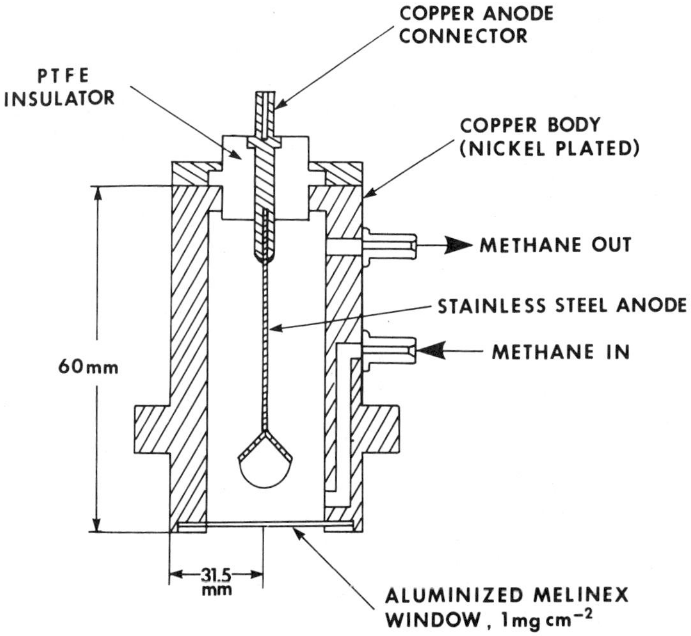
COPPER ANODE
CONNECTOR
PTFE
INSULATOR
COPPER BODY
(NICKEL PLATED)
METHANE OUT
STAINLESS STEEL ANODE
METHANE IN
60mm
31.5
mm
ALUMINIZED MELINEX
WINDOW, 1mg cm^{-2}

(a)

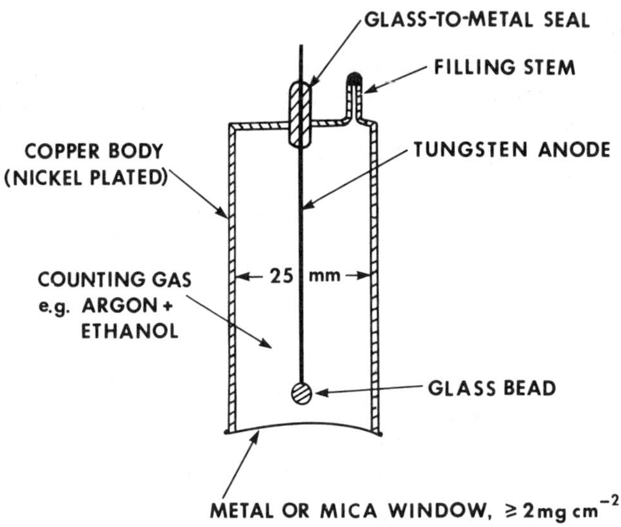
GLASS-TO-METAL SEAL
FILLING STEM
COPPER BODY
(NICKEL PLATED)
TUNGSTEN ANODE
COUNTING GAS
e.g. ARGON +
ETHANOL
25 mm
GLASS BEAD
METAL OR MICA WINDOW, ≥ 2mg cm^{-2}

(b)

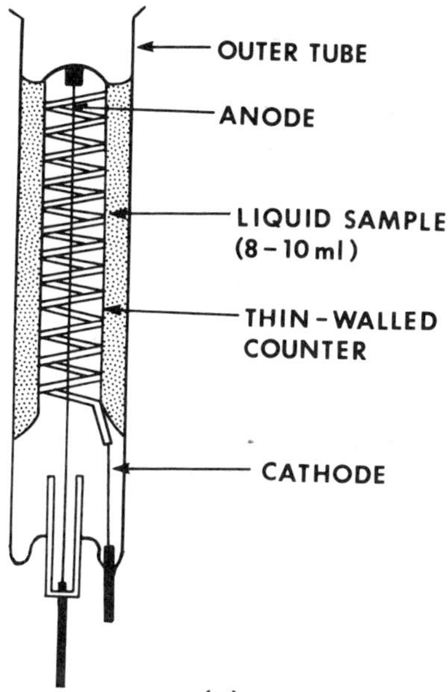

(c)

Figure 6 (a) End-window gas flow proportional counter. (From Ref. 28.) (b) Typical G-M counter. (From Ref. 28.) (c) Glass G-M counter for liquid samples.

While proportional counters are most commonly used for counting β particles, they can also be used for counting α emissions if fitted with a very thin window. In another version of the counter, the α source is placed inside the counting chamber [28].

Modern laboratory instruments are available incorporating proportional counters with automatic sample changers and printout systems, which are capable of accommodating up to 100 samples.

Geiger-Muller Counter

In the Geiger-Muller region, any ionization gives rise to uniform, large pulses. This means that although the G-M detector is unable to distinguish radiations of different energy it is ideal for use with electronic systems which are simple, cheap, and robust. As a consequence it has proved very popular as a general purpose β counter in spite of its other limitations [7,18]. A typical end-window G-M tube is shown in Fig. 6b. The anode is a quite substantial wire sealed into but insulated from the metal body of the tube. The end of the wire is coated with a glass bead to prevent unwanted point discharges. The tube is fitted at one end with a sealed mica window which is as thin as possible. Even so, the higher gas pressure used in G-M tubes means that this window must be thicker than that used in a gas-flow proportional counter, thereby causing a decrease in counting efficiency.

A particularly important part of G-M tube design is the nature of the gas filling. To ensure that each ionization event results in a finite pulse, a quenching agent, usually chlorine or ethanol, is introduced into the counting gas (normally argon) to bring the discharge to an end. During the time taken by the ionization and quenching processes the tube is unable to detect or respond to any other ionizing radiation. To ensure that this *dead-time* is a constant value it is usual to superimpose a known, electronically generated dead-time, triggered by the receipt of a pulse. Formerly, dead-times of 400 μsec were commonly used, but with modern equipment 100 μsec is more usual [7]. When the count rate is high it becomes necessary to perform a correction to make up for lost counts. While this is not difficult, it is inconvenient, and it is better practice to avoid where possible the need for correction by careful adjustment of sample size and activity and its distance from the tube.

If a dead-time correction *is* necessary then it can be calculated from

$$C_t = \frac{C_0}{1 - C_0 t} \tag{8}$$

where C_t is the true count rate (per second), C_0 the observed count rate (per second), and t the dead-time (seconds). This correction is an approximation but is adequate for most routine purposes.

G-M tubes can be made in forms other than the end-window type described above. The electrodes and gas can, for instance, be enclosed in a glass envelope. This gives a very robust tube but the thickness of the glass envelope restricts its use to higher energy β emitters. One particular use of this type of tube is an arrangement for counting liquid samples such as that shown in Fig. 6c, which has found use in soils work with isotopes such as ^{32}P. Although in the last few years most workers have used Cerenkov counting methods (see Sec. III.D) to determine ^{32}P in soil extracts and solutions, recent papers [31,32] have referred to the continued use of the liquid counter for this purpose. Although it was first described in 1948 [33], it is still commercially available in a number of versions, e.g., the Type M6H (20th Century Electronics, Ltd. Croydon, Surrey, England).

Absorption and Scattering of β Particles

Solid sources for β counting are commonly obtained by evaporating to dryness aliquots of dilute salt solutions or of slurried precipitates on metal planchets, or by filtering precipitates through suitable papers, membranes, or sinters (see Sec. IV). The resulting layer of solid material on its support is then counted. In these cases the solid material of the source itself gives rise to self-absorption, which reduces the count rate, and scattering, which can either reduce or increase the count rate.

Self-Absorption. This problem can be largely overcome by ensuring that the amount of solid material on the planchet, filter, etc., is sufficiently small so as to cause no significant absorption. If this is not possible it becomes necessary to calibrate the system using standards of known activity and known amounts of solid. Alternatively, if the radioisotope emits γ rays as well as β particles, it may be preferable to count the γ radiation, since this is much less affected by self-absorption.

If increasing weights of a material of given specific activity are made into sources of uniform density and cross-sectional area, and counted, the count rate increases with increasing mass (and thickness), but since self-absorption also increases, the relationship is not linear (Fig. 7). A source thickness is reached which is equal to the range of β particles in this medium. All particles emitted from below this depth are absorbed, and the count rate remains constant; this is known as an "infinitely thick" source. Such sources of equal mass can be counted and compared on the basis of specific activity

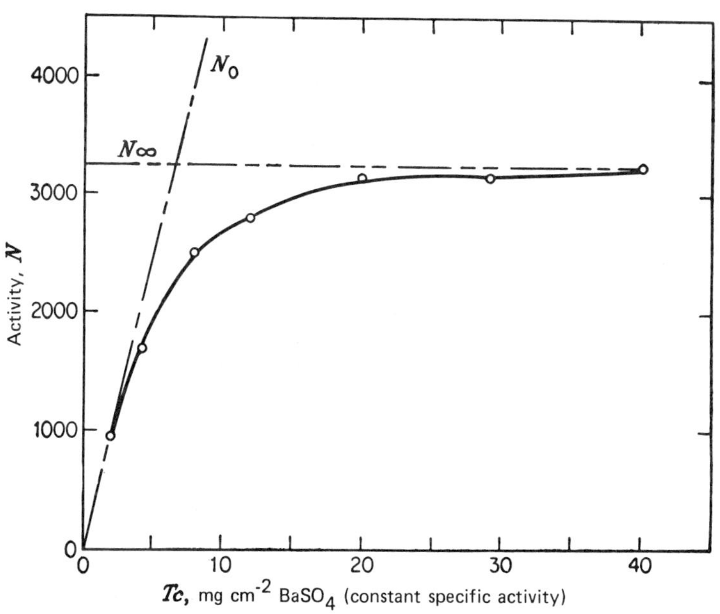

Figure 7 Self-absorption of beta particles. (From Ref. 23.)

rather than total activity. This technique can be particularly useful when working with heavy precipitates containing low-energy β emitters, e.g., barium carbonate containing ^{14}C or calcium oxalate containing ^{45}Ca.

Back-Scattering. Particles emitted from a source in a direction away from the detector may be scattered into the detector, due to collisions with nuclei in the source itself, the source mount, or the counter shield. Back-scattering increases with the thickness of the source mount and with increasing atomic number of source and mounting materials. The counting rate of a high-energy β source may thus be raised by 50-60% above that expected purely on the basis of counting geometry [2,34]. Thus it is important, in order to prevent errors, to prepare and count sample sources and standards under the same conditions.

Use of Absorbers. The introduction of an aluminum or polyethylene absorber between a β source and a detector will prevent all β particles below a certain energy from entering the detector. This simple principle has been used for many years to allow the determination of two different isotopes, with different E_{max}, present in the same source. Good examples are the determination of ^{89}Sr in the presence of ^{90}Sr, using a 100 mg cm^{-2} absorber [34], and ^{32}P and ^{33}P used in double-labeling experiments [35]. In suitably equipped laboratories this technique has been largely superseded by liquid scintillation counting [36].

C. Principles of Scintillation Counting

When an α or β particle, or a γ ray, passes through certain transparent materials, all or part of the energy is converted into a small light pulse, or *scintillation*. These materials may be crystals of inorganic salts, e.g., calcium fluoride or sodium iodide, or organic compounds such as anthracene, solutions of certain aromatic compounds in hydrocarbon solvents, or plastics. By recording the number of light pulses emitted, the activity of the source can be determined. Furthermore, as the intensity of each light pulse is proportional to the energy absorbed, if pulses of different intensities can be distinguished, then the energy spectrum of the radiation may be determined. The scintillator (or *phosphor*) is placed in optical contact with the cathode of a photomultiplier (PM) tube and enclosed in a light-proof container. Light photons from the scintillator eject electrons from the photocathode, and the size of the pulse emitted by the photomultiplier is proportional to the light energy received. Detailed accounts of these processes are given in Birks [19] and Crouthamel et al. [21].

Because of the distribution of energy within the β particle spectrum, plus the nature of energy transfer in scintillation systems and poor resolution of pulse size, it is found that the relationship between pulse size and energy is less well defined for β than for γ radiation. This restricts the use of scintillation counting as a spectroscopic technique for β particles, but the resolution of energies is sufficiently good to allow double-labeling experiments to be carried out using two β emitters with significantly different energies (e.g., ^{3}H and ^{14}C). In contrast, spectroscopic applications of γ ray detectors are well developed and widely used. They are discussed in Sec. III.E and Chap. 7.

Traditionally the solid scintillators used to detect β particles were materials such as calcium fluoride and anthracene, but more recently plastics have been produced containing a variety of organic scintillators. These materials can be cast or machined to shapes, sizes, and thicknesses appropriate for the particular radiation detection task involved, and have replaced gas ionization detectors in some commercially available β counters. Such scintillation counters have a very much shorter dead-time (∿1 μsec) than either G-M counters or proportional counters. Liquid scintillators are discussed in Sec. III.D.

A common form of α counter has a screen of microcrystalline zinc sulfide (activated with silver) over the face of a PM tube. This phosphor is readily available, as it is used in the manufacture of cathode-ray tubes, and this type of counter is frequently laboratory-made. The source is placed very close to the phosphor, giving nearly 2π geometry, and actual efficiencies achieved in practice are usually 40-45%. Counters may employ 25 or 50 mm diameter PM tubes for counting sources on planchets, or ones as large as 125 mm for counting sources of finely powdered soil or plant ash of similar size [37]. The large pulses produced by such α-scintillation detectors allow the discriminator level to be set high enough to eliminate spurious pulses from the PM tube and associated circuitry, and very low background counting rates, typically a few counts per hour, can be achieved.

A key component of the scintillation counting system is the photomultiplier tube. When light photons hit the photocathode a number of electrons are emitted, and this photoelectric current is amplified by a factor of 10^{6}-10^{8} by the PM tube [21]. Thermionic emission of electrons from the photocathode or dynodes will give rise to unwanted pulses, and this can lead to an unacceptably high level of background noise or counts. Much of this consists of low energy pulses which can be rejected by the use of a discriminator or bias in the associated electronics. The noise can also be reduced by cooling the tube to minimize thermionic emissions, but this is not always possible. The best solution to this problem, now incorporated in almost all liquid scintillation counters, involves the use of two PM tubes linked through a coincidence circuit. This is shown schematically in Fig. 8. The coincidence circuit is designed so that an output pulse is produced only when pulses arrive simultaneously (or, in practice within a very short time interval, ∿1 μsec) from both PM tubes. Pulses due to random thermionic noise are unlikely to be produced simul-

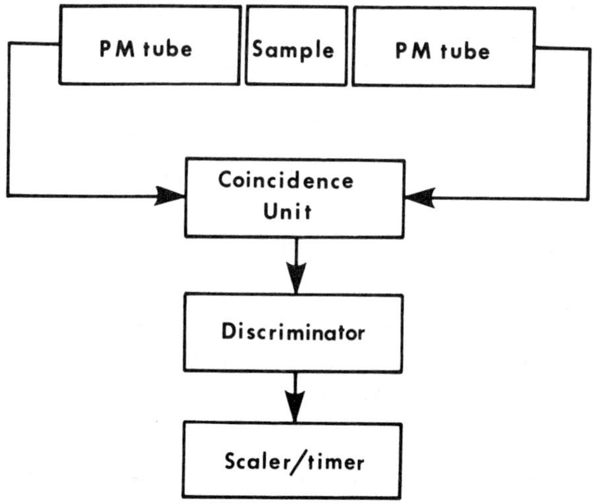

Figure 8 Coincidence counting system used in scintillation counters.

taneously by both PM tubes and thus nearly all are rejected. On the other hand, photon emissions caused by scintillations within the sample will arrive at both PM tubes simultaneously, giving rise to pulses in coincidence which are recorded.

D. Liquid Scintillation Counting

There are tow major reasons for the popularity of liquid scintillation counting in biological studies. First, the technique is ideal for counting the low-energy β radiation emitted by isotopes such as ^{3}H, ^{14}C. and ^{35}S which are used extensively in such studies. Indeed, work with ^{3}H would be almost impossible without this technique. Second, liquid scintillation counting lends itself to automation, and highly sophisticated counters are available which minimize the tedium of manual operation and the need for technical understanding by the operator. More detailed information on the technique than can be included here is available in a number of books [19,20,38-42].

In general terms the liquid scintillation source is a sample consisting of radioactive material, the organic phosphor, and a solvent, contained in a vial. The light flashes produced by the sample are detected by a PM tube and counted in the usual way. The intimacy of the sample and the detector (i.e., the organic phosphor) minimizes absorption problems and maximizes counting efficiency. Although the system is simple in essence, errors and problems can arise. The solvents used must be able to dissolve the organic phosphor and preferably the radioactive material also (although this is not absolutely essential). In addition, the solvents play an important role in the transfer of energy within the sample. They absorb the energy of the β radiation and then transfer it ot the organic phosphor [43]. Solvents which have been found to fulfil these criteria are aromatic compounds such as xylene, toluene, and, to a lesser extent, benzene.

A number of organic scintillants, known as primary solutes or fluors, are used; all are fluorescent, polyaromatic compounds. These molecules are excited by transfer of energy from the excited solvent molecules, and then give out this energy as photons of light by fluorescing (or *scintillating*). Commonly used primary solutes are 2,5-diphenyl oxazole (PPO) and 2-(4-biphenyl)-5-phenyl-1,3,4-oxadiazole (PBD):

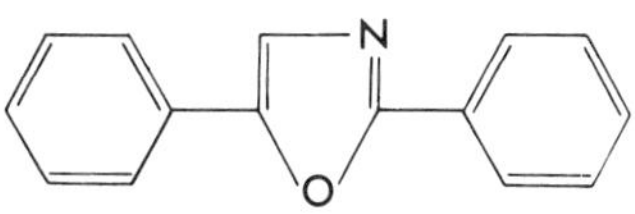

PPO

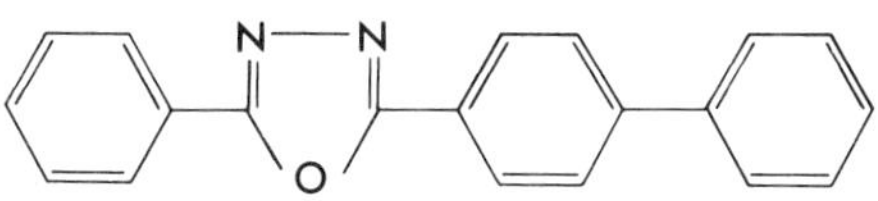

PBD

In some cases it is advantageous to include a secondary solute which absorbs the light emitted from the primary solute and re-emits it at a longer wavelength. This was originally done to produce a better response in the photomultiplier tube, but the greater sensitivity of modern photocathodes at shorter wavelengths now often makes this unnecessary. A secondary solute is useful, however, in other circumstances, e.g., when the scintillator vial is only partially transparent to the primary solute fluorescence [44]. A compound often used for this purpose is 1,4-bis-(5-phenyloxazol-2yl) benzene (POPOP):

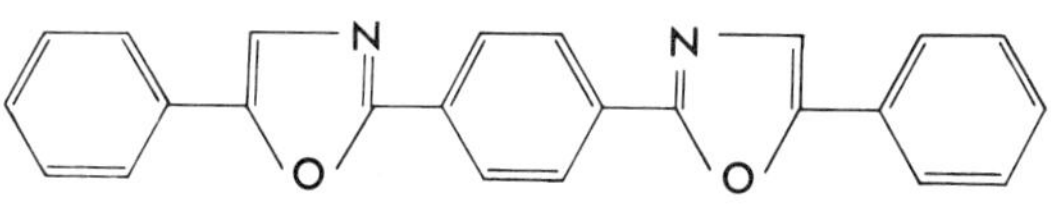

It is common practice to buy from manufacturers ready prepared "cocktails" of solvent plus primary and secondary solute suitable for the sample and counting system to be employed [39].

Many of the samples obtained from soil studies are either aqueous solutions or inorganic precipitates, neither of which dissolves readily in solvents such as toluene or xylene. One way of solving this problem is to incorporate a polar solvent such as dioxane, methanol, or ethanol into the cocktail [45]. Such mixtures are miscible with small amounts of water and can dissolve soluble inorganic salts. Another approach is to form a stable emulsion by incorporating a nonpolar surfactant into the liquid scintillant [46-49]. Methods of sample preparation in which up to an equal volume of aqueous solution can be incorporated with the scintillant are fully discussed by Fox [42].

Finely divided solid materials, such as barium carbonate or the material from a thin-layer chromatographic plate, can also be counted by incorporating a thixotropic gelling agent such as Cab-o-Sil (a finely divided silica gel) into the liquid scintillant [50,51]. This mixture is liquid when shaken or stirred but quickly sets to a firm gel on standing, thus giving a uniform distribution of the solid particles. In another widely used technique the insoluble sample is held on a solid support such as a glass fiber disk [42].

Whichever of these procedures is adopted, the final sample is placed in a vial to be counted. The vials are usually made of glass, which causes no problems at normal count rates. If, however, low count rates are being determined, the ^{40}K in the glass may give rise to an inconveniently high background count. To overcome this it is necessary to use either low-potassium glass or, more conveniently, polythene or Teflon vials, as long as these are not adversely affected by the solvents in the sample [52].

Apart from electronic considerations (discussed in Sec. III.C) the greatest problem in present-day liquid scintillation counting arises from a phenomenon known as quenching (this should be distinguished from the "quenching" of a G-M tube; see Sec. III.B). The possible causes of quenching are many, but a common one is that the sample to be counted contains compounds which absorb light and/or produce discolorations. This leads to a decrease in counting efficiency which can vary with each sample, and it becomes necessary to refer all samples to a standard state in order to obtain the true count rate. A number of methods can be used to overcome or correct for quenching [40,42,53]. For instance, the colored material can be removed by treatment with charcoal or some other adsorbent, but this is laborious, and not always successful. Another procedure is the *internal standard method*, in which the sample is counted, then a known amount of the same radioisotope is added and the sample recounted. Comparison of the count rate due to the standard added to the sample with the known unquenched count rate will give the loss of efficiency due to quenching. Neither of these methods readily lends itself to the high degree of automation found in modern liquid scintillation counters and other, more convenient techniques have been developed for this purpose.

In the *external standard method* [54,55] the sample is counted, then an external source is placed alongside the sample, which is then recounted. The external standard is usually a γ-ray source (e.g., ^{137}Cs) which generates Compton recoil electrons within the counting vial (see Sec. III.E) and these interact with the liquid scintillant in

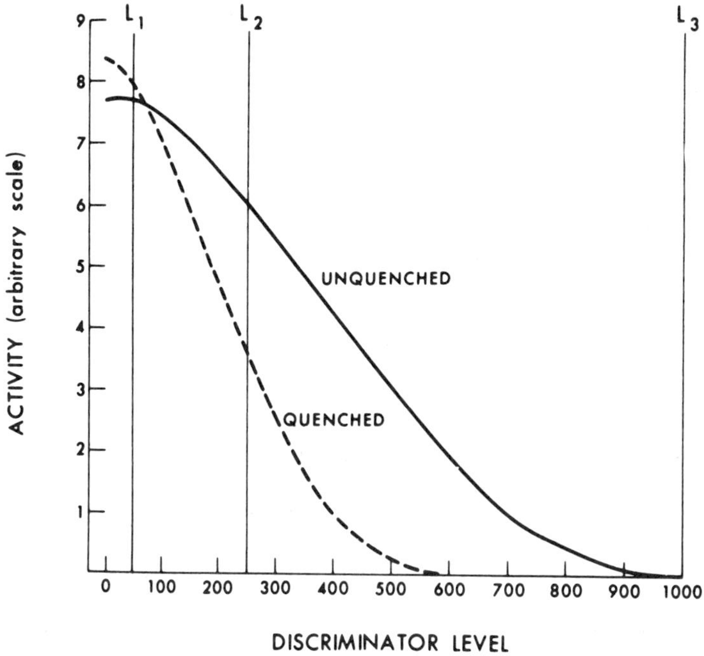

Figure 9 Unquenched and quenched pulse-height spectra for ^{14}C. (From Ref. 53.)

the normal way. Once again, comparison of the standard count rate against a calibration curve will enable the degree of quenching in the sample to be determined.

The method of choice for many workers is the *channels ratio* technique [56-58]. If a plot is made of counts against discriminator setting for β radiation, a curve of the type shown in Fig. 9 is obtained. It can be seen that while quenching reduces the overall count rate, the effect is greater on high-energy pulses. Thus, if counting is carried out in two channels (e.g., channel 1, L_1 to L_2, and channel 2, L_2 to L_3) then the ratio of counts in channel 1 to channel 2 will change as quenching occurs. By comparison with ratio values obtained for known levels of quenching the count rate can be corrected. The channels ratio count can be performed at the same time as the sample count and the equipment programmed to correct the observed sample count. The system is therefore ideal for use with automated equipment.

Erroneously high count rates can be caused by chemiluminescence of the sample. Cooling, addition of various compounds, and storage prior to counting are all employed to overcome this problem [10].

The current generation of microprocessor-controlled counters also includes features such as library programs for different counting requirements, automatic calibration, and on-line computing to produce information in the final form required.

Cerenkov Counting

High-energy β emitters dissolved in water have been found to emit light pulses even with no scintillant in the system. The light is known as Cerenkov radiation and arises when β particles travel at a velocity greater than that of light in the liquid medium [59]. The electron is found to lose some of its energy in the form of photons emitted at an angle to the direction of travel of the particle [5,60]. The mechanism is analogous to the production of a shock wave when an object travels at greater than the speed of sound in air. Cerenkov radiation is a threshold phenomenon and β-particle energies have to be greater than a certain value (related to the refractive index of the liquid medium) before light emissions occur. For water, E_{max} must be > 0.26 MeV but, in practice, worthwhile efficiencies are not achieved until $E_{max} \geqslant 1$ MeV. Even so, several isotopes which are very useful in soils work, e.g., ^{32}P, ^{24}Na, ^{42}K, and ^{86}Rb, can be counted by this method [61-64]. Although the number of photons emitted in this process is relatively small, the light pulses are easily detected by standard liquid scintillation counters.

The method has several advantages. Little or no sample preparation is required; the whole sample volume can be occupied by radioactive solutions, thereby decreasing the specific activity required; the sample is unadulterated by addition of solvent and scintillant and can be used for other purposes after counting. Also, the cost of conventional liquid scintillant is avoided, although small amounts of fluorescent compounds can be added to improve the counting efficiency [65]. The problems of quenching, standardization, etc., are much the same as those outlined for normal liquid scintillation counting and are overcome in similar ways. Since the emitted photons are highly directional they may only activate one PM tube and be discarded by the coincidence unit. This can be avoided by switching out the coincidence unit, but this practice can sometimes result in unacceptably high backgrounds.

Other Uses for Liquid Scintillation Counters

Having invested in an automatic liquid scintillation counter it is desirable to use it as effectively as possible. The system is ideal for counting α particles and, because of intimate mixing in the sample, very good efficiencies can be obtained. For γ radiation,

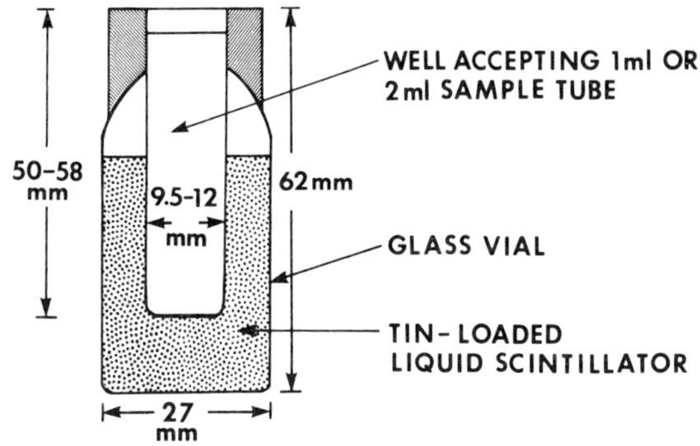

Figure 10 Vial (Koch-Light type) containing metal-loaded scintillator, for gamma counting in liquid scintillation counter. (Courtesy of Koch-Light Laboratories Ltd., Slough, Berkshire, England.)

however, the efficiency of the system is very low since most γ rays pass through the sample without interacting. Improved counting efficiencies can be obtained for γ rays by incorporating organometallic compounds (e.g., of Sn, Tl, Pb) in the liquid scintillant which improve its stopping power [10,66]. Such scintillant liquids are sufficiently expensive to deter the common practice of discarding the liquid after use. This can be overcome by using a vial with a central well for the sample, which is thus kept separate from the scintillant (Fig. 10) [67]. In this way the scintillant liquid can be used indefinitely. It should be pointed out, however, that the counting position of most automatic liquid scintillation counters is designed for use with β radiation. It is not adequately shielded for accurate use as a γ counter, thus an unacceptably high component of the count rate may be contributed by other sources in the autochanger of the counter.

E. Gamma-Scintillation Counting and Spectrometry

The effectiveness with which γ rays are absorbed increases rapidly with increasing atomic number of the absorbing material. A widely used type of γ detector consists of a large single crystal of sodium iodide (for iodine, Z = 53), activated with a low concentration of thallium, encased in a thin aluminum lightproof can and mounted in an integral assembly on a PM tube. Crystal sizes commonly range from 25 mm diameter × 25 mm high to 100 × 100 mm or even larger. Some versions of the detector have a re-entrant well cut into the crystal into which the source may be lowered to improve counting geometry (Fig. 11).

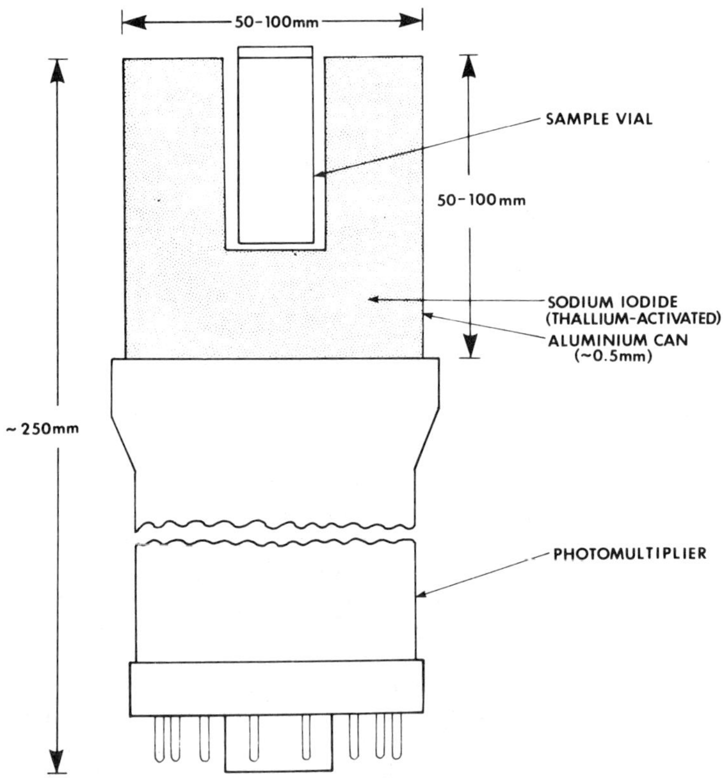

Figure 11 Sodium iodide (thallium-activated) well crystal-photomultiplier assembly, for gamma-scintillation counting.

When γ rays enter the crystal, they lose part or the whole of their energy by transfer to electrons. The size of the output pulse is proportional to the γ energy absorbed. In many studies the only requirement is to count the single γ-emitting tracer isotope present. For this purpose it is sufficient to record, within a single energy channel, all the pulses corresponding to total absorption, i.e., those in the photopeak. This is achieved by setting upper and lower discriminator levels to leave a "window" of the desired width. A conventional electronic scaler is used to record the events.

Large automated γ counters in current use normally include a number of preset channels for counting particular isotopes: mainly those used widely in medical research, e.g., ^{125}I. However, one or more channels may be adjusted to any desired energy. Crystal sizes are normally 50 × 50 mm or 75 × 75 mm, with wells up to 25 mm in diameter. The sample conveyors can accommodate several hundred samples, and because of the potentially high background activity caused by a large number of samples on the conveyor, heavy lead shielding of the crystal is normal. Typically, the crystal is surrounded by at least 75 mm of lead, with up to 200 mm on the side adjacent to the conveyor. As with liquid scintillation counters, microprocessor control of counting and data output is becoming a standard feature of the latest instruments.

Each γ-emitting radioisotope emits radiation of a characteristic energy. By determining the energy of the radiation it is usually possible to make an unequivocal determination of the identity of the isotope; furthermore, by separately recording radiation of different energies it is possible to determine quantitatively the separate activities present in a mixture. These principles are embodied in the *gamma spectrometer*, in which a scintillation detector or semiconductor detector (see Sec. III.F) is coupled to a pulse-height analyzer. The detector emits a pulse which is proportional to the γ energy absorbed, and this is converted from analog to digital form by a conversion circuit and recorded in the corresponding energy channel of the analyzer. The γ rays which are completely absorbed give rise to the photopeak, and those only partially absorbed to the Compton continuum (Fig. 12). In addition, γ rays in excess of 1.02 MeV in energy may result in *pair production*, i.e., an electron-positron pair is created: the reverse of the annihilation process described

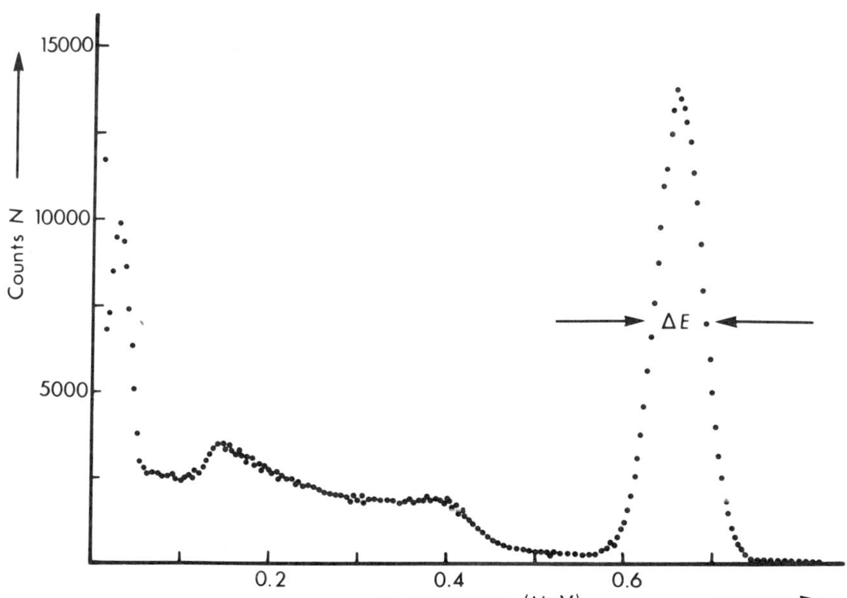

Figure 12 Gamma-ray spectrum, showing photopeak and Compton continuum. (From Ref. 5).

in Sec. II above. The positrons will inevitably be annihilated in turn by collision with other electrons, resulting in the production of 0.51 MeV annihilation radiation. The probability of this process occurring increases with increasing γ energy, and 0.51 MeV peaks are visible, above the Compton continuum, in the spectrum of the more energetic γ emitters.

The activity of a source is normally determined from the counting rate observed in the photopeak. Two or more γ emitters may be determined in the presence of each other if the photopeaks are adequately resolved. The resolution of detectors is usually expressed as a ratio of the full width of the ^{137}Cs photopeak at half the maximum height (or FWHM; shown as ΔE in Fig. 12) to the energy of the radiation causing the peak (667 keV). Thus if the FWHM for a scintillation detector = 50 keV, the resolution is $(50/667) \times 100 = 7.5\%$. Spectrometric applications of sodium iodide scintillation detectors are now primarily restricted to measurements involving only one or two radioisotopes. The applications of both scintillation and semiconductor detectors in γ spectrometry are discussed in detail by Crouthamel et al. [21].

F. Semiconductor Detectors

Semiconductor radiation detectors can be regarded as the solid-state equivalents of gas ionization detectors. When charged particles or γ rays lose energy by interactions with the detector material, electrons are raised in energy from the valence band to the conduction band. Under the influence of an applied electric field these electrons move through the material, resulting in a pulse [10,21].

Compared with gas detectors, semiconductor devices have certain advantages. First, the fact that they are solid means that they have a far greater stopping power than a gas-filled detector. Second, excellent energy resolution is obtainable, which may be as much as 10 times better than that of a gas ionization chamber or 20 times that of a sodium iodide scintillation detector [10]. Third, the ability to make semiconductor detectors in extremely small sizes greatly increases their possible applications as probes used in biological studies in vivo.

Composition of Semiconductor Detectors

The presence of pentavalent atoms such as phosphorus, arsenic, and antimony within the crystal lattice of tetravalent elements such as silicon or germanium results in surplus valence electrons which cannot take part in bonding, and thus imparts semiconducting properties to what would otherwise be an insulating material. Materials with such an excess of electrons are known as n type, to signify they contain carriers of negative charge. Trivalent elements such as boron, aluminum, and gallium result in a corresponding deficiency of valence electrons in the crystal, i.e., an "electron hole" or a positive charge. This hole can be filled by an electron from an adjacent atom, resulting in migration of the hole. Such semiconductors are known as p-type materials.

One type of detector consists of layers of p- and n-type materials with a voltage applied so that they are connected to negative and positive poles, respectively. Under this *reverse bias* potential the holes migrate toward the negative pole and the electrons to the positive, thus creating a depletion zone in the center. Radiation entering this zone induces the formation of electron-hole pairs (Fig. 13) which migrate to the poles, resulting in a current pulse whose magnitude is proportional to the energy of the radiation. Such a device is known as a noncompensated surface barrier detector, and has a depletion zone up to a few millimeters in width, thus making it suitable for counting charged particles or low-energy X rays.

Detectors with much greater depletion zone widths, which are thus suitable for use as γ detectors, are produced by diffusing lithium into germanium or silicon [21]. Lithium acts as an electron donor, and compensates for the holes caused by impurities in the material, thus raising its resistivity. Because of their high resolution, the Ge(Li) devices formed in this way are now widely used for γ spectrometry. This makes it possible to measure complex mixtures of γ emitters without prior separation (see Chap. 7).

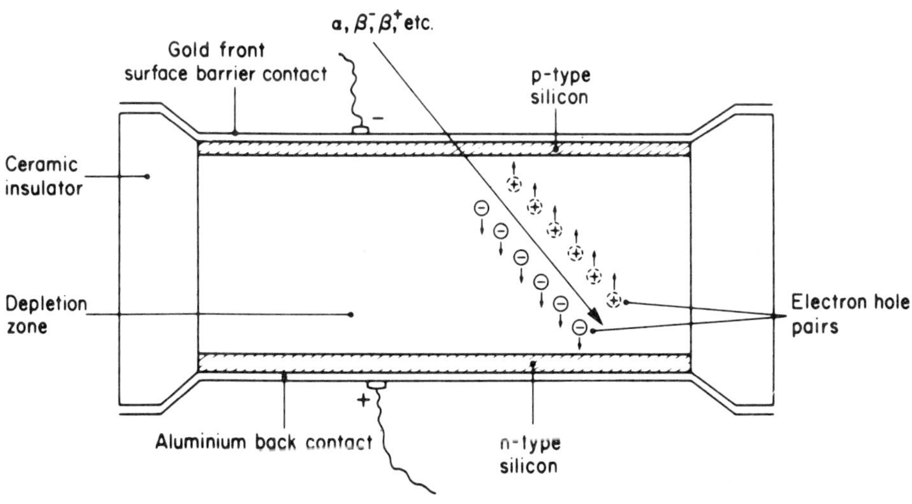

Figure 13 Cross-section of p-n junction semiconductor detector. [From Ref. 10, with permission from M. F. L'Annunziata, *Radiotracers in Agricultural Chemistry*, 1979, p. 67. Copyright by Academic Press, Inc. (London) Ltd.]

Alpha Spectrometry

Unequivocal identification of α-emitting isotopes can normally be made by determining the energy of the particles. The introduction of solid-state detectors has resulted in greatly improved resolution of α spectra compared with that obtainable with ionization chambers. However, good resolving power of the detector is useless if the energy spectrum is badly degraded by contamination of the source, and this is frequently the limiting factor in studies concerned with the analysis of radioisotopes present in matrices such as soil and plant ash. The presence of even a few micrograms of material on the source results in particles emerging with a range of energies from the maximum possible virtually to zero. Sources for α spectrometry, therefore, must be formed from carrier-free material after careful radiochemical purification to remove all extraneous matter. Procedures for this include electrochemical deposition, electroplating, and evaporation of solutions onto metal disks [34,68].

G. Autoradiography

Ionizing radiation will interact with and expose the silver halide in a photographic emulsion so that on development the areas subjected to the radiation show up as blackened. Thus a sheet of photographic paper can be used to give a facsimile, or an *autoradiograph*, of the distribution of radioactivity in a particular object.

Autoradiography is most suitable for use with low-energy β emitters, such as ^{3}H, ^{14}C, ^{33}P, ^{35}S, ^{45}Ca, etc., which produce well-defined images because of the short path length of the particles. High-energy β particles, with longer path lengths, and γ rays give much more diffuse images which are of limited use. Full details may be found in the book and review by Rogers [69,70].

Autoradiography has been used extensively in the study of nutrient uptake from soil by plant roots and to study the distribution of nutrients in plants after uptake [71-79]. Another important use is with paper and thin-layer chromatograms. Here it is used to monitor the distribution of activity (usually of ^{14}C) in particular compounds, following synthesis or degradation or some other biochemical process.

While autoradiography is essentially a qualitative technique, quantitative measurements have been made using ^{33}P and a microdensitometer to scan the film [74]. When quantitative results are required, a more common method is to recover the isotope from the sample and count it by one of the methods referred to in previous sections. The autoradiograph can be invaluable in pinpointing the areas of sample to be assayed.

An interesting combination of two techniques is shown in *fluorography*, in which the specimen for autoradiography is impregnated with scintillator, and the photographic film is affected by the light produced by the scintillator as well as by the direct action of the radiation. Fluorography is especially effective with tritium, because the low energy of the β particles from this isotope results in very poor penetration of the photographic emulsion. However, there is little to be gained by applying the technique to isotopes which emit more energetic radiation [10,69].

IV. RADIOCHEMICAL METHODS

A. Inorganic Separation Procedures

The manipulations which have to be performed to prepare a radioisotope in a form suitable for counting vary enormously in complexity. At one end of the scale, a direct nondestructive determination may be carried out on the original material. This is possible not only for γ emitters but also in some circumstances for high-energy β emitters, e.g., in the determination of ^{32}P in plant material [80-83]. Again, in studies of isotopic exchange between soil and solution (see Sec. V.B), the activity remaining in solution may be determined directly, after separation from the soil by centrifugation. Frequently, however, it is necessary to employ chemical separation methods to obtain the radioisotope in a sufficiently pure state. The extensive literature describing these methods has been reviewed by a number of authors [2,9,11-15]. Combustion of plant material and organic matter in soil has been used extensively to separate $^{14}CO_2$ and ^3H_2O [10,42], and automated equipment has been developed to transfer large numbers of samples into vials for liquid scintillation counting.

Precipitation Methods

Where chemical separation procedures are required, it is necessary to consider whether separations involving the use of isotopic carriers can be employed or whether carrier-free methods are necessary. If the former situation applies, precipitation of a radioisotope along with a macro-amount of the same (inactive) element is one of the commonest ways of separating the radioisotope from solution. Examples are the precipitation of ^{35}S as $BaSO_4$ and ^{45}Ca as calcium oxalate [23]. In situations where insufficient element is initially present, a quantity usually in the range 10-100 mg is added to the sample. To achieve satisfactory carrying of the radioisotope, it must first be uniformly mixed with the carrier in solution. Elements of variable valence, e.g., iodine and manganese, should also be taken through an oxidation-reduction cycle, to ensure that both active and inactive forms are in the same valence state.

When considering the effectiveness of a particular separation it is necessary to distinguish between *chemical* and *radiochemical* purity.

A precipitate of, say, 50 mg of barium sulfate containing 100 Bq (2.7 nCi) of ^{35}S would be radiochemically very impure if it contained half this activity of ^{32}P as a contaminant (particularly as the latter would be counted with much greater efficiency); yet the weight of phosphorus this activity represents is only 5×10^{-12} mg.

It is often difficult to achieve simultaneously both adequate chemical and radiochemical purity *and* a quantitative recovery. It is more desirable to achieve radiochemical purity with a reasonable yield, say in excess of 50%, and subsequently to determine that yield by a suitable analytical procedure, then to achieve 100% recovery and also carry down unwanted activities in the precipitate. Contamination of a precipitate can often be greatly reduced by the addition to the solution of milligram quantities of the inactive element isotopic with the unwanted activity, as a "holdback" carrier. The diluting effect of this addition greatly reduces the amount of unwanted activity which is coprecipitated.

Nonisotopic carrying, i.e., coprecipitation with ions of other elements, may be employed with carrier-free material. Several different processes may be employed: isomorphous replacement, anomalous mixed crystal formation, surface adsorption on the precipitate or internal adsorption, i.e., adsorption during the growth of crystals which traps the carried ions within the crystals [2]. These carrying reactions may also be employed to remove carrier-free impurities from a solution containing the wanted activity in the presence of its isotopic carrier. This process has become known as "scavenging," and a common example is the precipitation of iron (III) hydroxide to remove the ^{90}Y and ^{140}La daughter products of ^{90}Sr and ^{140}Ba, respectively [84].

Ion Exchange

This technique has the advantage of being applicable to both carrier-free and carrier-borne radioisotopes. It has been very widely used in radiochemical analysis, particularly for separations of complex mixtures of radioisotopes.

The two procedures used most frequently are ion exchange elution chromatography and selective adsorption [85]. In the former, a mixture of different ionic species is adsorbed on a column, and the individual species are then sequentially eluted with one or more reagents. In the latter, conditions are created so that one species is much more strongly adsorbed than the others. For detailed discussion of the theory of ion exchange processes and their applications in analytical separations, the reader is referred to books by Helfferich [86], Samuelson [87] and Rieman and Walton [88].

The great majority of ion exchange separations employ the synthetic cation and anion exchange resins based on polymers of styrene and divinylbenzene. Strong-acid cation exchangers (e.g.,

Dowex 50) contain sulphonic acid functional groups, and strong-base anion exchangers (e.g., Dowex-1) contain quaternary ammonium groups. In general, for solutions under 0.1 M the affinity for cation exchanger increases with charge on the cation and decreases with increasing size of the hydrated cation. This results in the following order of affinities [2]:

For cations,

$$Ba^{2+} > Sr^{2+} > Ca^{2+} > Mg^{2+}$$

$$Mg^{2+} > Mn^{2+} > Cd^{2+}$$

$$Cs^{+} > Rb^{+} > NH_4^{+} > K^{+} > Na^{+} > H^{+} > Li^{+}$$

For polyvalent anions, on a quaternary ammonium resin,

citrate > sulfate > oxalate

and for univalent anions,

$$I^{-} > HSO_4^{-} > NO_3^{-} > Br^{-} > Cl^{-} > HCO_3^{-} > H_2PO_4^{-} > OH^{-}.$$

Some ions which are only separated from each other with difficulty by elution from cation exchange columns may be much more readily separated by anion exchange in strong acid solution. For example, Fe^{3+} and Co^{2+} form anionic chloro-complexes in strong HCl, but Ni^{2+} does not and is not adsorbed [89].

The use of complexing agents often greatly improves the separation which can be achieved between two very similar ions, because the ion with the greater affinity for the resin also has the lesser tendency to form complex ions. Thus Ba^{2+} may be separated from Ra^{2+} by eluting the former ion from a cation exchange column with 0.01 M EDTA at pH 9 [68].

Some ions may be selectively adsorbed from solution by the use of chelating resins, which have a particular affinity for ions of a certain size or charge. One such resin—Chelex 100—has been employed in this way in the determination of copper in soils [90]. Certain insoluble inorganic substances act as ion exchangers, and these materials can also have a high selectivity for particular ions. A good example is ammonium molybdophosphate, which has been used to separate ^{137}Cs from solutions of environmental samples [34].

Solvent Extraction

Solvent extraction, or liquid-liquid distribution as it is now becoming known [91], shares with ion exchange the property of being applicable to carrier-free or macro-amounts of material. In its most common form, the process involves the extraction of a metal chelate

Table 2 Some Solvent Extraction Procedures Applicable to Radiochemical Analysis

Element (radioisotope)	Complexing agent	Solvent
Cobalt (^{58}Co or ^{60}Co)	2-Nitroso-1-naphthol	Acetone + isoamyl alcohol
Copper (^{64}Cu)	Dithizone	Carbon tetrachloride
Iron (^{59}Fe)	Cupferron	Chloroform
	Chloride (6 M HCl)	Diethyl ether
Manganese (^{54}Mn or ^{56}Mn)	Diethyldithiocarbamate	Chloroform
	Tetraphenylarsonium chloride	1,2-Dichloroethane[a]
Molybdenum (^{99}Mo)	8-Hydroxyquinoline	Chloroform
Selenium (^{75}Se)	2,3-Diaminonaphthalene	Cyclohexane
Zinc (^{65}Zn)	Dithizone	Carbon tetrachloride

[a]As permanganate ion-association complex.
Source: Refs. 91 and 161.

from aqueous solution into an immiscible nonpolar solvent, thus achieving a satisfactory separation from interfering ions. The widespread analytical applications have resulted in a large output of papers and a number of books, e.g., Refs. 91 to 94.

According to Coomber [85], for satisfactory separation of a substance from others by solvent extraction the following points should be satisfied:

1. The solubility of the extracted compound should be high.
2. The separation factor should be as large as possible.
3. The solvent should be immiscible, or nearly immiscible, with the aqueous phase.
4. The specific gravity of the solvent should differ as much as possible from the aqueous phase, to encourage rapid separation and minimize the formation of emulsions.

If adequate recovery is not achieved with a single extraction, the process may be repeated as many times as is necessary, and the extracts combined. Repeated extraction with small aliquots of solvent is preferable to one extraction with a single large volume.

Analytical separations may be readily carried out in separating funnels or stoppered or capped test tubes and centrifuge tubes. When several successive extractions are needed, or when many samples are to be extracted, the use of commercially available continuous-extraction apparatus is desirable [2].

Metal chelates are not the only species which can be extracted into organic solvents. Other species include ion-association compounds formed by large organic ions such as tetraphenylarsonium; nitrate and chloride complexes (solvates) formed with neutral solvents such as tri-n-butyl phosphate (TBP); and complexes between anions such as $(FeCl_4)^-$ and long-chain amines [91]. Some examples of complexes and solvents suitable for solvent extraction in radiochemical analysis are given in Table 2.

B. Preparation of Sources

Solid Sources for β-counting

These may be prepared either by (1) transferring a precipitate as a slurry onto a planchet and evaporating the liquid under an infrared lamp, or (2) evaporating a solution in the same way, or (3) preparing a flat pad of precipitate by filtration under suction. These techniques are discussed in detail by Overman and Clark [2].

Two alternative ways of applying method (3) are illustrated in Fig. 14a and b. Figure 14a shows a demountable assembly from which the filter paper, together with its pad of precipitate, may be transferred to a planchet. Figure 14b shows a perforated planchet which serves as a mounting for the filter paper during filtration. This avoids the risk of damage to the pad during transfer.

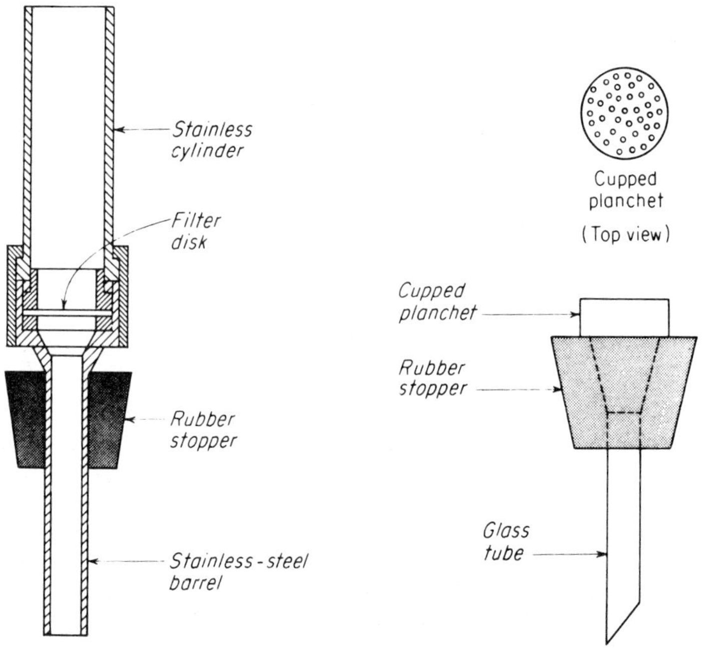

(a) (b)

Figure 14 Filtration devices for preparation of solid sources for beta counting. (From Ref. 2.)

Precipitates should be dried slowly, to avoid cracking. Curling of the filter paper, and crumbling or flaking (which results in lost activity or contamination of the counter) can be avoided by cementing with a dilute solution of polyvinyl acetate in methanol [84].

Other Sources

The lack of self-absorption problems in γ counting, and the consequent ability to count large samples, greatly simplifies source preparation. Either the original sample may be counted directly without processing, or, if concentration of the radioisotope is required to improve efficiency, a simplified procedure may be adopted such as adsorption of the radioisotope on ion-exchange material which is then transferred to a bottle and counted in a well-type sodium iodide scintillation detector [34]. Where a purification procedure is necessary, resulting in a solution of the radioisotope, this solution can be counted directly, thus eliminating the source preparation stages necessary for most β-counting techniques.

Reference has already been made in Sec. III to the preparation of sources for α spectrometry and liquid scintillation counting, and will not be discussed further here.

C. Isotope Dilution Analysis

Standard Procedure

In radioisotope dilution analysis, the quantity of an element or compound present in a sample is determined by adding a known amount of labeled element or compound. A procedure is followed which ensures uniform mixing of labeled and unlabeled material. Then, after a suitable separation, which need not be quantitative but which gives the element or compound in a pure form, the change in specific activity is determined. Thus, if the specific activities before addition to the samples and after recovery are S_a and S_r, respectively,

$$S_r = S_a \frac{W_a}{W + W_a} \tag{9}$$

where W and W_a are the quantities of the element or compound initially present and added to the sample, respectively. Rearranging Eq. (9) gives

$$W = W_a \left(\frac{S_a}{S_r} - 1 \right) \tag{10}$$

When the radioisotope added is carrier-free or if the quantity of inactive substance added is very much less than the quantity in the sample ($W_a << W$), then

$$W = \frac{A_a}{S_r} \tag{11}$$

where A_a is the activity added to the sample and S_r, as before, is the specific activity of the substance recovered from the sample.

The method may be applied to the determination of naturally occurring radioelements, as well as to inactive ones, provided that it is possible to distinguish between the radiation emitted from the radioisotope and that from the radioelement already present in the sample. For example, an artificial isotope of radium or polonium can be used to measure the recovery of another isotope of the same element because the radiations can be distinguished by α spectrometry [22,68]. Gamma-emitting isotopes may be determined in a similar manner by γ spectrometry [21]. In situations where the two sources of radiation cannot be easily distinguished, e.g., when both are β emitters of similar energies, it is possible to determine the two components by resolving the decay curve, provided that one half-life differs appreciably from the other. Applications of the method are in neutron activation analysis and the determination of fission products in environmental samples, but it is more common to apply the technique described in the next section.

With the exception of those heavy elements which have no stable isotopes, it is possible for the added isotopic yield tracer to be carrier-free or to contain added inactive element, as desired. For those separation procedures which work as well with carrier-free materials as with macroscopic amounts, such as solvent extraction and ion exchange, either approach may be employed. If precipitations are involved, milligram quantities of inactive carrier are required.

Reverse Isotope Dilution Analysis

Reverse isotope dilution analysis involves the addition of inactive element or compound to a sample containing the element or compound in labeled form. As the name implies, this is the reverse of the procedure described in the previous section.

A common application is in neutron activiation analysis (NAA), in which the activity in the sample is induced by irradiation by neutrons from a nuclear reactor or neutron source (see Chap. 7). A quantity W_a of carrier is added which greatly exceeds the quantity W in the sample. Typically, 10-50 mg may be added to a sample containing microgram or submicrogram amounts of the element. If the quantities of activity originally present in the sample, and recovered at the end of the chemical separation are A_s and A_r, respectively, and the weight of carrier recovered = W_r,

$$A_s = A_r \frac{W_a}{W_r} \tag{12}$$

The specific activity S_s of the element in the sample is determined by comparison with a standard irradiated under identical conditions. Thus W may be calculated, as

$$W = \frac{A_s}{S_s} \tag{13}$$

The neutron activation technique has been fully described in a number of books and reviews, e.g., Refs. 95 to 99 (see also the references in Chap. 7).

Chemical recovery may be determined either by using a radioactive yield tracer (with or without inactive carrier) or by using an inactive carrier alone. The choice depends very much on the availability of necessary instrumentation, and the variability in concentration of the element between samples. The yield of inactive carrier is normally measured by flame emission or atomic absorption spectrometry or a gravimetric method. The use of radioactive yield tracers may be preferred if a multichannel pulse height analyzer is available. Computer processing of the counting data may be employed to determine the individual components of the composite spectra obtained from the mixture of isotopes. However, when element concentrations vary widely, there is a problem of matching the quantity of added tracer to the induced activity, to prevent one activity being "swamped" by the other. In such situations, superior results may be achievable by determining the recovery of inactive carrier. The latter method is also more appropriate in laboratories which are well equipped for analysis of inactive elements, but which have only simple γ counters available.

Substoichiometric Isotope Dilution Analysis

The determinations of inactive elements by the procedures described in the two previous sections necessitate a chemical analysis as well as a measurement of radioactivity. This can cause difficulties if concentrations are very low. A technique known as substoichiometric isotope dilution analysis was pioneered by Růžička and Starý [100], as a means of eliminating the need for a chemical determination.

If the specific activity of the radioisotope added to the sample is

$$S_a = \frac{A_a}{W_a} \tag{14}$$

the specific activity after isotopic exchange with the element in the sample is

$$S = \frac{A_a}{W + W_a} \tag{15}$$

where W is the weight of element in the sample. By adding the same *substoichiometric* quantity of reagent required to bring about a precipitation of the element or to form a complex for solvent extraction, the same weight of element W_r ($< W_a$) can be separated from the radioisotope solution and the sample which has undergone isotope dilution. If the activities of these two fractions are A_r and A_x, respectively,

$$A_r = W_r S_a = W_r \frac{A_a}{W_a} \tag{16}$$

and

$$A_x = W_r S = W_r \frac{A_a}{W + W_a} \qquad (17)$$

Conbining these two equations,

$$A_x (W + W_a) = A_r W_a \qquad (18)$$

thus

$$W = W_a \left(\frac{A_r}{A_x} - 1 \right) \qquad (19)$$

i.e., the weight of element in the sample can be determined from the weight added and a ratio of two activity measurements.

D. Radiocarbon Dating

Radiocarbon dating has become a powerful tool for estimating the age of organic material in archaeological and geological studies [101], and is now being widely used for the dating of organic material found in soils.

The technique depends on the change in the specific activity of ^{14}C in the organic matter with time, due to the decay of the ^{14}C ($t_{1/2}$ = 5700 years). The isotope is produced in the upper atmosphere by the reaction $^{14}N(n,p)^{14}C$, the neutrons being generated by the interaction of cosmic rays with other atoms. The ^{14}C is oxidized to CO_2, moves into living tissue by photosynthesis and becomes incorporated into soil organic matter. When the organic material is no longer living, exchange with the pool of atmospheric CO_2 ceases and, in principle, a comparison of the specific activity of carbon from a soil sample to be dated with a contemporary sample of organic carbon will allow the age of the former to be calculated. In practice, complications arise because of changes in the specific activity of atmospheric carbon during the past few decades, due to combustion of fossil fuels and the extra production of ^{14}C as a result of the atmospheric testing of nuclear weapons.

Because of the low specific activity, the achievement of satisfactory measurements requires the counting of samples which contain as much carbon as possible, and the use of low-background counters. The organic carbon is combusted to CO_2 which may be counted directly, or after conversion to a hydrocarbon, in an anticoincidence proportional counter. Such gas counters may be operated at high pressure, to incorporate more sample, and the conversion of CO_2 to acetylene, C_2H_2, is an effective way of increasing the amount of carbon in the gas filling the counter [102].

The formation of acetylene is achieved by reaction of the CO_2 with lithium, to form lithium carbide, Li_2C_2, which is then reacted with water. By catalytic coversion of the acetylene to benzene and counting in a liquid scintillation counter, even more carbon can be counted than in a large high-pressure gas counter [103-105]. It is now considered that, if sufficient care is taken, the liquid scintillation method can compete successfully with any gas counting system now in operation [106].

E. Preparation and Analysis of Labeled Organic Substances

Preparation

Study of the synthesis and degradation of simple and complex organic molecules in the soil is greatly facilitated by tracer techniques using radio-actively labeled compounds. Simple compounds such as sugars, phenols, or amino acids, labeled either uniformly or in particular positions, may be obtained from commercial suppliers. If it is necessary to synthesize an otherwise unavailable ^{14}C-labeled compound, procedures may be found in Catch [107]. Unformly labeled plant material (mainly maize), produced by growing plants in a growth chamber in an atmosphere containing $^{14}CO_2$, has recently become obtainable from the Commissariat A L'Energie Atomique, Saint-Paul-lez Durance, France. The necessary techniques for operating growth chambers containing high specific activity $^{14}CO_2$ have been described by Sauerbeck and Fuhr [108] and Zeller et al. [109]. Plant material labeled with ^{14}C may also be prepared by growing plants in soil to which $NaH^{14}CO_3$ has been added [102], and Knight [110] has described solution culture methods for labeling with ^{59}Fe, ^{22}Na, ^{36}Cl, ^{65}Zn, and ^{45}Ca. Humic substances, soil polysaccharides, etc., are not commercially available and have to be prepared by researchers themselves.

Labeled plant material can be used directly in decomposition studies, as in Jenkinson's work [111-114], or allowed to decompose in soil to produce labeled organic matter. The specific activity of organic matter fractions extracted after this process tends to be rather low. Higher specific activity normally results from adding labeled organic compounds directly to the soil for incorporation by the microorganisms [115].

Since the radioisotope is present only in the newly formed material, the techniques described above result in nonuniform labeling of the soil organic matter fractions, even though the substance incorporated was itself uniformly labeled. An alternative approach has been to synthesize "model" polymeric substances, e.g., humic-like materials [102] or polysaccharides [116] from suitable labeled substrates. These materials do not behave sufficiently like their natural counterparts for studies of their reactions to be very valuable.

A more promising method is that known as tritium recoil labeling [117], in which the sample is mixed with a lithium salt and irradiated with neutrons in a nuclear reactor. The lithium undergoes the reaction

$$^{6}Li + n \rightarrow {}^{4}He + {}^{3}H$$

liberating "hot" tritium atoms which can react with organic molecules to replace normal hydrogen atoms, or react with multiple bonds. High specific activity may be achieved in this way, but the labeled materials require purification from the lithium salt and the degradation products resulting from radiation damage in the reactor.

Another approach involves isotopic exchange between $^{3}H_2$ and the more labile groups in the organic matter. This process is accelerated by supplying excitation energy in the form of an electrical discharge, or microwave or UV radiation. This work has been reviewed by Scharpenseel [102], who suggested that the technique could also be employed with $^{14}CO_2$. Tritium-hydrogen exchange can also be carried out in aqueous systems, using HTO and a catalyst, e.g., platinum, nickel, or aluminum chloride [118].

Analysis

The techniques most frequently employed for the separation and quantitative assay of labeled organic compounds have been recently discussed in detail by L'Annunziata [10] and will only be summarized here. Chromatographic methods, including paper and thin-layer, gel filtration, ion exchange and gas chromatography, and electrophoresis, have been used in this way. The separation procedures are the conventional ones used in nonradioactive work with the addition of an appropriate method of measuring the activity present.

The positions of radioactive compounds on a thin-layer chromatographic plate or a paper chromatogram may be determined either by autoradiography or by use of a scanner, which consists of a detector (under which the plate or paper moves) connected to a ratemeter and chart recorder. Alternatively, sections of a paper chromatogram, or the coating of a thin-layer plate, may be counted in a scintillation counter directly, or after digestion.

In liquid chromatography, including gel filtration and ion exchange methods, the effluent solution may either be collected in a fraction collector, followed by the determination of the activity in each fraction, or passed through a flow cell where the activity is continuously monitored with a suitable detector [119].

Gaseous or volatile substances may be assayed by gas radiochromatography. In a typical system, a gas chromatograph (see Chaps. 10 and 11) is modified to allow part of the gas stream leaving the column to pass through a combustion/reduction train in which the organic compounds are converted to CO_2 and H_2, and thence to a gas-flow proportional counter. The remaining part of the gas stream is monitored with a g.c. flame ionization detector. The outputs of the two detectors are normally recorded with a twin-pen chart recorder.

V. APPLICATIONS

A. Elemental Analysis of Soils, Plants, and Other Environmental Materials

Neutron Activation Methods

Applications of neutron activation analysis (NAA) to the determination of various elements in soils, fertilizers, and plant materials were first reported in the 1950s , and the literature up to the mid-1960s has been reviewed by Bowen and Gibbons [95] and Bowen [96]. References to this early work cited here have, in part, been abstracted from the latter review.

Soils and Soil Extracts. Determinations have been made of the essential trace elements Co, Cu, Mo, Ni, and Zn [120] and the rare earths Eu, La, and Sm [121-123] in Japanese soils, and of potassium in 19 soils from Taiwan [124]. Certain elements of concern as environmental pollutants have been determined at very low concentrations. For example, the limit of detection for Hg was 10^{-10} g in 100 mg soil samples, using a procedure involving solvent extraction and precipitation (as sulfide) of the ^{197}Hg activation product [125]. The limit for Cd achieved by Steinnes [126] was 5×10^{-6} g g^{-1} after separation by anion exchange and γ-counting the ^{115m}In daughter product, the method being slightly less sensitive than atomic absorption for this element.

Kronborg and Steinnes [127] have reported a method for the simultaneous determination of As and Se. Soil samples were irradiated for 2 days at a thermal neutron flux of 1.5×10^{13} neutrons cm^{-2} sec^{-1}, and after 7 days' decay fused, together with carriers, with sodium hydroxide and sodium peroxide. The elements were subsequently precipitated with thioacetamide, and the activities of ^{75}As ($t_{1/2}$ = 26.4 hr) and ^{75}Se ($t_{1/2}$ = 120 days) measured with a Ge(Li) detector connected to a multichannel analyzer. An interesting feature of this work was the technique employed to determine the chemical yield. After the ^{76}As had decayed to a negligible level the precipitates were irradiated for 30 sec, and the recovery of carriers determined by counting the ^{81m}Se ($t_{1/2}$ = 57 min) and ^{76}As activities. Selenium has also been determined by neutron activation to ^{75}Se, by Zmijewska and Semkow [128], in soils, waste dumps, and biological materials. These workers used solvent extraction and distillation methods to separate the desired activity, and achieved a limit of detection of 2 μg.

NAA was first applied to ammonium acetate extracts of soil by Bowen and Dymond [129]. They precipitated Ca, Sr, and Ba from the extracts as oxalates, irradiated the precipitates and, after radiochemical purification, separately measured the induced ^{87m}Sr ($t_{1/2}$ = 2.8 hr) and ^{139}Ba ($t_{1/2}$ = 1.4 hr) activities. Smith [130] varied the technique by irradiating the dried extracts prior to any chemical separation and subsequently measuring the ^{87m}Sr and ^{139}Ba activities simultaneously by γ spectrometry.

Bowen and Cawse [131] separated the soil solution from moist soil by centrifugation and determined Cu, Mn, and Zn by activating the solution obtained. A similar method was employed by Brune and Landström [132] to determine As, Br, Cd, Cu, Hg, Mo, Na, Sb, Se, U, and W. The suitability of this technique for the determination of phosphorus in the soil solution, when the concentrations are too low for satisfactory colorimetric analysis, was pointed out by Bowen [96]. One such application has been in the analysis of flooded rice soils [133].

Separation of the exchangeable form of an element may be carried out *after* irradiation, thus eliminating blank determinations. For example, Kline et al. [134] measured the rate of exchange of ^{64}Cu between irradiated soil and copper sulfate solution. Exchange was virtually complete after 1 hr, but the amount of copper released varied with the strength of the solution. Bowen [96] has suggested that radioelements in irradiated soils do not behave in the same way as the natural elements because of Szilard-Chalmers effects, i.e., the rupture of chemical bonds due to the recoil energy of nuclei formed through (n, γ) reactions. It seems likely that results obtained in this way are less realistic than those based on extraction before irradiation.

Plant Materials. Most applications to plant analysis have been concerned with essential trace elements or unwanted toxic elements. Selenium has been determined in tomato juice by activation to ^{81}Se ($t_{1/2}$ = 18 min) and rapid separation [135]. The more commonly used method has been to irradiate for longer periods and to determine the 120 day ^{75}Se. This approach has been applied to cereal diets [136,137], kale [138], and forage grasses [139].

Molybdenum has been determined in clover by activation to both ^{99}Mo ($t_{1/2}$ = 66 hr) and ^{101}Mo ($t_{1/2}$ = 14.6 min), with a limit of detection of 0.1 μg [140]. The uptake of this element from nutrient solution, relative to tungsten, has also been investigated [141]. The levels of Cu, Mn, and Zn have been determined in the seeds of 10 grass species [142], and iodine which, like selenium, is essential to animals though not to plants, has been determined in standard kale and flour samples by activation to ^{128}I [143].

A number of nonessential elements have also been studied. The list includes Cr [144] and W [145] in tomato; Eu, La, and Sm in rice [146]; U in clover [147,148]; Sr in alfalfa [149] and clover and ryegrass [150]; Sr and Ba in wheat and brazil nuts [130,151], and Hg used in fungicidal seed dressings [132].

Isotope Dilution Methods

Several elements have been determined in soils and other environmental materials by radiochemical isotope dilution procedures. A typical study was that by Brown et al. [152] who used ^{76}As as yield tracer in the determination of arsenic in soils and rocks. Their procedure involved the separation of arsenic as arsine gas which was then absorbed in iodine-potassium iodide-sodium bicarbonate solution and determined colorimetrically. This inevitably resulted in variable losses due to leakage, and chemical yields (determined by γ-counting with a single-channel analyzer) varied from 40-90%. Branquinho and Robinson [153] have similarly used ^{212}Pb in the determination of lead in water and dust samples, in the course of an investigation of lead pollution in Rio de Janeiro, Brazil. This short-lived tracer ($t_{1/2}$ = 10.6 hr) is a decay product of the gaseous isotope ^{220}Rn, which itself results from the successive decays of ^{228}Th and ^{224}Ra. The ^{212}Pb was conveniently "milked" from a ^{228}Th emanation source by collection on a gold cathode, from which it was removed with concentrated HCl.

The γ-emitting isotope ^{85}Sr has proved useful in the determination of stable strontium in soils and plant materials. Even in samples containing a significant activity of ^{89}Sr and ^{90}Sr from environmental fallout, chemical yields have been determined by γ-spectrometric measurement of ^{85}Sr activity without interference from the other isotopes, which are pure β emitters. The stable strontium was determined by flame photometry [84].

The naturally occurring α-active radioisotopes of the Uranium and Thorium Series are difficult to determine quantitatively in complex matrices such as soils and plant ash, because of the need to remove all unwanted material prior to preparation of a source for α spectrometry. This makes estimation of chemical yield essential. Suitable yield tracers for the thorium and radium isotopes are ^{229}Th and ^{225}Ra, which do not occur in nature [154]; ^{209}Po has been produced, for use as a tracer for ^{210}Po, by bombardment of lead with 27 MeV ^{3}He nuclei in a cyclotron [22,155]. ^{225}Ra and ^{209}Po have been used in this way in analyses of soil extracts and plant materials [22,68,130,151].

Substoichiometric Analysis

Růžička et al. [156-161] have applied substoichiometric isotope dilution techniques to biological materials and also automated them by using a Technicon AutoAnalyzer. Thus traces of phosphorus at concentrations down to 0.01 $\mu g\ ml^{-1}$ may be determined by the formation of a Zr:P:Mo complex which is retained in aqueous solution while the excess phosphomolybdic acid is extracted into a butanol-chloroform mixture. The ^{32}P activity added to the solution is determined with a flow counter connected to a ratemeter and chart recorder. The method has been applied to the measurement of the specific activity of ^{32}P in plants fertilized with labeled phosphate [156]. Using ^{203}Hg as tracer, and automated procedure for mercury in biological materials has also been devised, in which the mercury is extracted with dithizone and the γ activity measured in a flow cell mounted in a 75×75 mm NaI(Tl) well-type detector [159-161].

Other applications include the determination of iron in soil solution [133] and plant material [162] by separating an Fe-EDTA complex labeled with ^{59}Fe from the remainder of the iron by cation exchange; the determination of fluoride in plant material and water using ^{18}F tracer, by solvent extraction into trimethylchlorosilane [163,164]; and the determination of chloride, sulfate, and carbonate, using precipitation methods [165,166].

Direct Counting Methods

The decay products of ^{238}U and ^{232}Th include γ emitters as well as α emitters, and these natural radioelements (as well as ^{40}K) may be determined directly in soil by γ spectrometry. Vassilaki et al. [24] used a spectrometer incorporating a 75×75 mm NaI(Tl) detector and a system of gain stabilization to count 3 kg samples. The ranges of activities observed in surface soil were 20-80 Bq kg^{-1} (0.5-2 pCi g^{-1}) for ^{226}Ra plus daughters, 4-60 Bq kg^{-1} (0.1-1.5 pCi g^{-1}) for ^{232}Th plus daughters and 20-400 Bq kg^{-1} (0.5-9.7 pCi g^{-1}) for ^{40}K.

More recently semiconductor detectors have been used for nondestructive counting. The low-background system employed by Quam et al. [167] incorporated a Ge(Li) detector surrounded by an NaI(Tl) annular detector in anticoincidence. Samples up to 350 g in weight were counted, giving limits of detection of 0.03 and 0.09 $\mu g\ g^{-1}$ for U and Th, respectively, and 1.1 Bq (30 pCi) for ^{40}K. A limit of detection of 1 $\mu g\ g^{-1}$ for U in soil was obtained by Coles et al. [168] by measuring the 63 keV peak due to ^{234}Th, compared with 35 $\mu g\ g^{-1}$ when the 1.00 MeV peak of ^{234}U was measured. The latter procedure has also been applied to the determination of uranium in phosphate fertilizers [169].

The potassium content of potassic basic slag fertilizer (a mixture of potassium chloride and basic slag) has been monitored by counting the β activity of ^{40}K with four G-M counters arranged in parallel [170].

B. Measurement of Labile Ions

Phosphorus

The importance of phosphorus in plant nutrition has resulted in many studies of labile soil phosphorus, i.e., that fraction of the total soil phosphorus which can enter the soil solution by isionic exchange within an appropriate time span, e.g., one cropping season [171,172]. Various approaches have been used by different workers, but all are based on the principle of isotope dilution analysis.

When a surface containing ions is in contact with a solution containing isotopically labeled but chemically identical ions, exchange takes place between the two phases. If the inactive and active forms of the ion are M and M*, respectively, at equilibrium

$$\frac{[M^*]_{\text{surface}}}{[M^*]_{\text{solution}}} = \frac{[M]_{\text{surface}}}{[M]_{\text{solution}}} \tag{20}$$

This principle was first applied as long ago as 1922, in an investigation of the amount of isotopically exchangeable lead in insoluble lead salts [173]. It was first applied to soil phosphorus by McAuliffe et al. [174], who equilibrated carrier-free ^{32}P (as phosphate) in solution with the inactive ^{32}P-phosphate of the soil. At equilibrium

$$\frac{[^{32}P]_{\text{surface}}}{[^{32}P]_{\text{solution}}} = \frac{[^{31}P]_{\text{surface}}}{[^{31}P]_{\text{solution}}} \tag{21}$$

The ^{31}P-phosphate in solution was experimentally determined, as was the ratio of ^{32}P activities, thus the surface ^{31}P-phosphate, and hence the labile ^{31}P initially in the soil, could be calculated.

Any isotopic-exchange reaction follows first-order reaction kinetics. If the exchange is only with one soil form of phosphate, a logarithmic relationship will be obtained between the disappearance of the isotope from solution and time. It has been found, in practice, that the rate is usually a composite of a number of simultaneous reactions. Some of these reactions are slow, and complete isotopic equilibrium is not normally attained in a laboratory experiment, but after a certain period the rate of isotopic dilution decreases drastically. Thus an arbitrary time t is chosen for shaking, and the quantity of labile P which exchanges in this time has become known as the *E value*, or E_t. Russell et al. [171] showed that the initial rapid exchange was nearly complete at 48 hr but not at 4 hr, and Olsen and Dean [175] recommended 24 hr as a suitable shaking period.

It is not necessary to use carrier-free ^{32}P in the solution undergoing exchange wiith the soil. E values have been found to be relatively independent of the amount of phosphate added in the equilibrating solution, because phosphate sorbed by the soil remains exchangeable [171,176,177]. The experimental procedure is normally straightforward, usually involving counting the ^{32}P activity remaining in solution, either in a solution counter of the type shown in

Fig. 6c [171], or by evaporating an aliquot to dryness on a planchet for end-window β counting [175], or by Cerenkov counting in a liquid scintillation counter [64], and determining the ^{31}P (inactive P) in solution colorimetrically. A complication occurs in acid soils, for which it has been suggested that E_t values may be underestimated due to the presence of labeled complexes of phosphorus and aluminum in solution, which are detected by counting but not by conventional orthophosphate analysis [64,178]. According to Larsen [172], this complex formation is lowest in only slightly acid soils where the $H_2PO_4^-$ ions predominate and where problems of this type do not arise, variations in results can occur due to changes in temperature, in t or in the soil:solution ratio [31,179,180]. The satisfactory determination of the radioisotope is generally the least of the problems.

Larsen [181] and Gunnarsson and Fredriksson [182] determined the labile P in soils by growing plants in pots of soil to which ^{32}P-labeled phosphate fertilizer had been added, and comparing the specific activity of the phosphorus in the plants with that of the fertilizer. The labile P (the *L value*) was calculated using the standard isotope dilution equation [Eq. (10)]. The L value was found to be independent of the amount of P applied and also became independent of time, suggesting that isotopic equilibrium was attained. Russell et al. [176] found that the attainment of equilibrium was facilitated by very thorough mixing of the soil with the labeled P. If equilibrium is *not* attained the value is not a true L value, although it can provide an index of "available" P in the soil [172].

Identical E and L values are not normally obtained for a given soil because of the different environments in which isotopic exchange occurs. The E value refers to a soil suspension from which no phosphorus is removed, whereas the L value refers to a soil at or below field capacity in which plants are grown and from which some phosphorus is removed [172]. The E value can be easily and quickly measured, but the L value is more relevant to plant uptake.

Although reasonably satisfactory relationships have been found between E and L values and plant uptake for many soils, unsatisfactory results have been obtained for some soils, particularly those which "fix" large quantities of added phosphate [175,176,183]. The occurrence of high E and L values in soils from which plants are able to absorb only relatively small quantities of phosphate has been interpreted as evidence that the fixation process is due primarily not to the precipitation of phosphate in insoluble forms, but to labile phosphate being held by exchange sites of sufficient affinity for this ion to prevent its absorption by plants [176,184].

Fried and Dean [185] conceived the *A value* as the amount of available nutrient in a soil determined in terms of a standard, i.e., the same nutrient added to the soil as fertilizer. If a plant takes up the nutrient from the two sources in direct proportion to the amounts available, and if the respective contributions in the plant can be identified, then the A value can be calculated. The most suitable way of doing this is by isotopic ratio measurements; for phosphorus, ^{32}P is used and the mathematical expression is identical to that for the L value [Eq. (22)]. This has caused confusion between the two quantities—Larsen [172] has summarized the essential differences between them (Table 3).

By labeling the exchangeable phosphorus in soil with ^{33}P and fertilizer phosphorus with ^{32}P, Fardeau and Jappe [36] have compared L and A values in the same experiment. ^{32}P was determined by liquid scintillation counting and ^{32}P by Cerenkov counting. They found that L values were unaffected by whether or not fertilizer phosphate was present, but that the relationship between L and A values varied widely. The A value was expressed as a product of the L value and a coefficient K characteristic of the fertilizer used in the experiment.

For further discussion of these technqiues for assessment of labile P the reader is referred to the reviews by Fried [186], Larsen [172], and Jose and Krishnamoorthy [187], and the paper by Dalal and Hallsworth [183].

Other Ions

Calcium. The isotope ^{45}Ca has been used to determine both E and L values by techniques similar to those described above for soil phosphorus. The ^{45}Ca activity has typically been determined by precipitation as calcium oxalate and counting with an end-window G-M counter. Blume and Smith [186] studied 22 soils and found that for noncalcareous soils containing less than 50 mmol Ca kg^{-1} (100 meq kg^{-1}) results agreed with those obtained by extraction with ammonium acetate. For soils with higher exchangeable calcium, the results obtained by isotopic exchange were generally higher than those obtained by extraction, and where diffferences occurred the calcium remaining in the soil after extraction could be measured by isotopic exchange. The exchange method was also applicable to soils containing free $CaCO_3$. Newbould and Russell [189] found that the initial fast exchange was complete within 24 hr except where free $CaCO_3$ was present. They concluded that the E value and the quantity extracted by ammonium acetate were equally good measures of the labile pool, but as the L value exceeded the E value by 15-35%, some calcium which was not isotopically exchangeable was accessible to plants, although this represented a very small fraction of the total calcium in the soil.

Potassium. The short half-life of the isotope ^{42}K (12.4 hr) necessarily limits the duration of exchange studies. Conversely, the naturally occurring isotope ^{40}K has a half-life of 1.3×10^9 years, so that the specific activity even of enriched material is very low. In spite of these difficulties, both isotopes have been used in

Table 3 Distinction Between A and L Values

	A value	L value
Object	To measure the *availability* of soil phosphorus relative to a standard fertilizer source	To measure the *total quantity* of plant available soil phosphorus
Basic requirements		
1.	A minimum of reaction between source and soil: therefore, a short experiment	Sufficient reaction to give a steady state value: therefore, a long experiment
2.	^{32}P added with standard fertilizer source	^{32}P added with a minimum of carrier (to facilitate exchange with soil phosphorus)
3.	No mixing, ^{32}P placed	Maximum mixing, no placement
Results		
1. Pot experiments	Result dependent on degree of soil exploitation (equivalent to L value if all soil in pot is exploited)	Result independent of degree of soil exploitation
2. Field experiments	Can be used to characterize degree of exploitation of soil phosphorus	Not measurable

Source: Ref. 172.

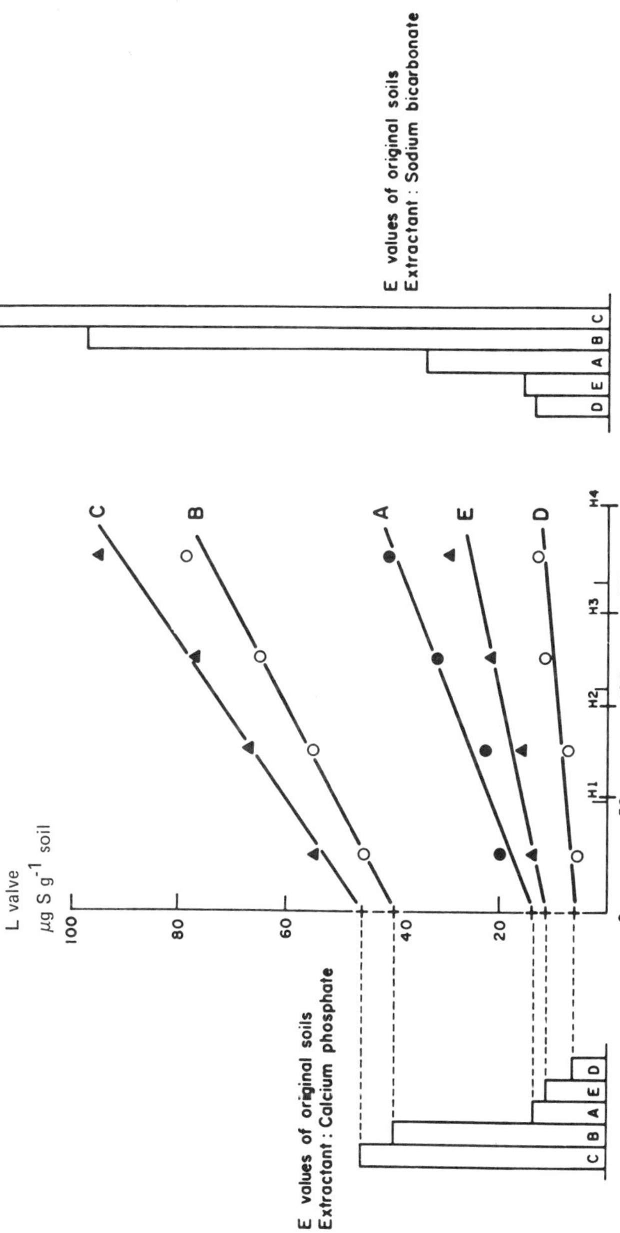

Figure 15 Increase in L values for sulfur in Australian soils with time, and comparison with initial E-values obtained with two extractants. (From Ref. 197.)

agronomic studies. For example, using ^{42}K, it has been shown that the quantity of potassium which is isotopically exchangeable is greater than that which can normally be extracted with M ammonium acetate [190] but very similar to the quantity removed by eight extractions with 0.01 M $CaCl_2$ [191]. The extensive literature in this field has been reviewed (up to 1967) by Bouat [192].

The use of rubidium as a tracer for potassium has been investigated, but significant differences occur between the two elements [192,193]. A more satisfactory method for long-term exchange studies with potassium has been described by Mercer and Gibbs [194], who used the stable isotope ^{39}K free from the other stable isotope ^{41}K, which occurs in natural potassium to the extent of 6.9%. Exchange between a solution containing ^{39}K and a soil results in both ^{39}K and ^{41}K entering the solution, and the increase in ^{41}K concentration is determined by neutron activation analysis (by activation to ^{42}K).

Sulfur. For sulfur, A values have been determined using ^{35}S as tracer. Nearpass et al. [195] applied this technique to 30 soils from the southeast United States, using cotton as the test plant and applying sulfur at six rates up to 16 $\mu g\ g^{-1}$ soil. The determination of ^{35}S activity involved precipitation as $BaSO_4$ and counting an infinitely thick source with an end-window G-M counter. Harward et al. [196] found that in 16 Oregon soils the A values were better correlated with plant yield than were measurements of S content in plants or extractable S plus S released on incubation. Probert [197] has made a detailed comparison of E and L values for sulfur in Australian soils. His data (Fig. 15) are a good illustration of the relationship between these two quantities for an available nutrient.

Trace Elements. The size of the labile pools of Fe, Mn, Zn, Cu, Co, and Se have all been investigated over the past 25 years, using ^{59}Fe, ^{54}Mn, ^{65}Zn, ^{64}Cu, ^{60}Co, and ^{75}Se tracers, respectively. Generally, experiments have been conducted with single isotopes, but Lopez and Graham [198,199] have counted ^{59}Fe, ^{54}Mn, and ^{65}Zn simultaneously in a multichannel γ spectrometer. The technique used by these workers, and by Rule and Graham [200], was to equilibrate soils with carrier-free isotope in a buffer solution containing 0.005 M diethylene triaminepentaacetic acid (DTPA), usually for 48 hr. Buffer solutions at pH 5 proved most satisfactory, as this minimized precipitation or sorption of the isotopes into less soluble compounds [199]. For Mn, very similar results were obtained for L values and the E values determined by DTPA extraction, but poor agreement was obtained for Fe and Zn [200].

An alternative to the technique of using a complexing agent in equilibration experiments is first to shake the soil with a buffer solution to establish equilibrium of the inactive isotope between soil and solution phases, and then to introduce the isotope into the solution in a carrier-free state so as not to disturb the equilibrium. The system is shaken for a further period and the extent of isotopic exchange is measured [201]. From the eidence of correlations between assessment of the labile pool and plant uptake, it appears that both variants of the technique produce satisfactory results, although in principle the inclusion of a complexing agent creates more artificial conditions in the experiment.

Manganese availability has been studied in several investigations. At least two first-order exchange reactions were found to occur between soil and solution [202,203] and all the soil Mn in equilibrium with that in solution could be extracted with pyrophosphate and $ZnSO_4$ [204]. Recently Salcedo and Ellis [205] have compared the correlations with plant uptake of extractable Mn in various extractants, E values (in 0.03 M H_3PO_4), and L values, and found they decreased in the order H_3PO_4 > steam/NH_4OAc > E value > L value > DTPA > $NH_4H_2PO_4$ > HCl.

Tiller et al. [201,206-208] have undertaken exchange studies with Co and Zn in Australian soils. Isotopically exchangeable Co was very highly correlated with NH_4OAc-extractable + hydroquinone – extractable Co in 25 soils. The technique allows quantitative study of sorption of Co by soils, and it was found that Mn minerals played a dominant role in the sorption of Co by many soils [201]. For Zn, the L values were highly correlated with those obtained by complexing agent extractions, but not with E values for alkaline soils [206]. It was concluded that soil Zn not in equilibrium with the soil solution, in the absence of plants, must be brought into solution by agents associated with the root environment, e.g., amino acids, complexing agents, or H^+. The role of these substances has been discussed by Rovira [209] and recently reviewed by Cheshire [210]. Using oats, maize, and orange as test plants, Shaw et al. [211] measured A values for Zn in the range 1.4–18 μg g^{-1} soil. Stewart et al [212] determined A and L values using maize, but obtained poor results when attempting to apply the technique to rice paddy soils because of precipitation of $ZnCO_3$ at the onset of flooding.

In Denmark, Gissel-Nielsen and Hamdy [213] have determined L values for Se in sandy loam soils, using ^{75}Se tracer. Italian ryegrass was used as the test plant and it was found that the L values were correlated with uptake of native soil Se.

C. Organic Matter Studies and Radiocarbon Dating

The literature on the use of labeled substances in organic matter is too extensive to be cited in full here, and only selected references are included. For more detailed information reference should be made to existing reviews and specialist publications [115,116,210,214-218].

Decomposition and Transformation of Organic Substances

Much of the work carried out with soil organic matter has involved the study of the decomposition or transformation of labeled compounds or plant material, and the evolution of $^{14}CO_2$. The techniques available for monitoring these processes and assaying the products have been reviewed elsewhere [10,219]. The measurement of $^{14}CO_2$ evolution, known as radiorespirometry, is particularly important because it makes possible the determination of the amount and rate of degradation [220].

When readily metabolized, ^{14}C-labeled compounds are added to the soil, there is an initial rapid decomposition and accompanying evolution of $^{14}CO_2$, followed by a much slower evolution of $^{14}CO_2$ as the metabolites and microbial tissues, into which the ^{14}C has been incorporated, are themselves degraded as part of the normal carbon turnover process. The behavior of glucose is typical in this respect. It has been shown that roughly 75% of this compound is liberated as $^{14}CO_2$ in the initial period and the remainder enters into microbial tissues and organic components [221-223].

Jansson and Persson [224] showed that while some of the incorporated ^{14}C was distributed throughout the organic matter, it was concentrated in the readily hydrolyzable fraction, and was probably carbohydrate material. However, substantial amounts of ^{14}C from glucose were shown to have been converted into aromatic compounds [225,226] and amino acids [227,228]. Other carbohydrates and polysaccharides such as starch and cellulose gave similar results, but the rates of decomposition and of transformation differed substantially from that of glucose [210,218,222,229].

Amino acids are also broken down very readily in the soil [230]. Labeling of specific carbon atoms in the tryptophan molecule has shown the carboxyl group to be less stable than the aliphatic side-chain, which was in turn less stable than the aromatic rings [220,230]. Proteins were found to be generally more stable than amino acids [231,232], particularly when they were complexed with other soil components. Such "protection" of proteins and other organic compounds by complexing with metals, clays, or humic substances in the soil, has been commonly observed [233-235].

The behavior of substituted aromatic carboxylic acids and phenols is of great interest in soil organic matter studies since they are components of both lignin and humic substances. Using compounds with specifically labeled carbon atoms, Martin and Haider [236] showed that, following the addition of p-hydroxybenzoic acid to soil, the carboxyl group was liberated as CO_2 even more rapidly than glucose and very little ^{14}C was incorporated into the soil. Evolution of CO_2 from labeled ring carbon atoms was slower than from glucose and there was more incorporation of the ^{14}C. Many other compounds, e.g., benzoic, cinnamic, caffeic, coumaric, ferulic, and vanillic acids and corresponding aldehydes and ketones have been studied in the same way [236-238]. All of these compounds were degraded fairly readily, and side chains (e.g., carboxyl, methoxyl, carbonyl, etc.) were found to be generally more susceptible to breakdown than aromatic rings. Similar studies have also been carried out on labeled lignin and "model" lignin polymers [239-241]. In these, the differences in rates of decomposition of rings versus side chains were not so clear cut.

The decomposition in the soil of labeled plant materials, including roots and root exudates, has been the subject of considerable study [111-114,210,218,242-248]. In temperate climates it has been shown that 60-70% of the carbon is lost from the soil in the first year but thereafter the rate of decomposition is much slower and 10-20% of the labeled carbon is still present in the soil after several years. Jenkinson and Ayanaba [114] found that in warmer climates the rate of decomposition was much more rapid (Fig. 16), and suggested that the decomposition curves can be explained by considering the plant material to consist broadly of two components with different half-lives. In the temperate climate of England these half-lives were 0.25 and 8 years, respectively. Cheshire [210] found a similar two-component system for rice straw but the most persistent part of the cellulose and xylan constituents was chemically indistinguishable from the more readily decomposable.

Jenkinson [243] reported that much of the labeled carbon persisting in the soil for several years was present in microbial tissue. Sterilization of the soil rendered this fraction decomposable, and Jenkinson suggested that the size of the soil biomass could be roughly estimated from the flush of CO_2 following sterilization. Studies of the decomposition of labeled microbial tissue have been amply reviewed by Webley and Jones [249].

Much work has been carried out using labeled compounds to elucidate the detailed biochemical reactions performed by various organisms isolated from the soil. These organisms are thought to be responsible for synthesis and decomposition in the soil and their use in pure cultures simplifies the chemical studies. Such work is beyond the scope of this chapter and readers are referred to the review by Haider et al. [115].

The breakdown of labeled pesticides has also been studied in culture systems and in the soil, and the rate and mode of decomposition determined [10]. The use of labeled compounds has also facilitated studies of the movement of pesticides and other compounds within soils and plants [10,250,251].

Most radioisotope work with soil organic matter has involved the use of ^{14}C, but other isotopes have also been used to good effect. For example, several studies have been carried out, using ^{35}S, to monitor the distribution of added sulfur through the soil organic fractions and to ascertain which fractions are most likely to be

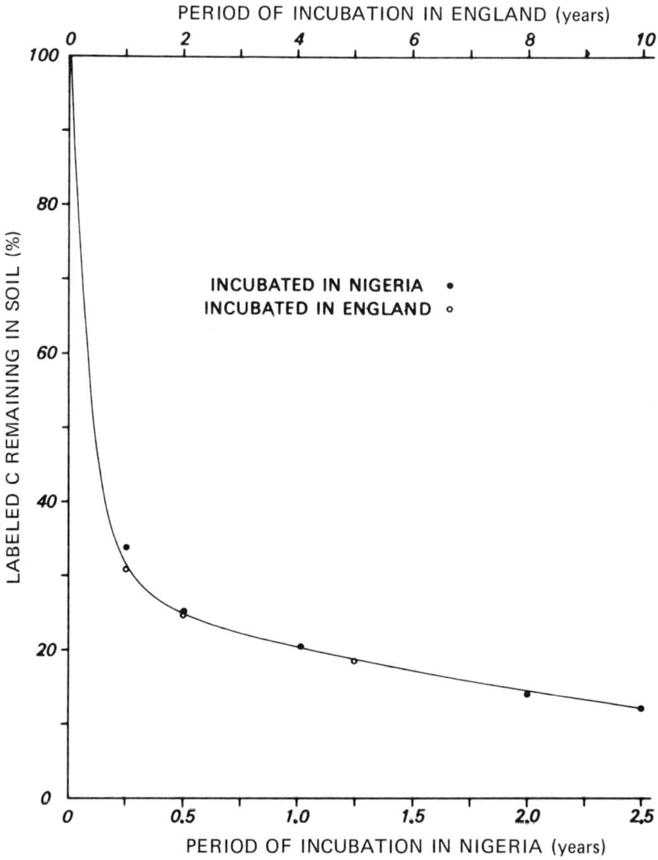

Figure 16 Decomposition of uniformly labeled ryegrass in soil in England and in Nigeria. (From Ref. 114.)

mineralized [252-255]. In contrast, ^{32}P has not been similarly used, probably because its half-life (14.8 days) is too short for incorporation and mineralization experiments. However, the use of ^{32}P does provide one of the few examples of isotope dilution analysis in soil organic matter work, i.e., the use by Martin [256] of labeled inositol phosphate in an attempt to measure the amount of this substance in soil extracts. This technique could find greater application in soil studies, particularly as gas and liquid radiochromatography equipment becomes more widely used, making it possible to check the purity of the added tracer compound and the product isolated for the determination of specific activity.

Multiple labeling of organic matter with ^{14}C, ^{3}H, ^{35}S, and possibly ^{32}P or ^{33}P, and also with the stable isotopes ^{15}N and ^{18}O, would in principle allow the simultaneous study of the transformations involving these elements. However, although such an approach appears to have great potential, it has, as yet, hardly been explored.

Radiocarbon Dating Studies

Radiocarbon dating has been used to attempt to determine the absolute age of soils. For existing topsoils (which contain much recently formed material) this has only been possible when the soil contains material such as charcoal or preserved wood which entered the soil at or around the date of formation; these materials can then be used for dating purposes. Absolute soil ages have also been obtained for buried soils, where incorporation of fresh ^{14}C ceased at the time of burial. Usually such processes are associated with volcanic or loessal activity and it has sometimes proved possible to correlate radiocarbon dates with geological dates. Studies using these techniques to date organic matter and charcoal taken from buried profiles in a chronosequence have given ages ranging from 1×10^3 to $> 40 \times 10^3$ years [257,258].

Radiocarbon dating has also been used to assess the mean residence time of organic matter in surface soils. Special care must be taken with "rejuvenating" materials such as fresh plant and microbial debris and the living soil biomass, to avoid misleadingly low values [214]. Attempts have been made to overcome these difficulties by chemical pretreatment to remove such material or by extraction and use of organic matter fractions thought to be free of "rejuvenating" substances [257,259-261]. Typically, values for mean residence time have ranged between 500 and 1500 years when rejuvenating effects were excluded [262,263]. Values obtained for mean residence times of organic matter in topsoils and the ages of buried soils have been reviewed elsewhere [264].

Recently, Jenkinson and Rayner [265] have proposed a compartmentalized model to account for the observed behavior and turnover of organic matter in topsoils. For the soil examined the compartments consisted of: decomposable plant material (half-life 0.165 year); resistant plant material (2.3 years); soil biomass (1.69 years); physically stabilized organic matter (49.5 years); and chemically stabilized organic matter (1980 years). The model predicted a radiocarbon age of 1240 years compared wiith the measured value of 1450 years.

VI. CONCLUSION

Radioisotopes have found wide applications in soil analysis and in the study of chemical, physical and microbiological processes in the soil. Suitable tracer isotopes for many of the elements of biological interest are either commercially available or can be produced to order in reactors or accelerator facilities in many parts of the world.

There are severe limitations, however. In soil studies there are no usable tracers for the major constituents Si, Al, and Ti; and oxygen and nitrogen, which are major elements in the soil-plant-animal system, only have radioisotopes with extremely short half-lives. Half-life considerations also restrict the use of the major elements K and Mg and the trace elements Cu and Mo, and there is no radioisotope of boron, an essential element in plant nutrition.

Detection systems have become more sensitive and more varied in recent years, and now a wide range of instruments is available for the detection of all types of radiation. The advent of microprocessors has resulted in rapid developments in automatic sample handling and data processing, but for some applications simple systems involving the use of long-established methods continue to demonstrate their usefulness.

In elemental analysis, the trend towards nondestructive methods has resulted in increasing use of instrumental neutron activation analysis (see Chap. 7), but procedures involving radiochemical separations have continued to be of importance, particularly those involving the various isotope dilution techniques. For measurements of labile ions and for investigations of biochemical transformations involving carbon compounds, the traditional tracer techniques remain as useful today as 30 years ago, and depend less on sophisticated instrumentation than on sound radiochemical procedures and an adequate understanding of the system being studied.

The continued use of radioactive tracer techniques on a wide scale seems assured because of their general usefulness, the comparative ease with which radioisotopes may be detected, and the general recognition that, provided approved safety procedures are employed, no significant hazard is involved. The list of processes to which the techniques may be applied is unlikely to be exhausted for a very long time, and the potential exists for significant new developments in the future.

ACKNOWLEDGMENT

The authors wish to express their thanks to Mr. A. H. Knight of the Macaulay Institute for Soil Research, Aberdeen, for many useful suggestions and comments during the preparation of this chapter.

REFERENCES

1. G. B. Cook and J. F. Duncan, *Modern Radiochemical Practice*, Clarendon Press, Oxford, 1951.

2. R. T. Overman and H. M. Clark, *Radioisotope Techniques*, McGraw-Hill, New York, 1960.
3. G. Friedlander, J. W. Kennedy, and J. M. Miller, *Nuclear and Radiochemistry*, 2nd ed., Wiley, New York, 1964.
4. B. J. Wilson (Ed.), *The Radiochemical Manual*, 2nd ed., Radiochemical Centre, Amersham, Buckinghamshire, England, 1966.
5. P. Marmier and E. Sheldon, *Physics of Nuclei and Particles*, Vol. 1, Academic Press, New York 1969.
6. H. A. C. McKay, *Principles of Radiochemistry*, Butterworths, London, 1971.
7. R. A. Faires and B. H. Parks, *Radioisotope Laboratory Techniques*, 3rd ed., Butterworths, London, 1973.
8. W. E. Burcham, *Nuclear Physics*, 2nd ed., Longmans, London, 1973.
9. D. I. Coomber (Ed.), *Radiochemical Methods in Analysis*, Plenum Press, New York, 1975.
10. M. F. L'Annunziata, *Radiotracers in Agricultural Chemistry*, Academic Press, London, 1979.
11. C. L. Gordon, Anal. Chem., *21*:736 *26*:176 (1954).
12. W. W. Meinke, Anal. Chem., *28*:736 (1956); *30*:686 (1968); *32*:104 (1960).
13. G. W. Leddicote, Anal. Chem., *34*:143R (1962); *36*:419R (1964).
14. W. S. Lyon, E. Ricci, and H. H. Ross, Anal. Chem., *38*: 251R (1966); *40*:168R (1968); *42*:123R (1970); *44*:439R (1972); *46*:431R (1974).
15. W. S. Lyon and H. H. Ross, Anal. Chem., *48*:96R (1976); *50*:80R (1978).
16. M. S. Smith, M. K. Firestone, and J. M. Tiedje, Soil Sci. Soc. Am. J., *42*:611 (1978).
17. D. A. Barber, M. Ebert, and N. T. S. Evans, J. Exp. Bot., *13*:397 (1962).
18. L. A. J. Venverloo, *Practical Measuring Techniques for Beta Radiation*, Macmillan, London, 1971.
19. J. B. Birks, *The Theory and Practice of Scintillation Counting*, Pergamon, Oxford, 1964.
20. E. D. Bransome, Jr. (Ed.) *The Current Status of Liquid Scintillation Counting*, Grune and Stratton, New York, 1970.
21. C. E. Crouthamel, F. Adams, and R. Dams, *Applied Gamma-Ray Spectrometry*, 2nd ed., Pergamon, Oxford, 1970.
22. K. A. Smith, Ph.D. thesis, Reading University, Berkshire, England, 1968.
23. C. L. Comar, *Radioisotopes in Biology and Agriculture*, McGraw-Hill, New York, 1955.
24. M. Vassilaki, L. Salmon, and J. A. B. Gibson, Geochim. Cosmochim. Acta. *30*:601 (1966).

25. R. S. Russell and F. B. Ellis, Nature (Lond.) *217*:582 (1968).
26. E. R. Mercer, in *Radioactivity and Human Diet* (R. S. Russell, Ed.), Pergamon, Oxford, 1966, p. 489.
27. H. H. Seliger and A. Schwebel, Nucleonics, *12*(7):54 (1954).
28. J. G. Cuninghame, in Ref. 9, p. 1.
29. K. Z. Morgan, in *Nuclear Instruments and Their Uses*; Vol. 1 (A. H. Snell, Ed.), Wiley, New York, 1962, p. 391.
30. R. W. Williams in *Methods of Experimental Physics*; vol. 5A (L. C. L. Yuan and C. S. Wu, Eds.), Academic Press, New York, 1961.
31. N. J. Barrow and T. C. Shaw, Soil Sci., *119*:190 (1975).
32. P. A. Sarangamath, B. N. Shinde, and S. Patnaik, Soil Scil. *124*:40 (1977).
33. N. Veall, Brit. J. Radiol., *21*:347 (1948).
34. Anon., *Methods of Radiochemical Analysis*, W.H.O., Geneva, 1966.
35. J. R. Robinson, Int. J. Appl. Radiat. Isotopes, *20*:531 (1969).
36. J. C. Fardeau and J. Jappe, in *Isotopes and Radiation in Soil-Plant Relationships Including Forestry*, IAEA, Vienna, 1972, p. 499.
37. R. C. Turner, J. M. Radley, and W. V. Mayneord, Brit, J. Radiol., *31*:397 (1958).
38. A. A. Noujaim, C. Ediss, and L. I. Wiebe (Eds.), *Liquid Scintillation: Science and Technology*, Academic Press, New York, 1976.
39. J. B. Birks, *Solutes and Solvents for Liquid Scintillation Counting*, Koch Light Labs, Ltd., Colnbrook, Bucks, England, 1969.
40. K. D. Neame and C. A. Homewood, *Introduction to Liquid Scintillation Counting*, Butterworths, London, 1974.
41. Y. Kobayashi and D. W. Maudsley, *Biological Applications of Liquid Scintillation Counting*, Academic Press, New York, 1974.
42. B. W. Fox, *Techniques of Sample Preparation for Liquid Scintillation Counting*, North Holland, Amsterdam, 1976.
43. G. Laustriat, R. Voltz, and J. Klein, in Ref. 20, p. 13.
44. J. B. Birks, in Ref. 20, p. 3.
45. G. A. Bray, in Ref. 20, p. 170.
46. M. S. Patterson and R. C. Green, Anal. Chem., *37*:854 (1965).
47. J. C. Turner, Int. J. Appl. Radiat. Isotopes, *19* : 557 (1968).
48. J. C. Turner, Int. J. Appl. Radiat. Isotopes, *20*:449 (1969).
49. C. B. Campbell and L. W. Powell, J. Clin. Path., *23*:304 (1970).
50. D. G. Ott, C. R. Richmond, T. T. Treyill, and H. Foreman, Nucleonics, *17*(9):106 (1959).

51. H. J. Cluley, Analyst (Lond.) *87*:170 (1962).
52. G. E. Calf, Int. J. Appl. Radiat. Isotopes, *20*:611 (1969).
53. C. T. Peng, in Ref. 9, p. 79.
54. T. Higashimura, O. Yamada, N. Nohara, and T. Shidei, Int. J. Appl. Radiat. Isotopes, *13*:308 (1962).
55. I. T. Takahashi and F. A. Blanchard, Anal. Biochem., *29*: 154 (1969).
56. L. A. Baillie, Int. J. Appl. Radiat. Isotopes, *8*:1 (1960).
57. E. T. Bush, Anal. Chem., *35*:1024 (1963).
58. A. W. Rogers and J. F. Moran, Anal. Biochem., *16*:206 (1966).
59. P. A. Cerenkov, Dokl. Akad. Nauk S. S. S. R., *2*:451 (1934).
60. E. H. Belcher, Proc. Roy. Soc. Ser. A., *216*:90 (1953).
62. A. T. B. Moir, Int. J. Appl. Radiat. Isotopes, *22*:213 (1971).
63. J. R. Brownell and A. Lauchli, Int. J. Appl. Radiat. Isotopes, *20*:797 (1969).
64. R. E. White and A. W. Taylor, J. Soil Sci., *28*:48 (1977).
65. R. H. Elrick and R. P. Parker, Int. J. Appl. Radiat. Isotopes, *19*:263 (1968).
66. H. Lundqvist, K. J. Johanson, and G. Jonsson, Int. J. Appl. Radiat. Isotopes, *27*:233 (1976).
67. J. Ashcroft, Anal. Biochem., *37*:268 (1970).
68. K. A. Smith and E. R. Mercer, J. Radioanal. Chem., *5*: 303 (1970).
69. A. W. Rogers, *Techniques of Autoradiography*, 3rd ed., Elsevier, Amsterdam, 1979.
70. A. W. Rogers, *Practical Autoradiography*, Review No. 20, Radiochemical Centre, Amersham, Buckinghamshire, England, 1979.
71. H. F. Wilkinson, J. F. Loneragan, and J. P. Quirk, Soil Sci. Soc. Am. Proc., *32*:831 (1968).
72. H. F. Wilkinson, J. F. Loneragan, and J. P. Quirk, Science, *161*:1245 (1968).
73. J. P. Baldwin and P. B. Tinker, Plant Soil, *37*:209 (1972).
74. K. K. S. Bhat and P. H. Nye, Plant Soil, *38*:161 (1973).
75. K. K. S. Bhat and P. H. Nye, Plant Soil, *41*:365 (1974).
76. K. K. S. Bhat and P. H. Nye, Plant Soil, *41*:383 (1974).
77. D. G. Lewis and J. P. Quirk, Plant Soil, *26*:445 (1967).
78. J. M. Walker and S. A. Barber, Science, *133*:881 (1961).
79. S. A. Barber, J. M. Walker, and E. H. Vasey, J. Agric. Fd. Chem., *11*:204 (1963).
80. J. N. Marais, S. Afr. J. Agric. Sci., *11*:423 (1968).
81. P. Newbould, R. Taylor, and K. R. Howse, J. Brit. Grassland Soc., *26*:201 (1971).
82. F. S. Chapin and D. F. Holleman, Int. J. Appl. Radiat. Isotopes, *25*:568 (1974).

83. F. Fric and V. Finocchiaro, Radiochem. Radioanal. Lett., *25*:187 (1976).
84. E. R. Mercer, K. B. Gunn, P. M. Lay, W. Harris, W. Downs, and M. G. Johnson, *U. K. Agricultural Research Council Letcombe Laboratory Annual Report 1968*, Supplement, pp. 1-19, HMSO, London, 1969.
85. D. I. Coomber, in Ref. 9, p. 175.
86. F. Helfferich, *Ion Exchange*, McGraw-Hill, New York, 1962.
87. O. Samuelson, *Ion Exchange Separations in Analytical Chemistry*, Wiley, New York, 1957.
88. W. Rieman and H. F. Walton, *Ion Exchange in Analytical Chemistry*, Pergamon, Oxford, 1970.
89. K. A. Kraus and F. Nelson, *Proceedings of the International Conference on the Peaceful Uses of Atomic Energy*, Geneva, 1956, vol. 7, p. 113.
90. R. G. McLaren and D. V. Crawford, J. Soil Sci., *24*:172 (1973).
91. J. Starý, in *Analytical Chemistry*, Part 1 (T. S. West, Ed.) Physical Chemistry Series I, vol. 12, Butterworths, London, 1973, p. 279.
92. J. Starý, *The Solvent Extraction of Metal Chelates*, Pergamon, Oxford, 1964.
93. G. H. Morrison and H. Freiser, *Solvent Extraction in Analytical Chemistry*, Wiley, New York, 1957.
94. Y. Marcus and A. S. Kertes, *Ion Exchange and Solvent Extraction of Metal Complexes*, Wiley, New York, 1969.
95. H. J. M. Bowen and D. Gibbons, *Radioactivation Analysis*, Clarendon Press, Oxford, 1963.
96. H. J. M. Bowen, in *Nuclear Activation Techniques in the Life Sciences*, IAEA, Vienna, 1967, p. 287.
97. P. Kruger, *Principles of Activation Analysis*, Wiley-Interscience, New York, 1971.
98. R. Dams, F. de Corte, J. Hertogen, J. Hoste, W. Maenhaut, and F. Adams, in *Analytical Chemistry*, Part 1 (T. S. West, Ed.), Physical Chemistry Series 2, vol. 12, Butterworths, London, 1976, p. 1.
99. R. Cornelis, J. Hoste, A. Speecke, C. Vandecasteele, J. Versieck, and R. Gibels, in Ref. 98, p. 71.
100. J. Růžička and J. Starý, Talanta, *8*:228 (1961).
101. W. F. Libby, *Carbon Dating*, 2nd ed., Univ. of Chicago Press, Chicago, 1963.
102. H. W. Scharpenseel, in *Soil Biochemistry*, Vol. 2 (A. D. McLaren and J. Skujins, Eds.), Marcel Dekker, New York, 1971, p. 96.
103. H. W. Scharpenseel and F. Pietig, Geoderma, *2*:273 (1968/69).
104. H. W. Scharpenseel and F. Pietig, Atompraxis, *16*:1 (1974).
105. G. W. Pearson, J. R. Pilcher, M. G. L. Baillie, and J. Hillam, Nature (Lond.), *270*:25 (1977).
106. H. E. Suess, Nature (Lond.), *272*:649 (1978).

107. J. R. Catch, *Carbon-14 Compounds*, Butterworths, London, 1961.
108. D. Sauerbeck and F. Fuhr, in *The Use of Isotopes in Soil Organic Matter Studies*, Pergamon, Oxford, 1966, p. 391.
109. A. Zeller, H. E. Oberländer, and F. Hiemer, in Ref. 108. p. 401.
110. A. H. Knight, New Phytol. *79*:573 (1977).
111. D. S. Jenkinson, J. Soil Sci., *16*:104 (1965).
112. D. S. Jenkinson, Soil Sci., *111*:64 (1971).
113. D. S. Jenkinson, J. Soil Sci., *28*:417 (1977).
114. D. S. Jenkinson and A. Ayenaba, Soil Sci. Soc. Am. J., *41*:912 (1977).
115. K. Haider, J. P. Martin, and Z. Filip, in *Soil Biochemistry*, vol. 4 (E. A. Paul and A. D. McLaren, Eds.), Marcel Dekker, New York, 1975, p. 195.
116. G. H. Wagner, in *Soil Biochemistry*, vol. 3 (E. A. Paul and A. D. McLaren, Eds.), Marcel Dekker, New York, 1975, p. 269.
117. R. Wolfgang, Ann. Rev. Phys. Chem., *16*:15 (1965).
118. J. L. Garnett, L. Henderson, and W. A. Sollich, *Proceedings of the Symposium on Tritium in Biological and Physical Sciences*, IAEA, Vienna, 1962, vol. 2, p. 47.
119. E. T. McGuinness and M. C. Cullen, J. Chem. Educ., *47*: A9 (1970).
120. Y. Yamada, Radio-Isotopes, Tokyo, *13*:32 (1964).
121. M. Shibuya, *Proceedings of the 4th Japan Conference on Radioisotopes*, 1961, p. 1156.
122. M. Shibuya and T. Nakai, *Proceedings of the 5th Japan Conference on Radioisotopes*, 1963, p. 228.
123. M. Shibuya, Radio-Isotopes, Tokyo, *13*:45 (1964).
124. C. H. Ke, C. Y. Lin, and H. S. Cheng, J. Chin. Chem. Soc., Taipei, *11*:193 (1964).
125. S. Nakano, J. Hanboh, and N. Urabe, Bull. Chem. Soc. Japan, *49*:2437 (1976).
126. E. Steinnes, Talanta, *24*:121 (1977).
127. O. J. Kronborg and E. Steinnes, Analyst. (Lond.) *100*: 835 (1975).
128. W. Zmijewska and T. Semkow, Chem. Anal. (Warsaw), *23*: 583 (1978).
129. H. J. M. Bowen and J. A. Dymond, Proc. Roy. Soc. B., *144*:355 (1955).
130. K. A. Smith, Plant Soil, *34*:369 (1971).
131. H. J. M. Bowen and P. A. Cawse, Soil Sci., *98*:358 (1964).
132. D. Brune and O. Landström, Radiochim. Acta, *5*:1 (1966).
133. C. M. Cho and H. Axmann, *Tech. Report Series No. 48*, IAEA, Vienna, 1965.
134. J. R. Kline, S. S. Brar, P. F. Gustafson, and R. H. Rust, *Proceedings of the International Conference on Modern Trends in Activation Analysis*, College Station, Texas, 1966, p. 319.

135. H. J. M. Bowen and P. A. Cawse, Analyst (Lond.) *88*:721 (1963).
136. D. M. Hadjimarkos, Nature (Lond.) *193*:178 (1962).
137. K. P. McConnell, *Proceedings of the International Conference on Modern Trends in Activation Analysis*, College Station, Texas, 1962, p. 137.
138. D. Hoede, R. Massee and H. A. Das, Radiochem. Radioanal. Lett., *23*:379 (1975).
139. K. A. Cook and E. R. Graham, Soil Sci. Soc. Am. J., *42*:57 (1978).
140. B. Van Zanten, D. Decat, and G. Leliaert, Talanta, *9*:213 (1962).
141. H. J. M. Bowen and P. A. Cawse, Nature (Lond.) *194*:399 (1962).
142. W. Höfner, E. Schaumlöffel, and L. Völker, Planta, *57*:608 (1961).
143. O. Johansen and E. Steinnes, Analyst (Lond.) *101*:455 (1976).
144. H. J. M. Bowen, Analyst (Lond.) *89*:658 (1964).
145. H. J. Bowen, Biochem. J., *77*:79 (1960).
146. H. J. M. Bowen, Int. J. Appl. Radiat. Isotopes, *4*:214 (1962).
147. M. H. Dean, A. Stimson, and D. E. Green, Anal. Chim. Acta, *35*:530 (1966).
148. D. Decat, B. Van Zanten, and G. Leliaert, Anal. Chem., *35*:845 (1963).
149. B. A. Loveridge, R. K. Webster, J. W. Morgan, A. M. Thomas, and A. A. Smales, *UKAEA Report AERE C/R 2657* (1959).
150. C. Myttenaere and R. Kirchmann, Physiol. Plant, *15*:656 (1952).
151. K. A. Smith, Plant Soil, *34*:643 (1971).
152. F. W. Brown, F. O. Simon, and L. P. Greenland, J. Res. U. S. Geol. Survey, *3*:187 (1975).
153. C. L. Branquinho and V. J. Robinson, J. Radioanal. Chem., *24*:321 (1975).
154. E. R. Mercer and K. A. Smith, *U. K. Agricultural Research Council Radiobiological Laboratory Report ARCRL 12*, HMSO, London, 1964, p. 82.
155. E. R. Mercer and K. A. Smith, *U. K. Agricultural Research Council Radiobiological Laboratory Report ARCRL 14*, HMSO, London, 1965, p. 80.
156. J. Starý and J. Růžička, Talanta, *18*:1 (1971).
157. A. Zeman, J. Lesko, J. Prejza, N. Groth-Andersen, and J. Růžička, cited in Ref. 156.
158. J. Růzička and M. Williams, Talanta, *12*:967 (1965).
159. J. Růžička and C. G. Lamm, Talanta, *15*:689 (1968).

160. J. Růžička and C. G. Lamm, Talanta, *16*:157 (1969).
161. C. G. Lamm and J. Růžička, Talanta, *16*:603 (1969).
162. E. Gundersen and E. Steinnes, Talanta, *18*:1167 (1971).
163. G. D. Wals and H. A. Das, Radiochem. Radioanal. Lett., *26*:353 (1976).
164. H. A. Das, W. H. Kohnemann, G. D. Wals, and J. Zonderhuis, J. Radioanal. Chem., *25*:261 (1975).
165. J. K. Johannesson, Analyst (Lond.) *86*:72 (1961).
166. J. K. Johannesson, Analyst (Lond.) *92*:766 (1967).
167. W. Quam, J. Houk, and R. Mohr, *U.S. Energy Res. Dev. Adm. Report, EGG 1183-2362* (Rev.), 1976.
168. D. G. Coles, J. W. T. Meadows, and C. L. Lindeten, *USAEC Report, UCRL 75619*, 1974.
169. G. Philipp, H. Pfister, and C. Leisgang, Radiat. Environ. Biophys., *13*:239 (1976).
170. R. F. Hall, J. Sci. Fd. Agric., *25*:637 (1974).
171. R. S. Russell, J. B. Rickson, and S. N. Adams, J. Soil Sci., *5*:85 (1954).
172. S. Larsen, Adv. Agron., *19*:151 (1967).
173. F. Paneth and W. Vorwerk, Z. Phys. Chem., *101*:445 (1922).
174. C. D. McAuliffe, N. S. Hall, L. A. Dean, and S. B. Hendricks, Soil Sci. Soc. Am. Proc., *12*:119 (1948).
175. S. R. Olsen and L. A. Dean, in *Methods of Soil Analysis* (C. A. Black, D. D. Evans, L. E. Ensminger, J. L. White, and F. E. Clark, Eds.), Am. Soc. Agron., Madison, 1965, p. 1035.
176. R. S. Russell, E. W. Russell, and P. G. Marais, J. Soil Sci., *8*:248 (1957).
177. M. Fried, in *Atomic Energy and Agriculture*, (C. L. Comar, Ed.), Am. Assoc. Adv. Science, Washington, D. C., 1957, p. 1.
178. R. E. White, L. O. Tiffin, and A. W. Taylor, Plant Soil, *45*:521 (1976).
179. P. Arambarri and O. Talibudeen, Plant Soil, *11*:364 (1959).
180. G. E. C. Mattingly and O. Talibudeen, *Report Rothamsted Experimental Station, Hertfordshire, England, 1960*, p. 246.
181. S. Larsen, Plant Soil, *4*:1 (1952).
182. O. Gunnarsson and L. Fredriksson, *Proceedings of the Isotope Technology Conference, Oxford, 1953*, vol. 1, p. 427. Cited in Ref. 171.
183. R. C. Dalal and E. G. Hallsworth, Soil Sci. Soc. Am. J., *41*:81 (1977).
184. R. K. Scofield, Soils Fert., *18*:343 (1955).
185. M. Fried and L. A. Dean, Soil Sci., *73*:263 (1952).
186. M. Fried, *Transactions of the 8th International Congress of Soil Science*, Bucharest, 1964, vol. 4, p. 29.
187. A. I. Jose and K. K. Krishnamoorthy, Soils Fert., *35*:620 (1972).

188. J. M. Blume and D. Smith, Soil Sci., *77*:9 (1954).
189. P. Newbould and R. S. Russell, Plant Soil, *18*:239 (1963).
190. M. E. Sumner and G. H. Bolt, Soil Sci. Soc. Am. Proc., *26*:541 (1962).
191. E. R. Graham and D. H. Kampbell, Soil Sci. *106*:101 (1968).
192. M. Bouat, Ann. Agron. *20*:89 (1969).
193. A. A. R. Hafez and P. R. Stout, Soil Sci. Soc. Am. Proc., *37*:572 (1973).
194. E. R. Mercer and A. R. Gibbs, in *Soil Chemistry and Fertility* (G. V. Jacks Ed.), Trans. Int. Soc. Soil Sci. Commissions II, IV, Aberdeen, ISSS, 1967, p. 233.
195. D. C. Nearpass, M. Fried, and V. J. Kilmer, Soil Sci. Soc. Am. Proc., *25*:287 (1961).
196. M. E. Harward, T. T. Chao, and S. C. Feng, Agron. J., *54*:101 (1962).
197. M. E. Probert, Plant Soil *45*:461 (1976).
198. P. L. Lopez and E. R. Graham, Soil Sci., *110*:24 (1970).
199. P. L. Lopez and E. R. Graham, Soil Sci., *114*:295 (1972).
200. J. H. Rule and E. R. Graham, Soil Sci., Soc. Am. J., *40*:853 (1976).
201. K. G. Tiller, J. L. Honeysett, and E. G. Hallsworth, Aust. J. Soil Res., *7*:43 (1969).
202. C. C. Weir and M. H. Miller, Can. J. Soil Sci., *42*:105 (1962).
203. A. S. J. Reid and M. H. Miller, Can. J. Soil Sci., *43*:250 (1963).
204. E. M. Romney and S. J. Toth, Soil Sci., *77*:107 (1954).
205. I. H. Salcedo and B. G. Ellis, Soil Sci., *127*:227 (1979).
206. K. G. Tiller, J. L. Honeysett, and M. P. C. de Vries, Aust. J. Soil Res., *10*:151 (1972).
207. K. G. Tiller, J. L. Honeysett, and M. P. C. de Vries, Aust. J. Soil Res., *10*:165 (1972).
208. K. G. Tiller and P. Wassermann, in *Isotopes and Radiation in Soil-Plant Relationships Including Forestry*, IAEA, Vienna, 1972, p. 517.
209. A. D. Rovira, Bot. Rev., *35*:35 (1969).
210. M. V. Cheshire, *Nature and Origin of Carbohydrates in Soils*, Academic Press, London, 1979.
211. E. Shaw, R. G. Menzel, and L. A. Dean, Soil. Sci., *77*:205 (1954).
212. J. W. B. Stewart, B. V. Friaz, and F. M. Lapid, in *Isotopes and Radiation in Soil-Plant Relationships Including Forestry*, IAEA, Vienna, 1972, p. 539.
213. G. Gissel-Nielsen and A. A. Hamdy, Z. Pflanzen. Bodenk., *141*:67 (1978).
214. E. A. Paul and J. A. van Veen, *Transactions of the 11th International Congress of Soil Science*, Edmonton, 1978, vol. 3, p. 61.

215. *The Use of Isotopes in Soil Organic Matter Studies*, Pergamon, Oxford, 1966.
216. *Isotopes and Radiation in Soil Organic Matter Studies*, IAEA, Vienna, 1968.
217. *Soil Organic Matter Studies* (2 vols), IAEA, Vienna, 1977.
218. E. Grossbard (Ed.), *Straw Decay and Its Effect of Disposal and Utilization*, Wiley, Chichester, Sussex, England, 1979.
219. *Tracer Manual on Crops and Soils*, Tech. Report Series No. 171, IAEA, Vienna, 1976.
220. J. Mayaudon, in *Soil Biochemistry*, vol. 2 (A. D. McLaren and J. Skujins, Eds.), Marcel Dekker, New York, 1971, p. 202.
221. J. Mayaudon and P. Simonart, Plant Soil, *9*:376 (1958).
222. J. Mayaudon and P. Simonart, Plant Soil, *9*:381 (1958).
223. K. S. Chahal and G. H. Wagner, Soil Sci., *100*:96 (1965).
224. S. L. Jansson and J. Persson, in Ref. 216, p. 111.
225. G. H. Wagner, in Ref. 216, p. 197.
226. J. P. Martin, K. Haider, W. J. Farmer, and E. Fustec-Mathon, Soil Biol. Biochem., *6*:221 (1974).
227. L. H. Sørensen, Soil Sci., *104*:234 (1967).
228. M. V. Cheshire, C. M. Mundie, and H. Shepherd, J. Soil Sci., *22*:222 (1971).
229. J. M. Oades and G. H. Wagner, Geoderma, *4*:417 (1970).
230. J. Mayaudon and P. Simonart, Suppl. Ann. Inst. Pasteur, *109*:224 (1965).
231. J. Mayaudon, in Ref. 216, p. 197.
232. P. Simonart, L. Batistic, and J. Mayaudon, Plant Soil, *27*:153 (1967).
233. J. P. Martin, J. O. Ervin, and R. Shepherd, Soil Sci. Soc. Am. Proc., *29*:397 (1965).
234. J. P. Martin, J. O. Erwin, and R. Shepherd, Soil Sci. Soc. Am. Proc., *30*:196 (1966).
235. G. D. Swincer, J. M. Oades, and D. J. Greenland, Adv. Agron., *21*:195 (1969).
236. J. P. Martin and K. Haider, in Ref. 217 (vol. 2), p. 23.
237. K. Haider and J. P. Martin, Soil Sci. Soc. Am. Proc., *39*:657 (1975).
238. L. Batistic and J. Mayaudon, Ann. Inst. Pasteur, *118*:119 (1970).
239. K. Haider and J. P. Martin, Soil Sci. Soc. Am. Proc., *31*:766 (1967).
240. J. P. Martin and K. Haider, Soil Sci., *107*:260 (1969).
241. T. Fukuzimi, H. Takatuka, and K. Minami, Arch. Biochem. Biophys, *129*:396 (1969).
242. D. S. Jenkinson, in *Experimental Pedology* (E. G. Hallsworth and D. V. Crawford, Eds.), Butterworths, London, 1964, p. 199.
243. D. S. Jenkinson, J. Soil Sci., *17*:280 (1966).

244. F. E. Clark and E. A. Paul, Adv. Agron., *22*:375 (1970).
245. F. Führ and D. Sauerbeck, in Ref. 216, p. 241.
246. J. W. Nyhan, Soil Sci. Soc. Am. Proc., *39*:643 (1975).
247. J. K. Martin, in Ref. 217 (vol. 1), p. 197.
248. D. Sauerbeck and B. G. Johnen, in Ref. 217 (vol. 1), p. 141.
249. D. M. Webley and D. Jones, in *Soil Biochemistry*, vol. 2 (A. D. McLaren and J. Skujins, Eds.), Marcel Dekker, New York, 1971, p. 446.
250. E. P. Lichtenstein, K. R. Schultz, and T. W. Fuhremann, J. Agric. Fd. Chem., *20*:831 (1972).
251. E. P. Lichtenstein, T. W. Fuhremann and K. R. Schultz, J. Agric. Fd. Chem.,*22*:991 (1974).
252. J. R. Freney, G. E. Melville, and C. H. Williams, Soil Biol. Biochem., *3*:133 (1971).
253. A. R. Till and P. E. May, Aust. J. Agric. Res., *22*:391 (1971).
254. J. R. Bettany, J. W. B. Stewart, and E. H. Halstead, Can. J. Soil Sci., *54*:309 (1974).
255. H. J. van Praag, Plant Soil, *39*:49 (1973).
256. J. K. Martin, N. Z. J. Agric. Res., *13*:930 (1970).
257. K. M. Goh, B. P. J. Molloy, and T. A. Rafter, Quatern. Res., *7*:177 (1977).
258. K. M. Goh and W. A. Pullar, Geoderma, *18*:265 (1977).
259. K. M. Goh, J. Soil Sci., *29*:340 (1978).
260. H. W. Scharpenseel, in Ref. 217 (vol. 2), p. 193.
261. H. W. Scharpenseel, C. Ronzani, and F. Pietig, in Ref. 216, p. 67.
262. C. A. Campbell, E. A. Paul, D. A. Rennie, and K. J. McCallum, Soil Sci., *104*:81 (1967).
263. E. A. Paul, C. A. Campbell, D. A. Rennie and K. J. McCallum, *Transactions of the 8th International Congress of Soil Science*, Bucharest, 1964, vol. 3, p. 201.
264. H. W. Scharpenseel and H. Schiffman, Z. Pflanzen. Bodenk., *140*:159 (1977).
265. D. S. Jenkinson and J. H. Rayner, Soil Sci., *123*:298 (1977).

7

Instrumental Neutron Activation Analysis

LEONARD SALMON and PETER A. CAWSE Atomic Energy Research Establishment, Harwell, Oxfordshire, England

I. INTRODUCTION

An analytical method is selected by an investigator for a variety of reasons, and a method which offers fairly accurate results on a wide variety of elements at submicrogram levels, without any need for chemical treatment, is not to be discarded lightly. Indeed, instrumental neutron activation analysis (INAA) offers all these features to the soil chemist, yet the literature relating to this application is sparse compared with that, for example, for related studies of the atmospheric environment and in geochemistry.

Like all good physical methods of elemental analysis, activation analysis depends on the measurement of a specific property which is unique to the element and whose magnitude is reliably related to the quantity of the element present in the sample. The principle is simply to induce radioactivity in the material of the sample and then to measure radiation specific to each element being determined. It happens that, in soil, the major elements of low atomic number are barely susceptible to activation by neutrons, whereas many minor and trace elements do activate readily. In instrumental activation analysis, a radiation spectrometric technique is used to discriminate between different elements by measurement of photons specific in energy to individual isotopes.

Neutrons are chosen as the irradiation source since the probability of activation is greater than with charged particles, which need to overcome a coulomb barrier, while available nuclear reactors provide particularly intense and isotropic sources of neutrons of low enough energy to avoid the complication of interfering secondary reactions. Similarly, γ rays, rather than the primary β radiation are chosen for the identification and assessment of the induced activity, since they exhibit discrete energies and are not subject to significant absorption within the sample.

There are many general reviews of activation analysis in the literature including those by Gibbons and Lambie [1], De Soete et al. [2], Kruger [3], and recent compilations by staff of the University of Ghent [4,5]. Many papers are published in the proceedings of regular international conferences such as *Modern Trends in Activation Analysis* [6] and *Nuclear Activation Techniques in the Life Sciences* [7].

There have been numerous technical developments beyond the use of neutrons. One such, subject to much current interest, is the use of high-energy photons as an activation source. Photon activation offers analysis for certain elements not determined easily by neutron irradiation (e.g., Pb, Cd, and Ti) and a similar instrumental approach employing γ-ray spectrometry can be used. However, the application of photons differs sufficiently in practical details from neutron activation to make a single review of the two methods unsuitable. The reader is referred to a general review of the photon activation technique by Hislop [8] and its specific application to soil by Chattopadhyay and Jervis [9].

The present chapter confines its attention to the use of neutrons in soil analysis, to the employment of instrumental methods alone, and to consideration of published applications of the technique.

II. PRINCIPLES OF NEUTRON ACTIVATION ANALYSIS

A. Theory

When material is bombarded with neutrons, charged particles, or photons, some may be converted by a nuclear reaction to a radioactive isotope of the same or another element. In the case of neutron irradiation, the conversion is usually to a heavier isotope of the same element. Excess energy is released in the form of prompt γ rays:

$$ {}^{w}_{z}A + {}^{1}_{0}n = {}^{w+1}_{z}A + \gamma \tag{1} $$

Usually, though not always, the new isotope is radioactive and measurement of its activity yields an estimate of the quantity of element A present in the irradiated material.

Each isotope of an element (active or nonactive) has a specific probability of capturing a neutron of particular energy. Nuclear reactors are generally used as the source of neutrons; the neutron energy involved in most reactors is low (thermal), and the capture probability remains constant. This capture probability has the dimension of area and is called the *capture cross section*, having units of *barns*, viz. 10^{-24} cm^{-2}. The amount of induced activity can be calculated [10] from the following equation:

$$ D_0 = \frac{m\theta\sigma N\phi}{w} \left\{ 1 - \exp\left[-\ln(2)\,\frac{t}{T_{1/2}} \right] \right\} \tag{2} $$

where

D_0 = disintegration rate in becquerels (dis sec^{-1})

m = mass of element undergoing irradiation

θ = fractional natural abundance of the relevant isotope

σ = absorption cross section (cm^2)

w = atomic weight of isotope

N = Avogadro's number

ϕ = irradiation flux (n cm^{-2} sec^{-1})

t = irradiation period

$T_{1/2}$ = half-life of active isotope, in same units as t

After a decay period T,

$$ D_T = D_0 \exp\left[-\ln(2)\,\frac{T}{T_{1/2}} \right] \tag{3} $$

A consequence of the relationship shown in Eq. (2) is that for low values of t the induced radioactivity rises almost linearly with t, but is limited to a saturation activity at an infinite value of t when decay and production rates of activity are balanced. Note that there will be half the saturation activity produced after an irradiation period of one half-life.

The relationship between mass and induced activity is, of course, linear. Some indication of the potential sensitivity of activation analysis can be gained from consideration of a hypothetical isotope with some typical nuclear parameters. The activity induced in one nanogram of a monoisotopic element of atomic number Z = 100 and a cross-section of 0.1 barn, when irradiated for one half-life in a neutron flux of 10^{14} n cm^{-2} sec^{-1}, is 300 becquerels ($\sim$8 nCi). Classical chemical methods would permit us to separate this element with a high degree of radiochemical purity, and this level of activity would then be measured with considerable precision and sensitivity. Indeed, neutron activation analysis has employed radiochemistry as its major discrimination technique almost since its inception (see Chap. 6).

The principle of instrumental neutron activation analysis is to avoid tedious and expensive chemical separation and to use a nuclear property, the energy of associated γ rays, as the discriminating factor. As will be shown in the next section, the resolving power of the detection systems used for this, though considerable, can hardly compete with the separation factors of 10^8 or more which are obtained by chemistry. Instrumental activation therefore relies on limiting its application to matrices in which most of the activation takes place among the minor and trace elements present. Examination of the activation cross sections of the elements in the periodic table shows very little relation between them and chemical properties (unlike parameters normally encountered in chemical discrimination methods), but activation susceptibility does tend to be low for elements of low Z, either because the cross section is small or because the resultant isotope has a very short half-life. So far as soil is concerned, C, N, O, H, and Si produce very little activity while Al only produces an isotope with a half-life of 2.3 min.

Although the linear relation between mass and induced activity is shown in Eq. (2) above, the establishment of the proportionality constants requires knowledge of a number of experimental and nuclear parameters. Such data are available, together with calculated values of induced activity, in tabular form in a compilation by Erdtmann [11].

In practice, a comparative method is used, in which a known quantity of the element to be determined is irradiated under similar conditions to the sample. Then all irradiation and nuclear parameters can be regarded as constants between sample and standard and the activity ratios can be considered equal to the mass ratios.

Normalization refers only to correction for differences in decay times between sample and standard:

$$\frac{\text{Mass of element in soil sample}}{\text{Mass of element in standard}} = \frac{\text{normalized activity in soil}}{\text{normalized activity of standard}} \tag{4}$$

The instrumental method permits discrimination not only by energy but also by decay rate. A judicious combination of irradiation and decay times can optimize the sensitivity with which any particular element can be determined. To do this rigorously in a complex matrix requires a suitable computer program simulating the pulse height distributions obtained for a variety of irradiation and measurement parameters such as those developed by Zikovsky and Schweikert [12], Guinn [13], and Law [14]. However, a common assumption is that a decay period of approximately one half-life will optimize the signal from a specific isotope. Because of the growth factor in Eq. (2), the irradiation period should be kept low in relation to the half-life, provided sufficient activity is induced to be measured with adequate precision.

At any one time only a few isotopes are visible in a pulse height spectrum, but by continuing measurements over a considerable period as many as 40 elements may be determined in suitable matrices.

B. Gamma-Ray Spectrometry

The general principle of γ-ray spectrometry has been discussed in Chap. 6, and is therefore stated only briefly here. It simply involves energy discrimination by use of a detector whose response to an interaction with a γ ray is related to the energy of that photon and thus behaves as a proportional counter. Analog signals from the detector may be translated to a digital form, and the resulting number used to select a memory channel of a pulse height analyzer, where the event is recorded. With present-day techniques, the memory of a suitable computer may be used instead. A large number of events received over a period of time will result in the buildup of a frequency histogram or pulse-height spectrum which relates quite closely to the initial distribution of photon energies from the radioactive source.

Currently, the most popular form of detector is the solid-state germanium (lithium drifted) type, which has considerable qualities of energy discrimination though somewhat inconveniently needs to be maintained at liquid nitrogen temperatures throughout its life. The principles of operation and other features of the spectrometric technique are described in the standard textbook by Crouthamel et al. [15].

Solid-state detectors may be conveniently thought of as ion chambers, where the number of ion pairs produced is proportional to the absorbed energy. Consideration of the energy required to produce an ion pair and the statistical nature of the ion production process shows that the resolving power, that is the width of a total absorption peak, can theoretically be as low as 1 keV at 1000 keV. In fact, detectors giving peaks less than 2 keV in width are routinely available commercially. It is not usual for these detectors to be available with useful volumes greater than 100-200 cm^3, and in some special applications requiring large samples, a scintillation (sodium iodide) detector is required simply to provide sufficient detection efficiency. Such detectors are available with a volume of a liter or more. However, the inefficiency of energy collection and its transfer in the scintillation process is such that the resolving power is some 30-40 times worse than that of a Ge(Li) solid-state detector.

The majority of γ rays entering the detector will be scattered due to the Compton effect, when the energy of the secondary electron is given by:

$$E - \frac{E}{1 + a(1 - \beta)} \qquad (5)$$

where

E_e = energy of secondary electron (keV)

E = energy of incident photon (keV)

$a = E/511$

β = angle of photon scatter

Figure 1 shows the relative probability of Compton interaction at different energies and compared with other processes. The dependence on energy is not great. The highest energy imparted to a scattered Compton electron is less than the primary energy of the incident photon and corresponds to a scattering angle of 180°.

$$E_e(\text{max}) = \frac{E}{1 + 0.5/a} \qquad (6)$$

Figure 2 illustrates a pulse-height spectrum from ^{24}Na. This isotope has two equally intense γ rays at energies 2754 and 1369 keV. The corresponding Compton continua are shown with maximum energies in accord with Eq. (6) above. The somewhat concave shape of the continua is due to the nonisotropic nature of the scattering process. The differential probability of scattering at specific angles is given by Klein and Nishina [16].

The important photon interaction process, from the analyst's viewpoint, is the photoelectric interaction in which all the energy is converted (albeit indirectly) into the output signal. The photoelectric absorption probability varies as Z^5, hence germanium ($Z = 32$) is preferred to silicon ($Z = 14$) as material for semicon-

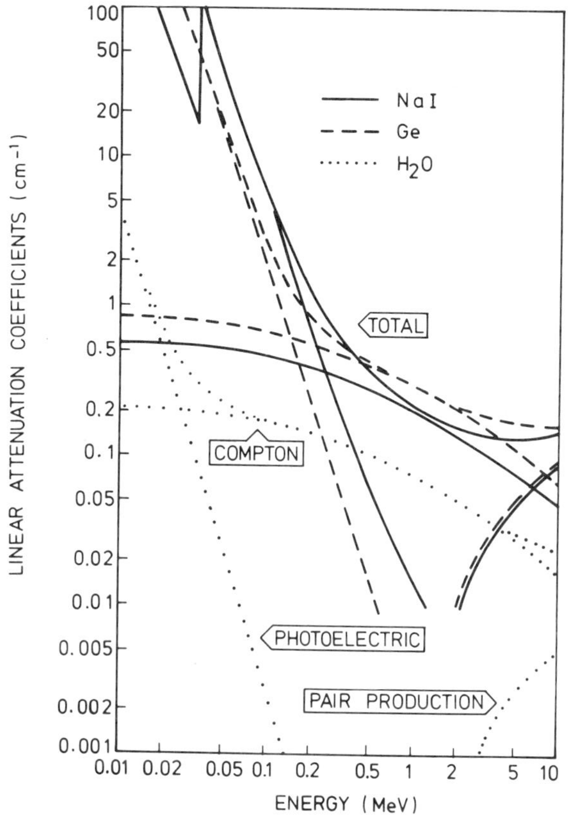

Figure 1 Absorption coefficients for γ-radiation in three media. (Courtesy of the Harshaw Chemical Co., Northants, England.)

ductor detectors of γ radiation. It appears from Fig. 1 that since the probability for photoelectric absorption falls rapidly with increasing energy, the size of the most energetic peaks will be very small. However, the scattering process reduces the initial energy of the photon, hence if the detector is sufficiently large to permit several interactions within its volume without the escape of the scattered photon, then its eventual photoelectric capture becomes distinctly probable as its energy drops. It is thus more accurate to refer to the observed peak as a "total absorption" peak. The appropriate two peaks are clearly seen in Fig. 2. Total absorption efficiencies for detectors of different sizes and materials are commonly tabulated in the literature either from experimental data such as those compiled by Heath [17] or from calculations using mathematical modeling techniques [18].

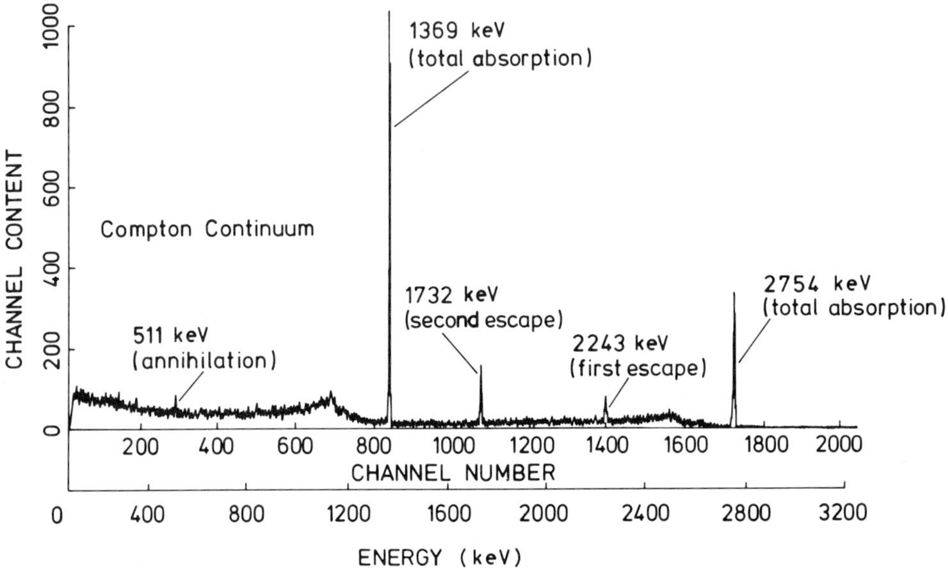

Figure 2 γ-Ray spectrum from ^{24}Na.

Other peaks are observed from high-energy photons due to the "pair production" phenomenon (see Chap. 6) when "escape" peaks occur at E − 511 and E − 1022 keV (see Fig. 2), where E is the energy of the total absorption peak.

Ideally, the peaks in a pulse-height spectrum should be of a symmetric Gaussian type. However, if the source itself is large or an absorbing material is interposed between the source and the detector, then some degradation may take place due to low angle scattering and the low-energy side of the peak may be relatively enhanced. Quite severe degradation may also take place if the amplification circuitry is unable to restore the pulse base line rapidly following the record of an event. Then if event rates are high, pulse heights are in error, resulting in peak broadening or shape distortion. The interpretation of such data is discussed in a later section.

The important relationships in a spectrometer are those between photon energy and peak position and between peak size and source intensity. The first relation is practically linear, though deviations from this are also discussed later. The property of linear addition in a spectrometry system is almost always obeyed at low event rates. At high rates, the total number of events recorded in a peak is subject to linear addition, but this may not be true of individual channels within the peak.

The resolving power of a detector is usually defined in terms of the full width at half the maximum height of a peak (FWHM) at a specific energy, e.g., 1.9 keV at 1330 keV (^{60}Co). The resolution varies as:

$$FWHM = k_1 + k_2E \tag{7}$$

where k_1 and k_2 are specific constants for the detection system.

In activation analysis, the detector used should be as efficient and of as high resolution as possible. The second factor is the more important, should economic considerations prevail. A useful Ge(Li) detector should have a resolving power of less than 2.0 keV at 1330 keV and have a total absorption efficiency at 667 keV of 10% of that of a standard 3 in. (75 mm) sodium iodide crystal. (The parameters are those generally in use commercially.)

Many laboratories find it useful to support an additional low efficiency detection system to measure sources of high activity, rather than perform additional irradiations to suit the single sensitive detector.

It has been demonstrated by Atkins et al. [19] that, for some limited applications, sodium iodide detectors are quite adequate devices for short-lived activation analysis. Also, a nonactivation application, the determination of natural radioactivity (radium, thorium, and potassium) in soil, is also best performed by a large sodium iodide detector [20].

There is probably some advantage in employing detectors particularly sensitive to low-energy photons or X rays. This has been demonstrated particularly in the estimation of a range of rare earths and platinum metals in rocks and lunar soils by Rosenberg and Wiik [21] and Gijbels and Govaerts [22]. Such detectors are virtually windowless Ge(Li) and Si(Li) or latterly made of intrinsic germanium.

III. PRACTICAL DETAILS OF ANALYSIS

A. Collection and Preparation of Samples

The application of instrumental neutron activation analysis to the determination of trace elements in soils requires special techniques of sample collection and preparation. Although most soil samples are taken by auger or spade, the use of a coring device of known cross-sectional area is preferred for studies on the input of elements from the atmosphere to ground, since the integrated deposition per unit area needs to be known. Thus Ragaini et al. [23] took soil core samples to a depth of 26 cm by driving a steel cylinder into the ground, employing an interchangeable plastic liner to avoid contamination by metals. A polypropylene corer, 32 mm in diameter, has also been used to avoid metal contamination [24].

Whereas for conventional analyses, samples > 10 g are commonly prepared from air-dried soil sieved through a 2 mm mesh, the requirement of the activation method for small samples, usually from 50 to 200 mg, demands the production of a very homogeneous specimen of soil with a final particle size below 300-500 μm diameter. This problem was also recognized during analysis of the standard Bowen kale powder [25].

For preparation of finely ground or sieved soil, various approaches have been described by different workers. Nevertheless, many reports in the literature do not provide details of the processing of soil samples prior to analysis by INAA. The danger of contamination by metals from conventional disk grinders and mills, and the difficulty of cleaning out such apparatus between samples to avoid cross-contamination, has led authors to the use of an agate ball mill for pulverization [26,27] or an agate pestle and mortar [28].

Tests with a vibrating agate ball pestle mill of 50 cm^3 capacity showed that 15 g of dry soil (< 2 mm) was pulverized in 20 min. to < 200 μm particle diameter. Furthermore, 70% of the sample passed a 150 μm sieve and 18% was < 53 μm [29]. Other grinding techniques include the use of diamonite mortars [30,31] and an iron mill [31]. In archaeological work [32], samples of 20 g of soil were freeze-dried before grinding in a vibrating mill to pass a 1 mm^2 nylon mesh when 200 mg samples were used for analysis by neutron activation. In other work [23], soil samples were air-dried and ground to <75 μm mesh prior to short-term irradiation of 25 mg and longer irradiation of 200 mg samples.

The dependence of trace element concentrations on soil particle size must be appreciated if specific size fractions are sieved out for analysis without homogenization. Evidence to show the effect of particle size on measured element concentration is mentioned in Sec. V.C.

For the packaging of samples for irradiation, practically all analysts recommend the use of clear silica tubes. Tubing should be cleaned with a powerful oxidant and thoroughly rinsed in tapwater followed by deionized water. After insertion of the soil sample, the tube should be sealed by fusion. For quite short irradiations (a few minutes) the authors have successfully used unsealed silica tubes plugged with aluminum foil. Similarly, polythene vials and sealed bags are useful if selected for their low inorganic content.

Liquid samples may arise from soil extracts and solutions (Sec. IV.B). These should be carefully evaporated onto suitable matrices (for example, Whatman 40 filter paper) and the matrix itself packaged and irradiated. It is possible to irradiate, for short periods, liquids sealed in quartz or even polythene tubes. However, the procedure is not recommended nowadays. An air space should be left in the vessel equal to at least half of its volume. Also the solution should not contain substances (e.g., nitric acid) subject to radiolysis.

Samples should always be removed from silica irradiation containers before measurement. However, for the measurement of isotopes having short half-lives it is quite practicable to measure irradiated soil or filter papers within an irradiated polythene container.

B. Irradiation of Samples

The majority of neutron activation analyses performed on soil have used a nuclear reactor as the source of neutrons. Not only is a high neutron flux available (up to 10^{14} n cm^{-2} sec^{-1}) for maximum sensitivity, but a reasonable degree of irradiation uniformity can be expected due to the isotropic distribution of neutrons within a reactor.

Nuclear research establishments frequently permit access to the larger reactors (e.g., DIDO heavy water reactor at A. E. R. E. Harwell, U.K.) where irradiation facilities exist which permit uniform, high intensity thermal neutron fluxes. However, many universities throughout the world provide "swimming-pool" reactors which use a light water moderator with enriched uranium fuel. These may even be deliberately designed for activation analysis, providing special irradiation facilities and operating at low cost. Such reactors have been marketed as TRIGA reactors by General Atomic Inc. [33] and SLOWPOKE by Atomic Energy of Canada Ltd. [34]. Nuclear power reactors rarely have useful irradiation facilities and operate at inconveniently high temperatures.

It is possible to use nonreactor sources of neutrons for analysis of rocks and soils, but they are so lacking in intensity that they are really only suitable for measurement of major elements. Plutonium-beryllium and radium-beryllium sources have been used in the past though effective thermal fluxes have not usually exceeded 10^5 n cm^{-2} sec^{-1} [35]. In recent years, however, ^{252}Cf sources have become available, yielding fluxes of up to 10^7 thermal n cm^{-2} sec^{-1}, depending on the quantity and physical arrangement of the isotope [36].

An alternative source is the neutron generator which, by deuteron excitation of a tritium target, produces 14 MeV neutrons from the $^3_1H(d,n)^4_2He$ reaction. Fluxes of 10^{10} n cm^{-2} sec^{-1} are possible [37], though again the irradiating beam is not isotropic and the high energy of the neutrons makes the device more suitable for special reactions on elements of low atomic number, rather than for the more usual application of multi-trace element analysis. A detailed review of neutron beams in activation analysis is given by De Soete et al. [2].

The relative intensities of thermal and epi-thermal neutrons vary considerably, depending on their source. Generally it is advantageous to employ as low energy a source of neutrons as possible. The cross-section for (n,γ) reactions is highest at low energies, thus yielding maximum sensitivity. Also, a wide variety of alternative reactions, e.g., (n,α), (n,p), $(n,2n)$, occur with generally increasing probability at higher energies. These may well interfere with the determination of one element due to the enhancement of the appropriate activity by an alternative neutron reaction with another element (see Sec. II.A).

It may happen that a particular element can only be conveniently measured by use of a high-energy neutron reaction. A well-known example is the measurement of nickel by $^{58}Ni(n,p)^{58}Co$, a more useful and sensitive reaction than $^{64}Ni(n,\gamma)^{65}Ni$, while oxygen can be determined by $^{16}O(n,p)^{16}N$. In general, however, it is in the interest of the analyst to use as thermalized a flux as possible. For this reason, large heavy water (i.e., well moderated) reactors available at some national laboratories are practically ideal. Elsewhere, a homogeneous thermal flux can usually only be obtained at the cost of source intensity, and a compromise has to be accepted between sensitivity and potential interference.

Another area in which some compromise is frequently necessary is the length of time of irradiation and the subsequent period of decay before measurement. To obtain the maximum level of induced activity would require an irradiation of several half-lives. However, the rate of induction of activity ceases to be linear before an irradiation period of one half-life has elapsed (Eq. (2)). Thus inducement of sufficient total activity should be controlled by selection of the sample weight and the neutron flux, while the irradiation period should be kept short in relation to the relevant half-life. Also, the decay period should be selected to optimize the relative level of activity it is desired to measure compared with all others present. If all other activities were uniformly distributed in quantity and half-life then the optimum time to make the measurement would be about one half-life of the activity of interest after irradiation. In most practical cases, however, one interfering activity is often dominant for a particular period of time. Such examples in the case of soil are the activities arising from aluminum, sodium and scandium. Thus decay periods between measurements are chosen to avoid the major sources of interference. It is possible to forecast the optimum conditions of irradiation and decay by means of suitable computer programs, which model the γ-ray distribution induced in soil of known approximate composition and hence derive the relative signal of particular interest [9,10,14].

The instrumental activation technique is normally employed for multielement analysis, when 20-30 elements are estimated. Conditions cannot be optimized for all these, and some general compromise needs to be adopted. In the authors' laboratory, two separate soil samples each of about 50-100 mg, are prepared and irradiated in the DIDO reactor, one for a few seconds and the other for 4-8 hr. These periods could be scaled up or down depending on the flux used and the weight of sample taken. The first sample is measured within

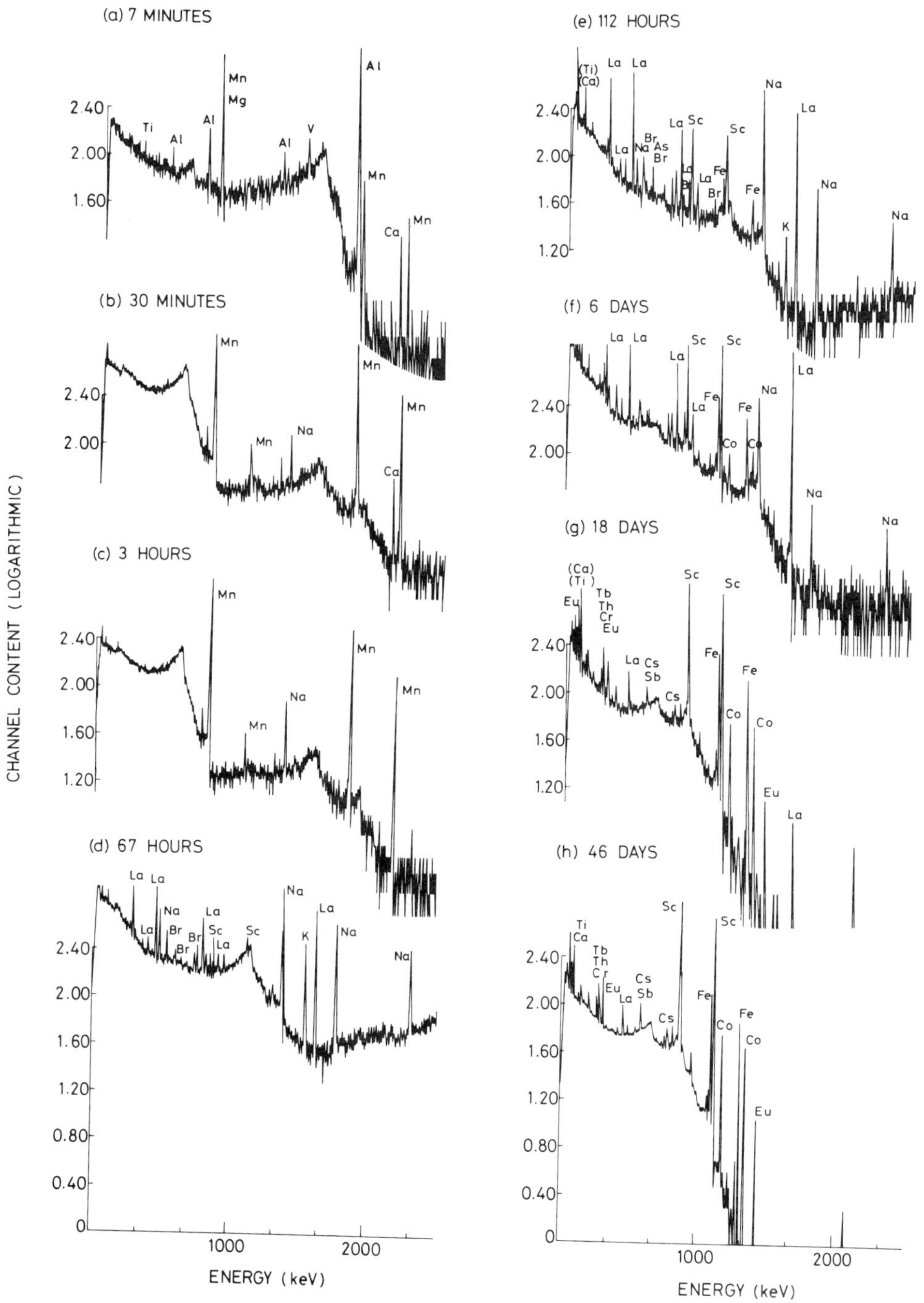

Figure 3 Sequential γ-ray spectra from irradiated soil.

5 min of irradiation and thereafter at intervals of 30 min, 3 hr, and (possibly) 14 hr. The second is measured after 2 days and 20 days.

Figure 3 displays the differing form of the spectra as the source decays and indicates the appearance of a variety of peaks from many isotopes.

Ideally all samples should be irradiated uniformly in an isotropic neutron flux unattenuated by the sample. In the case of small soil samples, the average capture cross section is sufficiently low [<< 1 barn (10^{-24} cm^2)] for attenuation to be quite negligible. It is, however, quite important to ensure that neither the irradiation container nor material in the reactor near the sample are of sufficiently high cross section to produce significant local flux changes in the sample. In reactors designed for activation experiments there are devices to rotate a number of samples in the reactor to ensure uniformity of irradiation [33]. This uniformity is particularly difficult to achieve in nonisotropic irradiation conditions as found in neutron generators and ^{252}Cf sources.

In order to monitor the neutron flux received by the sample, standards of all elements to be determined should be irradiated with the sample and be in a similar physical state to the sample. It is not practicable, however, to irradiate a large set of fresh standards for every irradiation made, and it is assumed that one set of standards can be irradiated and measured to cover a reasonable period of time (perhaps some weeks), during which the energy spectrum of the neutron source will not change. Small differences in irradiation time and absolute flux level can then be monitored by including one or two suitable reference standards with each set of samples irradiated. Such standards should be of high purity and of such concentration and nuclear characteristics that an accurate measurement of the neutron dose received can be made without perturbing that flux. The use of sodium salts (e.g., sodium carbonate) and pure iron wire is recommended for short and long periods of irradiation, respectively.

Master standards of elements to be determined will, of course, be required. Most workers employ freshly prepared dilute solutions of appropriate compounds. The compound chosen should be a soluble salt of the element concerned containing no other "activatable" elements and available in a chemically pure form. It should be diluted to a concentration comparable with that to be determined and a measured quantity sealed in a container of material such as silica. After irradiation under known conditions, the solution should be removed from the container and a source prepared physically resembling a soil sample.

C. Measurement of Radioactivity

The actual measurement of γ radiation is performed in the conventional manner of radiation detection. The detector is shielded so far as is possible from external radiation sources and the sample itself arranged in a well-defined and reproducible position relative to the detector.

The physical arrangement may be somewhat difficult with solid-state Ge(Li) detectors, since the associated crysostat is inevitably bulky (e.g., 0.8 m in height and 0.4 m in diameter), though in activation work it is usually unnecessary to provide a lead shield to cover the entire assembly. Most laboratories performing activation analysis routinely provide some form of sample changing mechanism. For such a system to be useful, it is essential that each sample be moved to a measurement position identical with its predecessors and that the storage system for samples awaiting measurement be sufficiently distant and well shielded for their activity to be insignificant to the detector. Figures 4a and b show such a system with automatically controlled doors to the shield. An accurate sample positioning mechanism driven by stepping motors is controlled by a small local computer and provides options in the measurement sequence.

The minimization of radiation background is not perhaps so important in activation analysis as in some other applications of γ-ray spectrometry. Nevertheless, some pains should be taken to shield the detector effectively with 100 mm of lead (selected for its low activity) and to plug gaps between the shield and the cryostat with bags of lead shot. Regular measurements of background should be made to discover occasional contamination and sources of varying external radiation present in some nuclear environments. The lead should be lined with successive layers of cadmium, copper, and aluminum sheet, to absorb interfering characteristic X rays from the lead and subsequent metals due to irradiation by the source being measured. The device supporting the sample should be of lucite or similar material.

If the irradiation and measurement conditions are correctly chosen, then there will be a sufficient event rate registered by the detector to provide an adequate spectrum in a reasonably short time. At too great a rate there will be spectrum distortion due to overloading of the pulse amplification equipment. If the source is of such activity that it needs to be measured at a considerable distance from the detector, then unwanted scattering effects are observed to distort the continuum of the spectrum. It is essential that the sample exhibits the same geometry as the standards used for comparison. Efficiency is very dependent on the solid angle subtended by a source onto the surface of the detector. It should be noted that the point of greatest sensitivity to small changes for a disk-shaped source appears to be a distance of one disk radius from the detector.

It is equally important that characteristics of the detection system remain unchanged between measurement of standard and sample. The position and shape of absorption peaks must remain constant. They should also exhibit the property of linear addition, that is, twice the activity should produce twice the size of the cor-

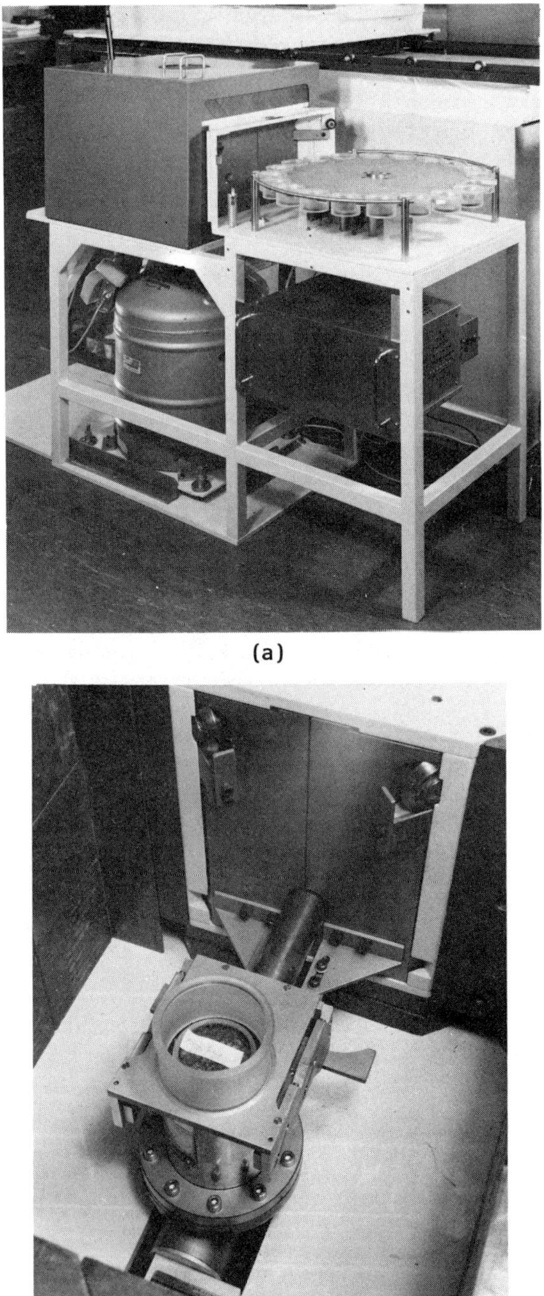

(a)

(b)

Figure 4 (a) General view of detector and sample-changing mechanism. (b) Local view of detector and sample-changing mechanism.

responding peak. The amplification characteristics (zero offset and channel width) should remain fixed with carefully chosen equipment. The detection equipment must be kept at a constant temperature. This will obviously be so for a solid-state detector cooled by liquid nitrogen. A scintillation detector, if used, should not only be thermostatically housed but should have some gain stabilization system fitted, e.g., that used by Keane [38], employing an internal reference standard. The peak shape is best maintained by measuring similar event rates with both sample and standard, but the amplification system should use a base-line restoration system before presenting an analog signal to a digital converter, and thus avoid some of the distortion caused by overlap of signals that occur during high event rate measurements.

The analog-to-digital converter (ADC) should be of the Wilkinson type employing a high-frequency (20-100 MHz) oscillator to cope with the high levels of activity that will inevitably be encountered at some time in activation work.

The pulse-height analyzer stores the information gained from the ADC by incrementation of the corrresponding address in its memory. It is now commonplace to use a computer as the controlling device for pulse-height analysis whereupon the analyzer memory is either treated as a computer peripheral or forms part of the memory itself. In the latter case the memory incrementation is nearly always performed as an "autonomous operation" without involving the computer's processing unit.

The pulse-height spectrum should be accumulated for a sufficient time to achieve acceptable statistical precision. Thereafter, depending on the level of prior information available on similar spectra, it will be necessary to examine, manipulate, and transport the data before a numerical examination is performed. The complexity of such operations which may be carried out using the analyzer depends on the capability of the built-in circuitry. These techniques, however, bear such a resemblance to computer operations that this circuitry is nowadays replaced by cheaper and more powerful "firmware" (prewritten computer code fixed within the system). This may be extremely powerful and provide digital and analog interactive displays with, to a limited degree, assessment of the identification and magnitude of individual spectrum components.

"Small" computer systems are now available which control simultaneously the acquisition of pulse height spectra from a variety of detectors and provide complete computer programs for the local interactive analysis of activation data. (The reader is referred to the current major manufacturers in this field for literature.) Alternatively, less sophisticated systems may provide the wherewithal for controlled data acquisition, but the numerical interpretation is performed with the aid of a large independent computer available at an institution or commercial facility.

A number of computer codes are available which can be used for activation analysis:—HEVESEY [39], GASP [40,41], GASPAN [42], HYPERMET [43], GAMANAL [44], SAMPO [45] are but a few. Although such computer codes are available for performing numerical analyses, it should be realized that quantities of data need to be recorded, transported, and presented to the computer in a suitable form. Storage and transport of data have been a stumbling block in the efficient analysis of pulse-height data for many years. Magnetic tape, though often used, is expensive and one pays for a considerably greater data capacity than is actually needed. Paper tape, though very cheap, is clumsy, unreliable, and voluminous. Cassette tape is attractively cheap and compact, but in the absence of an industry standard is rarely compatible with a major computing system. Perhaps the most convenient medium is that of a flexible magnetic diskette (floppy disk). This compact form of data storage is rapidly replacing tape and cards as a standard medium which can be read by most computers. Each disk, costing only a few dollars, can store over 100,000 channels of spectrum data, and a stack of a dozen 200 mm disks is only 20 mm deep.

Some installations may have a link permitting land-line transmission between a local data acquisition system and a central computer. This is naturally a preferred method of transfer, but the rate of transfer needs to be a minimum of 150 channels per second (2400 baud), and preferably a good deal higher, to be useful. Automatic error checking facilities are essential.

It is often convenient to retain a permanent graphical presentation of the spectra obtained, and most spectrometer systems either incorporate a pen recorder or have access to some computer-controlled system of graphics.

D. Interpretation of Spectra

The activation analyst will inevitably find the need for data evaluation and analysis by computing methods even though subjective judgment must still play a large part in the initial interpretation of activation data. The quantity of digital data produced is so large, even from one measurement, that some aid is necessary. Admittedly, much useful analysis has long been performed by manual examination of data plotted in analog form and with purely graphical methods used to assess the peak energies and their magnitudes. Such methods, however, apart from being laborious, inevitably introduce subjective error, and usually fail to utilize all available information. They certainly fail to provide a statistical assessment of the data.

The operations performed in the interpretation of spectra begin with a recognition of the major features, viz., the total absorption peaks. The energies of the peaks need to be measured with some accuracy using suitable calibration methods. Qualitative analysis of the spectrum constituents is then carried out by recognition of specific photon energies and their relative intensities. In the routine analysis of similar batches of soil samples, this assessment does not need to be made manually in most laboratories.

The quantitative assessment may be made a general one in which all components are included. In this case some routine procedure such as regression analysis, with large library of standard spectra, must be invoked. Alternatively, if specific elements are sought, then direct comparison between total absorption peaks in the sample and an appropriate standard must be used. If a direct reference standard is not available, then an efficiency calibration may be established with secondary standards of different energies.

The first requirement, that of finding significant peaks, has been subject to considerable investigation by Routti and Prussin [45], Robertson and Kennet [46], Mariscotti [47], Yule [48], Blok et al. [49], among others. The topic has been reviewed by Op de Beeck and Hoste [50] and by Phillips and Marlow [51]. A numerical procedure based simply on the rise and fall of successive channel contents is too crude even when put onto a formal statistical basis. It is found necessary to perform a "convolution" operation (e.g., calculate a smoothed second derivative) on successive groups of channels. Then, although the procedure is one in which statistical fluctuations are smoothed, it also filters out the contributions to the spectrum from continua and other slowly varying changes such as those caused by scatter peaks.

By comparing the convulate with the standard deviation at any point, to give a ratio R, we will locate peaks where R exceeds some arbitrary value. In practice, five standard deviations will detect a peak with considerable certainty. Above this value the test becomes insensitive and below, too many "spurious" peaks will be found.

Although location techniques may indicate the channel corresponding to the center of a peak, a more accurate figure is required to assess the corresponding γ-ray energy. For this, a peak "centroid" measurement is needed. The continuum beneath the peak is first subtracted, making use of located boundaries to the peak or by making assumptions about the peak width and shape. A trapezoidal shape for the continuum is normally assumed. The residual peak may then be treated as a symmetric Gaussian and a simple least squares fit is performed on the quadratic shape obtained from the logarithmic transform of the peak. Alternatively, a simple way to calculate the centroid is by moments [50]. In cases where the symmetry assumption is unreasonable, then a more elaborate nonlinear regression analysis needs to be performed using a suitable function describing peak shape, for example, those by Routti and Prussin [45] and Booker [52].

The conversion of peak position to energy is performed by measurement of the position of peaks of known energy, either from calibration standards measured independently but under identical conditions, or from known peaks used as internal standards within the spectrum itself. The relation between peak position and energy is very nearly linear:

$$E = (a + p)b \tag{8}$$

where E is the energy represented by the peak at position p channels and a and b are the calibration constants "zero offset" and "channel width" respectively. However, some small deviation from linearity is normal, and quadratic or higher polynomials may be invoked for the calibration line, depending on the number of accurately measured points that are available, [50,53].

A calibration line may alternatively be drawn from experimental points using a cubic spline interpolation [54].

The identification of isotopes from energies and relative intensities is mostly performed by manual search. Major compilations of γ-ray energies include those by Lederer et al. [55] and Erdtmann and Soyka [56]. The latter provides a very complete list of all γ rays in ascending energy order and is available in computer-compatible form for incorporation into any suitable program suite. Tables limited to neutron activation products are available (e.g., Adams and Dams [57] and Carder et al. [58]).

Computer methods of isotope identification vary in complexity. The simplest method is just to match a measured energy against that which is arithmetically nearest in a computer stored table of isotopes. Usually, an arbitrary level of energy uncertainty is chosen such that all isotopes within a probable energy level are offered for consideration. Clearly, the more comprehensive the table stored, the more complex the solutions offered. A degree of sophistication can be incorporated which checks on the half-life of the isotope in relation to the proffered γ-ray energy and the presence of confirmatory γ rays can also be noted [58]. A more stringent procedure by Gunninck and Niday [59], elaborated by Christensen et al. [60], generates a confidence index of identification by measurement of all associated γ rays.

In the limited case of activation analysis associated with relatively few isotopes in a relevant half-life range, it might be more practicable to employ a method of inverted search such as that employed in analysis of mass spectra. Here, rather than identify each peak in the spectrum separately, a library of candidate isotopes is compared with the spectrum to find whether or not relevant γ-ray energies are present and whether in the right proportions.

The quantitative assessment of identified isotopes may either be performed by individual peak area measurement or by linear regression analysis applied to the whole of the spectrum. The former technique is commonly employed, particularly in small computer systems.

The area of a peak may be simply calculated as the area above a baseline which can either be assumed straight or subjected to some polynomial extrapolation of the continuum on either side of the peak, as in Quittner [61].

It is often difficult to assess the peak boundaries accurately, due to local statistical fluctuations (though convolution methods may be used to smooth these effects), and a comparison of a peak with one from an identical standard may be made in relative terms, based on partial peak area measurement. The classical method is due to Covell [62], where a center channel is chosen and a symmetric range of k channels on either side of the center is selected which does not reach either peak boundary. As modified by Op de Beeck and Hoste [50], this is:

$$A_p = \sum_{i=-(k-1)}^{k-1} a_i - (k - 1/2)(a_k + a_{-k}) \tag{9}$$

The variance on this assessment is:

$$V_p = \sum_{i=-(k-1)}^{k-1} a_i + (k - 1/2)^2(a_k + a_{-k}) \tag{10}$$

Maximum estimates for components too low to be measured can therefore be made. Other useful methods of partial peak area measurement are reviewed by Baedecker [63] and by Op de Beeck and Hoste [50].

Where two or more peaks overlap, their area may be simply assessed by linear regression analysis if the position and width of the components are known accurately. This method has been further elaborated by Christensen et al. [60], who provide an iterative correction to the assessment of peak position.

Nonlinear fitting procedures for peak fitting have been in use for many years; the literature is considerable and has been reviewed by Phillips and Marlow [51], McNelles and Campbell [64], and Booker [52]. Peak shapes suggested vary from simple Gaussian to modifications with exponential tails on one or both sides of the peak and a variety of asymmetric distributions based on the supposed behavior of pulses emerging from the analog amplifier and shaping systems. None of these models are very difficult to describe mathematically, but their fitting to experimental data using the method of least squares depends on the controlled behavior of an iterative minimization procedure. This can be a delicate operation when dealing with experimental data containing information limited by the statistical nature of the nuclear disintegration process. Nevertheless, such numerical procedures are commonly used for resolving multiple peaks of unsymmetric shape and are employed by the programs listed earlier in this section.

For the analysis of large numbers of similar spectra, a nonsubjective means of analysis, that of multilinear regression, is suitable. Although developed for sodium iodide detectors where the majority of component spectra overlap [65,66], it is equally applicable to solid-state detection systems, provided a high degree of system stability is maintained and the computational power to handle fairly large matrices is available.

Other numerical procedures may involve corrections for minor differences in irradiation and measurement parameters between sample and standard. Gibbons et al. [67] have considered the corrections to be made from extended flux level variations during irradiation.

In cases where an isotope has been identified as present in a spectrum but no standard is available, then an estimate can be made of the original elemental mass by applying the equations of Sec. II.A and using the appropriate nuclear constants and estimates of absolute flux level. Absolute disintegration rates are needed and thus the peak area needs to be estimated from one or more multi-energy standards such as ^{226}Ra or ^{152}Eu. Efficiencies calculated for a series of known energies are fitted to an experimental curve or joined by a series of cubic splines. A suitable general function is a pair of power law curves [53].

Finally, an important area of data processing is the collated storage of analytical results such that they can be validated [68] and presented in a useful form either as analytical or derived data [69]. It is also profitable in such circumstances to have automatic mechanisms for associating analytical data with the sample from which it arose.

E. Sensitivity and Applicability

A simple statement of sensitivity for the determination of individual elements in soil by INAA cannot be made. The relative contributions of constituent elements determine the signal-to-background ratios of absorption peaks in γ-ray spectra, hence sensitivity varies with each element in every soil type. Figure 5 shows four γ-ray spectra of differing soil types (see Table 3, Sec. IV.A) which illustrate the differing relative magnitude of constituent peaks and hence their degrees of interference in the determination of minor peaks.

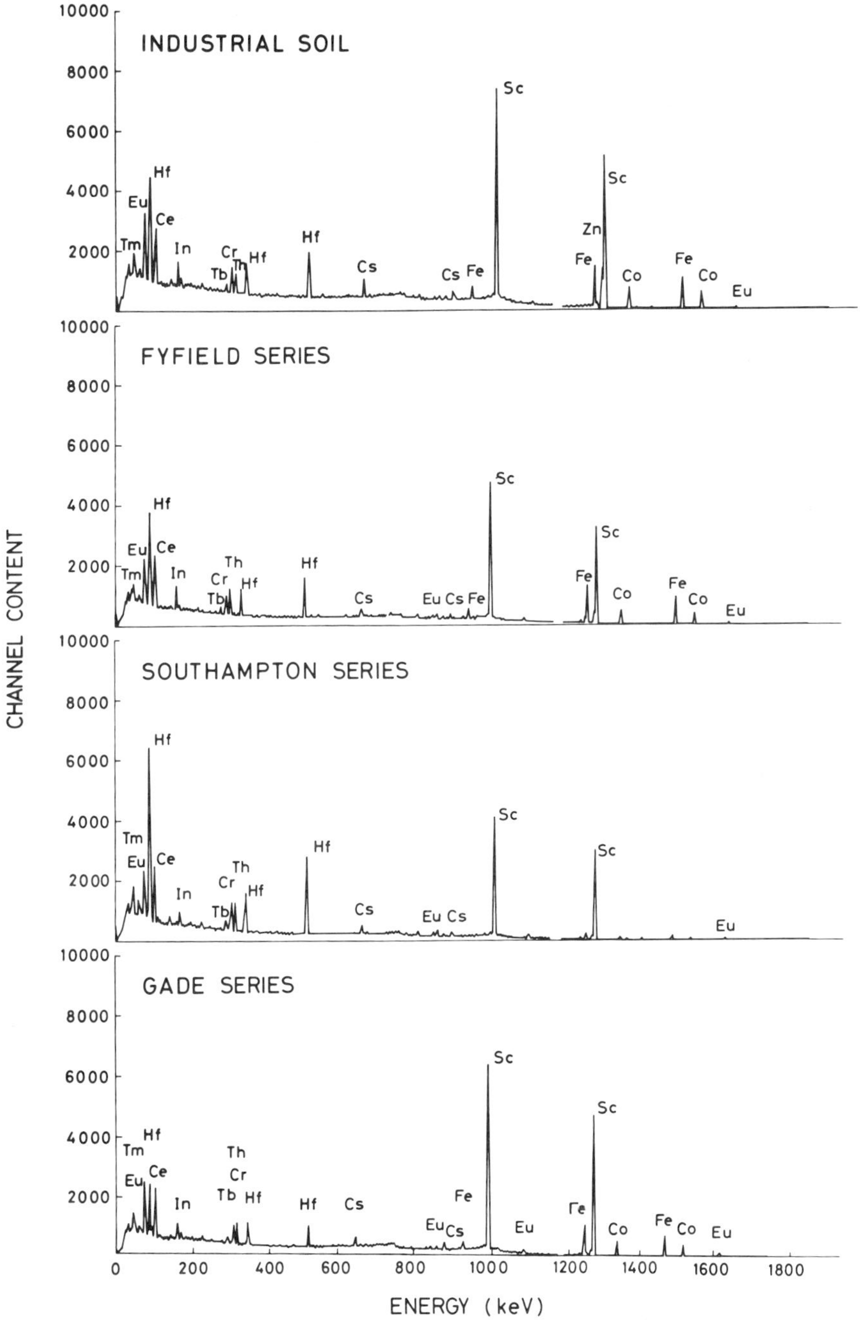

Figure 5 γ-Ray spectra from irradiated samples of differing soil type.

Detection (or minimum determination) levels can be expressed in a variety of ways, as discussed in Currie's classical review [70]. However, it is convenient to use the calculated standard deviation of the measured levels in the determination of a particular element as some indication of the minimum levels measurable. To give the reader some approximate minimum determination levels of particular elements in soil, Table 1 provides a list of *average* minimum levels for a group of U.K. soils.

The quoted levels are also some measure of the precision or reproducibility of results to be expected in soil analysis by INAA. The data are based on the statistical uncertainties associated with the radiation measurement, i.e., of "counting" statistics and detector resolving power. No account is taken of uncertainties in sampling or in preparation of the standards. Errors related to nonuniformity of irradiation flux may occur, but these should be very small in well-designed facilities. Likewise, errors due to geometrical dissimilarities in the measurement equipment should be negligible if reasonable

Table 1 Average Minimum Concentrations of Elements Detectable in a Group of U.K. Soils

Element	Concentration ($\mu g\ g^{-1}$)[a]	Element	Concentration ($\mu g\ g^{-1}$)[a]
Al	0.15%	K	0.15%
Ag	0.15	La	1.5
As	0.5	Mn	20
Au	0.005	Mo	5
Br	2.0	Na	300
Ca	0.4%	Ni	30
Cd	4	Rb	20
Ce	2	Sb	0.5
Co	0.35	Sc	0.25
Cr	2.5	Se	0.4
Cs	0.3	Sm	0.2
Cu	4	Tb	0.1
Dy	1	Th	0.3
Eu	0.02	Ti	1000
Fe	0.10%	V	10
Hf	0.15	W	0.3
Hg	0.6	Yb	0.3
I	10	Zn	4.0
In	0.05		

[a]Except where indicated.

care is taken in its design and operation. A more likely source of significant error is that introduced by distortion of spectrum shape due either to large differences in levels of measured radioactivity, or to instabilities in the measuring equipment itself.

Overall accuracy is not easily determined, but where analysis of similar material was examined using INAA [71], then 10-20% accuracy at least could be expected for most elements where the precision was good.

A review of this form of analysis of soil should include a statement of disadvantages. These are at least threefold:

1. The physics and economics of irradiation procedures demand that samples should not exceed a few grams and preferably be a good deal less. Thus a problem immediately exists with such a heterogeneous material as soil in the production of a suitably representative sample for analysis.
2. Access to a suitable irradiation source is a very limiting factor, especially if isotopes with a half-life of a few hours or even minutes are to be measured. Furthermore, it is unrealistic to expect a wide range of elemental analyses unless the irradiation source is a nuclear reactor.
3. A wealth of numerical data is provided with every sample measured. The laboratory is faced with either an excessive amount of manual computation on the part of its analytical staff or a fairly heavy investment in computing resources (both hardware and software) beyond the already expensive electronic equipment necessary to measure the samples and store the consequent digital data. For the investment to show a good economic return the throughput of samples needs to be quite high.

These problems are essentially organizational and economic. They do not detract from the basic scientific worth of the activation technique with its high specificity and sensitivity for a range of environmentally significant elements. It seems a reasonable conclusion, therefore, that in cases where the particular scientific advantages of the method are overwhelming, it is worth the trouble and labor of applying the method manually to a few important samples. At the other end of the scale, if the irradiation, measurement, and data handling resources are available then this powerful method should be used routinely.

Depending on the element and the irradiation and measurement conditions, levels of activity corresponding to submicrogram quantities can be determined.

A feature, not necessarily an advantage, of activation analysis is that the total element, irrespective of its chemical or physical composition, is measured.

Finally, some advantages of INAA may be summarized thus:

1. It is an economic multielement technique in the appropriate circumstances and well suited to automation and data processing operations.
2. The sensitivity is particularly good for a number of elements of interest both in plant and animal nutrition studies and in toxicity investigations related to pollution.
3. Trace element contamination by reagents and chemical processing is completely avoided.
4. Dissolution and other chemical operations are avoided in total element studies.

IV. BASIC STUDIES

A. Total Element Analysis of Soils

Basic investigations on the induced γ activities in a wide range of neutron irradiated soils have been reported over the last 20 years, and have concentrated on details of irradiation, instrumentation, and analysis of γ spectra rather than applications of the technique to various areas of research, which are discussed in detail in Sec. V.

In 1959, McNeilly and Eichler [72] examined the activities induced by irradiation of five U.S. soils of varied composition with both thermal and 14 MeV neutrons. They presented spectra obtained by a NaI(Tl) detector and identified peaks from ^{28}Al, ^{56}Mn, and ^{24}Na following thermal neutron irradiation, and an additional peak from ^{27}Mg from fast neutron irradiation.

Instrumental NAA of soils and rocks in Norway was achieved by Thoresen [73] in 1964, for Al, Br, Cu, Fe, K, La, Mn, Na, and Sc. In this case, separate irradiations were made at low and high neutron fluxes up to 10^{12} n cm^{-2} sec^{-1}. Analysis of γ-ray spectra was made by application of a spectrum stripping method to reveal the photopeaks of lower energy. Thoresen [73] commented that the amounts of Na in his samples limited detection of shorter-lived isotopes, and quoted approximate relative detection limits of 10^{-2} for Cu:Na and 1 for K:Na. For detection of Br, the amount of La was critical, thus the limit of detection was Br:La $\sim 10^{-1}$. Also in 1964, Kline [74] was able to obtain quantitative measurement by INAA of ^{60}Co, ^{51}Cr, ^{59}Fe, ^{140}La, and ^{46}Sc. Interference from ^{24}Na prevented analysis of isotopes with shorter half-lives than 15 hr.

A method for the rapid determination of total Al, Mn, and Na in Norwegian soils by INAA, with a precision of 1-2%, was published by Khera and Steinnes [75] in 1969. The authors gave attention to interfering nuclear reactions by (n,p) or (n,α) processes and found that correction factors were significant only for ^{28}Al.

With the advent of Ge(Li) detectors, Salmon and Creevy [76] developed an on-line computer system for INAA of air, water, and soil samples whereby γ-ray spectra were obtained from a NaI(Tl) detector for short-lived isotopes, and a Ge(Li) detector for longer-lived activities. The authors pointed out the inflexibility of pulse-height analyzers for accumulation of data from several detectors in a laboratory handling large numbers of routine determinations of numerous elements in many different media, and suggested replacement by a computer analyzer system.

More recently, in 1977, Naidenov and Travesi [77] presented INAA results for 43 soils from Bulgaria which were representative of a wide range of soil classes and profile depths. Altogether, 19 elements were determined following irradiation of samples with thermal neutrons and application of an Ortec 459 Ge(Li) semiconductor detector for low-energy photon spectrometry (LEPS) in addition to a 40 cm^3 Ge(Li) detector for energies > 200 keV. Use of the LEPS system on irradiated soils permitted analysis of low-energy peaks from ^{141}Ce, ^{177}Lu, ^{153}Sm, and ^{169}Yb, and reductions in interference from high-energy peaks allowed measurement of isotopes such as ^{147}Nd (91.4 keV peak) and ^{160}Tb (87.0 keV peak). For some elements, a check of the analysis was possible at both low and high energies, using the different detectors. The authors also showed that there was negligible interference from U in soil, in the analysis of elements whose activation products were also fission products.

Trace elements were measured in several Alaskan soils and clay minerals following irradiation with resonance (epithermal) neutrons to extend the analysis to elements with longer half-lives. Samples were sealed in quartz ampoules for irradiation over 40 hr at a total flux of 10^{12} n cm^{-2} sec^{-1} and were shielded by Cd foil to exclude thermal neutrons [78]. Quantitative data were obtained for Co, Cr, Fe, Ni, Ti, and Zn using a Ge(Li) detector. Ten additional elements were identified, including Ba, Cs, Ta, and U. This technique was claimed to represent an improvement on solely thermal neutron irradiation, where induced activities of elements with high cross-section or concentration in soil may obscure the γ-ray spectra of other elements that are present in trace amounts and have low activation cross-sections.

In 1973, total elemental analysis of soils from eastern Uruguay was reported by Buenafama [79]. Combined techniques of INAA and NAA with radiochemical separation were used, and both thermal and epithermal neutrons were applied for activation. Concentrations of 25 elements were thus determined by INAA, γ-ray spectra being measured by a Ge(Li) detector system; peak areas were evaluated by the Wasson method [63]. After radiochemical separation, Cu, Ga, Mo, Sr, W and Zn were determined. Some results obtained by

Table 2 Trace Element Content of a Basaltic Soil Determined by INAA

Element	Total Concentration ($\mu g\ g^{-1}$ dry soil)	S.D. (± %)
Br	22	23
Ca	8900	12
Co	33	1.2
Cr	220	4.2
Fe	69100	0.7
K	5900	3.4
Mn	1930	1.2
Na	4570	1.3
Sb	2.8	10
V	660	5.0

Source: Ref. 79.

Buenafama [79] for some elements of agricultural interest, including potentially toxic pollutants, are listed in Table 2, and represent the means of at least triplicate determinations.

Modern techniques of INAA have been applied by the authors [29] to the analysis of a range of contrasting soil types from southern England. These soils were sampled to 15 cm depth from arable, grassland, and woodland locations. The concentrations and the uncertainties found for 37 elements are shown in Table 3. Compared with other soils, the podzol (Southampton Series) showed considerably lower concentrations of As, Co, Ni, V, and Zn than others, probably the result of leaching under acid conditions. The excellent permeability to water of the brown earth (Fyfield Series loamy sand) was reflected in relatively low concentrations of Br, Cs, Na and rare earths.

Kerr [32] established that activation of soil samples with thermal neutrons combined with Ge(Li) spectrometry gave quantitative data for Al, Ba, Br, Ca, Cl, K, Mg, Mn, Na, Ti, and V; use of a low-energy photon detector provided additional results for Co, Dy, Eu, Sm, Th, and U. Cadmium shielding for core irradiations was only of value for Si, Sr, and Sm. Following cyclic neutron-activation (30 cycles, 5 sec irradiation and 5 sec count), Sc, Se, and Hf were also detected, but this technique was the most affected by the production of ^{28}Al by neutron capture, which interferes with the measurement of spectra from other isotopes.

Detection limits by activation analysis were compared with other methods of microanalysis for Cl, Co, Cu, Fe, Mn, Mo, Si, and Zn, and while it was recognized that B, an important trace element, cannot be determined by activation, sensitivity for the other elements was considered adequate [80].

B. Analysis of Soil Fractions

Measurements of trace elements in soil solutions and extracts by modern techniques of INAA are relatively few compared with work on the determination of total element content, although to most agriculturalists the exchangeable and solution "pools" of individual elements and the interelement ratios in these available fractions are vital in studies of plant nutrition and plant uptake.

Determination of Cl, Fe, K, and Na in neutron-irradiated soil solutions (simulated) was achieved by direct γ-ray spectrometry and counting with a NaI(Tl) detector, but with reliance on chemical separation by ion-exchange columns for analysis of Ca, Mg, P and S [82]. More recently, removal of soil solutions by a centrifugation method [83] followed by evaporation of 50-100 ml of sample onto filter paper (Whatman 41 grade) disks for irradiation has been demonstrated as a practical technique [29]; the composition of soil solutions extracted from a calcareous silty loam and analyzed by this method (Icknield Series) are shown in Table 4.

Soil water from lysimeters was analyzed for Al after removal of Cl by an anion exchange resin, to avoid interference by ^{38}Cl following irradiation of liquid samples by thermal neutrons [84]. The samples contained from 2 to 173 $\mu g\ ml^{-1}$ chloride. It was noted that a high Mg content of the soil water precluded the use of the 1040 keV γ-ray peak of ^{27}Mg as employed in metallurgical analysis. Production of ^{28}Al by an (n,p) reaction on ^{28}Si necessitated correction after separate colorimetric determination of Si concentrations. A NaI(Tl) detector was used for counting, and it was possible to reduce interference from ^{56}Mn by spectrum stripping.

For measurement of "plant-available" trace elements in soil, 0.5 M acetic acid is frequently used as an extractant. This reagent may easily be freed from metal contamination by distillation in quartz apparatus, and dilution with quartz-distilled water. An extractant/dry soil ratio of 40 is recommended, with a 2 hr shaking period [85]. As with soil solutions, analysis by INAA can be effected by evaporation of filtered extracts onto filter paper.

Other reagents frequently used as soil extractants in agricultural advisory work are KCl, NH_4OAc, NH_4Cl, 0.1 M HCl, 0.1 M HNO_3 or Morgan's reagent (NaOAc and acetic acid). Of these, NH_4OAc is best suited to analysis of extracts by INAA, unless only long-lived isotopes are sought and radioactive decay is allowed to eliminate interference from ^{24}Na and ^{38}Cl. Moreover, NH_4OAc can be prepared in a very pure state by bubbling ammonia gas through quartz-distilled acetic acid.

Khera et al. [86] determined ^{28}Al in irradiated extracts from 29 Norwegian soils; a well-type NaI(Tl) detector was used to measure the 1780 keV γ ray. The method had a sensitivity of 5×10^{-5} μg Al, and the overall results compared closely with those obtained

Table 3 Determination of Total Elemental Concentrations in Topsoils from Southern England ($\mu g\ g^{-1}$ dry soil, or % where stated)[a]

Element	Fyfield, Brown earth, Arable	Denchworth, Surface-water gley, Woodland	Southampton, Humus-iron podzol, Woodland	Gade, calcareous ground-water gley, Grassland
Al	1.4%	5.3%	1.1%	3.6%
Ag	0.96	2.4	1.5	1.3
As	7.3	16.7	4.1	6.6
Au	0.0036	0.024	0.065	0.0037
Br	3.1	14.4	6.2	16.2
Ca	0.96%	2.4%	2.1%	5.3%
Cd	3.9	6.6	< 6	< 3
Ce	32	59	37	41
Co	5.3	9.1	0.85	8.0
Cr	42	57	57	50
Cs	1.4	6.1	1.3	2.9
Cu	20	5.9	10.9	3.1
Dy	2.4	6.9	4.4	6.1
Eu	0.38	0.73	0.40	0.65
Fe	1.6%	2.8%	2.6%	1.94%
Hf	6.9	6.9	12.9	5.6
Hg	< 0.4	< 0.8	< 0.4	< 0.6
I	31	56	12.6	< 10
In	0.15	0.38	0.20	0.23
K	0.69%	2.1%	1.1%	0.54%
La	18.2	35	22	30
Mn	220	250	510	340
Mo	< 7	< 5	< 2	< 9
Na	410	3000	810	1800
Ni	45	81	< 20	69
Rb	< 13	82	24	48
Sb	0.65	< 0.9	0.98	< 0.6
Sc	2.4	7.6	2.3	5.3
Se	0.92	< 0.9	1.1	1.3
Sm	2.0	3.6	2.7	3.7
Th	2.6	6.9	4.3	4.2
Ti	1520	3900	5900	1300
V	51	102	30	58
W	< 0.3	1.8	1.8	0.94
Yb	0.71	1.4	1.9	1.1
Zn	59	78	11.6	80

[a]Names given are recognized soil series, and follow the Soil Survey of England and Wales classification [81].

Icknield, Rendzina, Grassland	Sulham, Peaty gley, Grassland	Westhay, Organic soil, Grassland	S.D. (%)
4.5%	3.4%	3.7%	< 5
1.8	2.4	1.6	< 10
10.2	13.7	16.5	< 5
0.029	0.018	0.057	< 20
28	10.0	89	< 10
4.2%	2.1%	3.2%	< 15
< 7	5.9	9.9	> 25
59	44	42	< 5
10.8	7.2	8.3	< 5
38	37	47	< 20
2.9	1.7	4.4	< 10
22	29	18.4	> 25
8.2	7.1	4.9	< 15
1.2	0.83	0.66	< 5
2.4%	2.2%	2.2%	< 5
4.7	4.6	3.0	< 5
< 0.7	< 1.7	< 0.6	> 25
93	< 12	24	> 25
0.35	0.49	0.16	< 20
1.92%	1.06%	2.1%	< 10
59	36	26	< 5
1430	350	490	< 5
< 4	< 6	< 4	> 25
1910	1340	1670	< 20
150	107	< 40	> 25
< 24	< 19	47	> 25
< 0.7	1.6	< 0.7	< 15
6.8	5.3	5.9	< 5
< 0.8	1.8	1.0	< 20
6.1	4.7	3.3	< 5
5.8	3.9	4.7	< 5
1770	2200	< 700	> 25
74	81	71	< 10
2.0	1.0	1.4	< 20
2.2	1.2	0.79	< 20
104	52	96	< 5

Table 4 Elemental Composition of Different Soil Fractions Determined by INAA[a]

	Concentrations ($\mu g\ g^{-1}$ dry soil)				
Element	Exchangeable fraction[b]		Soil solution[c]		% Uncertainty of determination
Al	69	(0.2%)	0.074	(0.0002%)	< 5
As	0.066	(0.7%)	< 0.0007	(< 0.007%)	< 5
Br	10.7	(38%)	0.0048	(0.02%)	< 10
Ca	17200	(41%)	98	(0.20%)	< 10
Cd	0.47	(< 7%)	< 0.0002	n.d.	> 25
Cl	< 40	n.d.	1.2	n.d.	> 25
Co	0.077	(0.7%)	0.00047	(0.004%)	< 5
Cr	0.17	(0.5%)	0.0049	(0.01%)	< 10
Cu	6.3	(29%)	0.22	(1%)	< 10
Fe	9.2	(0.04%)	< 0.4	(< 0.002%)	> 25

Hg	< 0.01	n.d.	0.0048	(> 0.7%)	> 25
K	240	(1%)	6.2	(0.03%)	> 25
Mg	520	(> 13%)	5.1	(> 0.1%)	< 20
Mn	49	(3%)	0.0093	(0.0007%)	< 5
Mo	< 0.02	n.d.	< 0.0004	n.d.	> 25
Na	155	(8%)	10.2	(0.5%)	< 5
Ni	1.4	(0.9%)	< 0.06	(0.04%)	> 25
Sb	0.029	(> 4%)	< 0.0006%	n.d.	< 20
Se	0.041	(> 5%)	0.17	(> 21%)	< 15
V	< 0.1	(< 0.1%)	0.0019	(0.003%)	< 20
Zn	8.1	(8%)	< 0.02	(< 0.02%)	< 5

n.d. Not determined; results for both soil fraction and total were below detection limit.

[a]Soil was a rendzina from Icknield Series, Chilton Oxon, U.K., sampled from permanent grassland, 0-15 cm depth.

[b]Extracted with 0.5 M acetic acid; 10 g of air-dried soil shaken for 2 hr with 400 ml extractant.

[c]Soil solution removed from soil by centrifugation at 41% moisture on a wet weight basis. Figures in parentheses denote percentage of total elemental concentration in soil.

by atomic absorption and spectrophotometric methods. In this study the extracts were irradiated in liquid form in a thermal neutron flux of about 1.5×10^{13} n cm^{-2} sec^{-1}. Subsequently Khera and Steinnes [87] determined ^{56}Mn and ^{24}Na (as well as ^{28}Al) after neutron irradiation of similar soil extracts.

Extraction of soil with 0.1 M HNO_3, followed by irradiation of the centrifuged extracts, was used by other authors [26]: for short-lived isotopes, 0.1-0.5 ml of the supernatant was irradiated in a polyethylene vial, while for long-lived ones 0.5 ml of the extract was sealed in a quartz vial prior to irradiation. Results were obtained for 20 elements, using a neutron flux of 2×10^{13} n cm^{-2} sec^{-1} for 50 hr, and a 45 cm^3 Ge(Li) detector coupled to an 8192 channel pulse-height analyzer.

Extraction of ^{46}Sc and ^{140}La from thermal neutron-irradiated soils by agitation with 0.1 M HCl has been used in pedological studies [88]; the extractable forms of these elements were compared with their total concentrations, to examine relative differences between soil-mineral associations.

The availability of Cu for plant uptake was examined in surface and subsurface soils by irradiation of samples with thermal neutrons, followed by extraction of a diffusible fraction with dilute $CuSO_4$ solution [74]. Soils were further extracted with acidified (0.1 M HCl) $CuSO_4$ solution, and finally the residual Cu fraction was also determined from the induced ^{64}Cu activity.

Clay mineral fractions were prepared for INAA by sedimentation of soil samples in distilled water in a lucite cylinder [78]. Glassy fractions from volcanic ash soils were obtained by centrifugation in high purity bromoform-bromobenzene and through ultrasonic scrubbing, centrifuging, and freezing procedures [89].

Vanadium in clays has been determined by paper chromatography of acid extracts, followed by neutron-irradiation of the relevant portion of the chromatogram [90]. This enabled measurement of ^{52}V with little interference from the induced activities of Al, Mn, and Na, and the method achieved better sensitivity than atomic absorption.

V. APPLICATIONS

The overall range of elements analyzed in soils by INAA has been very extensive; Table 5 indicates some of the main areas to which the technique has been applied. The cited papers are discussed in of Sec. V.A to V.E.

A. Plant and Animal Nutrition

The determination of minor and trace elements by INAA in life sciences research has been reviewed by Bowen [91] and Leddicotte

[92]. It is evident that INAA may be used to determine all trace elements, with the exception of B, that are essential to plants. Rahman [80] has reviewed the importance of this method of micronutrient analysis in research on soil-plant relationships in Bangladesh.

The essentiality of Co in plant and animal nutrition led Johansen and Steinnes [93] to apply INAA to routine determinations. Neutron-irradiated soils and plant ashes were stored for 4 weeks before measurement of ^{60}Co, using a well-type NaI(Tl) detector. The relative standard deviation was 7.2% for soils, and 6% for plant material, based on duplicate determination of 14 samples in each case. The authors point out that if immediate analytical data are not essential and long post-irradiation cooling times can be tolerated, the method is very well suited to handling large numbers of samples.

The identification of minerals and the reserve of major and trace elements present in soils on the Matadi rock formations in Zaire was determined with the assistance of INAA with respect to Al, K, Mn, Na, and Si [94]. However, chemical separation was made for Ba, Cu, and Sr prior to γ-ray spectrometry with a Ge(Li) detector.

In Thailand, routine analysis of rice soils for Al, Mn, and Na has been accomplished by INAA [95]. Other elements were measured following chemical separation of thermal neutron-activated soils, and included As, Cu, Mn and Zn.

More recently, activation of plant and fertilizer samples with 14 MeV fast neutrons to determine the macronutrient elements Ca, Cl, K, Mg, N, P, and Si has been carried out on a large scale in the USSR by Srapenyants and Saveliev [96]. A throughput of 250-500 plant and fertilizer samples per 8 hr shift has been achieved, and the system has accomplished analysis of more than 50,000 such specimens. Although these authors did not describe any application of their technique to analysis of soils or soil extracts for diagnostic purposes, other workers in the USSR [97] have applied INAA to analysis of cotton plant tissues and soils for seven elements. Thus by irradiation with a flux of epicadmium (epithermal: Cd shielded) neutrons the following isotopes were measured: ^{28}Al, ^{49}Ca, ^{38}Cl, ^{42}K, ^{56}Mn, ^{101}Mo, and ^{24}Na.

B. Environmental Pollution

Research to assess the extent and impact of soil pollution by operations such as power generation, smelting and refining of metals, and disposal of industrial waste has made extensive use of INAA. Frequently such work has been carried out on an interdisciplinary basis to include atmospheric, terrestrial, and aquatic measurements.

The application of sewage sludge to agricultural land for disposal and/or as fertilizer can cause problems because of the content of associated trace metals. The accumulation of metals in the soil

Table 5 Application of INAA to Analysis of Soil; References to Research Areas

Element positively measured	Plant and soil	General environmental studies	Environmental pollution	Forensic science	Archaeology
Ag		118	23		
Al	95, 97	114, 123	27, 103, 109	131, 132	
As		114, 118, 123	23, 27, 103, 109		
Au			23, 27, 103		
Ba		118	27, 103		137, 138
Br		114, 118, 123	27, 104, 109		
Ca	97		103, 109		
Cd			23		
Ce		114, 118, 123	27, 104, 109	132	137, 138
Cl	97		27		
Co		114, 118, 123	27, 103, 109	131, 132	137, 138
Cr		114, 118, 123	27, 109		137, 138
Cs		114, 118, 123	27, 103, 104, 109		137, 138
Cu		114	23, 103		
Eu		118	27, 103, 104	131	137, 138
Fe		114, 118, 123	27, 103, 109	131, 132	137, 138
Ga		118	104		
Hf		118	27, 103		137, 138
Hg			23		

In			23		
K		118		132	138
La		114, 118	27, 103, 104, 109	131, 132	137, 138
Lu		118			137
Mg		114	27, 103		
Mn	95, 97	114, 123	27, 103, 109	132	138
Mo	97	118			
Na	95, 97	114, 118, 123	27, 103, 109	132	138
Ni			23		
Rb		118	27, 103, 109		137, 138
Sb		118, 123	23, 27, 103, 104, 109		
Sc		118, 123	23, 27, 104, 109	131, 132	137, 138
Se		123	23, 27, 103		
Sm		118	27, 104	131	
Ta		118	27, 103		137, 138
Tb		118	27, 104		
Th		114, 118, 123	103, 104, 109	132	137, 138
Ti		114	27, 103		
U			103, 104		138
V		114, 123	27, 103, 109		
W			109		
Yb			27		
Zn	97	118, 123	23, 27, 109		
Zr			27	132	

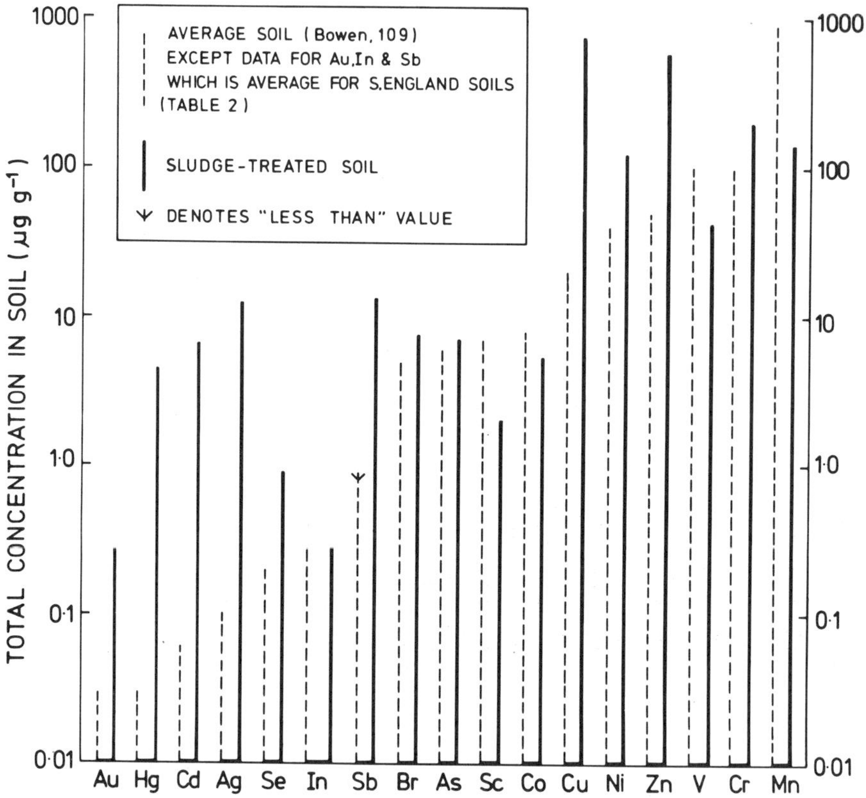

Figure 6 INAA of trace elements in sandy loam subject to long-term disposal of sewage sludge.

may be toxic to plants, or render vegetables and fruits unacceptable for human consumption. Contamination of ground water supplies by leaching of metals through soil may also require investigation. These problems demand a multi-element analysis technique. The determination of metals in sludges supplied from six cities in North Carolina has been achieved by INAA [98]. Results were obtained for Co, Cr, Fe, Hg, Sb, and Se, and it was suggested that Ag, Sn, and Zn could also be measured after a long decay time. The authors used both X-ray and γ-ray detectors.

Analysis of soil from grassland used for sewage sludge disposal was carried out by INAA, with the results shown in Fig. 6 [29]. Comparison with uncontaminated soil showed increases of two orders of magnitude in total concentrations of Ag, Cd, and Hg. For Cu, Sb, and Zn, an order of magnitude increase was observed.

In Japan, pollution by Cu of rivers used as sources of irrigation water has led to harmful effects on the growth of crops: INAA of Cu in plant tissue, soil and soil solution was made using a neutron flux of 5×10^{11} n cm^{-2} sec^{-1} and a 50 hr cooling period prior to γ-ray spectrometry [99].

The use of methyl bromide for soil fumigation to eradicate pests increases the uptake of Br by plants, with the result that marketed crops for human consumption may contain more than the acceptable level of Br. Although INAA has recently been used for the determination of ^{80}Br and ^{82}Br in vegetables by Suess and Staerk [100], it has so far found little application in the analysis of Br in soil or soil components. However, INAA would appear to be a promising technique to assess plant uptake factors for Br under varied conditions of soil management.

The use of arsenical pesticides in agriculture has led to a requirement for the determination of trace elements of As in plants and soils. Among the various techniques discussed by Talmi and Feldman [101], INAA is included. The induced γ activity of ^{76}As is measured with a Ge(Li) detector, and good sensitivity can be achieved in the absence of excessive ^{24}Na in the samples. Contamination of soils with chlorinated hydrocarbon pesticides has been examined by neutron irradiation of organic extracts of the soils to produce ^{38}Cl for measurement by the 2160 keV peak [102].

Emissions from coal-fired power stations into terrestial environment have been investigated by several workers in the United States. Lyon [103] determined 23 trace and major elements by INAA in soil profiles around a power plant at Memphis, Tennessee. Potential marker elements for fly-ash were identified and soil concentrations of Cu and Zn were 5 to 10 times higher than levels in average "world" soil. At Aiken, South Carolina, analysis of 29 elements was made in samples of fly-ash, soil, and vegetation collected at varying distances from a coal-fired power station [104]. Neutron activation was used to supplement atomic absorption analysis, plasma source emission spectrometry, and conventional spectrometry; 10 g samples were irradiated for 2 weeks with a ^{252}Cf source (100 mg), giving a flux of 10^7 n cm^{-2} sec^{-1}. Elements determined by γ-ray spectrometry were Br, Ce, Cs, Eu, Ga, La, Sb, Sc, Sm, Tb, Th, U, and Yb.

The use of Sr as an indicator for pollution around coal-fired power plants in the western United States prompted a comparison of techniques for analysis of soil and vegetation by Gladney, et al. [105]. INAA, epithermal neutron activation (IENAA) and atomic absorption (AAS) methods were applied. In the case of IENAA, higher energy (0.5 MeV) nonthermal neutrons were used to produce ^{87m}Sr and, although the detection limit was superior to INAA, the lowest detection limit was given by AAS; however, effort is required for sample dissolution in the latter method.

Crecelius et al. [106] measured As, Sb, and Pb contamination of soils at varying distances from a copper smelter: it was found that As and Sb were the best tracers of smelter stack dust. Dried soil was irradiated for 1 to 2 hr in a thermal neutron flux of 10^{12} n cm^{-2} sec^{-1}, and the ^{76}As and ^{122}Sb γ-ray peaks were counted after a cooling time of 2-4 days.

Near to gold mining and refining operations at Yellowknife, Canada, analysis by INAA of soil showed relatively high concentrations of As, Au, Co, Cr, and Sb [31]. The more organic soils showed higher levels of Co and Cr than mineral soils from the same general area. The ratios of As to Sb in soil were close to those reported for the mined ore.

More recently, the same gold refinery complex at Yellowknife was the subject of an epidemiological study by Jervis and Tiefenbach [107], but specifically on As pollution. INAA was used on samples of water, hair from children and workers, and on soil and plant specimens, using neutron irradiation at 10^{12} n cm^{-2} sec^{-1}; ^{76}As was measured after a 24 hr decay using a Ge(Li) detector. The authors commented that soils were the most dramatic indicator of local arsenic contamination and within 1-2 km of the smelters they measured two to three orders of magnitude more arsenic than normal terrestrial concentrations.

Pollution caused by urban metal refineries in Toronto, Canada, was examined by Jervis et al. [108]. Soil, vegetation, airborne particulates, dustfall and human hair were all analyzed. These authors used INAA to obtain information on pollution by As, Cd, Fe, and Sb, as these were associated with Pb emissions; in addition, data were obtained for Br (from vehicle exhaust and combustion sources) and Al, Ca, V, and Ti from soil and nonsmelter sources.

Pollution from metal mining, smelting, and refining industries in Kellogg, Idaho was assessed by multielement analysis of air particulates, soil and vegetation, employing INAA together with X-ray fluorescence analysis, since information on Pb levels was required [23]. High enrichments in soil and grass were reported for Ag, As, Au, Cd, Cu, Hg, In, Ni, Pb, Sb, Se and Zn; Sc was chosen as the normalizing element because it was considered that the only likely source of this element was the soil. An irradiation time of 2 min at 2×10^{13} n cm^{-2} sec^{-1} was used for short-lived isotopes, and 12 hr at 4×10^{12} n cm^{-2} sec^{-1} for long-lived ones. In this work, the depth distribution (profile) of metals in soil was measured to confirm that the parent material at the soil sampling points did not contribute to high levels of metals.

An interdisciplinary study of atmospheric and terrestrial pollution in the Swansea/Neath/Port Talbot industrial area of South Wales was initiated in 1971-1972, and analysis of soil and smelter waste by INAA was included [109]. The local soils had significantly higher concentrations of As, Cu, Ni, Se and Zn than uncontaminated soil [110], and the importance of resuspension and transport of waste deposits was examined by a comparison of element ratios with those in atmospheric samples.

The contamination of soil at the site of an old tin smelter in Lancashire, England was assessed by multielement analysis using INAA, supplemented by gamma-photon analysis for Pb and Sn [111]. Compared with the average analysis of surface soils [110], Ag, Sn, and W were enriched by three orders of magnitude, while an increase of 350- to 700-fold was detected for As, Cu, and Pb. Radish (roots) from the contaminated soil showed at least an order of magnitude increase in As, Br, Cd, Sn, Sb, and W concentrations above crops from normal soil, and As exceeded the statutory limit for foodstuffs.

The practical value of INAA for determination of 32 elements in soil beneath a landfill (refuse) site has been described by Van der Klugt et al. [27], and the heavy metals Ag, As, Co, Cr, and Zn were easily detected. The effect of leaching on downward migration of metals, and the potential for pollution of ground water, could be explored by an extension of this particular application.

In Canada, Kuja and Hutchinson [112] have used INAA in combination with atomic absorption to assess the toxicity to vegetation of mine wastes, with a view to eventual selection of tolerant species of grasses and sedges. In this work, elements analyzed in soil and plant specimens included Al, As, Cd, Cu, Ni, Pb, Se, and Zn.

Pollution of urban soils in the Glasgow area of Scotland by Br from automobile exhaust was used by Farmer and Cross [24] as a qualitative indicator of Pb pollution from the same source, using INAA to determine ^{82}Br. The soil samples were irradiated for 24 hr at a flux of 6×10^{12} n cm^{-2} sec^{-1} in the Harwell reactor DIDO. Natural concentrations of Br in soil were enhanced up to fivefold by pollution in urban areas. In a similar study [113], analysis was made of surface dust and vegetation close to highways, and also vehicle exhaust; altogether, 14 elements were measured including Br, which again proved a useful indicator of contamination levels.

C. General Studies on the Terrestrial Environment

Over the last 10 years, greater attention has been given to the cycling of both major and trace elements in the natural environment. Inputs of major elements from the atmosphere to plants and to the ground are of interest to agriculturists and hydrologists, while the influence of resuspension of soil dust on the qualitative and quantitative composition of the atmosphere is the concern of atmospheric chemists. The provision of data on the multielement composition of soil, plant, water, and atmospheric samples by the application of INAA has proved particularly suitable in this research area, as shown by the following examples. In all cases the total element content of soil has been measured.

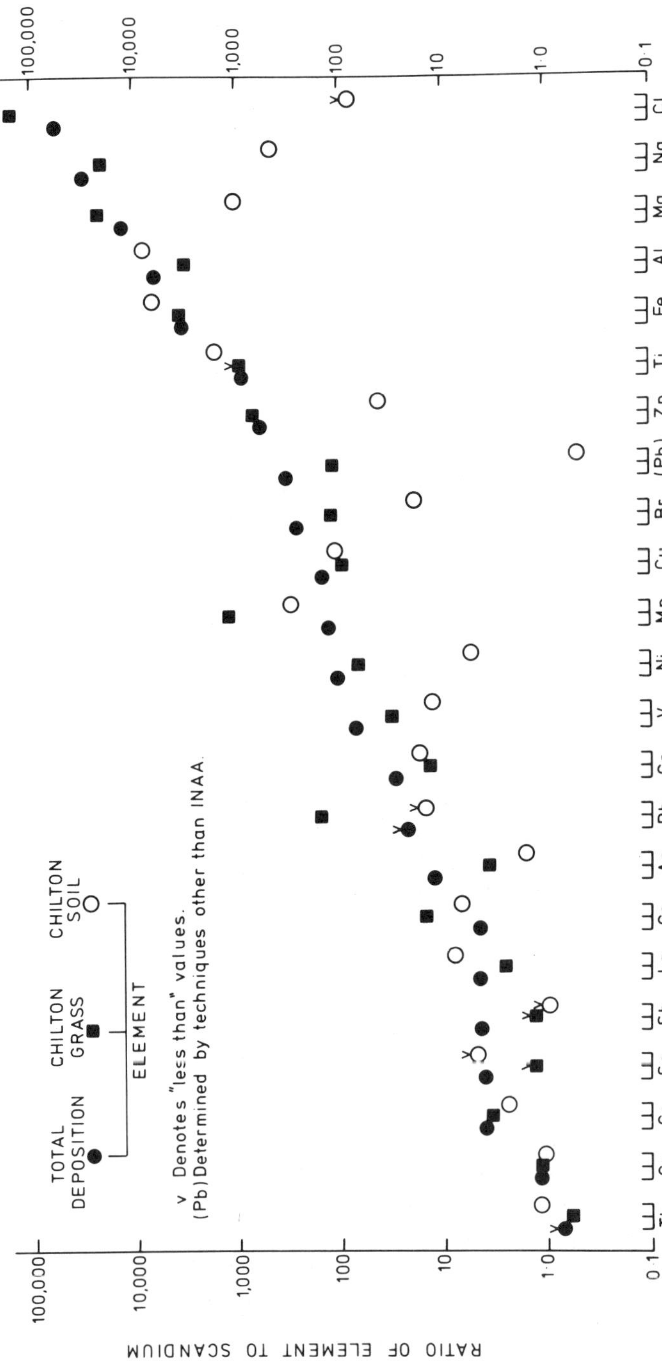

Figure 7 Ratios of elements to scandium in total deposition, grass and soil at Chilton, Oxon, England. (From Ref. 114.)

The multielement patterns of wet and dry (total) deposition, and of grass (*Festuca*) and surface soil from southern England were compared [114]. In this case, simple ratios of 23 elements to Sc rather than Al or Fe were used for comparison, because plants show negligible uptake and translocation of Sc; the results are shown in Fig. 7. It follows that any increase in element: Sc ratios relative to either soil or total deposition will demonstrate uptake by the roots or leaves. In fact, the results showed significant uptake of Ce, Cl, Mn, and Rb by grass, whereas for other elements analyzed, their ratios to Sc were similar in both total deposition and grass (Fig. 7). It was concluded that atmospheric deposition retained by the grass is a major factor influencing the composition of leaf tissues. Furthermore, it was shown that the ratios to Sc of As, Br, Cl, Co, Cr, Mg, Na, Ni, Sb, V, and Zn in the total deposition were greatly in excess of those in local soil, indicating industrial and maritime influences [114].

In Japan, Tanizaki and Nagatsuka [115] studied the distribution of 21 elements in lateritic soil, rainwater, and streamwater from the Ogasawara Islands by thermal NAA. The resuspension or scattering of soil was believed responsible for the occurrence of Al, Co, Cr, Cs, Fe, Rb, Sc, and V in rainwater and streamwater; however, the concentrations of As, Sb, W, and Zn found appeared to originate not from lateritic soil but rather from an industrial source.

The distribution of Cl, Br, and I in Norwegian forest soils has been studied by INAA [116] as part of a wider research program on trace elements of geochemical and geomedical interest. However, the presence of large quantities of ^{56}Mn after irradiation of soils with $< 80\%$ organic matter content precluded determination of I unless radiochemical separation was used.

Studies on the inorganic composition of the atmosphere have often been extended to include analysis of local surface soil, a potential contributor to airborne particulate matter. The enrichment of air particulates relative to soil may be examined by double normalization to an element that is not extensively used in industry, and is universally distributed in soil. Thus, an enrichment factor is derived, usually based on one of the normalizing elements Al, Ce, Fe, Sc, or Ti. For example, Gordon [117] reported analysis of 28 surface, subsoils, and deep soils from Maryland as part of an extensive research project on aerosol composition, enrichment, and sources. Some 20 elements were measured by INAA.

Concentrations of rare earth elements (REE) in Sahara soils and aerosols have been determined by INAA to examine their relative abundances, normalized to chondrites [118]. Light rare earth elements (Ce, Eu, La, Nd, Pr, and Sm) are enriched in urban aerosols relative to the heavier REE (Dy, Er, Gd, Ho, Lu, Tb, Tm, and Yb), and such enrichment is mainly found on the smaller particles in the atmosphere [118]. Other studies include the investigation of the origin of arctic haze bands that occur in the troposphere over Alaska [119], and measurement of the chemical composition of aerosols in the Sudan Gezira [120], northern Canada [121], and northern Nigeria [122].

In a survey of trace and major elements in air particulate, rainwater, and dry deposition at seven nonurban sites in the United Kingdom [123], analysis of the major soil type within 5 km radius of each sampling station was made. Surface soils (to 80 mm depth) were analyzed by INAA for total concentrations of 16 elements, supplemented by other methods for Ni and Pb. It was found that at certain locations the total deposition from the atmosphere of As, Br, Co, Cr, Na, Pb, and Zn was significant in relation to the total soil concentrations.

Particle size fractions in a surface profile of a buried loess soil from southeast Englane [124] were analyzed by INAA, using the Harwell reactor DIDO for activation at a flux up to 10^{14} n cm^{-2} sec^{-1} [29]. Preferential concentration was found on particles < 2 μm in diameter for the elements As, Ce, Cr, Co, Cs, Fe, La, Mn, Rb, Sc, Se, Th, and Zn. The reverse was true for Al, Na, and Ti, which were present at higher concentrations in the particle size range 2-50 μm diameter.

D. Forensic Science

The potential value of INAA in forensic investigations where the multielement composition of materials must be known for comparison and identification purposes, and a nondestructive method of analysis is preferred, was realized some 25 years ago. In addition, few alternative techniques could analyze extremely small specimens obtained from the scene of a crime, and accomplish this without dissolution of the sample, which may introduce trace elements as contamination from reagents. In an historic paper by Pro et al. [125], evidence was provided for the first time by activation analysis for use in the U.S. Federal Court. Their investigation was concerned with illicit distilleries, and involved analysis of whisky, soil, and paint. These materials were irradiated at 10^{12} n cm^{-2} sec^{-1}, followed by measurement of short- and long-lived γ-ray spectra with a NaI(Tl) detector. In eight soil samples, 13 isotopes were identified, including ^{198}Au, ^{131}Ba, ^{51}Cr, ^{239}Np, and ^{185}Os. Applications of INAA to forensic science have been reviewed by Bate et al. [126], Bryan and Guinn [127], Smythe [128], and Jervis [129]. Smythe [128] commented that the occurrence of trace elements in various drugs (plant-based) can often indicate the soils where they were grown, or the country of origin. Jervis [129] emphasized the improved resolution obtained by the Ge(Li) detector and gave examples from the analysis of hair, with comments on the extension of the technique to other media including soil.

In 1964, Bryan and Guinn [127] described developments in the United States of high flux neutron activation analysis, and differences in the elemental constituents of various media of forensic interest, including soil, were characterized. Furthermore, instrumental NAA was preferred to methods using chemical separation prior to γ-ray spectrometry on grounds of speed, preservation of specimens that might be required in court, and the ease of graphical presentation of results in legal proceedings.

In a comparison of forensic soil specimens by INAA [28,130], samples from Georgia, Texas, and Washington were irradiated in a thermal neutron flux of about 7×10^{12} n cm^{-2} sec^{-1}, for measurement of Ba, Ce, Cl, Cr, Cs, Hf, La, Na, Sc, and Sm using a NaI(Tl) detector. Additional elements were measured by atomic absorption, and with combined results for some 20 elements there was no difficulty in establishing the origin of the soils.

On the other hand, Kline and Brar [131] doubted the capability of INAA with NaI(Tl) detectors and 400-channel analyzers, to produce spectra that would characterize soils without other supporting evidence. Their conclusions were reached after analysis of 12 elements in soils from a world-wide collection. Samples were irradiated at a thermal flux of 2×10^{13} n cm^{-2} sec^{-1}. Short-lived isotopes ($T_{1/2} < 15$ hr) could not be measured owing to interference from ^{24}Na. The authors recognized that the use of low flux or Cd-shielded irradiations and higher resolution detectors for γ-ray spectrometry would improve the capability of INAA. Bate [132] discussed problems of comparison of soil analytical data for forensic work, and reported large variations in the content of some major and trace elements according to soil depth, with the exception of Co, La, Se and Th.

In conclusion, it seems essential that a wide multi-element spectrum is required before individual soil specimens can be characterized. It also appears desirable that results for more heavy metals are obtained for forensic work, since these elements are expected to be more variable in soils than the rare earths, Fe and Sc, and may be able to characterize urban locations particularly well.

E. Miscellaneous Studies

Further applications of INAA to soil research have been made in several areas including pedology, archaeology, and the measurement of tracer isotopes added to soil.

Pedological Studies

Garrec et al. [133] described the determination of Al, Fe, and Si in some French soils by INAA using 14 MeV neutrons. For the determination of Al, the reaction $^{27}Al(n,\alpha)^{24}Na$ was used, taking account of interference by $^{24}Mg(n,p)^{24}Na$. Silicon was determined by the production of ^{28}Al, and Fe by the induced activity of ^{56}Mn, both from (n,p) reactions. The results compared well with data obtained by activation of soils with thermal neutrons and separation of isotopes by ion-exchange resins.

The total concentrations of La and Sc in three soils from western Wisconsin were measured by INAA, for research on the effect of pedogenic processes on their translocation in depth profiles and on relationships of the two elements with clay, organic C and Fe content [88]. The authors concluded that since redistribution of Sc in soil development processes is less complex than that of La, it may be a useful marker for research on both soil morphogenesis and soil correlation.

In many parts of the world, soils have been derived from deposits of volcanic ash, and their identification and physico-chemical properties are of interest to pedologists, agriculturists and archaeologists. Thus the determination of trace elements in volcanic ash soils, pumice, and glassy fractions of ash has received attention. Borchardt et al. [134] applied INAA to measure 19 elements in samples from the Cascade Mountains area of the United States, using a flux of 7×10^{11} n cm^{-2} sec^{-1} and a 30 cm^3 Ge(Li) detector. The elements Ce, Fe, Hf, Na, Sc, Sm, and Th were determined with < 5% relative standard deviation. In a subsequent paper, Borchardt and Harward [30] established trace element correlations of volcanic ash soils from data obtained by INAA: the composition of volcanic glass from C horizons agreed with known volcanic ejecta from Mt. Mazama (Crater Lake, Oregon), and 15 stable elemental ratios unique to Mazama ash, e.g., Sc/Fe, La/Hf, Ce/Sc, Cs/Yb, were used to assist correlation of weathered soil horizons.

Iridium contents of soils were measured in the range 0.27-0.64 μg g^{-1} from the reaction $^{191}Ir(n,\gamma)^{192}Ir$ [135], for geochemical studies. A relatively short irradiation time of 30 hr at a flux of 5×10^{12} n cm^{-2} sec^{-1} was chosen to avoid excessive activation of major elements. Gamma-ray spectrometry was performed with a NaI(Tl) detector after 2-3 weeks cooling time.

For research on the geochemical behavior of Ti, the analysis of soil humic ash samples was achieved by INAA using 14 MeV neutrons [136]. An irradiation time of 20 hr was followed by 42 hr cooling before γ-ray spectrometry. The reaction $^{47}Ti(n,p)^{47}Sc$ was used for quantitative analysis, and spectra were obtained with a Ge(Li) detector. The results showed close agreement with a colormetric method, concentrations of 7 mg Ti g^{-1} ash being present.

Archaeological Research

INAA of clays and pottery from locations throughout the world has shown that measurement of Co, Cs, Hf, Sc, Ta, and Th could provide satisfactory clues on origins of articles [137]. Altogether, 15 isotopes were measured with a 30 cm^3 Ge(Li) detector coupled to a 4096-channel pulse-height analyzer, and grouping of the analytical data was studied statistically by cluster analysis.

Clay and shale from 20 sites in Egypt, and ancient pottery, were analyzed for 16 major and trace elements by instrumental thermal neutron activation [138]. Fluxes of 1-4 $\times 10^{14}$ cm^{-2} sec^{-1} were used for irradiation of these samples, and after appropriate cooling times the spectra of short- and long-lived activities were measured with a 35 cm^3 Ge(Li) detector. Differences were particularly marked in the concentrations of Ce, Co, Fe, Hf, Mn, Na, Rb, Ta, and Th.

Kerr [32] applied INAA to research at archaeological sites on the detection of buried soils, particularly the location of buried surfaces, by measuring variations in concentration of elements through the depth profile. Soil samples were obtained from Hadrian's Wall, near Carlisle, England, and were characterized by their relative levels of ^{28}Al, ^{29}Al, ^{139}Ba, ^{165}Dy, ^{42}K, ^{27}Mg, ^{56}Mn, ^{24}Na, ^{51}Ti, and ^{52}V following neutron activation.

Tracer Studies

The use of stable isotopes as tracers of transfer mechanisms in the terrestrial environment may be preferred to radioactive species, to avoid problems of safety in certain situations. In this connection, it has been demonstrated that the behavior of I in soils can be studied by means of stable ^{127}I tracer, with subsequent analysis by neutron activation at a flux of 2×10^{11} n cm^{-2} sec^{-1} to yield ^{128}I($T_{1/2}$ = 25 min) [139]. Benes et al. [139] also analyzed five of the most important soil types in Czechoslovakia by INAA, and established that increasing Mn and Al contents reduced the sensitivity of ^{128}I measurements.

The kinetics of Cs transport in a lake ecosystem was determined by the introduction of $^{133}CsCl$ as a tracer, and the concentrations in sediment and other media were measured by neutron activation to ^{134}Cs; absolute quantities of 10^{-8} g of ^{133}Cs were detectable [140]. This technique is promising for research on the behavior of ^{137}Cs fallout from atmospheric tests of nuclear weapons.

VI. CONCLUDING REMARKS

The value of INAA has been recognized and applied in a wide range of scientific disciplines that require soil analysis, including agriculture, research on natural and polluted environments, forensic science, and archaeology. Recent progress in the techniques of INAA resulting from improvements in general instrumentation and computer technology have permitted an extension of the range of elements which can be determined on a quantitative basis, and have allowed increases in sensitivity.

According to the published work throughout the world, it appears that research on environmental pollution has made greater use of modern techniques in INAA than other fields such as pedological investigations and research into problems of soil/plant and soil/plant/animal interrelationships. No doubt the remoteness of many agricultural research laboratories from reactor facilities and the complexity and cost of equipment generally required for INAA has been largely responsible. However, with an increase in the nuclear power program and demands for research on the fate of artificial isotopes in the terrestrial and marine environments, the availability of equipment for γ-ray spectrometry and data processing is also likely to increase. Such equipment can have a dual role i.e., the analysis of stable elements by INAA, given access to reactor facilities.

The application of INAA to problems of plant and animal toxicology has been relatively slow, and there are few integrated studies concerned with analysis of soils, plants, and animal tissues from specific areas unless there has been massive industrial pollution. Examples of the latter type of study are discussed in Sec. V.B.

The interrelationships of trace elements that are essential in plant and animal nutrition are such that an excess of one element may induce a deficiency of another (e.g., Mn/Fe, Cu/Mo). In this context, the comparison of multielement spectra obtained by INAA of soil and tissues would assist diagnosis. The deliberate addition of trace element fertilizers to land may also create problems of imbalance unless care is taken to avoid excessive use. Furthermore, modern methods of INAA could be usefully applied to analysis of soils and vegetation from regions where problems with livestock have been encountered, to detect the presence of As, Cu, Mo, Se, and Zn in particular.

ACKNOWLEDGMENTS

The authors wish to express their gratitude to Miss E. M. R. Fisher for providing much experimental detail and most of the analytical data published in this chapter, and to Mrs. M. J. Cann for considerable help in literature searches. Both Miss Fisher and Mrs. Cann are members of the Environmental and Medical Sciences Division at A.E.R.E. Harwell.

REFERENCES

1. D. Gibbons and D. A. Lambie, in *Comprehensive Analytical Chemistry*, vol. II c (C. L. Wilson and D. W. Wilson, Eds.) Elsevier, Amsterdam,
2. D. De Soete, R. Gijbels, and J. Hoste, *Neutron Activation Analysis*, Wiley Interscience, New York, 1972.

3. P. Kruger, *Principles of Activation Analysis*, Wiley-Interscience, New York, 1971.
4. R. Dams, F. De Corte, J. Hertogen, J. Hoste, W. Maenhaut, and F. Adams, in *Analytical Chemistry*, Part I (T. S. West, Ed.), Physical Chemistry Series II, Vol. 12, Butterworth, London, 1976, p. 1.
5. R. Cornelis, J. Hoste, A. Speecke, C. Vandecasteele, J. Versieck, and R. Gijbels, Ref. 4, p. 71.
6. J. Radioanal. Chem., *37* (1977). *Proceedings of the 5th International Conference on Modern Trends in Activation Analysis*, Munich, 1976. (Also earlier meetings at College Station, Texas, 1962 and 1965, at Gaithersburg, Maryland, 1968, and Paris 1973).
7. *Nuclear Activation in the Life Sciences*, Proceedings of the International Conference Vienna, 1978, IAEA, Vienna, 1979. (Also earlier meetings at Amsterdam, 1965, and Bled, 1972).
8. J. Hislop, Proc. Anal. Div. Chem. Soc., *15*:193 (1978).
9. A. Chattopadhyay and R. E. Jervis, Anal. Chem., *46*:1630 (1974).
10. H. J. M. Bowen and D. Gibbons, *Radioactivation Analysis*, Oxford Univ. Press, Oxford, 1963, p. 8.
11. G. Erdtmann, in *Kernchemie in Einzeldorstellungen* (K. H. Lieser, Ed.), Weinhem, New York, 1976.
12. L. Zikovsky and E. A. Schweikert, Nucl. Instr. Meth., *155*:279 (1978).
13. V. Guinn, unpublished work privately communicated, 1978.
14. D. V. Law, unpublished work, AERE Harwell, 1979.
15. C. E. Crouthamel, F. Adams, and R. Dams, *Applied Gamma-Ray Spectrometry*, Pergamon, Oxford, 1970.
16. O. Klein and Y. Nishina, Z. Physik, *52*:823 (1929).
17. R. L. Heath, USAEC Report IDO 16880, 1964.
18. B. J. Snyder, Nucl. Instr. Meth., *46*:173 (1967).
19. D. H. F. Atkins, E. M. R. Fisher, and L. Salmon, UKAEA Report, AERE-R6734, HMSO, London, 1972.
20. M. Vassilaki, L. Salmon, and J. A. B. Gibson, Geochim. Cosmochim. Acta, *30*:601 (1966).
21. J. Rosenberg and H. B. Wiik, Radiochem. Radioanal. Lett., *6*:45 (1971).
22. R. Gijbels and A. Govaerts, J. Radioanal. Chem. *16*:7 (1972).
23. R. C. Ragaini, H. R. Ralston, and N. Roberts, Environ. Sci. Tech., *11*:773 (1977).
24. J. G. Farmer and J. D. Cross, Water Air Soil Pollut., *9*:193 (1978).
25. H. J. M. Bowen and P. A. Cawse, in *Activation Analysis* (J. M. A. Lenihan and S. J. Thomson, Eds.), Academic Press, New York, 1965.

26. D. McKown, M. Kay, D. Gray, A. Abu-Samra, E. Eichor, and J. Vogt, in *Nuclear Methods in Environmental Research*, (J. R. Vogt, T. F. Parkinson, and R. L. Carter, Eds.), Univ. Missouri, Columbia, 1971, p. 150.
27. N. van der Klugt, P. Poelstra, and E. Zwemmer, J. Radioanal. Chem., *35*:109 (1977),
28. R. L. Brunelle, C. F. Hoffman, K. B. Snow, and M. J. Pro, J. Assoc. Off. Agric. Chem., *52*:911 (1969).
29. L. Salmon and P. A. Cawse, unpublished work, 1978.
30. G. A. Borchardt and M. E. Harward, Soil Sci. Soc. Am. Proc., *35*:626 (1971).
31. J. J. O'Toole, R. G. Clark, K. L. Malaby, and D. L. Trauger, in *Nuclear Methods in Environmental Research* (J. R. Vogt, T. F. Parkinson, and R. L. Carter, Eds.), Univ. Missouri, Columbia, 1971, p. 172.
32. S. A. Kerr, Ph.D. Thesis, Part II, Univ. Surrey, England, 1978.
33. H. Yule and V. Guinn, in *Radiochemical Methods of Analysis*, Proceedings of the Symposium, Salzburg, 1964, vol. 2, p. 111, IAEA, Vienna, 1965.
34. R. E. Jervis, R. G. V. Hancock, D. E. Hill, and K. Isles, J. Radioanal. Chem., *37*:463 (1977).
35. A. K. De and W. W. Meinke, Anal. Chem., *30*:1474 (1958).
36. Anonymous, *Radiation Sources 1977/78*, Radiochemical Centre, Amersham, Buckinghamshire, England, 1977, p. 60.
37. D. W. Downton and J. D. L. H. Wood, in *Proceedings of the 1968 Conference on Modern Trends in Activation Analysis* (J. R. DeVoe and P. D. LaFleur, Eds.) NBS Special Publication 312, National Bureau Standards, Washington, D. C., 1968, p. 1059.
38. J. R. Keane, Direct Information 1/65, Karlsruhe, 1965.
39. H. P. Yule, in *Proceedings of the 1968 Conference on Modern Trends in Activation Analysis* (J. R. DeVoe and P. D. LaFleur, Eds.), NBS Special Publication 312, National Bureau Standards, Washington, D.C., 1968, p. 881.
40. L. Salmon, in *Radiochemical Methods of Analysis*, Proceedings of the Symposium, Salzburg, 1964, vol. 2, p. 125, IAEA, Vienna, 1965.
41. L. Salmon and D. V. Booker, UKAEA Report, AERE-R8870, HMSO, London, 1979.
42. V. Barnes, UKAEA Report, PG-834(W), HMSO, London, 1968.
43. G. W. Phillips and K. W. Marlow, NRL Memo. Report 3198, Washington, D.C., 1976.
44. R. Gunninck and J. B. Niday, *Computerized Quantitative Analysis by Gamma-Ray Spectrometry*, vols. II and III, Lawrence Livermore Lab., Univ. California Report UCRL 51061, 1971.

45. J. T. Routti and P. G. Prussin, Nucl. Instr. Meth., *72*: 125 (1969).
46. A. Robertson and T. J. Kennet, Nucl. Instr. Meth., *100*: 317 (1972).
47. M. A. Mariscotti, Nucl. Instr. Meth., *50*:309 (1967).
48. H. P. Yule, Anal. Chem., *40*:1480 (1968).
49. H. P. Blok, J. L. DeLange, and J. W. Schottmann, Nucl. Instr. Meth., *50*:309 (1967).
50. J. P. Op de Beeck and J. Hoste, in *Analytical Chemistry*, Part 1 (T. S. West, Ed.), Physical Chemistry Series II, vol. 12, Butterworth, London, 1976, p. 151.
51. G. W. Phillips and K. W. Marlow, Nucl. Instr. Meth., *137*:525 (1976).
52. D. V. Booker, J. Radioanal. Chem., *48*:83 (1979).
53. Anonymous, *SCORPIO/SPECTRAN Version 2 User Manual*, CI-SE-905, Canberra Industries Inc., Meriden, Connecticut, 1977.
54. Anonymous, *System ND6600 Operation Instruction Summary*, Nuclear Data Inc., Schaumberg, Illinois, 1978.
55. C. M. Lederer, J. M. Hollander, and I. Perlman, *Table of Isotopes*, 6th ed., Wiley, New York, 1967.
56. G. Erdtmann and W. Soyka, Nucl. Instr. Meth., *121*: 197 (1974).
57. F. Adams and R. Dams, J. Radioanal. Chem., *3*:100 (1969).
58. W. Carder, J. D. Macmahon, and A. Egan, Talanta, *25*:21 (1978).
59. R. Gunninck and J. B. Niday, *Computerized Quantitative Analysis by Gamma-Ray Spectrometry*, vol. I, Lawrence Radiation Lab., Univ. California Report UCRL-51061, 1972.
60. G. C. Christensen, M. J. Koskelo, and J. T. Routti, CERN, Geneva Report HS-RP/015, 1977.
61. P. Quittner, Nucl. Instr. Meth., *76*:115 (1969).
62. D. F. Covell, Anal. Chem., *31*:1785 (1959).
63. P. A. Baedecker, Anal. Chem., *43*:405 (1971).
64. L. A. McNelles and J. L. Campbell, Nucl. Instr. Meth., *127*:73 (1975).
65. L. Salmon, Nucl Instr. Meth., *14*:193 (1961).
66. L. Salmon, in *Application of Computers to Nuclear and Radiochemistry* (G. D. O'Kelley, Ed.), Proceedings of the Symposium, Gatlinburg, Tenn., 1962, USAEC Report NAS-NS-3107, 1963, p. 165.
67. D. Gibbons, L. E. Fite, and R. E. Wainerdi, in *Proceedings of the 5th Conference on Nuclear Reactor Technology*, Gatlinburg, Tenn., USAEC Report TID-7629, 1961.
68. C. D. Carter and D. V. Booker, UKAEA Report AERE-PR/EMS1, HMSO, London, 1974.

69. C. D. Carter, AERE, Harwell, unpublished document, 1974.
70. L. A. Currie, Anal, Chem. *40*:586 (1968).
71. P. A. Cawse, AERE Harwell Report R8191, HMSO, London, 1976.
72. J. H. McNeilly and E. Eichler, Tech. Report CWLR 2299, U. S. Army Chemical Center, Maryland, 1959.
73. P. Thoresen, Acta Chem. Scand., *18*:1054 (1964).
74. J. R. Kline, Ph. D. Thesis, University Minnesota, Minneapolis, 1964.
75. A. K. Khera and E. Steinnes, Agrochim., *13*:524 (1969).
76. L. Salmon and M. G. Creevy, in *Nuclear Techniques in Environmental Pollution*, IAEA, Vienna, 1971.
77. M. Naidenov and A. Travesi, Soil Sci., *124*:152 (1977).
78. R. P. Murmann, R. W. Winters, and T. G. Martin, Soil Sci. Soc. Am. Proc., *35*:647 (1971).
79. H. D. Buenafama, J. Radioanal. Chem., *18*:111 (1973).
80. S. M. Rahman, Nucl. Sci. Applicat., *6*:42 (1972).
81. B. W. Avery, J. Soil Sci., *24*:324 (1973).
82. W. Zmijewska and J. Minczewski, in *Plant Nutrient Supply and Movement*, Tech. Report Series No. 48, IAEA, Vienna, 1965.
83. H. J. M. Bowen and P. A. Cawse, Soil Sci., *98*:358 (1964).
84. U. K. Misra, E. R. Graham, W. J. Upchurch, and D. M. McKown, Soil Sci. Soc. Am. Proc., *37*:193 (1973).
85. R. L. Mitchell, in *Trace Elements in Soils and Crops*, MAFF Tech, Bull. 21, HMSO, London, 1971.
86. A. K. Khera, E. Steinnes, and A. Øien, Acta Agric. Scand., *20*:33 (1970).
87. A. K. Khera and E. Steinnes, Geoderma, *5*:251 (1971).
88. J. R. Kline, J. E. Foss, and S. S. Brar, Soil Sci. Soc. Am. Proc., *33*:287 (1969).
89. G. A. Borchardt, Ph.D. Thesis, Oregon State University, Corvallis, Oregon, 1970.
90. H. Jaffrezic, A. Decarreau, J. P. Carbonnel, and H. Deschamps, J. Radioanal. Chem., *18*:49 (1973).
91. H. J. M. Bowen, in *Nuclear Activation Techniques in the Life Sciences*, IAEA, Vienna, 1967.
92. G. W. Leddicotte, Isotope Radiat. Tech., *5*:200 (1968).
93. O. Johansen and E. Steinnes, Acta Agric. Scand., *22*:104 (1972).
94. G. Van Compernolle, M. Krivanek, C. Masozera, and F. R. Lumu, in *Peaceful Uses of Atomic Energy in Africa*, IAEA, Vienna, 1970.
95. M. L. A. Nilubol, P. Sington, and D. Chamnirokasarnt, *Soil Analysis by Neutron Activation Method*, Thailand Atomic Energy Commission Report AEC-36, Bangkok, 1970.

96. R. A. Srapenyants and I. B. Saveliev, J. Radionanal. Chem., *38*:247 (1977).
97. R. Rustamov, Sh. Khatamov, I. I. Orestova, and A. A. Kist, Atomnaya Energiya, *34*:476 (1973).
98. J. N. Weaver, A. Hanson, J. McGaughey, and F. J. Steinkruger, Water Air Soil Pollut., *3*:327 (1974).
99. M. Shibuya and Y. Kamemoto, Mizushori Gijutsu, *4*:17 (1963).
100. A. Suess and H. Staerk, J. Radioanal. Chem., *37*:905 (1977).
101. Y. Talmi and C. Feldman, in *Arsenical Pesticides* (E. A. Woolson, Ed.), Am. Chem. Soc. Symposium Series 7, ACS, Washington D.C., 1975.
102. H. Gounchev and T. Dimchev, Pochvozn. Agrokhim., *8*:77 (1973).
103. W. S. Lyon, *Trace Element Measurements at the Coal Fired Steam Plant*, CRC Press, Cleveland, Ohio, 1977.
104. J. H. Horton, R. S. Dorsett, and R. E. Cooper, Savannah River Laboratory, Report DP-1475, Aiken, South Carolina, 1977.
105. E. S. Gladney, L. E. Wangen, and R. D. Aguilar, Anal. Lett., *10*:1083 (1977).
106. E. A. Crecelius, C. J. Johnson, and G. C. Hofer, Water Air Soil Pollut., *3*:337 (1974).
107. R. E. Jervis and B. Tiefenbach, in *Nuclear Activation Techniques in the Life Sciences*, IAEA, Vienna, 1978.
108. R. E. Jervis, J. J. Paciga, and A. Chattopadhyay, Trans. Am. Nucl. Soc., *21*:95 (1975).
109. N. J. Pattenden, AERE Harwell Report R7729, HMSO, London, 1974.
110. H. J. M. Bowen, *Trace Elements in Biochemistry*, Academic Press, New York, 1966.
111. P. A. Cawse, I. S. Jones, and G. W. Cox, AERE Harwell Report R8103, HMSO, London, 1975.
112. A. J. Kuja and T. C. Hutchinson, in *Ann. Report Slowpoke Nuclear Reactor* (W. Paul and R. G. V. Hancock, Eds.), Univ. Toronto, Toronto, 1978.
113. T. W. Oakes, A. K. Furr, D. J. Adair, and T. F. Parkinson, J. Radioanal. Chem., *37*:881 (1977).
114. P. A. Cawse, in *Inorganic Pollution and Agriculture*, Proc. ADAS Conf., London 1977, HMSO, London, 1980.
115. Y. Tanizaki and S. Nagatsuka, Nippon Kagaka Naishi, 5:667 (1977).
116. J. Lag and E. Steinnes, in *Isotopes and Radiation in Soil-Plant Relationships Including Forestry*, IAEA, Vienna, 1972.
117. G. E. Gordon (project director), *Atmospheric Impact of Major Sources and Consumers of Energy*, Progress Report

1975, Univ. Maryland Dept. Chemistry, College Park, Maryland.
118. K. A. Rahn, *The Chemical Composition of the Atmospheric Aerosol*, Tech, Report Univ. Rhode Island Dept. Oceanography, Kingston, R.I., 1976.
119. K. A. Rahn, R. D. Borys, and G. E. Shaw, Nature (Lond.), *268*:713 (1977).
120. S. A. Penkett, D. H. F. Atkins, and M. H. Unsworth, Tellus, *31*:295 (1979).
121. W. I. Gizyn and T. C. Hutchinson, in *Ann. Report. Slowpoke Nuclear Reactor* (W. Paul and R. G. V. Hancock, Eds.), Univ. Toronto, Toronto, 1978.
122. F. Beavington and P. A. Cawse, Sci. Tot. Environ., *10*:239 (1978).
123. P. A. Cawse, AERE Harwell Report R7669, HMSO, London, 1974.
124. A. H. Weir, J. A. Catt, and P. A. Madgett, Geoderma, *5*:131 (1971).
125. M. Pro, H. Schlesinger, and M. Cohan, J. Assoc. Off. Agric. Chem., *48*:459 (1956).
126. L. C. Bate, J. F. Emery, G. W. Ledicotte, W. S. Lyon, and M. J. Pro, Int. J. Appl. Radiat. Isotopes, *14*:549 (1963).
127. D. E. Bryan and V. P. Guinn, *Application of Neutron Activation Analysis in Scientific Crime Detection*, USAEC Report GA-5556, San Diego, 1964.
128. L. E. Smythe, Atomic Energy Australia, *9*:2 (1966).
129. R. E. Jervis, Radiat. Tech., *6*:57 (1968).
130. C. M. Hoffman, R. L. Brunelle, K. B. Snow, and M. J. Pro, in *Proceedings of the 1968 Conference on Modern Trends in Activation Analysis*, vol. 1, (J. R. DeVoe and P. D. LaFleur, Eds.), National Bureau Standards, Special Publ. 312, Washington, D.C., 1968.
131. J. R. Kline and S. S. Brar, Soil Sci. Soc. Am. Proc., *33*:234 (1969).
132. L. C. Bate, J. Radioanal. Chem., *15*;193 (1973).
133. J. P. Garrec, A. Fer, and A. Fourcy, C. R. Acad. Sci. Paris, *268*:3021 (1969).
134. G. A. Borchardt, M. E. Harward, and R. A. Schmitt, Quatern. Res., *1*:247 (1971).
135. G. Stefanov and L. Daieva, Isotopenpraxis, *4*:146 (1972).
136. P. Bornemisza-Pauspertl and M. Szilagyi, Radiochem. Radioanal. Lett., *12*:271 (1972).
137. O. Birgul, M. Diksic, and L. Yaffe, J. Radioanal. Chem., *39*:45 (1977).

138. S. K. Tobia and E. V. Sayre, in *Recent Advances in Science and Technology of Materials*, vol. 3 (A. Bishay, Ed.), Plenum Press, New York, 1974.
139. J. Benes, J. Frana, and A. Mastalka, Collect. Czech. Chem. Commun., *39*:2783 (1974).
140. T. E. Hakonson and F. W. Whicker, Health Phys., *28*:699 (1975).

8

Analysis of Nitrogen Isotope Ratios by Optical Emission Spectrometry

VICTOR MIDDELBOE The Royal Veterinary and Agricultural University, Copenhagen, Denmark

I. INTRODUCTION

The ideal tracer for nitrogen would, of course, be an appropriate radioisotope of that element. Nitrogen 13 (^{13}N), with a half-life of 10 min, has been used in isolated instances in experiments related to soil (see, e.g., Gersberg et al. [1]). All other known radioisotopes of nitrogen have a half-life considerably shorter than that of

^{13}N, which makes their use as tracers in soil investigations quite impracticable. Meanwhile, all naturally occurring nitrogen consists of two stable isotopes (^{14}N and ^{15}N), having an almost constant abundance ratio. In principle, therefore, stable nitrogen having any significantly unnatural isotopic ratio can be used as a tracer.

In tracer experimentation employing stable isotopes, the term "percentage isotope abundance" is often used. In the case of stable nitrogen, which is comprised of two isotopes only the relationships between the isotope abundances and the isotopic ratio for any given sample are as follows:

$$a_{15} = \frac{100}{r + 1} \text{ atom \%} \tag{1}$$

$$a_{14} = \frac{100}{r^{-1} + 1} \text{ atom \%} \tag{2}$$

where

a_{15} = percentage abundance of ^{15}N in sample

a_{14} = percentage abundance of ^{14}N in sample

r = $^{14}N/^{15}N$ ratio in sample nitrogen

The natural $^{14}N/^{15}N$ ratio is 272:1, which is equivalent to about four nitrogen-15 atoms in every 1000 nitrogen atoms. Consequently, a supply of nitrogen containing, say, 10 nitrogen-15 atoms per 1000 nitrogen atoms would be considered "labeled." The practical use of nitrogen labeled in this way depends on the accuracy with which small differences in $^{14}N/^{15}N$ ratios can be measured. Up to now, the commonly used methods of measurement have been (1) optical emission spectrometry, and (2) mass spectrometry (see Chap. 9). The subject of the present chapter is the measurement of $^{14}N/^{15}N$ ratios in soil and other environmental samples using optical emission spectrometry, including discussion of the procedure for the liberation of elemental nitrogen, N_2, from the sample, which is frequently the most difficult part of the complete analysis.

II. BASIC PRINCIPLES AND TECHNIQUES

A. Discovery of Nitrogen-15

The existence of the rare, stable isotope of nitrogen (^{15}N) was finally proven by Naudé in 1929 [2]. A tube containing nitric oxide was prepared and white light was transmitted through the tube parallel to the axis. In the absorption spectrum thereby obtained, isotopically shifted (i.e., wavelength displaced) bandheads of low intensity were observed at certain spectral positions, which coincided accurately with those that could be predicted by assuming the existence of nitrogen-15.

Isotopic shifts take place in molecular spectra mainly because the vibrational energy of a molecule (in any given quantum state) is a function of the masses of its constituent atoms. In fact, to a first approximation, the vibrational energy is inversely proportional to the square root of the reduced mass. The reduced mass μ, for a diatomic molecule, is defined as follows:

$$\mu = \frac{m_1 m_2}{m_1 + m_2} \tag{3}$$

where m_1 and m_2 are the masses of the two constituent atoms.

In the case of the UV bands studied by Naudé, wavelength shifts of 0.1 to 0.2 nm occur when the ^{14}N atom in the NO molecule is replaced by a ^{15}N atom. The observed isotopic shifts correspond to about 10 times the resolving power of the spectrographic equipment used by Naudé. However, he was not able to measure the natural $^{14}N^{18}O/^{15}N^{16}O$ intensity ratio with any degree of accuracy; but this was accomplished 2 years later by Urey and Murphy [3]. The result of the latter determination, combined with Bleakney and Hipple's value for the natural $^{16}O/^{18}O$ ratio [4], led to a natural $^{14}N/^{15}N$ ratio of 275, which is within 1% of the presently accepted value [5].

B. Measurement of $^{14}N/^{15}N$ Ratios by Optical Emission Spectrometry

In 1930, several bands in the emission spectrum of the N_2 molecule were isotopically analyzed by Herzberg [6], whereby the existence of the newly discovered ^{15}N isotope was verified. However, the value obtained for the natural $^{14}N/^{15}N$ ratio proved, with the advent of mass spectrometric determinations, to be rather inaccurate. Nevertheless, 20 years later (1950), reasonably accurate, semi-routine analyses of $^{14}N/^{15}N$ ratios in N_2 using optical emission spectrography were reported by Hoch and Weisser [7]. Two hours were needed per optical analysis, due to the use of photographic emulsion for recording the spectra. The quantity of nitrogen required for one analysis was about 50 μg. A series of ^{15}N abundances, ranging from natural (0.37 atom %) to 99.8 atom %, were analyzed in samples supplied in part by Clusius [8]. This was the beginning of routine optical emission analysis of $^{14}N/^{15}N$ ratios in N_2. The next major steps forward were (1) the introduction of photoelectric light detection by Broida and Chapman [9], and (2) the introduction of the monochromator by Meier and Müller [10]. As a result of these improvements the optical analysis time could be reduced to about 10 min, and the quantity of N_2 required was only 5 to 10 μg (in a 2 cm^3 discharge tube).

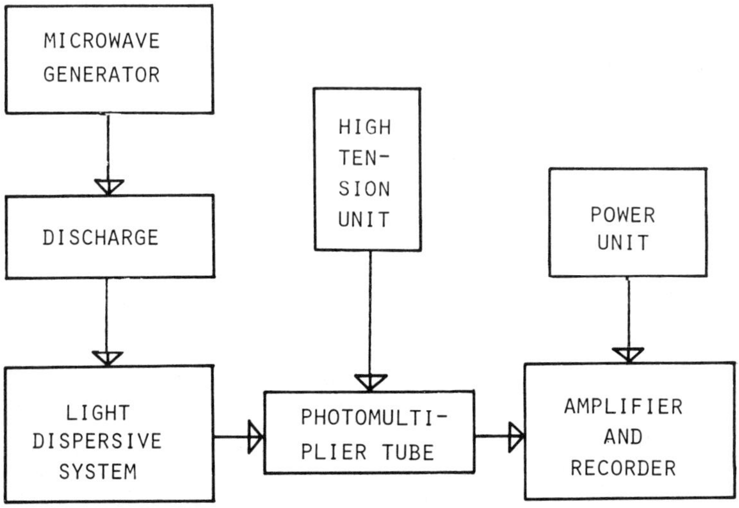

Figure 1 Block diagram of an optical emission spectrometer used for the analysis of $^{14}N/^{15}N$ ratios.

The main components of an optical emission spectrometer used for the measurement of $^{14}N/^{15}N$ ratios are shown in Fig. 1. Power from a microwave generator (also called a high-frequency transmitter) excites the N_2 in the discharge tube to emit light. The light dispersive unit is usually a prism or a line grating. The photomultiplier tube transforms light quanta into electronic signals, which are subsequently amplified and recorded. The spectral analysis of $^{14}N/^{15}N$ ratios in N_2 is based on an electronic transition ($C\,^3\Pi$-$B\,^3\Pi$), which gives rise to a group of bands belonging to the so-called second positive nitrogen system situated in the ultraviolet region (Fig. 2). Except in the case of very high ^{15}N abundances, the band at 298 nm (2977 Å), arising from the $2 \rightarrow 0$ transition, has been found to give the best results [10-12]. This band, like most of the others, appears in triplicate, as can be seen in Fig. 2. The three replicates, with diminishing peak intensities toward the $0 \rightarrow 0$ transition, correspond to emission by the three isotopic N_2 molecules $^{14}N^{14}N$, $^{15}N^{14}N$, and $^{15}N^{15}N$, respectively. These three isotopic molecules will be referred to as N_2-28, N_2-29, and N_2-30, according to their mass numbers.

During discharge the ^{14}N and ^{15}N atoms rapidly become randomly distributed among the three types of N_2 molecule, since these molecules continously undergo dissociation and recombination. Thus, according to the statistics of the binomial distribution, the following relations will be valid shortly after ignition of the discharge:

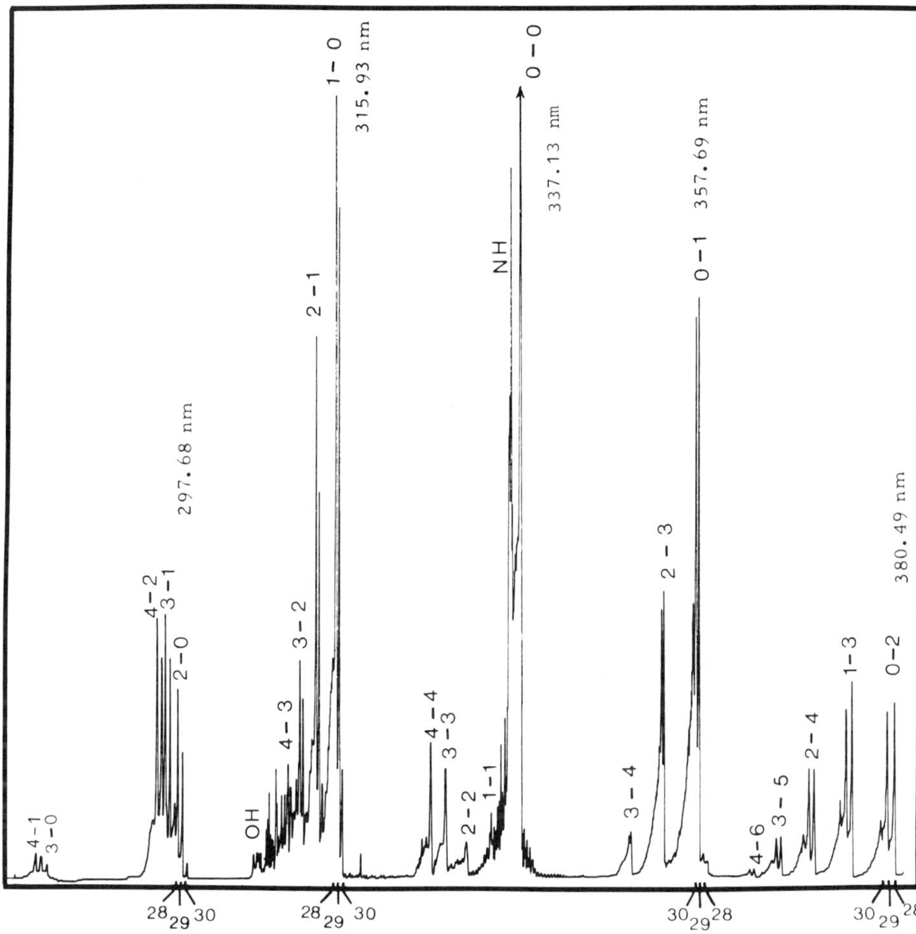

Figure 2 Spectrum of elemental nitrogen ($C^3\Pi$-$B^3\Pi$) showing light intensity vs. wavelengh in the region 280-380 nm. The numbers with a dash between them are vibrational quantum numbers. The nitrogen-15 abundance is 27 atom %, 28, 29, and 30 are N_2 mass numbers.

$$r = 2R \tag{4}$$

$$r^{-1} = 2R' \tag{5}$$

where r = $^{14}N/^{15}N$ ratio, R - N_2-28/N_2-29 ratio, R' = N_2-30/N_2-29 ratio.

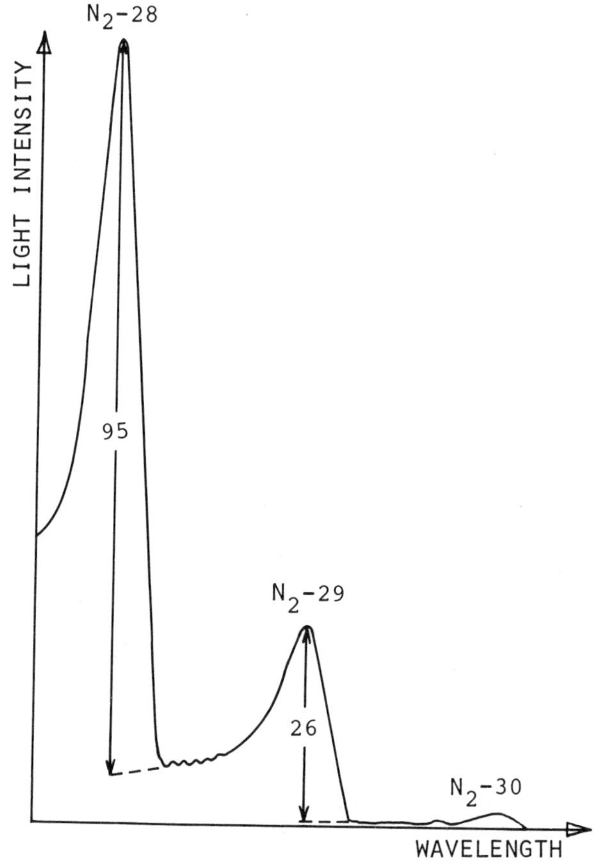

Figure 3 Spectrum of elemental nitrogen ($C^3\Pi$-$B^3\Pi$, $2 \rightarrow 0$) in the wavelength region 297.5-299 nm.

It is seen, by substituting Eq. (4) into Eq. (1), that the percentage ^{15}N abundance, a_{15}, in a nitrogen sample may be calculated as follows:

$$a_{15} = \frac{100}{2R + 1} \text{ atom } \% \tag{6}$$

where R is the population density of the $^{14}N^{14}N$ molecules relative to that of the $^{14}N^{15}N$ molecules.

A detailed spectrum of the $2 \rightarrow 0$ transition in the second positive nitrogen system is shown in Fig. 3.

The isotopic shift from N_2-28 to N_2-29 is 0.61 nm, and the shift from N_2-29 to N_2-30 is 0.57 nm. The N_2-28/N_2-29 peak-height ratio, assumed to be equal to the corresponding population density ratio R, is directly observable. For the particular sample from which the spectral section shown in Fig. 3 was obtained, it will be seen that R = 95/26 = 3.65, which, according to Eq. [6], corresponds to 12.0 atom % ^{15}N.

C. Comparison with Mass Spectrometry

Recently, a brief review appeared in which a comparison was made between the use of mass spectrometry and optical emission spectrometry in ^{15}N-aided research [13]. The following two main conclusions were drawn. Mass spectrometry is indispensable for the analysis of the extremely small variations in ^{15}N abundance that occur in nature, and optical emission spectrometry is uniquely applicable to isotopic analysis of nanomole quantities of nitrogen. These circumstances are largely due to the following two facts: (1) the accuracy of a mass spectrometer is superior to that of an optical spectrometer in the analysis of $^{14}N/^{15}N$ ratios, and (2) the nitrogen sample is consumed during analysis by mass spectrometry but conserved (in the discharge tube) during analysis by optical spectrometry.

Actually, the above-mentioned conclusions refer to two rather limiting cases, namely, minute differences in isotopic composition or very limited amounts of available nitrogen. Frequently the situation is intermediate and the choice of method not so clear-cut. We will consider below three typical experimental situations, namely (1) field experiments, (2) laboratory or greenhouse experiments, and (3) the use of ^{15}N-depleted nitrogen (i.e., almost pure ^{14}N) as the tracer. In each case the two most important considerations are the cost of the labeled N compound and the cost of the isotopic analyses (instrumentation and operation). The cost per gram of labeled nitrogen depends, progressively, on the extent to which the nitrogen isotope ratio has been altered relative to the natural ratio. The degree of alteration is usually expressed in terms of ^{15}N enrichment (depletion), i.e., the difference between the ^{15}N abundance of the labeled nitrogen sample and that of a natural nitrogen sample. The ^{15}N enrichment (depletion) is normally given in units of atom per cent excess (deficiency).

Field Experiments

Relatively large quantities of labeled nitrogen are needed in carrying out field experiments, and until about 10 years ago the level of ^{15}N enrichment considered to be desirable was rather high, in fact, the order of 10 atom % excess. Consequently, the cost of the labeled material often proved to be a serious deterrent. Meanwhile, the Joint FAO/IAEA Division, Vienna, who pioneered widespread field application of ^{15}N labeling in agronomic research, demonstrated in many instances that the level of ^{15}N enrichment mentioned above could be reduced by one, or even two, orders of magnitude without any appreciable loss of useful information [14]. However, the exploitation of *low level* ^{15}N *enrichment* makes it necessary, in most cases, for the isotopic analyses to be performed by mass spectrometry, which is considerably more expensive than optical emission spectrometry. Nevertheless, experience has shown that the overall savings on the labeled material, throughout a period of years, are substantial enough to justify the use of mass spectrometry in field experiments on a broad scale.

Laboratory or Greenhouse Experiments

The amount of nitrogen required for most laboratory or greenhouse experiments is small compared to that needed in field experiments. Therefore, the cost of the labeled material is usually relatively modest, even if the desired level of ^{15}N excess is high. Consequently, optical emission spectrometry is likely to be the analytical method of choice for pure laboratory or greenhouse experimentation, since the necessary optical equipment is considerably less costly to acquire, operate, and maintain than a mass spectrometer.

Use of Nitrogen Depleted in ^{15}N

At present, kilogram quantities of almost pure ^{14}N compounds are being produced commercially and these products are offered for sale at a somewhat lower cost than tracer material enriched in ^{15}N to an equivalent degree, i.e., N compounds containing 0.36 atom % excess of ^{15}N [15]. However, if the place where the tracer material is to be used is situated far from the isotope separation plant and shipping expenses are taken into account, the lesser cost of the ^{15}N-depleted compound can be more or less matched by transporting fairly highly enriched material, which is less bulky, and then diluting it locally with the corresponding natural N compound. Whether or not the use of optical emission spectrometry will suffice, when ^{15}N*-depleted material* is applied, will depend largely on the degree to which the tracer nitrogen becomes diluted with natural nitrogen during the course of the experiment; if the isotopic dilution factor involved is expected to be at least 10, and the relative error is to be kept below 25%, there is no alternative to mass spectrometry.

The same general procedures for the isolation of N_2 from the N-containing sample are applicable, irrespective of the particular method of $^{14}N/^{15}N$ measurement used. Consequently, the application of a certain sample preparation procedure has relatively little influence on the respective advantages and limitations of optical emission and mass spectrometry [13].

Finally, it is important to note that the $^{14}N/^{15}N$ ratios obtained by the use of mass spectrometry are accepted directly as "true" values, whereas those observed in optical emission spectrometry are usually considered as apparent values, i.e., the latter are normally corrected by the aid of a calibration curve in order to obtain true values. Experience shows that care must be taken in establishing the calibration curve. Spectroscopically pure N_2 samples (with known ^{15}N abundances) are not suitable for this purpose, presumably due to the absence of the small amounts of gaseous impurities normally present after preparation of samples of environmental material [16]. More consistent results are obtained by isotopic analysis of a set of ammonium chloride standards (with known ^{15}N abundances), samples of which are put through the same preparation procedure as the experimental samples. Furthermore, it is important to use the same type, and even the same batch, of glass in making the discharge tubes for the calibration samples and the experimental samples, otherwise variations in optical properties of the glass are likely to cause significant errors in the establishment of the calibration curve [17].

D. Principles of Sample Preparation

The aim of the sample preparation is to liberate the nitrogen contained in the sample as N_2. In principle, the sample nitrogen should be liberated completely and the N_2 produced should be uncontaminated by alien nitrogen or foreign gases. The original chemical procedure designed for use in $^{14}N/^{15}N$ analysis by optical emission was developed by Faust [18,19] in the 1960s.

Basically, there are two main approaches to sample preparation. In one procedure, called the Rittenberg method, the sample nitrogen is all converted to the ammonium form by conventional Kjeldahl digestion of the soil (or plant) sample [20], and N_2 is liberated from the solution containing the ammonium ions by mixing it with a solution of sodium hypobromite in a Rittenberg flask. In the alternative procedure, called the Dumas method, N_2 is liberated directly from the sample by heating it in the presence of copper oxide.

The Rittenberg method was originally designed for use in mass spectrometry, and the procedure is described more closely in Chap. 9. The chief modification encountered in applying the method to optical emission spectrometry is a reduction of the sample size by at least one order of magnitude.

The apparatus originally developed by Faust [19], designed for the preparation of discharge tubes containing N_2 liberated by the Dumas method, is shown schematically in Fig. 4. The Rasotherm glass discharge tube containing copper oxide and calcium oxide is attached by the aid of black wax (picein) to one end of a T piece, which can be rotated about a horizontal axis (the greased joint in Fig. 4). With the T piece in the horizontal position, the sample is placed just inside the free end of the T piece and this end is then closed by a cap having a ground glass joint. By opening the top valve (see Fig. 4) the T piece and discharge tube are evacuated, via a cold-trap, by means of a diffusion pump backed by another cold-trap and a rotary pump. Natural nitrogen adsorbed on the surface of the chemicals and the inner wall of the discharge tube is eliminated by heating under vacuum. Upon cooling, the sample is dropped into the discharge tube by turning the T piece to the vertical position. Finally, the discharge tube is sealed by flame and pulled off the vacuum line.

The closed discharge tube containing the sample and the chemicals is placed in a furnace at 550°C for 3 hr. The N_2 liberated from the sample should exert a pressure in the discharge tube that falls in the range 0.2-0.5 kPa (2-4 torr) in order to facilitate the production of a bright discharge. Most of the alien gases, such as water vapor and carbon dioxide, liberated together with the nitrogen, are absorbed by the calcium oxide in the discharge tube during a predischarge rest period of about 12 hr. However, one gas that remains unabsorbed is carbon monoxide, the spectrum of which overlaps and interferes somewhat with the section of the N_2 spectrum used in the optical analysis of $^{14}N/^{15}N$ ratios.

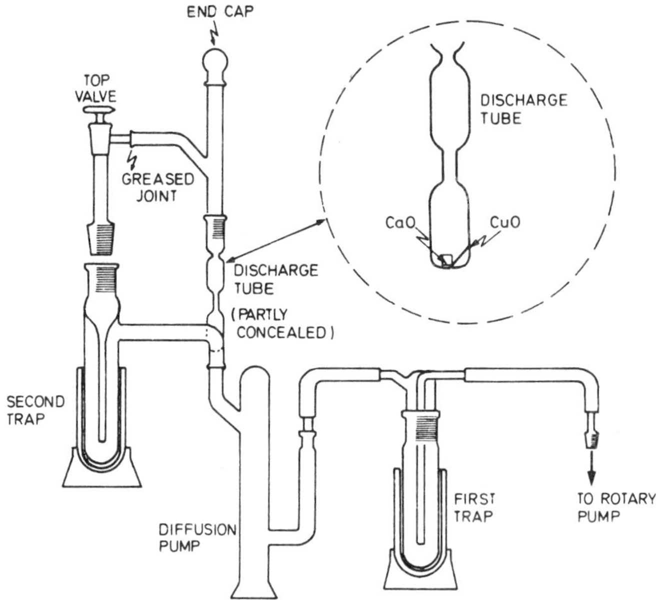

Figure 4 Faust's apparatus for the preparation of discharge tubes. (From Ref. 21.)

Many variations and further details in sample preparation are described in a fairly recent, comprehensive review by Fiedler and Proksch [22]. For example, the copper oxide in the discharge tube is replaced by metallic copper, when the sample nitrogen is present in the form of nitrates. Furthermore, in the case of very restricted amounts of sample nitrogen, of the order of 1 μg or less, isotopic analysis can be effectuated by the use of one or more noble gases to sustain the discharge, as demonstrated by Goleb and Middelboe [23].

E. Commercially Produced Instruments

The first series of commercial instruments designed specifically for the analysis of $^{14}N/^{15}N$ ratios by optical emission spectrometry were produced in the 1960s by VEB Statron, Fürstenwalde am Spree, German Democratic Republic. These instruments were based on the original work published by Meier and Müller in 1965 [10]. In 1973 a similar instrument was marketed by Japan Spectroscopic Company (JASCO), Tokyo, and at about the same time an advanced, fully automatic $^{14}N/^{15}N$ optical emission analyzer was launched commerically by VEB Statron under the name of Isonitromat 5200.

The latest version of the original nonautomatic instrument is called the NOI-5. This is the first version in which the electronics are transistorized, but otherwise there are no major changes. As previously, the high frequency of the microwave generator is 27 MHz, and the light dispersive medium is a 56° prism made of rock salt. The prism must be kept dry at all times (silica gel), which can present a problem in humid countries. Instead of the calibration discharge tubes originally supplied with the instrument, the NOI-5 is supplied with a set of ammonium chloride standards that have mass spectrometrically determined ^{15}N abundances ranging from natural abundance to 8 atom % ^{15}N. Ammonium chloride standards containing higher ^{15}N abundances are available as accessories.

The JASCO instrument is called the NIA-1. Essentially, it differs only from the NOI-5 in the following two respects: (1) the high frequency of the microwave generator is reduced to one-half, and (2) the prism has been replaced by a line grating. The first change is of little consequence in practice, but the second can be of practical importance in regions of high humidity, since the hygroscopic character of the light dispersive medium has been eliminated. The NIA-1 was thoroughly investigated by Guiraud and Buscarlet in 1974 [24]. They found, by comparison with mass spectrometry, that the range from natural abundance to 30 atom % ^{15}N can be adequately covered by a calibration curve consisting of three consecutive linear portions, or by one polynome of the fifth degree covering the whole range.

The Isonitromat 5200 is designed to accommodate 180 sample solutions containing the sample nitrogen as ammonium or any other form convertible to N_2 by the Rittenberg method. The samples are processed and isotopically analyzed automatically, and the resulting ^{15}N abundances are printed out. The quantity of nitrogen required for a duplicate analysis is 120 μg, and the instrument is able to handle one sample every 6 min. The optical part of the instrument is very similar to the NOI-5 except for the scanning mechanism. The dispersed light is scanned by the use of a vibrating exit slit, as opposed to mechanical rotation of the prism. The Isonitromat 5200 is roughly 2.5 times as expensive as the NOI-5 or the NIA-1, and the price is comparable to that of a lightweight mass spectrometer.

The precision of all the above-mentioned optical $^{14}N/^{15}N$ analyzers is very much the same. The standard deviation on a duplicate analysis is at best approximately 0.01 atom % in the case of samples in the range from 0 to 1 atom % ^{15}N. In the case of higher ^{15}N abundances the relative uncertainty is about 1%. However, if the analytical uncertainty in the complete sample preparation procedure is included, the relative overall standard deviation may be 2-5% [22].

F. Some Individual Analytical Modifications

The following brief descriptions of a few arbitrarily selected individual modifications, which have been devised in various laboratories, may serve to illustrate the versatility inherent in the optical emission method of $^{14}N/^{15}N$ analysis.

Lloyd-Jones et al. [25] avoid the distillation of ammonia, subsequent to Kjeldahl digestion of plant material, by precipitating the ammonium ions in the digest with Nessler reagent. The precipitate is treated with hydrochloric acid to give a solution of ammonium chloride, aliquots of which are dried and used later for the preparation of discharge tubes by the Dumas method.

In place of the commercial sample preparation apparatus made of glass (see Fig. 4), Keeney and Tedesco [26] assembled one constructed of metal and Teflon parts. The sample nitrogen is released by the Rittenberg method and the discharge tubes are made from Pyrex or Vycor tubing. In each analysis the pressure in the discharge tube is adjusted to 400 Pa (3 torr), either by removal of excess N_2 or, conversely, by addition of argon.

Meyer et al. [27] suggest (1) repeated use of a silica discharge tube permanently installed for a period of time, and (2) careful regulation of the partial pressures of both N_2 and argon. Carry-over from the previous sample is practically eliminated by evacuating for about 15 min under excitation by the microwave generator (of the NOI-5 instrument). Subsequently, sample nitrogen, liberated by use of the Rittenberg method, is ignited in the discharge tube for about 1 min, whereupon the pressure is reduced to 25 Pa (0.2 torr) by evacuation of excess N_2. Finally, argon is admitted to attain a total pressure of 150 Pa (1.1 torr), and the optical measurement of the $^{14}N/^{15}N$ ratio is then carried out.

Lazeeva et al. [28] introduced a novel sample preparation procedure. The sample is pyrolyzed under vacuum and the combustion products are passed across nickel oxide at a temperature of 1000°C. Carbon dioxide, water vapor, and other condensible by-products are frozen out in a liquid nitrogen trap, and the N_2, together with any remaining contaminants, flows through a purifying unit containing copper at 600°C and copper oxide at 300°C. Finally, the N_2 is admitted to the discharge tube via a second cold-trap.

Faust and Reinhardt [29] have investigated the use of molecular sieves, in place of calcium oxide, for the absorption of alien gases produced during the preparation of N_2 by the Dumas method. With 10 μg of N_2 in a 2 cm^3 discharge tube, the most promising results were obtained using 10-15 mg of molecular sieve grains (ca. 2 mm diameter) having a pore diameter of $3\text{-}5 \times 10^{-10}$ m (3-5 Å). The apparently improved quality of the N_2 discharge, brought about by use of this technique, is probably partly due to absorption of carbon monoxide by the molecular sieve and partly due to a reduction of the residual water vapor in comparison with that left unabsorbed by calcium oxide.

A modification of the Dumas method that permits the use of frozen aqueous samples, as opposed to the normal dehydrated samples, was developed in 1977 by Middelboe [30]. The frozen sample is placed in the usual position inside the T piece (Figs. 4 and 5),

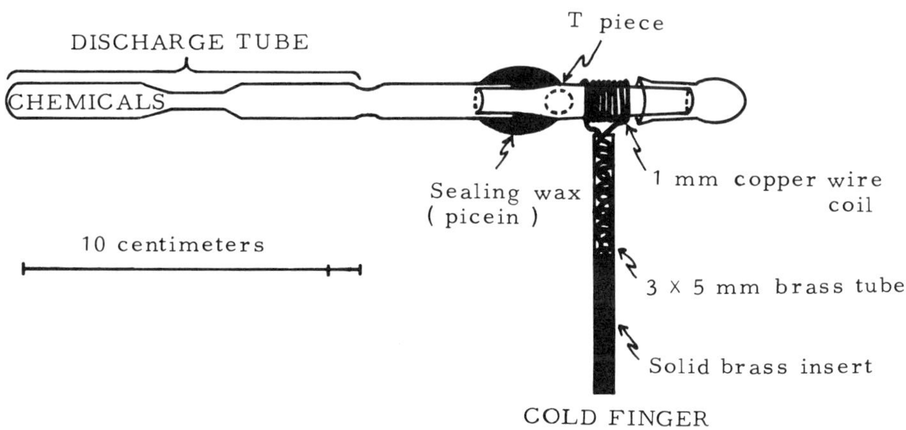

Figure 5 Part of sample preparation apparatus (see Fig. 4) modified for use of the "frozen sample" technique. (From Ref. 30.)

and it is kept at about $-100°C$ during evacuation and degassing of the discharge tube. Finally, having tipped in the frozen sample, the discharge tube is sealed before the sample begins to thaw. Naturally, the quantity of calcium oxide in the discharge tube must be sufficient to absorb all the water present, e.g., 30 mg of calcium oxide is found to be adequate in the case of a 1 μl aqueous sample.

Recently Goulden and Salter [31] published a paper describing an automated optical emission instrument for $^{14}N/^{15}N$ analysis. The new instrument is reported to be three times as fast as the Isonitromat 5200. The sample must be presented as a solution containing the test nitrogen in the form of ammonium chloride, from which ammonia can be readily released in the hot soda-lime reactor incorporated in the instrument. The ammonia is split catalytically, and the hydrogen together with water vapor is retained by a molecular sieve, while the N_2 flows through to the discharge tube. A dual wavelength spectrometer automatically compares the N_2-28 and N_2-29 peak intensities, computes the ^{15}N abundance, and prints out the result. The precision of the instrument apparently compares well with that of the currently available commercial instruments.

Finally, Burridge and Hewitt [32] have shown that the discharge tube may be filled directly with ammonia instead of using the Rittenberg method to release N_2 from the ammonium solution. In their procedure a frozen aqueous solution containing ammonium ions, to which caustic soda has been added, is attached to a line, evacuated, and allowed to thaw. The ammonia gas liberated in this way is expanded into a permanently installed discharge tube, while water vapor is retained by a carefully balanced cooling system. On excitation of the ammonia at 2450 MHz the emission spectrum of N_2 is obtained with only small interferences due to hydrogen and other gases. Very recently, this procedure has been taken one step further, particularly with regard to simplification of the analysis [33].

G. ^{13}C-Analog of ^{15}N-Analyzer

Preliminary investigations have demonstrated the feasibility of developing an optical emission method for the analysis of $^{12}C/^{13}C$ ratios [34,35]. The proposed method is analogous to that described above for the optical analysis of $^{14}N/^{15}N$ ratios.

Carbon dioxide would seem to be the most convenient carbon-containing gas for isotopic analysis, since (1) this gas is frequently the direct object of interest and (2) organic compounds are readily converted to carbon dioxide. However, a high-frequency discharge in pure carbon dioxide produces a spectrum that is dominated by emission bands due to carbon monoxide, and the diffuseness of these bands renders them ill suited for isotopic analysis. In contrast, ionized carbon monoxide produces slender bandheads, almost as sharp as atomic spectral lines, and bandtails with a relatively low intensity (degraded toward increasing wavelengths).

The spectrum of ionized carbon monoxide is obtained by using helium as carrier gas, at about 100 Pa (ca. 1 torr), mixed with carbon dioxide at a partial pressure one or two orders of magnitude lower than the helium pressure. The most useful part of the spectrum is that called the first negative system of ionized carbon monoxide with its bands situated in the ultraviolet region. In particular, the $0 \rightarrow 1$ transition corresponds to a band headed at 230 nm. This band undergoes an isotopic shift of 0.3 nm (to a lower wavelength), when the light is emitted by $^{13}CO^+$ instead of $^{12}CO^+$. The isotopic shift is of the same magnitude (in the opposite direction) as the N_2-28 to N_2-29 wavelength shift of the $0 \rightarrow 1$ band in the second positive nitrogen system, so in emission spectrometric analysis the optical dispersion required for carbon-13 is roughly the same as for nitrogen-15.

Up to now the main stumbling-block in optical $^{12}C/^{13}C$ analysis has undoubtedly been contamination of the sample, during discharge, by alien carbon compounds adhering to the inner wall of the tube. However, this problem can be overcome by using a flow-through discharge system, which facilitates (1) removal of alien carbon by oxidation and (2) elimination of unwanted adsorbed carbon oxides by isotopic exchange with the sample gas. On completion of the clean-up process described, the flow-through rate is either slowed down or stopped completely and the appropriate section of the ionized carbon monoxide spectrum is then analyzed.

At present the accuracy of the optical emission method in the analysis of natural and nearly natural ^{13}C abundances is about ±0.03 atom %, but some improvement may be expected as a result of further development. The radioactive isotope carbon-14 is, of course, generally applicable as a tracer of carbon. Nevertheless, the production of a ^{13}C-analog of the commercial ^{15}N-analyzer would no doubt stimulate ^{13}C-^{14}C double-label experiments and, furthermore, provide an alternative to the mass spectrometer in situations, such as human medicine, where ^{13}C is used because the radiation from ^{14}C is unacceptable or undesirable.

III. EXAMPLES OF APPLICATIONS

A. Introduction

In this section attention is focused on problems related to the practical application of $^{14}N/^{15}N$ analysis by optical emission spectrometry. The emphasis here is placed mainly on the limitations and advantages of the technique.

Up to now about 2000 papers have been published on agronomical experiments in which nitrogen-15 has been used as a tracer, and in recent years an increasing fraction of the isotopic analyses involved have been performed by use of the optical method [36]. A few examples will serve to illustrate some of the more important tracer problems that are frequently encountered. The examples have been selected, not primarily for the intrinsic value of the results, but as a basis for discussing certain points of tracer methodology associated with the application of isotopically labeled nitrogen.

The following examples include: the determination of A-values in a greenhouse experiment, the analysis for total-N by isotopic dilution in submilliliter samples of soil water, and the evaluation of asymbiotic nitrogen fixation in soil or sediment.

B. Evaluation of Soil Nitrogen Fertility

The well-known A-value was introduced by Fried and Dean [37] as a concept by which soil fertility could be assessed quantitatively. According to this concept, the amount of available soil nitrogen is mathematically defined as follows:

$$A_N = B_N \frac{1 - y}{y} \text{ kg ha}^{-1} \tag{7}$$

where

A_N = A-value for nitrogen

B_N = rate of fertilizer-N applied

$$y = \frac{\text{atom \% excess N-15 in plant-N}}{\text{atom \% excess N-15 in fertilizer-N}}$$

= fraction of plant-N derived from fertilizer

Evidently, the fertility of the soil is expressed in terms of an equivalent rate of the applied fertilizer.

A greenhouse experiment was carried out in pots containing nine different soils, and the respective A-values were determined at three rates of ammonium sulphate application (50, 100, 200 kgN ha^{-1}). There were three replicates for each combination of soil and treatment, and the fertilizer nitrogen was labeled with 1 atom % excess ^{15}N. The result was, essentially, that for each soil the A-value appeared to be independent of fertilizer rate [38]. The total quantity of labeled fertilizer used in the experiment contained approximately 0.1 g excess ^{15}N, which cost about \$10 (U.S.). The samples of plant nitrogen were prepared by the Rittenberg method and the $^{14}N/^{15}N$ ratios were measured by mass spectrometry. Thus, ^{15}N depleted fertilizer could have been used, but the consequent cost reduction would have been extremely small.

The determining factor in the calculation of the A-value is the fraction y in Eq. (7). In this experiment the fraction of plant nitrogen derived from the fertilizer was never less than one tenth. Therefore, the atom per cent excess ^{15}N in the plant was not less than one tenth of that in the fertilizer. Consequently, if the initial ^{15}N enrichment of the fertilizer had been 5 atom % excess (instead of 1 atom % excess) none of the samples to be analyzed would have contained less than 0.5 atom % excess ^{15}N, and their enrichments could presumably have been measured by optical emission spectrometry with a relative standard deviation of 2-3%. The cost of the labeled fertilizer would still have been very modest.

C. Analysis of Subsoil Water for Total-N

Hevesy's isotope dilution technique [30] may be applied to the determination of an unknown quantity of natural nitrogen as follows. A known amount of labeled nitrogen, called the spike, is mixed with the unknown nitrogen, which, being natural, contains 99.6 atom % ^{14}N. Next, the isotopic ratio in the uniform mixture is measured and, finally, the result is calculated according to the following equation:

$$n_x = n_s \frac{a_m - a_s}{99.6 - a_m} \text{ mol} \tag{8}$$

where

n_x = moles of N in the unknown

n_s = moles of N in the spike

a_s = ^{14}N abundance in the spike (mass-spectrometrically determined)

a_m = ^{14}N abundance in the mixture of spike and unknown

A high sensitivity in total-N analysis is achieved when (1) the spike contains practically pure nitrogen-15, and (2) the isotopic ratio in the nitrogen mixture is determined by the use of optical emission spectrometry. However, high sensitivity requires the determination of low ^{14}N abundances. Consequently, in the 2 → 0 transition (see Fig. 3), the N_2-30 band will be the most intense and its tail will overlap the less intense N_2-29 band, whereby accurate measurement of the N_2-29 peak height becomes impracticable. This problem is overcome by basing the optical emission analysis on the 0 → 2 transition, in which the isotopic shift is reversed with respect to wavelength (see Fig. 2). According to the previous Eqs. (2) and (5), the ^{14}N abundance in the mixture, a_m may be calculated as follows:

$$a_m = \frac{100}{2R' + 1} \text{ atom \%} \qquad (9)$$

where R' is the N_2-30/N_2-29 peak-height ratio observed.

The isotope dilution technique has recently been applied to trace analysis for total-N in a number of subsoil water samples [30]. The samples contained various concentrations of NH_4-N (0-77 μg ml^{-1}), NO_3-N (0-16 μg ml^{-1}) and organic-N (0-3 μg ml^{-1}).

The pH of the sample was adjusted to approximately 4.5, and an aliquot of 100 μl was used per analysis for total-N. The spike was added and the aliquot was concentrated by evaporation, under vacuum at 50°C, to a volume of 1 μl. Evaporation to complete dryness had to be avoided in order to prevent loss of nitrogen, either in the form of nitric acid fumes or as ammonia gas, toward the end of the drying process.

Discharge tubes containing N_2 were prepared according to the "frozen sample" technique described in Sec. II.F (see Fig. 5). Contamination by alien nitrogen, e.g., natural N_2 dissolved in the water or adsorbed on the surface of the ice, becomes significant when dealing with nanomole quantities of sample nitrogen. A number of blanks were analyzed and the contamination was found to be 0.11 ± 0.03 μg N per discharge tube.

Due to the statistical uncertainty of the blank value, the lower limit of detection* is about 2 nmol N (i.e., ±0.3 μg total-N per ml in this example). For total-N contents above 10 μg ml^{-1} the relative accuracy of the analysis is 2-3%. The amount of water needed for a duplicate analysis is only 0.25 ml.

D. Assessment of Nitrogen Fixation in Soil or Sediment

The rate of asymbiotic nitrogen fixation, or nitrogenase activity, in soil or sediment can be assessed by the acetylene reduction method. However, the appropriate conversion factor, according to Hardy et al. [39], has been observed to vary between 2 and 25. The "true" conversion factor for any given system can be evaluated by the simultaneous application of acetylene and ^{15}N-labeled N_2.

A recent investigation, carried out by Potts et al. [40], provides an illustrative example. Nitrogen fixation rates were determined by acetylene reduction as well as ^{15}N labeling at an intertidal mud- and sandflat on the coast of an island off Germany. Sediment cores 5 mm deep were transferred to 7 ml serum bottles and sea water was added to make a final volume of about 2 ml. Gases were intro-

---

*The "lower limit of detection" is defined as an amount, a single measurement of which carries a standard deviation equal to that amount.

duced or removed by the use of a syringe inserted through the rubber seal of the serum bottle. Labeled N_2, with an abundance of about 80 atom % ^{15}N, was generated in the field by application of the Rittenberg method.

Ethylene and acetylene concentrations were measured by gas chromatography. Total-N contents and ^{15}N abundances were determined by the use of a CHN-analyzer in connection with a combined gas chromatograph and mass spectrometer. The conversion factor, that is, the ratio of acetylene reduction rate (nanomoles C_2H_4 mg N^{-1} hr^{-1}) to nitrogen fixation rate (nanomoles N_2 mgN^{-1} hr^{-1}), was found to be 5.4 ± 0.8. This observed value is significantly greater than the theoretical ratio of 3. The overall standard deviation of the observed value corresponds to a relative error of 15%.

Whether or not essentially the same result could have been obtained using optical emission spectrometry (instead of mass spectrometry) remains an open question. In the experiment performed, the order of magnitude of the crucial ^{15}N measurement was 0.1 atom % excess. Hence, if the analytical error in optical ^{15}N measurement were assumed to be ±0.01 atom %, the relative analytical error on the observed conversion factor would be on the order of 10%, which compares favorably with the overall relative error mentioned above.

IV. SUMMARY

The use of optical emission spectrometry for the analysis of $^{14}N/^{15}N$ ratios in N_2 gas has become an established technique in tracer experiments related to nitrogen in soils. Commercial instrumentation is available for sample preparation as well as optical analysis, and many studies of pertinent technical details have been reported in the literature. The list of references given below is not exhaustive, since reference is made mainly to pioneer work on the one hand and selected recent publications on the other. However, the latter will enable the interested reader to find many further literature references.

The advantages and limitations of optical emission spectrometry, as opposed to mass spectrometry, may be summarized as follows. The optical technique is cheaper and simpler, and it permits the isotopic analysis of considerably smaller amounts of nitrogen. Conversely, the precision is limited to ±0.01 atom % ^{15}N, which excludes the use of depleted ^{15}N tracers in most cases, and reduces the maximum acceptable dilution with natural nitrogen in any case. Furthermore, natural variations in ^{15}N abundance cannot be detected by optical emission spectrometry. Nevertheless, a host of worthwhile tracer investigations related to soil analysis can be adequately accommodated using the optical method for the determination of $^{14}N/^{15}N$ ratios.

REFERENCES

1. R. Gersberg, K. Krohn, N. Peek, and C. R. Goldman, Science, *192*:1229 (1976).
2. S. M. Naudé, Phys. Rev., *34*:1498 (1929).
3. H. C. Urey and G. M. Murphy, Phys. Rev., *38*:575 (1931).
4. W. Bleakney ana J. A. Hipple, Phys. Rev., *47*:800 (1935).
5. G. Junk and H. J. Svec, Geochim. Cosmochim. Acta, *14*:234 (1958).
6. G. Herzbery, Z. Phys. Chem., *B9*:43 (1930).
7. M. Hoch and H.-R. Weisser, Helv. Chim. Acta, *33*:2128 (1950).
8. K. Clusius, Helv. Chim. Acta, *33*:2134 (1950).
9. H. P. Broida and M. W. Chapman, Anal. Chem., *30*:2049 (1958).
10. G. Meier and G. Müller, Isotopenpraxis, *1*:53 (1965).
11. A. I. Gorbunov and P. A. Sagorec, Zh. Fiz. Khim., *29*:1442 (1955).
12. J.-P. Leicknam, V. Middelboe, and G. Proksch, Anal. Chim. Acta, *40*:487 (1968).
13. V. Middelboe, in *Soil Organic Matter Studies*, vol. 2, IAEA, Vienna, 1977, p. 205.
14. D. A. Rennie and M. Fried, in *Proceedings of the International Symposium on Soil Fertility Evaluation*, vol. 1, Indian Agricultural Research Institute, New Delhi, 1971, p. 639.
15. F. E. Broadbent and A. B. Carlton, in *Nitrogen in the Environment*, vol. 1 (D. R. Nielsen and J. G. MacDonald, Eds.), Academic Press, New York, 1978, p. 1.
16. L. Karlsson and V. Middelboe, in *Isotopes and Radiation in Soil-Plant Relationships including Forestry*, IAEA, Vienna, 1972, p. 211.
17. C. P. Lloyd-Jones, J. Adams, and D. N. Salter, Analyst (Lond.), *100*:891 (1975).
18. H. Faust, Isotopenpraxis, *1*:62 (1965).
19. H. Faust, Isotopenpraxis, *3*:100 (1967).
20. J. M. Bremner and A. P. Edwards, Soil Sci. Soc. Am. Proc., *29*:504 (1965).
21. Anonymous, *Tracer Manual on Crops and Soils*, IAEA, Vienna, 1976, p. 108.
22. R. Fiedler and G. Proksch, Anal. Chim. Acta, *78*:1 (1975).
23. J. A. Goleb and V. Middelboe, Anal. Chim. Acta, *43*:229 (1968).
24. G. Guiraud and L. A. Buscarlet, Int. J. Appl. Radiat. Isotopes, *26*:187 (1975).
25. C. P. Lloyd-Jones, G. A. Hudd, and D. G. Hill-Cottingham, Analyst (Lond.), *99*:580 (1974).
26. D. R. Keeney and M. J. Tedesco, Anal. Chim. Acta, *65*:19 (1973).

27. G. W. Meyer, B. D. McCaslin, and R. G. Gast, *Soil Sci.*, *117*:378 (1974).
28. G. S. Lazeeva, A. A. Petrov, and E. P. Stolbova, in *Primenenie stabil'nogo izotopa ^{15}N v issledovanijach po zemledeliju*, Kolos, Moscow, 1973, p. 16.
29. H. Faust and R. Reinhardt, Isotopenpraxis, *11*:321 (1975).
30. V. Middelboe, Int. J. Appl. Radiat. Isotopes, *29*:753 (1978).
31. J. D. S. Goulden and D. N. Salter, Analyst (Lond.), *104*:756 (1979).
32. J. C. Burridge and I. J. Hewitt, *Anal. Chim. Acta*, *118*:11 (1980).
33. H. Saaby Johansen and V. Middelboe, Appl. Spectrosc., *36*:221 (1982).
34. V. Middelboe and H. Saaby Johansen, Appl. Spectrosc., *32*:511 (1978).
35. H. Saaby Johansen and V. Middelboe, Appl. Spectrosc., *34*:555 (1980).
36. R. D. Hauck and J. M. Bremner, Advan. Agron., *28*:219 (1976).
37. M. Fried and L. A. Dean, Soil Sci., *73*:263 (1952).
38. Z. Aleksic, H. Broeshart, and V. Middelboe, Plant Soil, *29*: 474 (1968).
39. R. W. F. Hardy, R. C. Burns, and R. D. Holsten, Soil Biol. Biochem., 5:47 (1973).
40. M. Potts, W. E. Krumbein, and J. Metzger, *Proceedings of the 3rd International Symposium on Environmental Biogeochemistry*, Wolfenbuettel, Germany, 1977.

9

Analysis of Nitrogen Isotope Ratios by Mass Spectrometry

ANTHONY HAYSTEAD Hill Farming Research Organisation,
Penicuik, Midlothian, Scotland

I. INTRODUCTION

The function of the mass spectrometer in nitrogen isotope analysis is in most cases identical with that of the optical emission spectrometer (Chap. 8) and, while each instrument has unique advantages, in many instances either would perform the task equally well. Both instruments exploit a physical property of the nitrogen molecule, N_2, to determine the relative amounts of each of the three possible species, $^{14}N_2$, $^{14}N^{15}N$, and $^{15}N_2$. In optical emission spectrometry, N_2 molecules are separated on the basis of their vibration properties, while in mass spectrometry, charged ions are separated according to their mass to charge (m/e) ratio. The terms *mass spectrometer*, *mass spectroscope*, and *mass spectrograph* all refer to instruments performing this function, the difference being limited to the way in which the ions are detected and measured.

A further similarity between nitrogen isotope analysis using a mass spectrometer and analysis using an optical emission spectrometer lies in the procedures used to convert soil or biological material to elemental nitrogen. Since N_2 is the material analyzed in both instruments, techniques developed for one can frequently be adapted for the other.

Historical aspects of the discovery of the isotopic nature of nitrogen and the properties of the various isotopes are described in some detail in Chap. 8, and are not discussed further here. This chapter describes the principles on which the most commonly encountered mass spectrometers operate and the advantages that isotope mass spectrometry has over optical emission spectrometry, and discusses briefly those applications to which mass spectrometry is best suited. A detailed review of practical aspects of nitrogen isotope analysis has recently been published by Fiedler and Proksch [1], and Hauck and Bremner [2] have discussed the applications of tracer nitrogen work in soil science and plant analysis. As far as possible, this contribution will try to complement these excellent authors. Commercially available mass spectrometers are not discussed in any detail in this chapter but, inevitably, the limited range with which the author is familiar is mentioned most frequently. This fact does not imply an endorsement of any particular instrument or the deficiencies exist in others. From the after-sales service point of view, however, there are strong arguments for buying a locally made product.

II. MASS SPECTROMETERS

The development of the mass spectrometer followed the discovery of "positive rays" (in fact, positively charged ions) which were seen to emerge from the perforated cathode of a partially evacuated discharge tube (Fig. 1). The forerunners of the modern mass spectrometer, developed by Thompson [3] and Aston [4], contained all the fundamental elements present in modern instruments. Figure 2 is a schematic diagram of a mass spectrometer. In the *source*, positively charged ions are generated in a gas or vapor and restricted into a narrow beam. The beam of ions is further restricted and accelerated electrically into the *analyzer* where the beam is separated into its

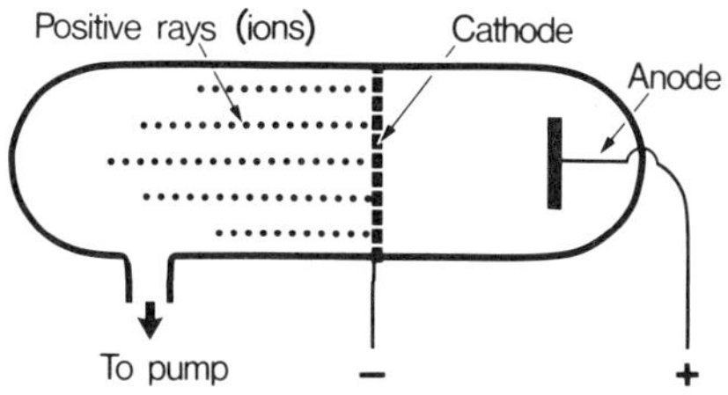

Figure 1 Positive rays in a partially evacuated discharge tube.

components in electric and/or magnetic fields. The separated components of the beam are allowed to impinge on the *collector* assembly which in turn is connected to some sort of voltage measuring device.

It is not within the scope of this chapter to discuss in detail the historical developement of the isotope mass spectrometer; interested readers are directed to review material by the founders of mass spectrometry [5-7] and also the texts of Duckworth [8] and McDowell [9].

A. Operating Principles

Mass spectrometer developement, particularly over the last 25 years, has been extremely rapid and at the present time a number of types of mass spectrometer are commercially available. Not all types are suitable for routine isotope analysis, however; in the main, single and double collector magnetic sector instruments and, to a lesser extent, quadrupole mass spectrometers have been used in isotopic tracer studies. A detailed discussion of mass spectrometer types can be found in Farmer [10]. All isotope mass spectrometers have certain features in common; for example, they all operate under high vacuum conditions and have the basic ion source-analyzer-collector arrangement shown in Fig. 2. Major differences between magnetic sector, double focusing, quadrupole, cycloidal focusing, and time of flight mass spectrometers lie in the analyzer section, in that the ion beam is separated into its components in different ways.

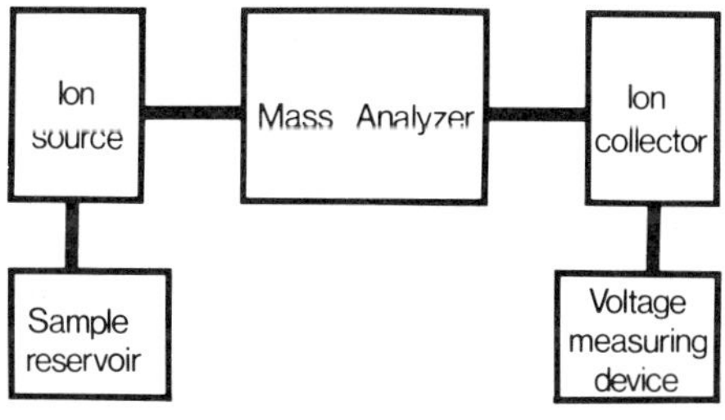

Figure 2 Schematic diagram of a mass spectrometer.

Several types of source are used in commercially available instruments. Chemical ionization sources in particular are extensively used in high-resolution mass spectrometry of labeled and unlabeled organics and metabolic intermediates. Isotopic ratio analysis of inorganic gases, however, is almost invariably carried out using electron impact ionization.

The resolution and precision of an isotope ratio mass spectrometer to a large extent determine how well it performs. Resolution, the ability of the instrument to separate the ion spectrum, is expressed in terms of the peak width (ΔM in atomic mass units, $^{16}O^+$ = 16 AMU) at 0.1 or 0.5 of the peak height (h). Resolving power (R) is often described as

$$R = \frac{M}{\Delta M_{0.1 \text{ or } 0.5}} \tag{1}$$

where M is the value of the peak in atomic mass units and ΔM is the width of the peak. Unfortunately there are several other definitions of resolution and resolving power, so a prospective mass spectrometer purchaser is well advised to check exactly how R has been calculated for a particular mass analyzer system. It should also be borne in mind that resolution is affected by the collector aperture or slit width of a magnetic sector instrument and by field parameters in a quadrupole. In both cases resolution can be increased or decreased to a certain extent with reciprocal changes in sensitivity.

Precision defines the ability of a mass spectrometer to differentiate reliably between samples of varied isotope abundance. Quantitatively, precision or internal reproducibility is expressed either as a relative standard deviation for a set of measurements or a minimum detectable difference in isotopic abundance. Both of these numerical estimates of precision are affected by sample size (relative to the optimum for the inlet system) and also the isotopic abundance range of the samples. The figures quoted by most manufacturers refer to an optimum sample size close to natural abundance; only rarely is it possible to determine how the instrument will perform at high enrichment.

Magnetic Sector Mass Spectrometers

Figure 3 illustrates diagrammatically how one of the simplest types of mass spectrometer, a 180° single collector instrument, operates. The whole of the source-analyzer-collector assembly is contained within an evacuated stainless steel enclosure which is continuously pumped, so that during sample admission the pressure equilibrates between 10 and 10^{-2} mPa (ca. 10^{-4}-10^{-7} torr). Low pressure must be maintained in the mass spectrometer so that the mean free path of the ions (the distance an ion can move between collisions) is sufficiently large with respect to the geometry of the analyzer

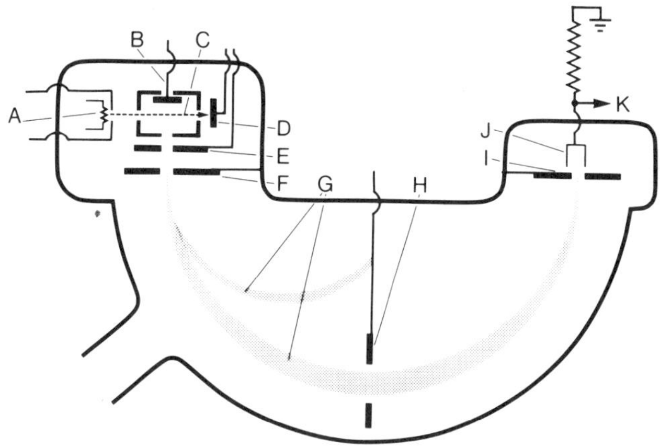

Figure 3 180° magnetic mass spectrometer: A, filament; B, ion repeller; C, electron beam; D, electron trap; E and F, beam focusing and ion accelerating plates; G, ion beams; H, alpha slit; I, collector slit; J, collector plate; K, output.

assembly. If this requirement is not met, then scattering due to ion-ion and ion-molecule collisions prevents sharp beam focusing. In the instrument shown in Fig. 3, positive ions are generated in the material to be examined by electrons which are accelerated from a heated filament into a space called the ion chamber. Ions emerge under the influence of voltages on plate E and the ion repeller and are further accelerated and collimated by a much larger electric field applied between E and F. A permanent magnet encloses the analyzer such that the homogeneous magnetic field is perpendicular to the diagram and the ion beam is deflected through 180° onto a collector plate J. In a constant magnetic field, the radius of the flight path of an ion is a function of its kinetic energy and its mass. The relationship between the radius of flight (r), the accelerator voltage between B and C (V), the mass (m), charge (e), and velocity (v) of an ion is as follows:

$$\frac{1}{2}mv^2 = eV$$

therefore

$$v^2 = \frac{2eV}{m} \tag{2}$$

If the flux density of the magnetic field through which the ions pass is B and the source, alpha, and collector slits are arranged so that only a fraction of the ion beam arrives at the collector, the system will select ions with a narrow range of flight radius, the relationship between r and B being

$$r = \frac{mv}{Be} \tag{3}$$

Combining Eq. (2) and (3) and eliminating v,

$$\frac{e}{m} = \frac{2V}{B^2 r^2} \tag{4}$$

An analysis of a mixture of ions of different e/m values (in fact m values since ions with a single positive charge predominate) can hence be made by varying V or B and measuring the current passing from source to collector as different parts of the ion spectrum enter the collector aperture. In practice, the voltage produced by the ion current across a very high resistance [10^{10}-10^{12} Ω (ohms)] is measured using a dc amplifier and voltmeter or potentiometric recorder.

In modern instruments the ion current is often first amplified by an electron multiplier and subsequently by an operational amplifier using field effect transistors (FETs). These first stages of signal amplifications are highly susceptible to outside influences and to reduce noise the components are usually enclosed in a metallic, often evacuated, enclosure.

Quadrupole Mass Spectrometers

The principle on which the analyzer of the quadrupole mass spectrometer operates was described simultaneously in Germany and the United States in 1953 [11,12]. A full understanding of the principle requires a fairly complex mathematical description of the trajectory characteristics of ions passing through the quadrupole field, which can be found in a number or recent review articles on the subject [13-15]. O'Hanlon [71] provides a nonmathematical description of the mass filter action of a quadrupole. Figure 4 is a diagrammatic representation of a quadrupole mass spectrometer, from which it is immediately apparent that the ion source and collector sections are identical to those in the magnetic sector mass spectrometer; only the analyzer is different. The quadrupole field into which the ions are accelerated is generated by applying dc and radio frequency (rf) voltages to four metal rods held equidistant and parallel to each other. The resolution of the mass analyzer depends to a large extent on the precision of the quadrupole assembly, which in consequence is manufactured to very rigorous standards. Singly charged ions entering the quadrupole field undergo oscillations which are dependent on field parameters and on the mass of the ion. Under any given set of conditions only ions having a particular mass will emerge and impinge on the collector plate. Mass spectra

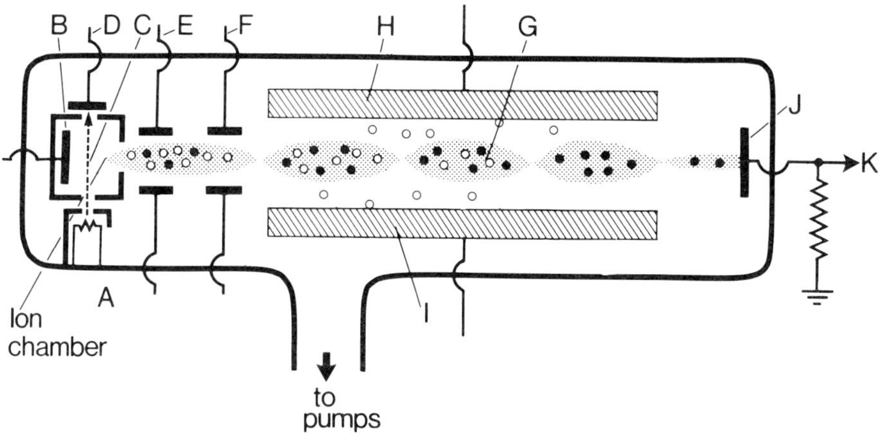

Figure 4 Quadrupole mass spectrometer: A, filament; B, ion repeller; C, electron beam; D, electron trap; E and F, ion beam focusing plates; G, ion beams; H and I, two of the four quadrupole rods; J, collector; K, output.

are obtained by varying the dc and rf components of the voltage supply to the quadrupole assembly and so changing the mass filter characteristics of the quadrupole. Conventional quadrupole mass spectrometers have the unique advantage of extremely high scanning speeds, but for isotopic analysis do not have the precision of magnetic sector mass spectrometers [16]. Schmidt and Favre [17] have designed a system analogous to a double beam spectrophotometer, in which sample and reference molecular beams are alternately allowed to enter the ion source of a conventional quadrupole mass spectrometer. After equalizing the major isotope ion currents, the ratio of the minor isotope ion currents is electronically measured. The precision obtainable with this system approaches that of many magnetic sector mass spectrometers. Quadrupole mass spectrometers which can be used for isotopic nitrogen ratio analysis are manufactured in the United States by UTI (Sunnyvale, Cal.). One instrument, the Q-30C, has a resolving power of 600 to 300 AMU (see Sec. II.A) and a precision of ±0.25% at an ion current ratio of 10.

B. Multiple Collector Systems

In the magnetic sector instruments described in Sec. II.A, ion beams are focused separately onto the collector electrode. If two ion currents, say m/e 28 ($^{14}N_2$) and m/e 29 ($^{14}N^{15}N$), are to be compared they must be measured sequentially. A loss of precision can occur in this procedure because the conditions under which ions are generated, the pressure and composition of the sample, and the characteristics of the measurement system are all susceptible to fluctuations during analysis. In a multiple collector instrument such effects are minimized because two or three ion beams are simultaneously collected and measured. The two ion beams in a dual collector instrument can be compared electronically and the ratio computed directly. Modern instruments designed for this specific purpose frequently make a series of automatic comparisons of an N_2 sample with a reference gas and are able to measure differences in ^{15}N abundance down to 0.00001 atom % ^{15}N.

C. High Vacuum Technology

Three main factors dictate the maximum operating pressure of the mass spectrometer. First, as discussed in Sec. II.A, analyzer resolution is dependent on the mean free path of the ions being sufficiently large with respect to the geometry of the flight path in the analyzer. Second, at pressures greater than about 10 mPa (10^{-4} torr), space charge effects upset the linearity of the ionization process. Third, the sensitivity of the mass spectrometer at any particular point in the mass spectrum is dependent on the ion current due to residual gases in the source. Ideally the background ion current at any particular value of m/e should be negligible with respect to the sample ion current. To achieve this objective the partial pressure of the molecular species giving rise to the ion being measured should be in the region of 10^{-1}-10^{-2} μPa (ca. 10^{-9}-10^{-10} torr) or below. It is not, however, sufficient simply to attach to the mass spectrometer a pumping system capable of removing the gas in question extremely quickly, since this would necessitate high rates of sample admission to maintain a pressure in the source sufficient for an analysis to be made. In some mass spectrometers a pumping system is used which pumps efficiently enough to give reasonable background ion currents but slowly enough to maintain instrument sensitivity, so that small samples can be admitted to the source. In other instruments, pumping system efficiency is maximized, and the sample is admitted directly into the ion chamber so that the pressure at the point of ion generation can be one or two orders of magnitude greater than that in the analyzer. Using such a system the objective of negligible backgrounds for the larger residual ion currents (for example at m/e 28, the $^{14}N_2$ peak) is attained.

The low pressures required in the mass spectrometer cannot be attained using mechanical, oil sealed, rotary pumps. Combined rough and high vacuum pumping systems must be used. The vapor (or diffusion) pump backed by a mechanical rotary pump is the type most commonly used for isotope mass spectrometers. Other systems,

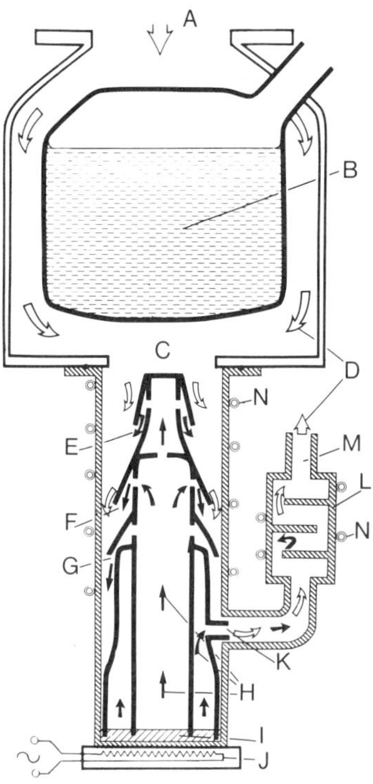

Figure 5 High-vacuum pumping system comprising a cold trap and an oil vapor pump. A, outlet from mass spectrometer; B, liquid nitrogen reservoir; C, inlet to vapor pump; D, residual gas flow; E, F, and G, stage 1, 2, and 3 jets; H, fluid vapor; I, pump fluid; J, heater; K, ejector jet; L, condenser; M, outlet to rotary pump; N, cooling water supply.

for example, ion pumps backed by sorption pumps, have been used, but have not found the general commercial acceptance of the vapor pump/rotary pump combination. Vapor pumps operate by creating jets of oil or mercury vapor which in turn transfer momentum to residual gas molecules, compressing them in the region of the rotary pump inlet. Figure 5 shows a typical pumping system comprising a rotary pump, a three-stage vapor pump, and a cold trap capable of maintaining a pressure in the region of 10^{-1} μPa (10^{-9} torr) in a mass spectrometer. The ultimate vacuum in a continuously pumped system is limited by (1) leaks into the system from outside, (2) backstreaming of oil vapor into the mass spectrometer from the pumps, (3) the rate at which substances are desorbed from the inside of the evacuated system (outgassing), and (4) the efficiency of the pumping system. Leak tightness is obtained by reducing demountable joints to a minimum and by using precision flanges with soft metal gaskets (copper, gold, or aluminum) where joints are essential. Ingress of oil vapor (backstreaming) is reduced by incorporating baffles (which may be refrigerated) or cold traps between the vapor pump and the analyzer assembly. Outgassing at room temperature can be a very slow process, depending on the nature of the gases being evolved and the material from which the evacuated system is constructed. The rate of outgassing increases with temperature, and for this reason the whole source-analyzer collector assembly can usually be heated (baked) to 200-350°C. In this way the ultimate vacuum of the system is attained as quickly as possible.

A number of isotope ratio mass spectrometers currently available employ turbomolecular high vacuum pumps rather than vapor pumps. In fact, the operating principle of the two is analogous. In the vapor pump, momentum is transferred to the residual gases by a supersonic vapor jet, while in the turbomolecular pump, a turbine rotating at 30,000-60,000 rpm is used. The turbomolecular pump has the advantage that there is no pump fluid backstreaming. It is, however, more expensive to purchase and has a higher maintenance requirement. The MAT 251 manufactured by Finnigan MAT used an automated turbomolecular pumping system, which the manufacturers claim gives the instrument greater sensitivity (due to less scattering by heavy residual gas molecules from the pump fluid) and increased user convenience, since refrigerant is not required.

D. Sample Inlet Systems

In most isotope mass spectrometers the sample is continuously admitted into the pumped system. Clearly, if the mass spectrometer is to be maintained at a low pressure during analysis, a means must be devised to admit sample gas (1) at a constant rate and (2) in extremely small amounts. This is usually done by introducing a flow restrictor (or leak) into the inlet of the mass spectrometer. The way in which gas flows through such a leak can be either *molecular* or *viscous*, depending on the pressure at the entrance side of the leak and the construction of the leak itself.

Molecular flow occurs when the pressure at the entrance to the leak is low and the mean free path of the sample molecules is large relative to the dimensions of the leak. Under molecular flow conditions lighter molecules pass into the source more readily and the sample reservoir becomes enriched in the heavier components of the mixture. In the source itself, however, movement of gases is also molecular and lighter gases are pumped away more quickly. Hence, at any point in time, the relative numbers of chemically identical ions generated in the source reflect accurately the partial pressures of the parent molecules in the sample reservoir. Isotope analysis using a typical molecular leak requires the sample reservoir to be large and sample pressure to be low, usually between 10 and 100 Pa (ca. 10^{-1}-1 torr). Molecular leaks are manufactured by sealing a plug of porous material into the sample inlet tube close to its connection with the mass spectrometer. Sintered silicon carbide, quartz, and glass are often used as refractive, chemically inert, microporous materials for this purpose.

At higher pressures, gas moves along a tube in much the way that water flows in a pipe, and flow is said to be viscous. A viscous leak is usually constructed by crimping a capillary inlet tube close to its point of entry to the mass spectrometer. Clearly, as the pressure falls across a constriction of this type, gas flow changes from viscous to molecular, so the overall characteristics of the leak depend on its construction and operating pressure. For example, a relatively long and narrow connection to the sample reservoir will restrict the backflow of heavy molecules accumulated at the constriction, reducing changes in the sample reservoir. Viscous inlets are generally preferred for isotope ratio measurements, because it is more convenient to handle gaseous samples in small volumes at relatively high pressures (5-10 kPa (40-80 torr)), but many commercially available isotope mass spectrometers use molecular leaks. Two gas handling systems are shown in Fig. 6, one for a double molecular leak suitable for isotope ratio comparisons (Fig. 6a) and one suitable for a viscous inlet into a dual collector mass spectrometer (Fig. 6b). Gas handling systems can be constructed from glass, using greaseless high vacuum stopcocks, but all-metal construction using packless (bellows) or diaphragm valves gives longer trouble-free service and has the advantage that these systems can be periodically baked.

E. Interpretation of Ion Current Measurements

When a sample of combined nitrogen is converted to N_2, using one of the recognized procedures (see Sec. III), the proportions of the three molecular species: $^{14}N_2$, $^{14}N^{15}N$, and $^{15}N_2$ will be binomially distributed. In a sample containing fractional amounts a and b of ^{14}N and ^{15}N atoms respectively, by definition a + b = 1, and the relative amounts of the three species will be

$$(^{14}N_2):(^{14}N^{15}N):(^{15}N_2) = a^2:2ab:b^2 \tag{5}$$

Since there is no significant difference in the ionization potential of the three species, the ratio of molecular ions with m/e values 28, 29, and 30, and hence the ion currents i_{28}, i_{29}, and i_{30} measured at the collector, will be in the same proportion as the original diatomic molecules. By measuring any two of the three ion currents, the third

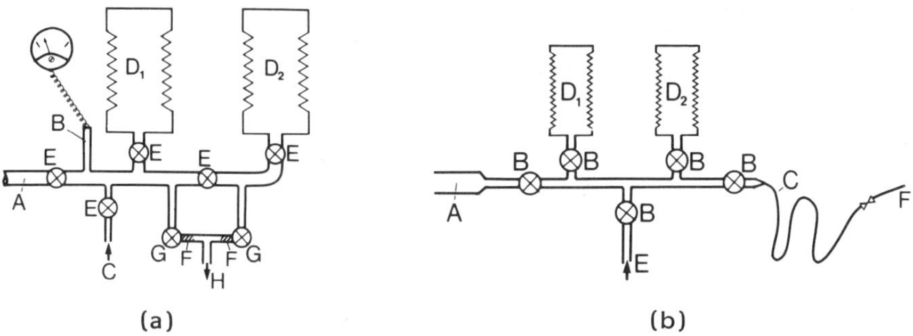

Figure 6 Inlet and gas handling system suitable for mass spectrometer with a molecular leak. A, outlet to vapor pump; B, Pirani gauge; C, sample inlet; D_1 and D_2, stainless steel bellows reservoirs (200-1000 ml); E, toggle-operated bellows gas-handling valves; F, porous leaks; G, diaphragm inlet valves (zero dead space); H, mass spectrometer inlet. (b) Inlet and gas handling system suitable for mass spectrometer with a viscous leak. A, outlet to rotary pump; B, toggle-operated bellows gas-handling valves; C, capillary leak with crimp (0.5 mm × 0.1 mm i.d.); D_1 and D_2, stainless steel bellows adjustable-volume reservoirs (50-150 ml); E, sample inlet; F, connection to mass spectrometer.

can be calculated from Eq. (5). The isotopic composition of a sample of nitrogen is usually expressed in terms of its content of the minor isotope, ^{15}N, either in absolute terms as a ^{15}N-abundance in atom percent (%) ^{15}N:

$$\text{Atom \% } {}^{15}\text{N} = \frac{\text{number of } {}^{15}\text{N atoms}}{\text{total number of N atoms}} \times 100 \tag{6}$$

or, alternatively, with respect to the ^{15}N-abundance of atmospheric N_2 (0.3663 ± 0.0004 [18]) as an enrichment or depletion from this norm:

$$|\text{atom \% } {}^{15}\text{N of sample} - 0.3663| = \text{atom \% } {}^{15}\text{N excess or depletion}$$

Small variations in ^{15}N abundance, for example, variations in the ^{15}N content of naturally occurring soil components, are often measured in terms of parts per thousand differences ($\delta\,^{15}N$) from a reference gas, usually atmospheric N_2:

$$\delta\,{}^{15}\text{N} = \frac{R_{\text{sample}} - R_{\text{reference}}}{R_{\text{reference}}} \times 1000 \tag{7}$$

where $R = {}^{15}N/{}^{14}N$.

Using this system, one $\delta^{15}N$ "unit" is equivalent to 3.7×10^{-4} atom % ^{15}N. Some authors [74,75] use a slightly different definition of $\delta^{15}N$ in that they use ^{15}N concentrations in their calculation rather than R. Close to natural abundance, the difference in $\delta^{15}N$ values calculated using the two procedures is very small. Equation (6) can be written in the form:

$$^{15}N \text{ abundance} = \frac{(^{14}N^{15}N + 2^{15}N_2) \times 100}{2^{14}N_2 + 2^{14}N^{15}N + 2^{15}N_2} \tag{8}$$

or in terms of ion currents:

$$^{15}N \text{ abundance} = \frac{(i_{29} + 2i_{30}) \times 100}{2(i_{28} + i_{29} + i_{30})} \tag{9}$$

Samples of very low or high ^{15}N content produce low ion currents at i_{30} and i_{28}, respectively, which are more difficult to measure accurately. Depending on the ^{15}N content of the sample, either the i_{28}/i_{29} or i_{29}/i_{30} ratio is measured and the ^{15}N abundance calculated from the equations:

$$\text{Atom \% } ^{15}N = \frac{100}{2R_1 + 1} \tag{10}$$

where $R_1 = i_{28}/i_{29}$,

$$\text{Atom \% } ^{15}N = \frac{200}{R_2 + 2} \tag{11}$$

where $R_2 = i_{29}/i_{30}$.

In practice the determination of either of these ratios depends on the characteristics of the mass spectrometer which is being used. For example, if residual gases in the source give rise to significant ion currents, then background values must be subtracted from the sample currents. Similarly, if atmospheric contamination is likely to be significant, corrections must be made to the sample ion currents to allow for the components due to atmospheric N_2. Fiedler and Proksch [1] have given a detailed account of mass spectrometric analysis of N_2 using a single collector magnetic sector instrument. In critical studies sample pressure may affect the measured ratios. If it is not possible to standardize sample pressures, then this phenomenon must be taken into account. Mook and Grootes [19] and Kerven and Saffigna [76] discuss corrections which must be applied to measured ion current ratios in high precision determination of isotope abundances. Statistical procedures for high precision determination of ^{15}N values are described by Steele et al. [77].

III. SAMPLE PREPARATION

It would be possible to determine the ^{15}N abundance of soil-derived material by converting it into any stable low molecular weight compound of nitrogen. Ideally, however, the compound used should be gaseous at room temperature and should contain only 1 or 2 atoms of nitrogen. Elemental nitrogen, N_2, meets these requirements and in consequence is used almost universally. The N_2 molecule is very stable in conventional ion sources, the contribution from atomic ions at m/e 14 and 15 being very small. In addition, since only nitrogen atoms are present, the mass spectrum of N_2 is not confused by ions originating from other isotopic elements. N_2O, for example, produces two species at m/e 46, $^{15}N^{15}N^{16}O$ and $^{14}N^{14}N^{18}O$, which are not separated in single focusing mass spectrometers. The principal requirements of the procedure used to prepare N_2 from experimental material are as follows:

1. Quantitative conversion to pure N_2
2. Random pairing of N atoms in the formation of N_2
3. Complete removal of volatile compounds of carbon (particularly CO) and phosphorus which may be formed
4. Complete removal of water vapor

Procedures in current use are based on either a direct oxidation of experimental material to N_2 with copper oxide (Dumas combustion) or a three-step procedure in which material is first converted to NH_3 by Kjeldhal digestion, and the NH_3 steam-distilled into acid solution and finally oxidized to N_2 using alkaline lithium or sodium hypobromite solution [20]. Although this latter procedure is clearly laborious and time consuming, it has been the method chosen by most workers preparing samples for mass spectroscopy and, until quite recently, Dumas combustion techniques have been associated with optical emission spectroscopy. Methods and sources of error in Kjeldahl digestion and steam distillation are described in detail by Bremner [21], so only the final step, the conversion of NH_3 to N_2, will be discussed here. Procedures in current use are of two main types: (1) those using a Rittenberg flask (Fig. 7), and (2) those based on the apparatus of Ross and Martin [22] (Fig. 8). In both systems, NH_3 as a dry ammonium salt or in solution is mixed under vacuum with an excess of hypobromite reagent. In Rittenberg flasks dissolved N_2 is removed from the reactants before mixing by either evacuating the flask on a separate pumping system for a few minutes, or, alternatively, a rather time consuming cycle of (1) freezing in liquid N_2, (2) pumping down on the mass spectrometer inlet pumping system, (3) closing off the flask and thawing to allow the solutions to evolve dissolved N_2, and (4) refreezing and pumping away the N_2. Two or three such cycles are required to remove completely dissolved N_2.

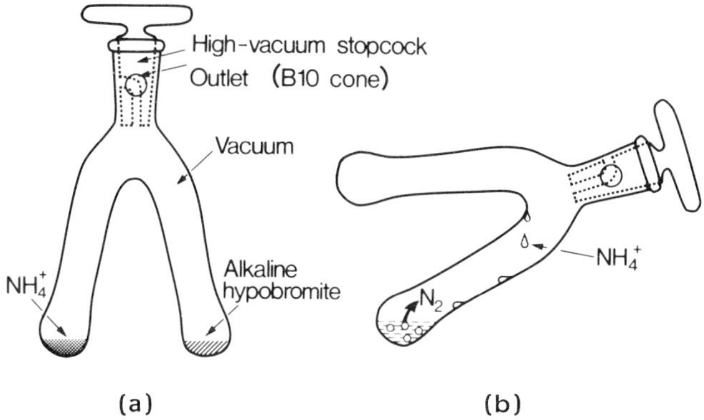

Figure 7 Rittenberg flask. (a) Upright position for reagent addition and degassing. (b) Tilted to add NH_4^+ to the alkaline hypobromite.

Smith and Chalk [78] have described an apparatus similar to that of Burris [57] in which N_2 generated in Rittenberg flasks is transferred to a small sample tube using a Toepler pump. In the Ross and Martin apparatus, samples are evaporated to dryness in a glass tube and the reagent is degassed in the reservoir by repeated pumping and filling with helium. VG isogas (Middlewich, Cheshire, England) currently supply a stainless steel sample preparation system in which degassed oxidant is added to evacuated sample vials through a rubber septum.

There are two attractively simple alternatives to these rather complex systems. (1) Oxidant and sample are injected into a previously evacuated container through a rubber septum. "Vacutainers" (Beckton-Dickinson, Rutherford, N.J.) have been used for this purpose [23]. After filling the containers with helium to atmospheric pressure, samples of the gas can be injected directly into the mass spectrometer through a cold trap [23]. (2) Sample and degassed oxidant are mixed in a tube immersed in liquid N_2. The tube is then attached to the mass spectrometer and after evacuation the contents are allowed to warm up and generate N_2 [24].

The direct oxidation of material to N_2, CO_2, and H_2O by heating at 600-1200°C with copper oxide combines all three steps of the Kjeldahl/Rittenberg procedure and in consequence ought to save considerable time. In practice, however, it is relatively difficult to obtain quantitative conversion to N_2 in a single oxidation step and it is usually necessary to "clean up" the gas produced by further oxidation and reduction reactions. Figure 9 shows a Dumas combustion system constructed in the author's laboratory and based on an apparatus in use at the Agricultural Research Council Letcombe

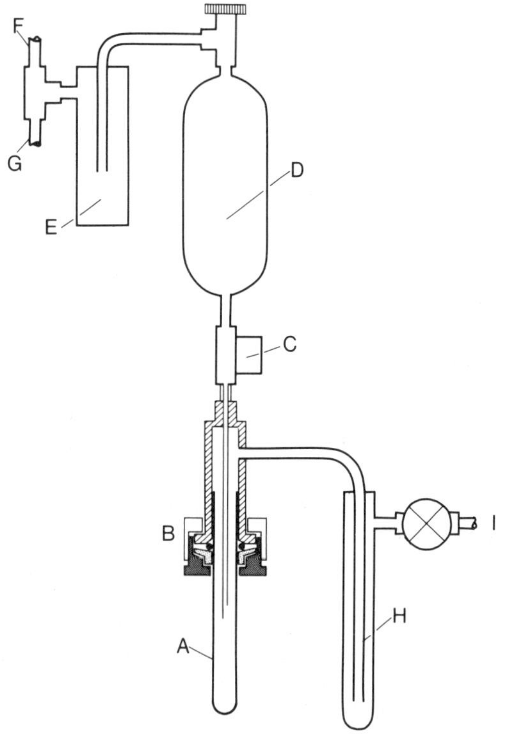

Figure 8 Stainless steel N_2 generation system based on Ross and Martin's [22] glass apparatus: A, 10 mm o.d. glass tube; B, rapid vacuum seal; C, PTFE diaphragm valve (Nupro Corp., Cleveland, Ohio; D, hypobromite reservoir; E, trap; F and G, pump and helium outlets: H, cold trap; I, inlet to mass spectrometer gas-handling system.

Laboratory [25]. An interesting feature of this apparatus is the way in which molecular sieve 5A pellets are used to transfer N_2 from the oxidation or reduction train to the mass spectrometer. The principle used is that of the sorption pump in which the molecular sieve takes up gases when cooled in liquid N_2 and, in the case of nitrogen, releases it again at room temperature. Fiedler and Proksch [1] have described a system in use at the International Atomic Energy Agency laboratory, Vienna, in which dry biological material is combusted with an oxidant (CuO) and a CO_2 and H_2O absorbent (CaO) in an evacuated glass tube. When oxidation is complete the tubes are loaded into an automatic sample admission device attached to the mass spectrometer.

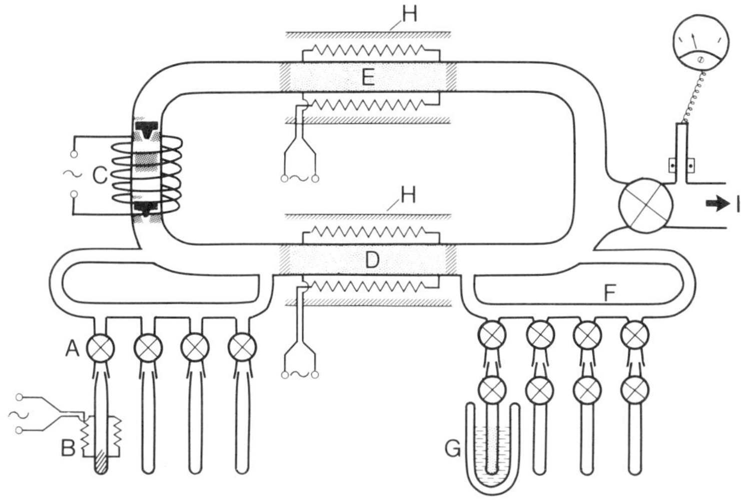

Figure 9 Glass Dumas combustion apparatus: A, Inlet manifold; B, electrical oxidation furnace (780°C) around silica sample tube; C, "one-way" solenoid-operated circulation pump; D, copper granules; E, copper oxide wire; F, collection manifold; G, collection tube containing molecular sieve 5A immersed in liquid nitrogen; H, furnaces at 900°C; I, outlet to vapor pump.

The device breaks the glass tubes in turn and expands the gas into the mass spectrometer inlet system.

Commercially available elemental and nitrogen analyzers utilize a Dumas combustion, usually followed by a gas chromatographic separation of the combustion products (see Chap. 4). The possibility of interfacing an elemental analyzer with a mass spectrometer is very attractive because it means that the whole procedure of total-N and isotopic-N analysis could be fully automated. Barsdate and Dugdale [26], using a Coleman nitrogen analyzer, collected the bubble of N_2 produced for volumetric determination of nitrogen, and transferred it with a Toepler pump to a mass spectrometer. The relatively poor precision obtained (±0.01 atom % ^{15}N) in these studies was mainly due to the unsuitability of the time-of-flight mass spectrometer used for isotopic abundance measurements. Japanese workers [27] have described an interfaced system in which N_2 emerging from the thermal conductivity detector of a CHN analyzer is automatically expanded into a series of evacuated reservoirs and admitted to an isotope mass spectrometer. A similar system using a jet separator to concentrate the N_2 prior to expansion has also been described [28].

IV. APPLICATION OF MASS SPECTROMETRY TO SOIL ANALYSIS

A. Introduction

Qualitative descriptions of nitrogen cycles in natural and agricultural ecosystems are common in the literature. There is no doubt, however, that the absence of a radioisotope of nitrogen which can be conveniently used in tracer studies has slowed down the quantification of such models. Over the last 10 years or so, high precision isotope mass spectrometers which can be easily operated and maintained by biologists and soil scientists have become available. In consequence, the rate of publication of isotopic tracer studies of the N cycle has dramatically increased. Table 1 indicates the type of mass spectrometer now available in the United Kingdom. Complete listings can be obtained from scientific instrument review journals. Instruments manufactured by Nuclide Corporation (642 East College Av., State College, Penn.) are extensively used in the USA and worldwide but at present no agency exists in the United Kingdom. The AEI (formerly Kratos) MS10 is no longer available as either a kit or a complete instrument from its manufacturers. Spares are available however, and a direct replacement (RGA 10) is manufactured by Vacuum Scientific Services Ltd. (44 Ellesmere St., Manchester, England).

The unique advantage of the mass spectrometer is its ability to measure extremely small differences in isotopic abundance between experimental material and a nitrogen standard, and it is when this advantage is being exploited that the additional expense of mass spectrometry is justified. Examples of such applications are:

1. Analysis of low enrichment material derived from experiments in which the tracer has been extensively diluted or was initially of low enrichment.
2. Analysis of material from experiments in which ^{15}N-depleted material is used.
3. Measurement of variations in natural abundance.
4. Determination of isotope fractionation in soil-N transfers.
5. Determination of N-isotope abundances in soil gases other than N_2, e.g., N_2O.

The fact that conventional mass spectrometers are able to measure routinely differences in isotope ratio down to $0.05\delta^{15}N$ units does not, however, mean that differences of this order can be interpreted simply as a bulk movement of N from one pool to another. At this level of precision, differences in isotopic abundance can result from the fact that chemically identical but isotopically different species do not react at identical rates, that is to say, a fundamental assumption underlying all tracer studies is invalid.

A comprehensive review of the application of mass spectrometry to soil analysis is beyond the scope of this chapter. Detailed informa-

Table 1 Mass Spectrometers for Determination of Nitrogen Isotope Ratios Available in the United Kingdom in 1982.

Manufacturer	Model	Type	Minimum detectable difference in abundance	Approximate 1982 Cost (E)
Vacuum Scientific Services Ltd., Manchester, England	RGA 10[a]	Magnetic sector	0.001 Atom % ^{15}N	8,000
VG Isogas Ltd. Middlewich Cheshire, England	MM622	Magnetic sector	0.0001 Atom % ^{15}N	27,000
	MM602E	Magnetic sector	0.00001 Atom % ^{15}N	45,000
Finnigan MAT, Paradise, Hemel Hempstead, Herts., England. (manufactured in Bremen, FRG)	MAT 251	Magnetic sector	Better than 0.00001 atom % ^{15}N	65,000
	Delta D	Magnetic sector	0.00006 Atom % ^{15}N	34,000
	Delta E	Magnetic sector	0.000015 Atom % ^{15}N	48,000
UTI, Sunnyvale, Cal. (British agent: Nanotech, Manchester, England).	UTI Q-30C	Quadrupole	0.001 Atom % ^{15}N	53,000

[a]Replacement of AEI MS 10 instrument.

tion of this type can be obtained in the reviews of Hauck and Bremner [2] and Hauck [29,30]. The remainder of this chapter will be devoted to a discussion of a few selected topics of particular interest, in which mass spectrometry is an indispensable analytical tool.

B. Isotope Discrimination in Biological N-Transfer Processes

From a practical point of view the limit of precision in ^{15}N-tracer studies is set by variation in ^{15}N abundance occurring between soil-N fractions of different chemical composition, from different depths in the soil profile, and from different locations [31,32]. These variations in ^{15}N abundance which usually fall in the range $\pm 10\,\delta\,^{15}N$ units (± 0.0037 atom % ^{15}N) originate from the isotope fractionation which occurs when nitrogeneous components of the soil undergo chemical and biological transformations. Delwiche and Steyn [33] and, more recently, Blackmer and Bremner [34] have shown that biological enzyme systems responsible for major soil-N fluxes react more readily with ^{14}N-containing molecules than with the equivalent ^{15}N species. Fractionation occurring as a result of such processes is expressed numerically as a fractionation factor. In the simple case of the nitrogen fixation reaction: $N_2 \rightarrow 2NH_4^+$, in which substrate supply is essentially infinite and not subject to isotope depletion, the fractionation factor β [35] is given by:

$$\beta = \frac{^{15}N/^{14}N \text{ in air}}{^{15}N/^{14}N \text{ in fixed N}} \tag{12}$$

Reactions occurring in the soil solution, denitrification, for example, in which the substrate is limited and subject to isotope depletion, require a more complicated analysis, and a kinetic fractionation factor must be derived [35]. Table 2 lists a number of soil processes and indicates the extent to which fractionation occurs during reaction. An examination of Table 2 shows that fractionation during N_2 fixation occurs to a lesser extent than during nitrification and denitrification, so it is apparent why soil organic and mineral N fractions are, in general, enriched relative to atmospheric N_2. A steady-state model of isotopic fractionation in field soils and laboratory microcosms has been described by Shearer and co-workers [72]. The predictions of this model agree well with experimental data when the fractionation factors of Delwiche and Steyn [32] are used. More recently, Marriotti et al. [73] have reviewed isotopic fractionation phenomena in soils and have presented data from laboratory incubation experiments to support their theoretical analysis of nitrification and denitrification.

If ^{15}N-tracers are to be used to study nitrogen transfers between soil, plant, and animal in grazed environments, a further com-

Table 2 Microbiological Reactions Resulting in Fractionation of Nitrogen Isotopes

Process	Formal reaction	Organism	Fractionation factor
N_2 fixation	$N_2 \rightarrow 2NH_4^+$	*Azotobacter*	1.0039
NH_4^+ assimilation	$NH_4^+ \rightarrow$ amine-N, amide-N	*Azotobacter vinelandii* *Pichia terricola*	1.0148 1.0018
Nitrification	$NH_4^+ \rightarrow NO_3^-$	*Nitrosomonas europea*	1.0260
Denitrification	$NO_3^- \rightarrow N_2O + N_2$	*Psuedomonas denitrificans*	1.0173
Denitrification	$NO_3^- \rightarrow NO_2^-$	Mixed soil population	1.013

Source: Refs. 33 and 34.

plication must be taken into consideration. Considerable isotopic fractionation occurs in the digestive tracts of ruminant animals [36]. This effect is not entirely due to the rumen microflora, as there is evidence of a similar effect in monogastric animals, including man [36]. In cattle eating pasture enriched by $0.76\delta^{15}N$ units relative to atmospheric N_2, urine and fecal nitrogen gave values of -1.49 and $+3.11\delta^{15}N$ units, respectively. In many pasture systems urine-N is extensively volatilized as NH_3 [37]. Depending on the extent to which the NH_3 is reabsorbed by plant shoots and subsequently re-eaten or returned to the soil, a change in the ^{15}N abundance of nitrogenous soil components will occur. The possible value of the natural abundance of ^{15}N in plant and soil material as a qualitative or quantitative indicator of the source from which the nitrogen was derived, is at the present time a contentious issue. Virginia [79] analyzed plant and soil nitrogen from a natural ecosystem and concluded that seasonal and spatial variations in natural abundance are substantial and that it is unlikely that quantitative estimates of nitrogen fixation can be made from this type of data. In a simpler arable situation, on the other hand [80], results have been more promising.

C. Movement of N Between Soil and Atmosphere

Gaseous Loss Processes

The two major soil-N fluxes involving gaseous reactants, N_2 fixation, and denitrification, are perhaps the most difficult to measure accurately. The importance of the two processes, however, in building up and depleting soil-N, make it essential that they be quantified in studies of the soil nitrogen cycle.

Denitrification. Gas analysis by mass spectrometry can be used to measure the gaseous products of denitrification (N_2O and N_2), but unless soils are incubated in an N_2-free atmosphere [38] it is difficult to measure the production of small amounts of N_2 accurately. Gas analysis of this type is best carried out using single collector magnetic sector instruments designed for the purpose. Double focusing mass spectrometers, in which the ion beam is analyzed sequentially in magnetic and electrostatic fields, are often used for this sort of analysis because the resolution obtainable permits the separation of ions differing by small fractions of a mass unit (for example, $^{14}N_2$ and $^{12}C^{16}O$), so producing mass spectra which are easier to interpret.

Measuring denitrification in a realistic situation, for example, a waterlogged soil exposed to the air, is possible if isotopically labeled substrates ($^{15}NO_3^-$ or $^{15}NO_2^-$) are added to the soil and the production of labeled N_2 and N_2O measured against a background of atmospheric N_2 [29, 39]. The interpretation of mass spectra produced in this sort of experiment presents certain problems, however, because account must be taken of (1) fragmentation of N_2O in the source [40] and (2) nonrandom distribution of nitrogen isotopes in N_2 derived from two sources. Hauck [29] has described equations which can be used to deduce the ^{15}N content of N_2 evolved into air from measurements of i_{28}, i_{29}, and i_{30}. Using these equations, the relative contribution of the added substrate and unlabeled soil-N to the N_2 evolved can also be calculated.

The disappearance of isotopically labeled substrate can be used to estimate denitrification by measuring the loss of total soil ^{15}N during incubation with labeled NO_3 or NO_2^-. This simple approach is useful if comparative estimates of rates of denitrification are required. For example, McKenzie and Kurtz [41] have used this method to investigate the effects of soil pretreatments on subsequent denitrification. The technique can be made more sensitive if loss of labeled N from the extractable NO_3^- and NO_2^- pools is measured, particularly if the evolved gases are simultaneously measured and isotopically analyzed [39].

Focht [16] has drawn attention to the potential value of interfaced gas chromatograph/mass spectrometer systems in denitrification studies. Using interfaced systems, the gaseous produces of denitrification can first be separated on a molecular sieve column and isotopic analyses performed on the separated components. In the system described by Focht [16], the quadrupole mass spectrometer used gave data of rather low precision (±0.01 atom % ^{15}N). This limitation is not, however, inherent in quadrupole mass spectrometers. Schmidt and Favre [17] and Green [42], for example, have described quadrupole instruments capable of precise isotopic abundance determinations.

Ideally, isotopic measurements of denitrification would be made in the field under conditions as close to ambient as possible. These requirements are closely met in the studies of Rolston et al. [43] who measured concentration gradients of N_2O and $^{15}N_2$ in the soil profile using buried soil atmosphere sampling devices. A comparison of the direct measurements of gaseous loss (calculated from the concentration gradients) with a classical "difference" estimate of denitrification gave rates of loss of added NO_3^--N of 44.9 and 68.4%, respectively. The authors point out that since the denitrification zone in the field was very close to the soil surface, calculations based on a diffusive model may have underestimated N loss. Ryden et al. [44] have measured N_2O evolution from microplots in the field by covering them with a canopy through which air is passed, and collecting the evolved N_2O on molecular sieve. In the laboratory, the N_2O is released and measured by gas chromatography. Portable canopies of this type could be used to collect isotopically labeled N_2O and, if flow rates were sufficiently low, the N_2O-free air emerging from the molecular sieve trap could be analyzed for ^{15}N enrichment at m/e 29 and 30. A further possibility would be to measure N_2O concentration and isotopic abundance from soils treated with acetylene, which has been reported to inhibit the reduction of N_2O to N_2 [45-48].

Ammonia Volatilization. Gaseous loss of N as NH_3 from grazed pasture soils has been reported to be of the same magnitude as denitrification from arable soils [49]. Keeney and MacGregor [50] have studied cycling of [^{15}N]-urea applied to ryegrass-white clover pastures but did not attempt to collect evolved NH_3. If fluxes of NH_3-N from the soils are indeed significant in grazed ecosystems, this would seem to be an area in which considerable information could be obtained using mass spectrometry.

Nitrogen Fixation

The incorporation of elemental nitrogen into biological systems is most simply measured by exposing the test material to an atmosphere containing ^{15}N-labeled N_2. Subsequent conversion of the isotopically enriched products of fixation to N_2 for analysis in a mass spectrometer permits a quantitative determination of N_2 fixation to be made. If the results from this sort of experiment are to be meaningful, however, the enclosed test system must be maintained under conditions identical to those from which the system was taken. N_2-fixing organisms which can be tested in this way, for example, aqueous suspensions of microorganisms [51,52], incrustations of blue-green algae [53], and lichens [54] tend not to be those of general agricultural significance. The method has, however, been used to determine N_2 fixation by detached nodules of legumes and nonlegumes [52] and by free-living soil bacteria [55]. When this approach is used to measure N_2 fixation by whole plants growing in soil it is difficult to maintain constant partial pressures of CO_2 and O_2 for sufficiently long periods to measure ^{15}N incorporation (see, for example, Ross and Martin [56]). The use of $^{15}N_2$ in enclosed systems to measure nitrogen fixation in a range of biological systems has been reviewed by Burris [57,58]. In particular, he describes the use of Saran bags to enclose plant/soil cores in atmospheres of $^{15}N_2$-labeled air.

In most situations in which N_2 fixation is to be measured on whole plants growing in undisturbed soil, either the acetylene reduction technique (see Chap. 10) or an indirect isotopic method must be used [59]. It is possible indirectly to estimate N_2 fixation by adding labeled fertilizer to a nonfixing test crop (often an ineffectively nodulated or uninoculated legume) and a nitrogen-fixing species growing in the same soil. The test crop derives all its nitrogen from the soil, so the nitrogen it takes up will reflect the ^{15}N abundance of the available soil-N pool integrated over the growth period. The nitrogen fixer, on the other hand, will contain less ^{15}N because a proportion of its nitrogen is derived from the atmosphere. The extent of this isotope dilution can be used to determine the amount of nitrogen fixed during an increment of growth as follows:

If ΔN_f, ΔN_{nf} are the increases in total plant nitrogen for the fixing and nonfixing plants, respectively,

$$N \text{ fixed} = \Delta N_f(1 - X) \quad (13)$$

where

$$X = \frac{^{15}N \text{ enrichment of } \Delta N_f}{^{15}N \text{ enrichment of } \Delta N_{nf}}$$

If plants are grown from seed, calculating ΔN presents no problem. If this method is used to determine N_2 fixation during regrowth of a defoliated or perennial crop, then calculating ΔN is clearly more difficult and subject to greater errors [60]. An early description of this approach was published by Walker et al. [61], and the idea has been "rediscovered" at intervals up to the present time [62-64]. Fried and Broeshart [62,71] transform the proportions of nitrogen derived from fertilizer, soil, and atmosphere calculated from the measured isotopic abundances into fertilizer equivalents (A values in their terminology), to simplify the arithmetic in cases where the nitrogen-fixing and control plants have received different amounts of fertilizer nitrogen. The technique has been used by the author and others to determine N_2 fixation and also in attempts to measure N transfer from legume to grass in pasture [60,65,66]. An estimate of fixed-N transfer is obtained by comparing the enrichment of the test crop growing alone with the same crop growing in close association with a legume. Several possible sources of error should be taken into consideration when this sort of indirect measurement is made, particularly when a new soil or association of plant species is being investigated. Examples of such errors are: (1) plant roots of different species may discriminate differently between isotpes of nitrogen; (2) the N_2 fixer and control species may exploit different rooting zones, so N available to one may not be available to the other; (3) root and shoot turnover in the two species may differ; (4) if only above-ground organs are samples it is possible that partitioning of labeled metabolites between root and shoot may be different in the two species; (5) uptake of nitrogen by the two species may occur at different times during the growth period if their patterns of growth are markedly different.

Legg and Sloger [64] (see also Refs. 67 and 68 for methods employed) have compared the ^{15}N dilution tcchniquc with plant nitrogen accumulation and acetylene reduction in nodulating and non-nodulating lines of soybean. Their data show a close correlation between total plant nitrogen accretion and the isotopic determination of nitrogen uptake from soil and atmosphere. The acetylene reduction technique, however, gave considerably lower values. The authors suggest that weekly assays of acetylene reducing activity used in their studies may not be sufficient to determine total nitrogen fixing activity. In addition, the assays were carried out using washed, decapitated root systems [68] which would almost certainly show reduced activity.

D. Future Developments in Analysis of Nitrogen Isotope Ratios

It is unlikely that further increases in mass spectrometer precision will be of real value in soil and plant analysis, since existing instruments already have sufficient precision to determine variations in natural abundance of ^{15}N and, furthermore, the major sources of error in practical systems originate in the preparation of gaseous nitrogen and its subsequent admission to the mass spectrometer.

Several alternative analytical techniques exist for the determination of nitrogen isotope ratios in organic and biological systems. Irradiation by protons [69] and gas chromatography [70] have been suggested, but very little further evaluation of these techniques has been published. Quite recently, diagnostic vacuum gauges operating on the quadrupole principle came onto the market. The resolution of some of these systems is adequate for useful quantitative ratio determinations at m/e 28 and 29. Such systems would be robust, relatively inexpensive, and simple to operate. An extremely simple gas handling and inlet system, combined with an insertion type quadrupole gas analyzer, is currently being evaluated in the author's laboratory.

One major advantage of the mass spectrometer has so far not been mentioned. While it is certainly true that mass spectrometers are expensive instruments, and by virtue of the fact that they run continuously can be quite expensive and time consuming to maintain, they have a unique advantage over the optical emission spectrometer in that they are capable of an extremely wide range of analytical functions. Although purpose-designed instruments exist for gas analysis, isotopic abundance determinations, organic analyses, etc., almost any of these instruments can be used (usually with some loss in precision) in a variety of other analytical roles. Double-collector instruments, for example, in addition to being able to determine isotope ratios in mixtures of H/D, $^{16}O/^{18}O$, and $^{12}C/^{13}C$, can be used in the scanning mode as single collector gas analysis instruments to determine rates of exchange of O_2, CO_2, N_2O, etc., in biological systems.

REFERENCES

1. R. Fiedler and G. Proksch, Anal. Chim. Acta, *78*:1 (1975).
2. R. D. Hauck and J. M. Bremner, Adv. Agron., *28*:219 (1976).

3. J. J. Thomson, Phil. Mag., *13*:561 (1907).
4. J. W. Aston, Phil. Mag., *38*:709 (1919).
5. J. J. Thomson, *Rays of Positive Electricity and Their Application to Chemical Analysis*, Longmans, Green and Co., London, 1913.
6. F. W. Aston, *Mass Spectra and Isotopes*, Arnold, London, 1933.
7. A. O. Nier, Rev. Sci. Instr., *11*:212 (1940).
8. H. E. Duckworth, *Mass Spectroscopy*, Cambridge Univ. Press, London, 1958.
9. C. A. McDowell, *Mass Spectrometry*, McGraw-Hill, New York, 1963.
10. J. B. Farmer, in *Mass Spectrometry* (C. A. McDowell, Ed.), McGraw-Hill, New York, 1963, p. 7.
11. W. Paul and H. Steinwedel, Z. Naturforsch., *8a*:448 (1953).
12. R. F. Post, University of California Lab. Report No. UCRL-2209, Berkeley, Cal., 1953.
13. J. Yinon and F. S. Klein, Vacuum, *21*:379 (1971).
14. P. H. Dawson and N. R. Whetten, Dyn. Mass Spectrom., *2*:1 (1971).
15. A. Frigerio, Boll. Chim.-Farm., *112*:715 (1973).
16. D. D. Focht, in *Nitrogen in the Environment*, vol. 2 (D. R. Nielsen and J. G. MacDonald, Eds.), Academic Press, New York, 1978, p. 429.
17. T. W. Schmidt and J. A. Favre, Int. J. Mass Spectrom. Ion Phys., *20*:379 (1976).
18. G. Junk and H. J. Svec, Geochim. Cosmochim. Acta, *14*:234 (1958).
19. W. G. Mook and P. M. Grootes, Int. J. Mass Spectrom. Ion Phys., *12*:273 (1973).
20. D. B. Sprinson and D. Rittenberg, J. Biol. Chem., *180*:707 (1949).
21. J. M. Bremner, in *Methods of Soil Analysis* (C. A. Black, Ed.), Part 2, Am. Soc. Agron., Madison, Wis., 1965, p. 1256.
22. P. J. Ross and A. E. Martin, Analyst (Lond.), *95*:817 (1970).
23. T. Preston, personal communication, 1978.
24. D. Halliday, personal communication, 1979.
25. E. R. Mercer, personal communication, 1973.
26. R. J. Barsdate and R. C. Dugdale, Anal. Biochem., *13*:1 (1965).
27. O. Tsiyi, M. Masugi, and Y. Kosai, Anal. Biochem., *65*:19 (1975).
28. M. Potts, E. E. Krumbein, and J. Metzger, in *Environmental Biogeochemistry and Geomicrobiology*, vol. 3 (W. E. Krumbein, Ed.), Ann Arbor Science Publ. Inc., Ann Arbor, Mich., 1978, p. 753.

29. R. D. Hauck, in *The Use of Isotopes in Soil Organic Matter Studies*, Pergamon, Oxford, 1966, p. 447.
30. R. D. Hauck, in *Nitrogen-15 in Soil-Plant Studies*, IAEA, Vienna, 1971, p. 65.
31. L. D. Owens, Soil Sci. Soc. Am. Proc., *24*:372 (1960).
32. H. H. Cheng, J. M. Bremner, and A. P. Edwards, Science, *146*:1574 (1964).
33. C. C. Delwiche and P. L. Steyn, Environ. Sci. Technol., *4*:929 (1970).
34. A. M. Blackmer and J. M. Bremner, Soil Biol. Biochem., *9*:73 (1977).
35. J. Y. Tong and P. E. Yankwich, J. Phys. Chem., *61*:540 (1957).
36. K. W. Steele and R. M. Daniel, J. Agric. Sci. Camb., *90*:7 (1978).
37. O. T. Denmead, J. R. Simpson, and J. R. Freney, Science, *185*:609 (1974).
38. R. D. Hauck, S. W. Melsted, and P. E. Yankwich, Soil Sci., *86*:287 (1958).
39. C. M. Cho and L. Sakdinan, Can. J. Soil Sci., *58*:443 (1978).
40. C. T. Chen, USAEC Report, TID-21648, 1965.
41. E. McKenzie, Jr. and L. T. Kurtz, Soil Sci. Soc. Am. J., *40*:534 (1976).
42. L. A. Green, Diss. Abstr. Int. B., *35*:5272 (1975).
43. D. E. Rolston, M. Fried, and D. A. Goldhamer, Soil Sci. Soc. Am. J., *40*:259 (1976).
44. J. C. Ryden, L. J. Lund, and D. D. Focht, Soil Sci. Soc. Am. J., *42*:731 (1978).
45. T. Yoshinari and R. Knowles, Biochem. Biophys. Res. Commun., *69*:705 (1976).
46. W. L. Balderston, B. Sherr, and W. J. Payne, Appl. Environ. Microbiol., *31*:504 (1976).
47. A. M. Blackmer and J. M. Bremner, Soil Biol. Biochem., *10*:187 (1978).
48. A. M. Blackmer and J. M. Bremner, Soil Biol. Biochem., *11*:313 (1979).
49. O. T. Denmead, J. R. Freney, and J. R. Simpson, Soil Biol. Biochem., *8*:161 (1976).
50. D. R. Keeney and A. N. MacGregor, N. Z. J. Agric. Res., *21*:443 (1978).
51. V. A. Dugdale and R. C. Dugdale, Limnol. Oceanogr., *7*:170 (1962).
52. W. D. P. Stewart, *Nitrogen Fixation in Plants*, Athlone Press, Univ. of London, 1966.
53. E. A. Shtina, E. M. Pankeratova, G. N. Perminova, A. N. Tretjakova, and L. A. Yung, *Transactions of the 9th International Congress of Soil Science*, Adelaide, 1968, vol. 2, p. 151.

54. G. Bond and G. D. Scott, Ann. Bot. Lond., N.S., *19*:65 (1955).
55. C. C. Delwiche and J. Wijler, *Plant Soil*, *7*:113 (1956).
56. P. J. Ross and A. E. Martin, Nature (Lond.), *204*:444 (1964).
57. R. H. Burris, Meth. Enzym., *24*:415 (1972).
58. R. H. Burris, in *The Biology of Nitrogen Fixation* (A. Quispel, Ed.), North-Holland/American Elsevier, Amsterdam, 1974, p.9.
59. R. W. F. Hardy, R. D. Holsten, E. K. Jackson, and R. C. Burns, Plant Physiol., *43*:1185 (1968).
60. A. Haystead and A. G. Lowe, J. Brit. Grassl. Soc., *32*:57 (1977).
61. T. W. Walker, A. F. R. Adams, and H. D. Orchiston, J. Brit. Grassl. Soc., *9*:249 (1956).
62. M. Fried and H. Broeshart, Plant Soil, *43*:707 (1975).
63. D. A. Phillips and J. P. Bennett, Argon. J., *70*:671 (1978).
64. J. O. Legg and C. Sloger, *Proceedings of the International Conference on Stable Isotopes* (E. R. Klein and P. D. Klein Eds.), Oak Brook, Ill., 1975, p. 661.
65. I. Vallis, E. F. Henzell, and T. R. Evans, Aust. J. Agric. Res., *28*:413 (1977).
66. A. Haystead and C. Marriott, Ann. Appl. Biol., *88*:453 (1978).
67. J. H. Smith, J. O. Legg, and J. N. Carter, Soil Sci., *96*:313 (1963).
68. C. Sloger, D. Bezdicek, R. Milberg, and N. Boonkerd, in *Nitrogen Fixation by Free Living Organisms* (W. D. P. Stewart, Ed.), Cambridge University Press, Cambridge, 1975, p. 271.
69. E. Ricci and J. H. Gibbons, Trans. Am. Nucl. Soc., *12*:509 (1969).
70. O. Wernberg, C. G. Lamm, and T. Nielson, J. Chromatogr. Sci., *9*:373 (1971).
71. J. F. O'Hanlon, *A User's Guide to Vacuum Technology*, Wiley, New York, 1980.
72. G. Sherer, J. Duffy, D. H. Kohl and B. Commoner, Soil Sci. Soc. Am. J., *38*:315 (1974).
73. A. Marriotti, J. C. Germon, P. Hubert, P. Kaiser, R. Letolle, and A. Tardieux, Plant Soil, *62*:413 (1981).
74. M. Fried and H. Broeshart, Plant Soil, *62*:331 (1981).
75. M. Fried and V. Middelboe, Plant Soil, *47*:713 (1977).
76. G. L. Kerven and P. G. Saffigna, Int. J. Mass Spectrom. Ion Phys. *23*:333 (1977).
77. K. W. Steele, M. J. Grinstead, A. T. Wilson, and C. B. Dyson, Int. J. Mass Spectrom. Ion Phys., *28*:313 (1978).
78. C. J. Smith and P. M. Chalk, Anal. Chim. Acta., *104*:245 (1979).
79. R. A. Virginia, PhD. Thesis, University of California (1980).
80. N. Armarger, A. Marriotti, F. Marriotti, J. C. Durr, C. Bourguinon, and B. Lagacherie, Plant Soil, *52*:269 (1979).

10

Gas Chromatographic Analysis of the Soil Atmosphere

KEITH A. SMITH The Edinburgh School of Agriculture, Edinburgh, Scotland

I. INTRODUCTION

The soil atmosphere may be significantly modified in its composition, with respect to the air above the soil, as a result of the respiratory activity of plant roots and soil microorganisms. The concentration of oxygen decreases, and the concentration of carbon dioxide increases, particularly when gaseous exchange with the air above the soil is impaired. Furthermore, other substances such as oxides of nitrogen (resulting from the decomposition of nitrate by denitrification processes), gaseous hydrocarbons, and volatile sulfur compounds, may be present in significant quantities. The development of suitable gas chromatographic methods for soil atmospheric analysis, to study these changes, has been of considerable importance in soil science and related areas of microbiology and plant physiology. These methods, and their applications, are considered in this chapter.

The first analysis of the free air in the soil was apparently by Boussingault and Lewy in 1853 [1]. They inserted a pipe into the soil to a depth of 0.3-0.4 m and aspirated 2.5-10 liters of soil air through barium hydroxide solution, and weighed the carbonate formed, to determine the carbon dioxide content of the soil air. For a further 100 years or more the measurement of soil gases continued to rely on absorptiometric/manometric procedures, and, in spite of the experimental problems involved, many of the important chemical and biochemical processes in the soil involving gaseous substances had been investigated long before the advent of modern instrumental methods of analysis. However, there were often limitations due to lack of sensitivity which could not be overcome.

The availability, since the end of the 1950s, of gas chromatographic methods for the analysis of permanent gases and volatile compounds has resulted in very considerable progress in research in this area. The change from traditional chemical methods of analysis, which were not specific for some gases, to a system with which most, if not all, the substances of interest could be readily separated, and the accompanying increases in sensitivity obtainable by gas chromatography, have made possible a wide range of new investigations in soil science, as in so many other disciplines.

Gas chromatographic methods are applicable both to investigations of the soil atmosphere as it occurs in the field, and also to laboratory studies of the evolution of gases from, and their uptake by, soil. Such studies form an integral part of many investigations which have as their ultimate objective the understanding of natural processes in the field.

II. GAS CHROMATOGRAPHY

A. Principles

A brief outline of the principles involved in gas chromatography is given below, but for a more comprehensive discussion the reader is referred to books solely devoted to this subject, including those by

Keulemans [2], Jeffery and Kipping [3], Giddings [4], and Szepesy [5] and a number of important theoretical papers [6-11].

Chromatography is essentially the separation of the components of a mixture resulting from differences in their partition between a stationary phase with a large active surface area and a moving phase which flows over the stationary phase. Depending on the state of the moving phase, a distinction is made between liquid and gas chromatography. Separations which would be very difficult to achieve by any other means are possible by chromatographic methods because even small differences between the components in their partition between the phases are multiplied many times during their passage through the chromatographic system.

Gas chromatography may be conveniently classified into two types: gas-liquid and gas-solid chromatography. In the former, the stationary phase is an involatile liquid coated either on an inert support material in a packed column, or on the internal wall of an open tubular (capillary) column. In the latter type, the column is packed with an adsorbent of small particle size, or the inner surface of a capillary column is coated with a thin adsorbent layer. For separations of gaseous mixtures, solid adsorbents (see below) have been more widely used than liquid stationary phases.

At constant temperature, pressure, and carrier gas velocity, the rate at which a component travels along the column is related to the partition coefficient K:

$$K = \frac{C_s}{C_g} \tag{1}$$

where C_s and C_g are the concentrations in the stationary and gaseous phases, respectively. The more strongly bound components stay for a longer time in the stationary phase and thus travel along the column more slowly than the more weakly bound components. In an actual column the zone occupied by a component broadens as it moves along, due to diffusion of molecules in the gaseous phase.

If K is independent of concentration and the adsorption isotherm is as in Fig. 1a, then the elution peak has a symmetrical, Gaussian shape. However, if the adsorption isotherm is of the Langmuir type, which is generally the case for solid adsorbents (Fig. 1b), then the chromatographic peaks have sharp fronts with trailing backs, although the degree of asymmetry may often be quite small.

By analogy with the theory developed to explain the separation of components in a distillation column, the resolving power of a chromatographic column was formerly described in terms of the *number of theoretical plates* to which it was equivalent. By dividing the length of the column by the number of theoretical plates, *the height equivalent to a theoretical plate* (HETP) could be determined. It is considered to be more appropriate to replace the term "theoretical

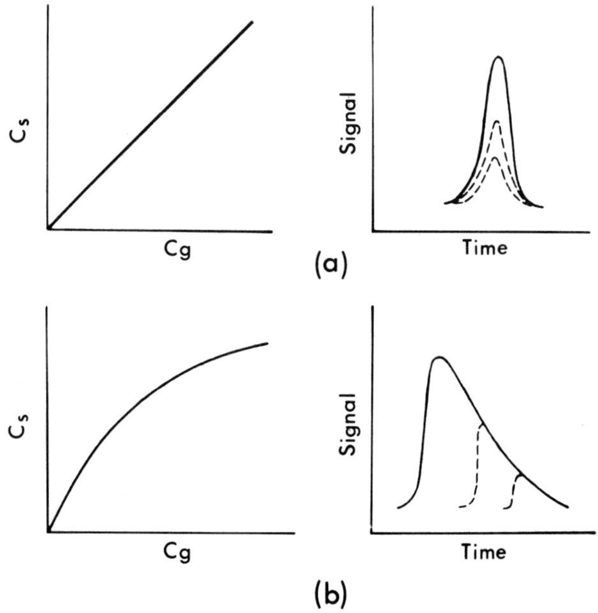

Figure 1 (a) Linear adsorption isotherm—symmetrical elution peak (position of peak maximum constant). (b) Nonlinear (Langmuir) isotherm—distorted elution peak "tailing" (position of peak maximum depends on quantity of sample). (From Ref. 5.)

plate number" by *efficiency* and "plate height" by *relative peak broadening* [5].

The separation of the components in a sample depends on the relative values of the partition coefficients. The closer the values are to each other, the longer is the column required to give an adequate separation. Reducing the temperature has the effect of increasing the separation. In practice, the variables which are most commonly exploited to achieve satisfactory separations are:

1. The material used as stationary phase
2. The temperature
3. The length and diameter of the column
4. The flow-rate of the moving phase (the carrier gas)

The essential parts of all gas chromatographs consist of:

1. A carrier gas system, generally in the form of a cylinder supply at pressures up to 15 MPa (150 bar), with a two-stage regulator to reduce the pressure and a flow-rate regulating valve

2. A device for introduction of the sample into the carrier gas stream
3. A column, to separate the components of the sample
4. A detector, to indicate the elution of each component, and produce a response proportional to the quantity of the component present
5. A recording system, to provide a permanent record of the detector response

B. Commercial Instruments

The column, the detector, and sometimes the injection device must be enclosed in thermostats. In the simpler types of laboratory instrument the detector and column may be within the same enclosure, but in the more complex and expensive instruments it is common for the detector(s) to be separately housed and maintained, if desired, at a different temperature from that of the column. A wide range of commercial instruments is now available from manufacturers in several countries. Many instruments are of modular construction, and it is usually possible to purchase only those modules (e.g., particular types of detector and their associated electronics) required for the type of analysis to be undertaken.

For many applications the choice of instrument is not critical, because the substances to be measured are easily within the detection capabilities of any commercial instrument. In this situation the decision is more likely to be made on the basis of price and the availability of after-sales service than on technical features. However, in those applications where the common detectors have inadequate sensitivity (see Sec. II.C), only one or two manufacturers may offer the particular device which is required. McNair [12] has reviewed the commercial instruments available, and gives details of the various features included in each model. For information on the latest developments, manufacturers' literature should be consulted, particularly on aspects relating to automation (see Sec. II.F).

It frequently happens that the chromatographer has a perfectly serviceable gas chromatograph, which has been used for some years on one particular application, and that the chromatographer now wishes to use for a new analysis, which requires a different type of detector. In most situations it is perfectly feasible to fit a detector produced by one manufacturer to a chromatograph made by another, with the only drawbacks being cosmetic ones if, for instance, holes have to be cut in the instrument case. However, in the author's experience, manufacturers have widely differing attitudes toward the worker who wishes to do this. Whereas one company may be only too pleased to cooperate, another may insist on selling an unwanted g.c. oven and gas inlet controls along with the detector and associated electronics. It is to be hoped that the introduction of microprocessor-controlled instruments will not result in greater reluctance to supply individual modules, or in greater technical problems when making hybrid assemblies.

C. Columns

Gas chromatographic methods commonly used for the measurement of the constituents of the soil atmosphere have generally been adopted from methods originally developed for other purposes. There has been little or no basic research into new types of column for this particular application. As was mentioned earlier, solid adsorbents have been more widely used than liquid stationary phases for separations of permanent gases. The most widely used adsorbents are:

1. Molecular sieves 5A and 13X, which are synthetic zeolites with channels in the crystal lattice capable of occluding small molecules. They are the only commonly used packing materials that will separate oxygen from nitrogen at ambient or higher temperatures.
2. Porous polymer beads, first developed by Hollis [13] and marketed under the name Porapak, and followed by similar materials such as Polypak, Phasepak, and Chromosorb 101 and 102. These materials are copolymers of styrene and divinylbenzene or ethylvinylbenzene, and are available in different grades which vary in their retention characteristics for particular substances. Porapak Q has been widely used for separating carbon dioxide and nitrous oxide from the major constituents of air samples.
3. Alumina. This has been widely used for the separation of hydrocarbon gases, often after partial deactivation by addition of water or silicone oil [14] or inorganic salts [15], to reduce tailing of the elution peaks. Alumina is a polar adsorbent on which the order of separation depends on the molecular structure. Thus, for example, unsaturated hydrocarbons are more strongly retained than saturated ones with the same number of carbon atoms.
4. Silica gel and activated charcoal. These materials have been used for the separation of gases such as carbon dioxide and nitrous oxide. However, the properties of both these adsorbents have been found to vary considerably with source of supply and the method used in preparation, making it very difficult to reproduce retention times from one column to another [3]. They have now been largely superseded by the porous polymers, which have given much more reproducible results.

Two comparatively new types of carbon adsorbent are now finding applications in gas analysis: the graphitized carbon black materials such as the Carbopack range, and the nongraphitized carbon molecular sieve, first prepared by Kaiser [16,17] by pyrolysis of polyvinylidene chloride. The latter material is now available under the name Carbosieve B, and has proved to be very useful for the separation of complex mixtures of inorganic gases and hydrocarbons [18,19]. Carbopack B has been developed to separate hydrogen sulfide, sulfur dioxide, and methanethiol at concentrations in the range 10^{-9}-10^{-6} ml ml^{-1} [20].

Separation of Oxygen from Argon

Although the separation of the two major components of air, oxygen and nitrogen, from each other is easily accomplished on molecular sieves, the separation of oxygen from argon is much more difficult, and at ambient temperature or above the two gases elute as a single peak. Their separation involves the use of molecular sieve 5A or Porapak Q columns operated at subambient temperatures, which necessitates either access to one of the few commercial chromatographs equipped with a subambient thermostat facility, or the use of cooling baths or Dewar flasks containing dry ice-acetone or similar mixtures. On molecular sieve 5A at $-72°C$, argon is eluted first, followed by oxygen, and the nitrogen is retained on the column [21]. In contrast, on Porapak Q nitrogen is eluted first, followed by oxygen, then argon [22].

Oxides of Nitrogen

The principal problem associated with the determination of nitrous oxide by gas chromatography is its separation from carbon dioxide. In recent years Porapak Q has been the most widely used material for the separation of these two gases, and satisfactory results have been obtained using columns of several different sizes at various temperatures, e.g., 3.7 m × 2 mm i.d. at 30°C [23], 5.5 m × 1.6 mm i.e. at 25°C [24], 1.5 m × 6 mm (4 mm i.d.) at 55°C [25], and 1.5 mm × 3 mm at 25°C [26]. Carbosieve B has also been used successfully [27,28,146]. If the carbon dioxide concentration is very high relative to that of nitrous oxide, as is frequently the case, it is important to ensure that the peaks are well separated on the chromatogram. Otherwise, it is possible that the nitrous oxide peak may be obscured by the tail from the carbon dioxide peak. Passage of the sample through a precolumn of sodalime [26] or ascarite [29] will either remove carbon dioxide completely, or so reduce the peak size that this ceases to be a major problem.

The determination of nitric oxide by gas chromatography is complicated by its rapid oxidation to nitrogen dioxide in the presence of molecular oxygen. In samples which contain no oxygen the gases from which separation is most difficult are nitrogen and carbon monoxide. Molecular sieve columns have been widely used for this separation, but problems with tailing of the nitric oxide peak can occur [3]. Quinlan and Kittrel [30] compared several adsorbents and concluded that Chromosorb 102 and Carbosieve B, both at subambient temperatures, were the most satisfactory. Nitric oxide can be irrevocably adsorbed on "active" sites on a charcoal column, but if these sites are saturated by prior treatment with the gas, then this column material can be used for its separation at ambient temperature [31]. If the temperature is raised desorption from the active sites occurs.

Nitrogen dioxide also presents problems because of its reactivity. It reacts with the aromatic rings of the polymer comprising Porapak Q, to form nitric oxide and water, also nitration products of the aromatic rings [32]. The reaction between nitrogen dioxide and water to form unstable nitrous acid, which then decomposes to yield nitric oxide, has been used as the basis of a chromatographic method. Part of a molecular sieve 5A column was wetted with water, and nitrogen dioxide injected onto the column was converted to nitric oxide and eluted as a distinct peak [33]. Smith et al. [34], in attempting to apply this technique to soil atmosphere samples, found that it would not work in the presence of oxygen; presumably this was due to the re-oxidation of nitric oxide back to nitrogen dioxide. They modified the procedure by trapping out nitrogen dioxide in acetone-dry ice, flashing out the oxygen with helium, and then vaporizing the frozen nitrogen dioxide and transferring it to the chromatograph. Other workers [35] have abandoned attempts to measure this gas by gas chromatography and have resorted to alternative chemical methods. It is clear that reliable results can only be obtained by chromatographic techniques when great care is taken to overcome problems of the type outlined here.

Ammonia

Ammonia is another nitrogen gas which is readily determined by nonchromatographic methods, and which has not been very successfully determined in soil atmosphere samples by gas chromatography. Smith and Clark [29] reported that of many packings tried, only Carbowax 600 on firebrick washed with dodium hydroxide had proved successful. The strong retention of ammonia on some gas-solid chromatographic packings is indicated by the work of Burford and Bremner [24,36] who could not obtain a peak for the gas after injecting it onto Porapak Q columns at 25 and 100°C, even though phosphine, hydrogen sulfide, and (at the higher temperature) water were all eluted. However, the manufacturers' literature indicates that ammonia is retained less strongly than water on all the Porapaks, and that Chromosorb 105 is also a suitable material. No reference can be found to the use of carbon molecular sieve or graphitized carbon for the separation of ammonia, but these adsorbents may well prove to have advantages for this purpose.

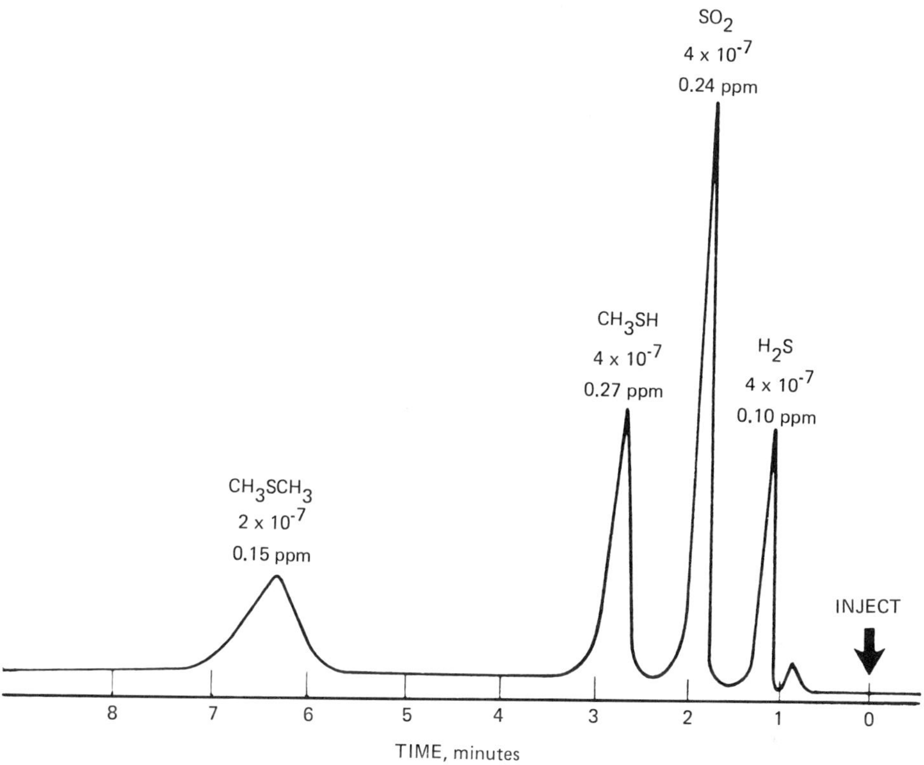

Figure 2 Separation of low molecular weight sulfur compounds on polyphenyl ether/orthophosphoric acid on Teflon support. (From Ref. 37, reprinted with permission from R. K. Stevens, J. D. Mulik, A. E. O'Keefe, and K. J. Krost, Anal. Chem. *43*:827 (1971). Copyright by the American Chemical Society.)

Sulfur Gases

Reactive sulfur compounds, sulfur dioxide and hydrogen sulfide in particular, have proved difficult to determine by gas chromatography at concentrations of the order of 10^{-6} ml ml^{-1}, due to adsorption on column walls or solid supports or irreversible reaction with the stationary phase. Glass, stainless steel, and even Teflon (polytetrafluoroethylene) tubing have all been found to exhibit undesirable retention of SO_2 below 10^{-5} ml ml^{-1}, but FEP Teflon (fluorinated ethylene-propylene) is free of this effect [37]. Powdered Teflon appears to be the one material sufficiently inert to serve as support, and one stationary phase has been identified which combines the two qualities of inertness and the ability to separate low molecular weight sulfur compounds: polyphenyl ether (five-ring polymer) to which orthophosphoric acid had been added [37]. A typical separation on this material is shown in Fig. 2. Recently several solid adsorbents have become available which make possible a number of difficult separations of sulfur gases: the graphitized carbon, Carbopack B, to which reference has already been made; a specially treated silica gel, Chromosil 310; and Supelpak-S, which is a modified Porapak QS [38].

D. Detectors

Gas chromatograph detectors are devices which can indicate the presence of a sample component eluting from the column in the carrier gas stream, and produce an electrical signal proportional to the concentration of the component. A wide variety of physical properties has been employed as the basis of these devices. For satisfactory performance, detectors should have high sensitivity to the components of the sample, a rapid response which increases linearly with concentration, and a high signal-to-background noise ratio. They should be easy to operate and reliable in constant use. A useful property associated with certain types is that of selectivity, only responding to components with particular properties.

Some of the detectors in current use for the determination of gaseous substances are listed in Table 1 and are described below. However, as the properties of the electron capture and flame photometric detectors are discussed in detail in Chap. 11 because of their great importance in pesticide analysis, their consideration in this section is restricted to discussion of particular applications in gas analysis.

Table 1 Some Detectors Used for Determination of Gases

Detector	Application
Thermal conductivity (Katharometer)	All gases (major constituents of mixtures)
Flame ionization	Hydrocarbons (trace concentrations)
Helium ionization	All gases (trace concentrations)
Electron capture	Nitrous oxide; oxygen; halogenated compounds
Ultrasonic	All gases (wide concentration range)
Flame photometric	Sulfur and phosphorus compounds

Thermal Conductivity Detector

In this detector, the carrier gas passes through holes drilled through a metal block and flows over heated resistance filaments or thermistors suspended in the channels. Pure carrier gas passes over the filaments in one channel of the detector, and the carrier gas containing the sample constituents passes over the filaments in the other channel. Changes in composition of the gas result in changes in thermal conductivity and thus in the temperature of the filaments and hence their resistance. The filaments are connected in a Wheatstone bridge circuit which is balanced when pure carrier gas flows through both channels. As a sample constituent elutes from the column into the detector, the bridge will be unbalanced and result in a deflection on the chart recorder. The sensitivity of the detector is dependent on the difference between the thermal conductivities of the carrier gas and the sample. As hydrogen and helium have much higher thermal conductivities than all other gases, these are the most appropriate carrier gases. In view of the possible hazards associated with emission of hydrogen into the laboratory atmosphere, helium is to be preferred in spite of its higher cost.

For the determination of the major constituents of permanent gas mixtures, the thermal conductivity detector is still widely used. Because of its relative cheapness and reliability, this detector is a logical choice for analytical tasks where very high sensitivity is not a primary requirement. Its main application in soil atmospheric analysis has been for the measurement of nitrogen, oxygen, and carbon dioxide, and occasionally argon and methane. Under ideal conditions, a limit of detection as low as 10^{-6} ml ml^{-1} may be obtained with both filament and thermistor detectors [39], but for routine operation concentrations above 10^{-4} ml ml^{-1} are most conveniently measured.

Flame Ionization Detector

The flame ionization detector (FID) is probably the most widely used of the selective detectors. It consists of a hydrogen-air burner and two electrodes at a potential difference of 100-300 V. Usually the burner jet is one electrode and the other, the collector electrode, commonly takes the form of a cylindrical tube surrounding the flame. At the high temperature of the flame some ionization of hydrogen occurs, producing an ion current between the electrodes, i.e., the standing current of the detector. Gases eluting from the column pass into the jet, and the combustion of any organic compounds present greatly increases the ion current, which, after amplification, is recorded. Baseline drift during temperature programmed analysis is overcome by the use of dual detectors with opposite polarities, one of which receives carrier gas and column effluent and the other carrier gas only.

For studies of hydrocarbons and other volatile organic species in the soil atmosphere, the FID has great advantages over other detectors. Apart from a small response to high concentrations of oxygen and nitrous oxide, the detector is insensitive to the presence of inorganic gases and water vapor, and in principle any inorganic carrier gas can be used. Nitrogen and argon are the gases most commonly used in practice. The most modern detectors have a limit of detection for hydrocarbon gases of 10^{-9}-10^{-8} ml ml^{-1}, i.e., a few parts per billion, with a linear dynamic range of 10^6 to 10^7. Thus the upper limit of concentration satisfactorily determined with the FID is of the order of 10^{-2} ml ml^{-1}, and samples containing higher concentrations will have to be diluted before analysis.

Helium Ionization Detector

The helium ionization detector (HID) depends on the increase in conductivity resulting from the ionization of gases by collision with metastable helium atoms. A radioactive source (or, in the device formerly made by Beckman Instruments, an electrical discharge) produces electrons by ionizing the helium carrier gas. These electrons, after acceleration, excite helium atoms to their metastable state. The excitation potential of helium (19.8 eV) is higher than the ionization potentials of all other gases except neon, thus the HID is a nonselective detector. It has a smaller linear dynamic range than the FID, 10^4 compared with 10^6-10^7, and is suitable for measuring concentrations in the 10^{-7}-10^{-6} ml ml^{-1} range.

Ultrasonic Detector

The recent introduction of the ultrasonic detector has provided a powerful new tool for the study of all gases at trace or high concentrations. It is much more sensitive than the thermal conductivity detector and has a linear response over the whole range from 10^{-6}-1 ml ml^{-1}. The principle on which it operates is the measurement of changes in the speed of sound traveling through the gases eluted from the column. When using hydrogen or helium as carrier gas the response of the detector to sample components is proportional to their respective molecular weights. Nitrogen, air, and argon may also be used as carrier gases and carrier impurities do not affect the detector response.

The ultrasonic detector has been applied to the study of nitrous oxide emissions from soils [66,147]. Although it is not as sensitive as a heated electron capture detector for this purpose (see below) and prior concentration of the gas is required for concentrations below 10^{-6} ml ml^{-1} [55], the ultrasonic detector has the advantages of great stability and durability. However, it is extremely expensive.

Electron Capture Detector (ECD)

The electron capture detector (ECD) is known for its applications in anlaysis of substances such as halogenated pesticides, but two specific applications in soil atmospheric studies have been developed in recent years. The detector has a high sensitivity to nitrous oxide at high temperatures [40], and at 350-400°C this is sufficient to allow the direct determination of normal atmospheric concentrations of about 0.3×10^{-6} ml ml^{-1} without the need for prior concentration. The versions of the detector which have a tritium (^{3}H) source are unsuitable for such high temperatures, and the types which incorporate a ^{63}Ni source must be used. The other use is for the measurement of low levels of oxygen, particularly oxygen in aqueous solution. The procedures for this determination are discussed in a later section.

A disadvantage of the ECD is its somewhat limited linear dynamic range compared with detectors such as the FID. This is inherent in the principle of operation of the ECD. Unlike ionization detectors which produce an increase in ion current in the presence of detectable substances, the ECD exhibits a decrease from an upper limit set by the standing current. This situation is somewhat similar to that encountered in spectrophotometric determinations involving absorption of light, and Wentworth et al. [41] have derived a relationship between response and sample concentration which is linear even in the region approaching detector saturation. If I_b is the standing current and I_e the current during the passage of an electron-capturing compound, $(I_b - I_e)/I_e = KA$, where K is a constant and A is the sample concentration. An electronic amplifier system which converts I_e to $(I_b - I_e)/I_e$, and thus gives a response which is a linear function of sample concentration, has been described [42]. Commercially available instruments which incorporate electronic linearizing systems have recently been employed in soil atmospheric studies [148,149]. Procedures for improving linearity are also discussed in Chap. 11.

Flame Photometric Detector (FPD)

The specificity and high sensitivity of this detector for sulfur compounds has been a major asset in studies of sulfur-containing gases in the soil atmosphere and the environment generally. The sensitivity for sulfur compounds is 10^4 times as great as that for other substances, so only the major constituents of the soil atmosphere are likely to produce a response comparable with that from trace levels of sulfur gases. However, this detector too has an inherently nonlinear response, the signal-concentration relationship being given by the expression

$$I = I_0[S]^n$$

where I is the signal, [S] the sulfur concentration, and I_0 and n are constants under given experimental conditions [43]. In general, the values of n reported have been between 1.5 and 2, but n varies with the flame conditions and is thus affected by flow rates of hydrogen and fuel gases. All the factors affecting the flame conditions must be kept constant to ensure any degree of analytical accuracy [43]. As with the electron capture detector, linearizing circuits are now available which convert the flame photometric detector signal to a linear function of concentration.

E. Calibration

The concentrations of the various constituents in a sample are conveniently determined by comparisons of peak height (or peak area) with that obtained from a known volume of a gas mixture of known composition, having first established the relationship between peak height (or area) and detector response. Cylinders of compressed gases of the desired composition may be obtained from commercial suppliers, together with a certificate of analysis. This is normally the most satisfactory method for gas concentrations of 10^{-2} ml ml^{-1} or more, but an alternative is the use of gas-proportioning pumps, which can produce accurate dilutions of one gas in another, down to concentrations of about 3×10^{-3} ml ml^{-1} [3]. For trace levels, commercial mixtures may also be obtained, but it is possible to achieve reasonably satisfactory results at negligible cost by injecting small volumes of the pure gas into larger vessels of known volume. For example, mixtures of hydrocarbons in nitrogen at concentrations of the order of 10^{-6} ml ml^{-1} can be readily prepared by injecting 1 ml of pure hydrocarbon into a 1 liter flask filled with nitrogen and, after equilibrium has been reached, injecting 1 ml of the diluted gas into a second flask. However, an advantage of using comparessed gases as standards is that the cylinder may be connected directly to a gas sampling valve fitted with a loop of suitable volume.

Better reproducibility of results is usually possible with a sampling valve than with a syringe, particularly if the carrier gas pressure is much above atmospheric. Improved results with syringe injection may be achieved by injecting the sample into a chamber forming part of the sample loop of a sampling valve [3].

F. Automation

Sample Injection

When large numbers of samples have to be analyzed, a degree of automation can greatly reduce operator time. In principle, the automation of gas sample injection and analysis is not difficult, and such systems have become widely used in industrial process control applications and in the monitoring of the levels of air pollutants. In these situations the supply of sample gas is unlimited, and is frequently at pressures above atmospheric. Thus a continuous flow of gas can be maintained through a gas sampling valve, and the contents of the sample loop can be automatically injected at any desired time interval. In contrast, the samples obtained from gas sampling probes buried in the soil are normally too small to allow the purging of a sampling valve to be carried out in the normal way, and each one (usually in a syringe) must be kept separated from the others. An automatic system for the injection of such samples has been described by Smith and Harris [44]. The loop of a motor-driven gas sampling valve is evacuated by a pump and then coupled, by the operation of a rotary switching valve, to each of 16 sample syringes in turn. The gas in the syringe expands into the loop, and the contents of the loop are then injected. A later development of this system accommodated 40 sample syringes [15,45].

General Chromatograph Operation

Whereas automated equipment such as the sample injection system described above has to be purpose-built, automation of the operation of the gas chromatograph itself is becoming a common feature of the latest commercial instruments. The incorporation of microprocessors allows parameters such as oven temperature and carrier gas flow rate and the timing of events such as the switching of a valve to be set to values stored in the microprocessor memory and called up by means of a small BASIC program. In the most sophisticated instruments, visual display units and keyboards allow interactive operation. Questions can be displayed on the screen which prompt the operator to enter the required instructions to program the instrument for a particular analysis. Several complete programs can be stored in the instrument's memory and, if necessary, more can be stored on floppy disks. The range of facilities offered is increasing so rapidly that any summary of the "state'of the-art" will become outdated in months rather than years, and readers are urged to consult manufacturers' literature to establish what features are currently available.

Data Acquisition and Processing

Automatic computation of chromatographic data can be achieved either by using purpose-built digital integrators or by passing the analog output signal of the chromatographic detector via an analog-to-digital converter to a computer. Dedicated integrators incorporate several standard programs which will allow, for example, the calculation of either peak heights or areas, and the percentage of the total sample represented by each peak; by incorporating response factors or normalizing to internal or external standards, results can be printed out in suitable concentration units. Instruments such as the Perkin-Elmer Sigma 10 Chromatograph Data Station and the Hewlett-Packard 3390A Reporting Integrator conveniently provide as output both an analog chromatogram and the quantitative numerical data on the same paper chart. Integrators are normally able to cope with sloping baselines and unresolved or tailing peaks, and "spikes" on the baseline due to noise can be ignored.

When the chromatograph is interfaced to a computer it becomes possible, in addition to performing the same operations as can be undertaken by the integrator, to undertake statistical evaluation of replicate analyses and store results for subsequent retrieval without manual transfer of data at any stage. This has the benefit of eliminating errors, as well as saving operator time.

III. ANALYSIS OF SOIL ATMOSPHERES

A. Sampling

Reference has already been made to the nineteenth century work of Boussingault and Lewy [1]. Their techniques of withdrawing a sample via a pipe inserted to the required depth is fundamentally the same as the methods in use today; the only real difference is that now, with the availability of gas chromatography, small samples of the order of 1-5 ml are adequate for most analyses.

Two main types of gas sampling device have been used. For instantaneous sampling, a tube is driven into the soil so that it fits tightly against the soil, and gas samples are withdrawn via an aperture at the lower end up through the tube into a syringe or some other receptacle. The alternative method involves the creation of a cavity in the soil, usually by the insertion of a small container, which is connected to the soil surface by a capillary tube. The gas in the container is allowed to come to equilibrium with the surrounding soil atmosphere by diffusion, and thus a sample withdrawn from the container is representative of the soil atmosphere. Examples of both types of sampling device are shown in Fig. 3.

The former type of probe has been used for an investigation of the levels of natural gas in subsoils above gas-bearing strata. A steel tube was driven 2-3 m into the ground by a sledge hammer, then withdrawn 0.2-0.3 m so as to expose part of the hole, then gas samples were sucked from the subsoil and taken for chromatographic analysis [47]. This method was very similar to that used as long ago as 1915 by Russell and Appleyard [48]. Tackett [49] used a copper tube 1.6 mm in diameter which could be pushed into the soil. To the upper end was attached a tubing union sealed with a septum. The gas entered the tube through a slot in the side. The side opening has the advantage of being less susceptible to plugging by soil when the

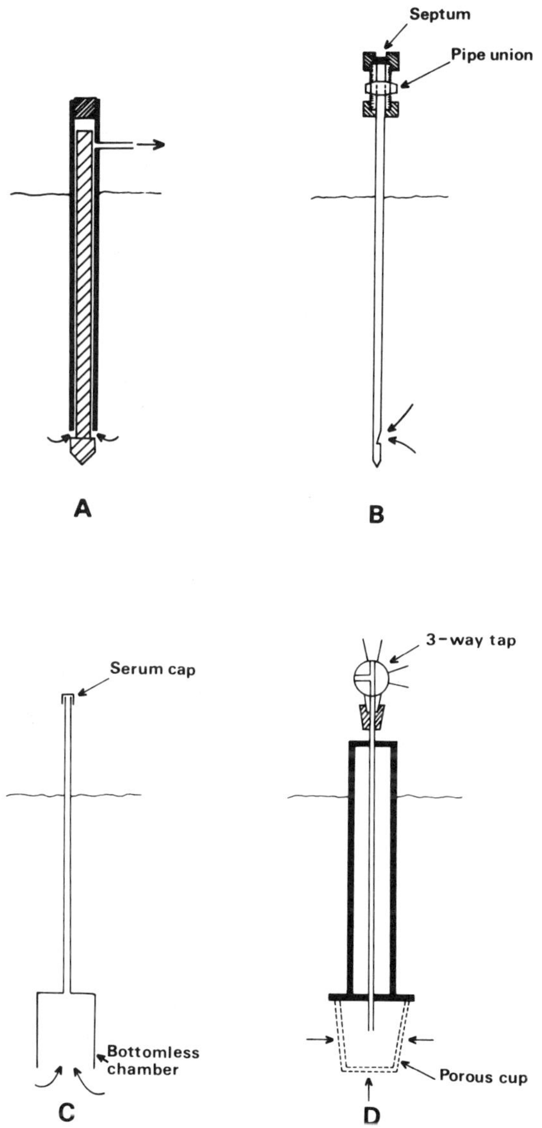

Figure 3 Devices used for sampling the soil atmosphere. Arrows indicate the direction of gas flow. A variety of different combinations of the above- and below-ground parts shown here has been used by different workers. (From Ref. 46.)

probe is inserted. A modification of Tackett's probe has been described in which the gas sample is sucked into a glass ampoule, which is then sealed off by melting with a portable gas burner [50]. In this condition the sample can be stored for several days, if required, before analysis by gas chromatography or mass spectrometry.

Several devices based on the diffusion reservoir principle have been developed to provide samples for gas chromatographic analysis. Reservoirs have been made from small open-ended metal cans [51], pipe unions [49], glass tubing [52], and sintered bronze filter cups [53]. In all these designs the reservoir is connected to the soil surface by a metal or glass capillary tube. To install the reservoir it is necessary to make a hole with an auger to the required depth, insert the apparatus, and then backfill the hole around the sampling tube with soil or some other material, tamping down firmly to provide a seal against the downward diffusion of air.

Burford and Stefanson [54] have compared the results obtained for the nitrous oxide content of the soil atmosphere by sampling from permanently installed reservoirs with those obtained from probes inserted into the soil immediately before sampling. No significant differences were observed.

If the concentrations of the gases of interest are too low for direct analysis, it is necessary to trap them out from a large volume of air. This approach has been particularly valuable for studies of nitrous oxide. Dowdell and Crees [26] collected the nitrous oxide diffusing to the soil surface by covering the soil with a hood and circulating the gases from under the hood through a column of molecular sieve 5A for 18 hr. The adsorbed gases were regenerated by addition of water to the molecular sieve in a sealed flask; the gaseous atmosphere in the flask was then sampled by syringe. More recently Bremner and Blackmer [55] have trapped out nitrous oxide from 300 ml air samples on Porapak Q at $-135°C$ and regenerated the gas into a small volume to inject into the gas chromatograph. Gas samples taken in the field may not be analyzed for some hours, or even days, and considerable care is required to ensure that the composition does not change significantly during the intervening period. Sealing the gas in a glass ampoule [50] would seem to be by far the most reliable method, but the risk of contamination by gasses from a blow-torch would necessarily limit the range of constituents which could be determined in this way. The high affinity of hydrocarbons for stopcock greases and also their rapid adsorption by the rubber diaphragms of disposable plastic syringes [25] makes necessary the use of all-glass syringes lubricated with glycerol or glycerol-mannitol-starch mixtures [3,25], when these gases are being measured. The alternative is to use gas-tight glass syringes with Teflon plungers manufactured specifically for gas chromatographic work. However, this is likely to be an extremely expensive solution when several dozen syringes are needed to sample a well-replicated field experiment. Some adsorption of inorganic gases by, and diffusion through, plastic syringes was reported by Burford [56], who had to make corrections for these effects. Glass syringes are to be preferred if samples have to be stored for considerable periods before analysis.

Instead of extracting gases from undisturbed soil, it is possible to remove soil samples from the field and subsequently extract the gases for analysis. Core samples have been sealed with wax [57] and oil [58] and encased in metal cans which were capped and soldered in the field [59].

The sampling of gases from incubation vessels in laboratory experiments is generally a straightforward operation. Samples can be withdrawn by syringe through a septum or a serum cap and usually can be analyzed within a short time. However, major problems may be experienced in this type of experiment, due to a variety of causes. The creation of leaks, following repeated sampling through the same rubber closure, is a common problem, and adsorption of the gases of interest by the rubber may occur [60]. The latter effect can be minimized by using the "Mininert" type of bottle closure in which the underside of the rubber septum is only exposed to the gases in the bottle at the moment of sampling [34,61,62]. During the intervening periods it is sealed off by a gas-tight Teflon plug. Repeated sampling of small incubation vessels can result in the removal of a significant fraction of the gases present, and this can have effects on the analytical results which are often ignored. The creation of a partial vacuum within the vessel will cause air to enter through any leaks by mass flow as well as diffusion. A less obvious problem is that the gas in the sampling syringe will be at a reduced pressure, and on removal of the needle from the septum, air can enter the syringe. Thus the presence of a small quantity of oxygen in the gas mixture injected into the chromatograph may provide false evidence of the presence of oxygen in the incubation vessel. The existence of a partial vacuum is often readily detectable when samples are withdrawn with a relatively large syringe. However, when 1 ml or smaller syringes with tightly fitting plungers are used, the friction of the plunger in the barrel is often quite sufficient to resist any inward movement under the influence of the pressure of the atmosphere; under these circumstances it is possible for the operator to take samples which are considerably less than atmospheric pressure without being aware of it. This contamination problem can be overcome by using syringes which can be closed by a valve. In analyses of the total gas mixture present, the occurrence of a reduced quantity of sample can then be detected by the reduced total area of the peaks in the chromatogram. However, in experiments in which trace gases only are being determined this check cannot be made; but provided that the experimental system used is leak-tight, then the quantity removed on each successive sampling occasion can be calculated from the known volumes of the vessel and the syringe by applying Boyle's law. Alternatively, the use of low concentrations of krypton as an internal standard, in conjunction with a helium ionization detector [24,36] provides an elegant solution to the problem of estimating sample size.

B. Column-Detector Systems for Major Constituents

In many investigations it is required to measure the concentrations of all the major constituents of the soil atmosphere, and a number of multiple column and/or multiple detector systems have been devised to provide this information from the analysis of a single sample. Several of these systems have been designed to allow the simultaneous measurement of low or trace levels of other substances, such as oxides of nitrogen and hydrocarbons. Methods specifically designed for these gases are described elsewhere in this chapter, but the general-purpose separation procedures in which they, and oxygen, nitrogen and carbon dioxide are all determined, may conveniently be discussed at this point. Figure 4 shows five different column/detector arrangements which have been described for analysis of inorganic gases and methane with the aid of thermal conductivity or ultrasonic dtectors, and Fig. 5 shows three systems which have been used for simultaneous measurement of hydrocarbon gases with a flame ionization detector and other permanent gases by thermal conductivity detector. The gases measured are listed in Table 2.

The system shown in Fig. 4A can be used to determine some or all of the following gases: oxygen (+ argon), nitrogen, carbon dioxide, nitrous oxide, and methane. By appropriate adjustments of temperature, column lengths and relative flow rates through the two columns the relative positions of the gases from each column can be altered on the chromatogram. Thus oxygen and nitrogen may be eluted before nitrous oxide and carbon dioxide [29], or after [46,63]. Alternatively carbon dioxide may come between oxygen and nitrogen [49]. In the system used in the author's laboratory a 1.6 m × 4 mm i.d. Porapak Q column is connected in parallel with a 1.6 m × 1.5 mm i.d. molecular sieve 5A column, and the sample is split so that only a small fraction passes through the molecular sieve. At constant attenuation, a 1% concentration of carbon dioxidc gives a similar peak height to atmospheric concentrations of nitrogen (Fig. 6). Many soil atmosphere samples have carbon dioxide concentrations of this order, and this system redices the need for alteration of attenuator settings in the middle of an anlsysis. = An alternative method of dealing with large and small peaks on the same chromatogram is to link the two imputs of a dual pen recorder in parallel, with different voltage-deflection settings. Thus those peaks which go off-scale on themoresensitive setting are usually a suitable size for measurement on the other [15]. The use of wide range digital integrators is a more satisfactory, thought generally a more expensive solution.

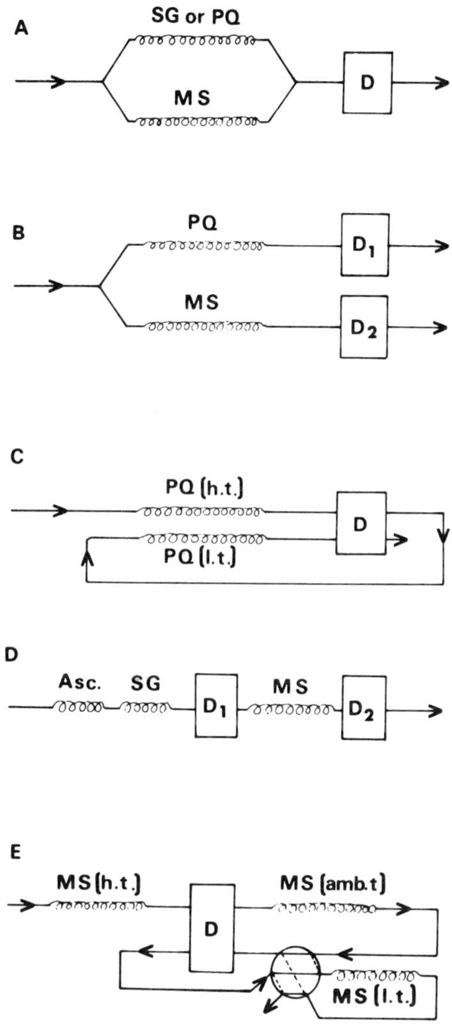

Figure 4 Multiple-column/detector systems used for determination of the major constituents of the soil atmosphere. SG, silica gel; MS, molecular sieve 5A or 13X; PQ, Porapak Q; Asc., Ascarite; h.t., high temperature; amb.t., ambient temperature; l.t., low temperature; D, detector. System C has also been used with different packings at the same temperature (see text and Table 2). (From Ref. 46.)

An alternative version of the two-column, one-detector system is shown in Fig. 4C; the columns may be at the same temperature, using different packing [58,64] or at different temperatures with the same packing. Obermiller and Charlier [22] devised a system in which one

Table 2 Gases Measured with Multiple Column/Detector Systems Shown in Figs. 4 and 5

System	Columns	Gases	References
4A	Silica gel, mol. sieve 5A	O_2, CO_2, N_2	49
	Silica gel, mol. sieve 5A	O_2, N_2, N_2O, CO_2	29
	Silica gel, mol. sieve 5A	CO_2, O_2, N_2, CH_4	63
	Porapak Q, mol. sieve 5A	CO_2, O_2, N_2	46
4B	Porapak Q, mol. sieve 5A	CH_4, CO_2, N_2O; O_2, N_2	68
4C	Porapak Q, Porapak Q	CH_4, CO_2, N_2O; N_2, O_2, Ar	22, 25, 65, 66
	Silica gel, mol. sieve	CO_2, O_2, N_2	58
	HMPA, mol. sieve 13X	CO_2, O_2, N_2, CH_4, H_2S	64
	Porapak Q, mol. sieve 5A	CO_2, O_2, N_2	73

4D	Ascarite/silica gel, mol. sieve 5A	H_2, N_2, N_2O, CH_4	71
4E	Mol. sieve 5A, mol. sieve 5A, mol. sieve 13X	N_2, N_2O, CO_2, Ar, O_2^2	56
5A	Porapak Q, mol. sieve 5A alumina/NaI	CO_2, N_2O; O_2, N_2; CH_4, C_2H_6, C_2H_4, C_3H_8, C_3H_6, iso-C_4H_{10}, n-C_4H_{10}, C_4H_8	15
5B	Porapak R, mol. sieve 5A	C_2H_4, CO_2, O_2	69
	Porapak Q, mol. sieve 5A	CH_4, C_2H_4, CO_2, O_2, N_2	70
5C	Porasil B, mol. sieve 5A, mol. sieve 5A	CH_4, C_2H_6, C_2H_4, C_3H_8; O_2, N_2, N_2O, CO_2	67

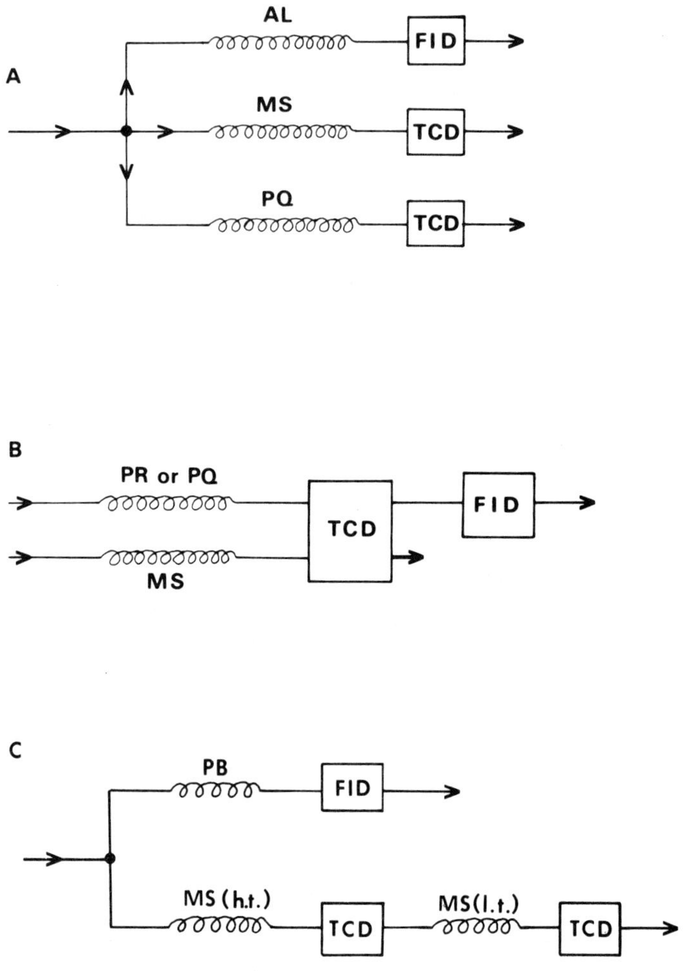

Figure 5 Multiple column/detector systems for simultaneous analysis of major inorganic constituents and trace hydrocarbons in the soil atmosphere. Al, alumina deactivated with NaI; MS, molecular sieve 5A; PB, Porasil B; PQ, Porapak Q; PR, Porapak R; FID, flame ionization detector; TCD, thermal conductivity detector; h.t., high temperature; l.t., low temperature. (After Smith, Ref. 46.)

Porapak Q column was maintained at above ambienc temperature to separate carbon dioxide and nitrous oxide from a composite peak of nitrogen, oxygen, and argon, and a second Porapak Q column was cooled to −78°C to separate these last three gases from each other. This system has been applied to soil atmospheric analysis by Smith and

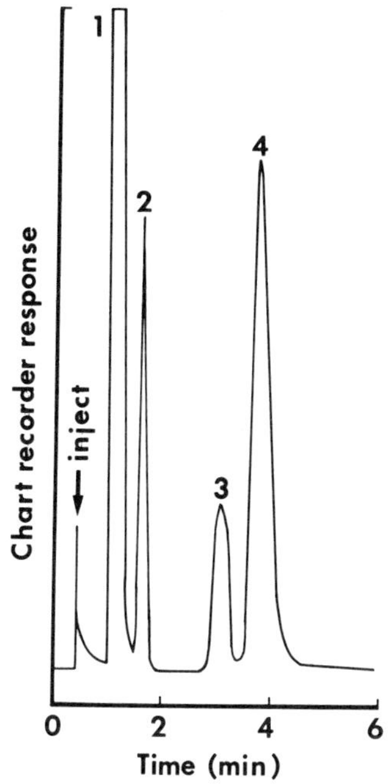

Figure 6 Separation of gases on column system shown in Fig. 4A. Sample is split so that only a small fraction passes through MS 5A column. 1, $O_2 + N_2 + Ar$; 2, CO_2; 3, $O_2 + Ar$; 4, N_2.

Restall [25] and Blackmer et al. [65]; Blackmer and Bremner [66] have modified it by replacing the thermal conductivity detector with an ultrasonic detector. Burford's system (Fig. 4E) [56] was more elaborate, involving a third column in a cold bath at −98°C on which argon and oxygen could be separated, following the separation of nitrogen, carbon dioxide, and nitrous oxide on two other columns at higher temperatures. In later work and third column was omitted [67].

An arrangement in which two parallel columns are connected to two separate thermal conductivity detectors (Fig. 4B) has been described [68]. This arrangement has been extended to include a third column coupled to a flame ionization detector to determine trace hydrocarbons (Fig. 5A) [15]. Another method for simultaneous analysis of the main constituents of the soil atmosphere and trace hydrocarbons involves the use of a Porapak Q or R column with thermal conductivity and flame ionization detectors in series (Fig. 5B) [69,70].

Laskowski and Moraghan [71] used two columns and two detectors in series (Fig. 4D) to separate hydrogen, nitrogen, nitrous oxide, and methane. Van Cleemput [72] arranged three columns in series, with a single detector. In this system oxides of nitrogen and carbon dioxide were retained on a Carbowax column at liquid air temperature while oxygen and nitrogen passed, via a Porapak Q column, to one of molecular sieve 5A where they were separated. The cold column was then heated in boiling water to drive the adsorbed gases onto the Porapak Q column.

Satisfactory measurements of oxygen concentrations can normally be obtained in soil atmosphere investigations without having to resort to cryogenic separation from argon, because the argon contribution to the single peak obtained at ambient or elevated temperatures can be calculated to a sufficient level of accuracy, and the oxygen content obtained by difference. The nitrogen, oxygen, and argon contents of CO_2-free dry air are N_2, 0.7806; O_2, 0.2100; and Ar, 0.0094 ml ml^{-1}, respectively. The composition of a soil atmosphere sample will differ from this in that the carbon dioxide level will usually be above the 3×10^{-4} ml ml^{-1} found in unpolluted air, the oxygen may be depleted, and the sample will be almost saturated with water vapor (which contributes about 0.012 and 0.023 ml ml^{-1} at 10 and 20°C respectively). The consumption of oxygen and evolution of carbon dioxide by respiring organisms causes a net reduction in pressure in the gaseous phase of the soil, because of the greater solubility of carbon dioxide. The result is an inward flow of air and, at equilibrium, an increase in the concentrations of nitrogen and argon. But none of these processes will change the ratio of these two gases to each other (83:1); the only way that such a change can be made is when nitrogen is evolved in the soil as the result of denitrification of nitrate. Calculations show that an unrealistically high rate of denitrification would have to occur in the field before the ratio of introgen to argon changed sufficiently to produce a significant error in the calculated concentration of argon, and thus in the calculation of oxygen in the composite oxygen-argon peak [15]. Table 3 shows the errors in the calculated oxygen concentration resulting from the extreme and unlikely situation in which 100 kg NO_3^--N ha^{-1} has been totally denitrified to N_2 before any has escaped from the profile. It is evident that in most circumstances, particularly for very variable field situations, such errors are quite acceptable, as they are likely to be well within the other errors associated with the experiment. To cite one example, Burford [67] has estimated that the maximum error in oxygen concentrations, using this approach in this study of the soil atmosphere after heavy applications of organic manure, was less than 0.002 ml ml^{-1}.

C. Determination of Dissolved Gases

When studies of the composition of the soil atmosphere are carried out over extended periods, there may well be occasions when the soil is saturated, and any gases present in the profile are in solution, not in the gaseous phase. If the concentrations of those dissolved gases

Table 3 Maximum Likely Errors in Determination of Oxygen Concentrations by Subtraction of Calculated Argon Contribution to Joint Peak

Concentration measured by g.c. (%)		Oxygen concentration		
O_2 + Ar	N_2	True	Calculated [a]	% error in O_2 value
15	80	14.12	14.04	0.57
10	85	9.07	8.98	0.99
5	90	4.01	3.92	2.24

[a]At 10% air-filled porosity (N_2 increased by denitrification of 100 kg NO_3^--N ha^{-1})

Table 4 Solubilities of Gases in Water at Different Temperatures

Temperature (°C)	Solubility (milliters of gas at STP per milliter of water)				
	O_2	N_2	CO_2	N_2O	C_2H_4
0	0.049	0.024	1.71	1.28	0.26
10	0.038	0.019	1.19	0.88	0.17
20	0.031	0.015	0.88	0.61	0.13
30	0.026	0.013	0.67	0.45	0.09

Source: *Handbook of Chemistry and Physics*, 37th ed., Chemical Rubber Publishing Co., Cleveland, Ohio, 1955.

which are of interest can be determined in samples of soil water, and the corresponding equilibrium gas phase concentrations are calculated from Henry's law, it is possible to interrelate data obtained under waterlogged conditions with those obtained when the soil is drained [53,74]. This is of considerable value in investigations of the concentrations of physiologically active gases to which plant roots and microorganisms are exposed. Table 4 shows the solubilities of a number of gases at different temperatures.

A number of workers in several different scientific fields have successfully applied gas chromatographic methods to the determination of dissolved gases. In the method of Swinnerton et al. [75] a water sample (1 or 2 ml) was transferred to a glass chamber divided into two compartments by a fritted glass disk, through which the carrier gas could be diverted by operating a four-way switching valve. The passage of the gas through the fritted glass resulted in the formation of a stream of bubbles which removed the dissolved gases from solution. An alternative extraction method involves the equilibration of a water sample of known volume with a known volume of air, nitrogen, or helium in a sealed bottle by vigorous shaking, and sampling the gas phase with a syringe. Concentrations of oxygen, nitrogen, nitrous oxide, carbon dioxide, and ethylene have been determined in this way [25,46,53,76,77]. However, shaking the sample with another gas creates problems of contamination with air when the method is applied to measurements of dissolved oxygen or nitrogen, and exceptional care has to be taken when flushing both the bottle and the sampling syringe. These difficulties are not very great during the analysis of gases which are either very minor constituents of the atmosphere, or not normally present, e.g., carbon dioxide, nitrous oxide, and ethylene. The alternative methods which could be applied to oxygen analysis, apart from stripping with carrier gas, to which reference has already been made, include:

1. Outgassing the sample under reduced pressure in a modified Van Slyke apparatus [78,79] or a Toepler gauge [80] and transferring the evolved gases to a gas chromatograph
2. Pumping an extractant gas continuously through the water sample in a closed circuit by means of a peristaltic pump, followed by transfer of an aliquot of the gas to the chromatograph [81]
3. Direct injection of the water sample into the chromatograph, followed by separation of gases and water vapor

The first two methods should be suitable for soil water analysis but, as yet, no such application has been described. The only disadvantage of such techniques is that they require additional ancillary apparatus coupled to the gas chromatograph, and may be time consuming. A technique for the measurement of dissolved oxygen in water samples by direct injection was described by Smith et al. [45]. The

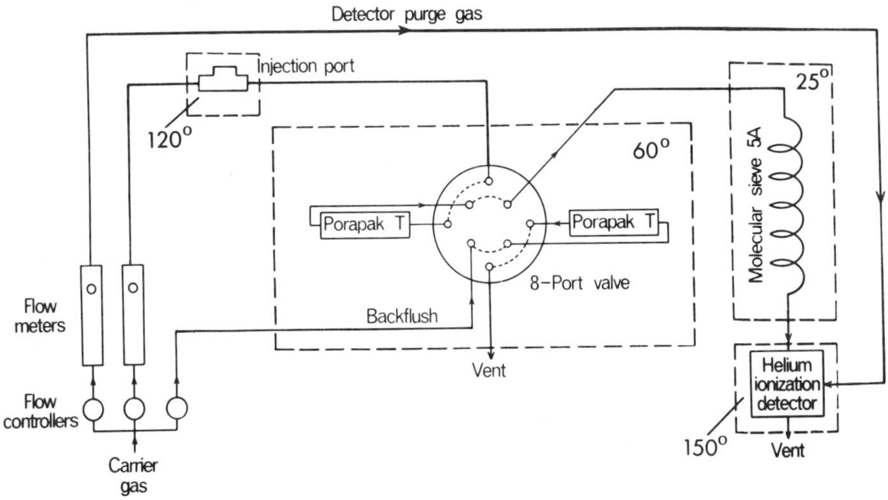

Figure 7 Gas flow diagram for gas chromatographic measurement of dissolved oxygen, using a helium ionization detector. (From Ref. 45.)

procedure involved the injection of 10 μl samples through an injection port heated to 120°C. The water vapor produced was adsorbed on a short Porapak T precolumn at room temperature, and the residual gases passed to a 1.8 m × 3 mm molecular sieve 5A column at 25°C, and thence to a helium ionization detector. Operation of an eight-port switching valve allowed a second Porapak T precolumn to be used while the first one was backflushed with helium to remove the accumulated water (Fig. 7). A very similar system, except for the use of a thermal conductivity detector, has been independently developed by Yamaguchi and Komatsu [82], and Hall [83] has now described a version in which an electron capture detector replaces the HID. This has the advantage that the ECD does not respond to nitrogen, so nitrogen may be used as carrier gas. Figure 8 shows the chromatograms obtained with the HCD and ECD systems.

It was first established by Henry in 1803 that the mass of gas dissolved by a given volume of solvent, at constant temperature, is proportional to the pressure of the gas with which it is in equilibrium. Shortly afterward, in 1805, Dalton found that Henry's law was obeyed by each gas in a mixture independently of the pressure of the others; in other words, each constituent dissolved according to its own-partial pressure. Dalton's work has been substantiated by much experimental evidence since then, for gases of low solubility, provided the pressures are not too high or the temperatures too low. However, Maharajh and Walkley [84] published data (obtained by a "helium-stripping gas chromatographic technique") indicating that for mixtures of gases, in-

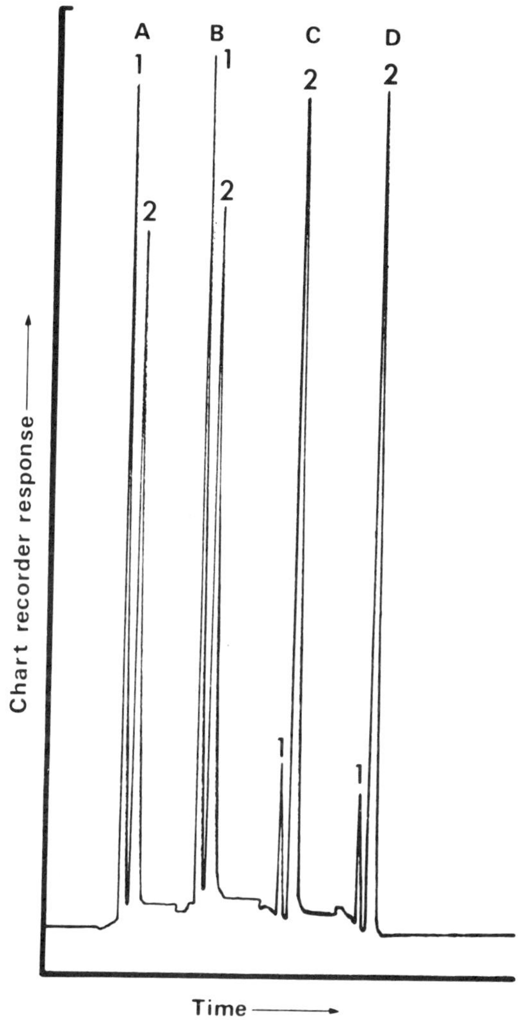

Figure 8a Chromatograms obtained from successive injections of 10 μl water samples directly into a chromatograph equipped with a helium ionization detector. A, B, water saturated with air; C. D, water equilibrated with mixture of 2.8% O_2, 97.2%; (1) O_2; (2) N_2 (From Ref. 45.)

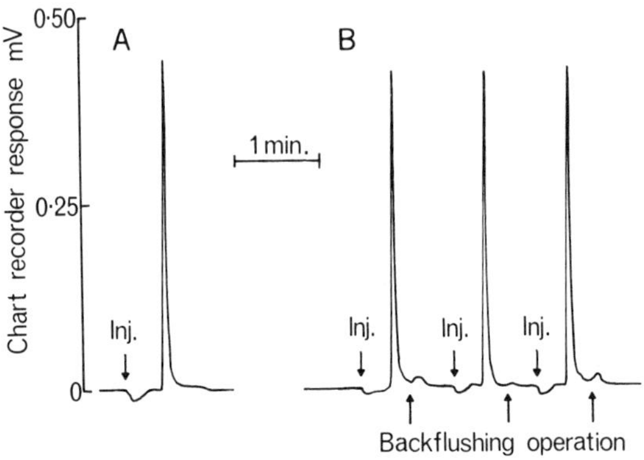

Figure 8b Chromatograms obtained from injection of 10 μl water samples directly into chromatograph equipped with electron capture detector. All samples saturated with air. (A) single injection; (B) successive injections using a backflush valve. (From Ref. 83, reproduced from the *Journal of Chromatographic Science*, by permission of Preston Publications, Inc.)

cluding oxygen-nitrogen, there was a mutual lowering of saturation solubility, compared with those of the pure gases. This work was rapidly critized by several workers [85-87] and Wilcock and Battino [88] reported that the 3% deviation from Henry's law which they found was no greater than the highest precision obtainable by gas chromatography. They commented that the precision of other workers may be as poor as ±10%.

It must be concluded that at present there is no good reason to doubt the general validity of Henry's law, and that any small deviations which may occur are likely to be small compared with the inherent variability of biological systems. A similar comment applies to the precision of measurement of dissolved gases by gas chromatography; in some areas of research a variation in results by ±10% of the value may be unacceptable and thus alternative techniques must be considered. However, for investigation of biological systems which are changing with time, for example, the development of anaerobic conditions in a previously well-aerated environment, such precision is often acceptable and well within the variability dictated by other factors.

IV. APPLICATIONS

A. Soil Oxygen and Carbon Dioxide

The use of gas chromatographic methods such as those described in Sec. III.B has greatly extended the available information about the state of soil aeration under field conditions. Recent studies have investigated the effects of drainage and irrigation [76], compaction by wheels [89], heavy applications of organic manures [67], and soil type [76] on the oxygen and carbon dioxide concentrations in the profile to which plant roots and microorganisms are exposed. In a comparison of different cultivation treatments, lower levels of oxygen and higher levels of carbon dioxide were found in unploughed plots into which cereals had been direct-drilled than in conventionally ploughed plots [70,89]. However, variations in seasonal rainfall have been shown to have greater effects than differences in cultivation [90]. In wet upland soils under afforestation, aeration was found to be very poor throughout the growing season, being particularly bad in midsummer when respiratory demand for oxygen was at its highest [91,92].

The composition of the soil atmosphere varies greatly over short distances, and this has been attributed to the effects of soil structure on gaseous diffusion. In fine-textured, aggregated soils in particular, at high moisture contents gaseous diffusion will occur primarily through the larger, interaggregate pores. If neighboring sampling probes are inserted (1) into the center of a wet aggregate and (2) close to a macropore between aggregates, respectively, there are likely to be major differences in the oxygen and carbon dioxide concentrations from the two environments. These effects make it necessary to maximize the number of replicate sampling locations. Usually the upper limit is set by practical considerations such as the number of analyses which can be handled. Automatic injection techniques are particularly valuable in this context, because time lost between successive analyses can be kept to a minimum. Provided that the analysis time per sample can be kept down to a few minutes, then throughputs of the order of 100 samples per day are feasible.

Increasing numbers of applications of oxygen and carbon dioxide analysis are being made in laboratory studies. An example is the measurement of carbon dioxide evolution in soil respiration, which has been greatly facilitated by the employment of gas chromatography [24,93]. Many investigations of soil microbiological processes require either strictly aerobic or strictly anaerobic conditions, and the ability to check rapidly the composition of the atmosphere in the incubation vessel has proved to be a major advantage when attempting to establish the necessary experimental conditions [94,95].

Gas chromatography is not the only instrumental method used for soil aeration studies. For instance, the platinum electrode has been employed by a number of groups since the early 1960s to measure oxygen diffusion rates. However, McIntyre [96] has shown that although the electrode works satisfactorily in saturated conditions, in which a constant current-concentration relationship is obtained, this relationship does not apply to unsaturated media. Thus the method gives unreliable results over the important range of conditions in which aeration may be impeded but the soil is not completely waterlogged. An alternative method which gives a direct analysis of the concentration of soil gaseous oxygen is that based on a portable polarographic oxygen analyzer [97]. Ten milliliter volumes of air are drawn through a cell of less than 1 ml volume from 75 ml reservoirs in the soil, and individual measurements take less than 1 min. This appears to be a more reliable technique than the one based on the electrode. Selection between gas chromatography and polarography as the most suitable method for soil oxygen is likely to be influenced by whether other gases are also to be determined.

B. Oxides of Nitrogen

Considerable losses of fertilizer nitrogen from the soil result from the microbial reduction of nitrate to nitrous oxide and molecular nitrogen under anaerobic conditions, and the agricultural importance of this phenomenon has led to its extensive investigation. A further impetus to research has been the claim that nitrous oxide in the upper atmosphere may bring about reactions leading to the destruction of the ozone layer [98,99]. In earlier studies of nitrous oxide concentrations, in laboratory systems and in field soils, thermal conductivity detectors were employed. Burford and Stefanson [54] studied the occurrence of nitrous oxide in generally well-aerated Australian soils, using a system with a limit of detection of 3-4 $\times$ 10^{-6} ml ml^{-1} in a 3 ml sample [56]. They attributed the small concentrations found to the presence of anaerobic microsites within the aerobic profile. Essentially the same conclusions were reached as a result of investigations in the United Kingdom [53,77] and were reinforced by work in California [100,101]. Rolston et al. [101] investigated the quantities of nitrous oxide evolved from laboratory columns and field plots heavily fertilized with nitrate and maintained at high water contents. This work was done with the aid of a helium ionization detector [102]. Gas samples were taken from several depths in the soil profile, and the total flux of nitrous oxide calculated from standard diffusion equations. Other studies of gaseous fluxes have involved enclosing the soil surface with a canopy and concentrating the gas by adsorbint it on a molecular sieve and releasing it in a small volume for analysis [26,150,151]. Other workers have sampled the atmosphere under the canopy directly and determined the nitrous oxide with an electron capture detector [28,103,152] or with an ultrasonic detector after prior concentration from 300 ml air samples [55].Flühler et al. [104] determined nitrous oxide concentrations which were mainly below 1 $\times$ 10^{-6} ml ml^{-1} by interfacing a gas chromatograph with a mass spectrometer. They concluded that the presence of the gas was a good indicator of the development of anaerobic zones within the soil matrix.

Bremner and Blackmer [55] found that the ratio of nitrous oxide to nitrogen evolved during denitrification was greatly dependent on nitrate concentration and pH. They also found that nitrous oxide was evolved from aerobic soils when ammonium ion was being nitrified to nitrite and nitrate [105], and concluded that although this process was unlikely to be a major cause of loss of fertilizer nitrogen, it could well contribute significantly to the fertilizer-derived nitrous oxide entering the stratospheric ozone layer. The aerobic formation of nitrous oxide has been confirmed in a number of other reports [106,148,153,154].

The formation of nitric oxide in soils has been attributed to the chemical decomposition of nitrite formed by microbial reduction of nitrate under anaerobic conditions [107]. Nelson and Bremner [35] found that nitric oxide was released predominantly by acid soils. It could only be detected under anaerobic conditions, because spontaneous oxidation to nitrogen dioxide occurred if any oxygen was present. Similar observations were made by van Cleemput and Baert [108].

C. Sulfur Gases

Cruciferous plants are characterized by high contents of sulfur-containing compounds such as methionine and methylcysteine sulfoxide, from which sulfur-containing volatiles may be released by chemical or enzymatic degradation. Lewis and Papavizas [109] investigated the volatiles resulting from the decomposition of crucifer residues in soil and identified methanethiol, dimethyl sulfide, and dimethyl disulfide. These products, along with a number of nonsulfur compounds, were separated on a column of Carbowax 20 M and measured with a flame ionization detector.

Banwart and Bremner used a flame photometric detector in conjunction with a variety of column packing materials, to determine the sulfur gases produced by microbial transformations in soils. All columns were made from Teflon tubing to minimize adsorption of the gases [110]. The following compounds were identified as products of the decomposition of sulfur-amino acids added to soils under aerobic or waterlogged conditions: methanethiol, dimethyl sulfide, dimethyl disulfide, carbonyl sulfide, carbon disulfide, ethanethiol, ethyl methyl sulfide, and diethyl disulfide [111]. The first five gases mentioned were evolved from soils treated with animal manures, sewage sludge, and plant materials [112], and small amounts of dimethyl sulfide, sometimes accompanied by traces of other gases, from unamended or sulfate-treated soils [113]. Hydrogen sulfide was not detected. Elliott and Travis [114] investigated which sulfur gases were included in the malodorous volatile products of bovine manure. Gas samples from the headspace above anaerobically incubated columns of manure were separated on Porapak Q or Chromosorb 101 and found to include hydrogen sulfide and carbonyl sulfide. The same two gases were also detected in the soil atmosphere below cattle feedlots into which, presumably, manure had infiltrated. The concentration of carbonyl sulfide was 0.1 ng S ml^{-1} (i.e., about 7.5×10^{-8} ml ml^{-1} of carbonyl sulfide).

Lovelock et al. [115] measured the emission of dimethyl sulfide by soils, living and decaying plant leaves, and marine algae. They found that although the compound was evolved from soil in flasks plugged with glass wool, hydrogen sulfide was produced instead if the flasks were completely sealed. As the oxygen concentration fell very little in the sealed system, the inhibition of dimethyl sulfide production was thought to be due to suppression of microbial activity by carbon dioxide, rather than to the development of anaerobic conditions. These authors suggested that dimethyl sulfide, rather than hydrogen sulfide, was the gaseous compound of most importance in the sulfur cycle in the environment.

Surface soils have the capacity to adsorb considerable quantities of gaseous atmospheric pollutants, either by physico-chemical processes or as the result of uptake by microorganisms. The rates of uptake may be conveniently studied by following the disappearance of a gas from the atmosphere above a soil in a closed system, by taking successive samples for gas chromatographic analysis. Experiments with sterilized and unsterilized soils have shown that sulfur dioxide, hydrogen sulfide, and methanethiol are taken up by physico chemical processes [61], whereas dimethyl sulfide, dimethyl disulfide, carbonyl sulfide, and carbon disulfide are partly taken up by microorganisms [116]. Sulfur dioxide has a much greater affinity for soil surfaces, wet or dry, than any of the other sulfur gases studied.

D. Hydrocarbons

The anaerobic decomposition of organic matter in waterlogged soil results in the formation of methane, or "marsh gas," in considerable quantities. The first systematic investigation of this process was as long ago as 1884 [117]. The only evidence of the emanation of other hydrocarbon gases from soils, prior to the application of gas chromatography to soil atmosphere studies, was reported by Neilson Jones [118]. He found that the effects on plant growth brought about by exposure to gases extracted from a forest soil were characteristic of those caused by trace levels of ethylene. The first mention of the use of gas chromatography for determining methane from flooded soils seems to have been by Koyama [119], but no experimental details were given. The earliest papers describing the determination of other gaseous hydrocarbons in soil air by gas chromatography were all concerned with the application of this method of analysis to geochemical prospecting for underlying petroleum or natural gas deposits. Smith and Ellis [59] used a 9 m × 6 mm column packed with 30% Carbowax 400 on 30-60 mesh Chromosorb, followed by a 6 m × 6 mm column of Silicone SF1017 on Chromosorb, to separate C_1 to C_5 hydrocarbons, both saturated and unsaturated compounds. It was found that the dominant peaks from soil, and decomposing vegetation and plant roots, were due to unsaturated compounds, particularly butene-1. Alkanes were also present, although in smaller quantities, and it was concluded that the presence of saturated hydrocarbons in the soil atmosphere was not necessarily attributable to an underlying petroleum deposit. A similar conclusion was reached by Juranek [120] and Neglia and Favretto [47]; the latter workers found that ethylene was present in soil air although this gas was not present in natural gas.

The development of the flame ionization detector has made possible the measurement of gaseous hydrocarbon compounds at concentrations in the 10^{-9}-10^{-8} ml ml^{-1} range. Using this detector in conjunction with columns of alumina partially deactivated with sodium iodide, Smith et al. [15,25,74] found that methane, ethane, ethylene, propane, propene, n- and iso-butane and butene were all present in the soil under anaerobic conditions. The presence of ethylene was of interest because of its effects on plant growth, and the alumina column was found to be particularly useful for measurements of this gas because it enabled a good separation to be achieved between ethane and ethylene at temperatures sufficiently high to keep the retention times for higher hydrocarbons tolerably short. Thus temperature programming could be avoided and the time taken between successive analyses kept to a minimum. Other types of column used to measure ethylene in the soil atmosphere have included activated alumina [121], Porapak Q [36,69] and Porapak R [70].

Extensive studies of ethylene have shown that it is present in a wide range of soils as a result of microbial activity [15,25,53,76,95, 122]. The quantity formed was found to be related to organic matter content and was greatly affected by wetting and drying and whether the soil had been recently cropped [25,95]. The effects of temperature, moisture, and aeration on ethylene concentrations in the atmospheres of field soils were compatible with the concept of its formation within anaerobic zones distributed through a generally aerated soil profile [76]. Evolution of ethylene was stimulated by additions of methionine and glucose [122] and the fermentative degradation products of carbohydrates, particularly ethanol [123]. The simultaneous formation of a mixture of C_1 to C_4 hydrocarbons was generally observed in field studies [15] and in laboratory incubation experiments [25], possibly due to a similar mechanism. Changes in relative and absolute concentrations of these gases with time in field samples are shown in Fig. 9. It has been claimed that ethylene has a major inhibitory effect on soil respiration [124], but a recent study employing gas chromatographic methods to determine not only ethylene but also oxygen consumption and carbon dioxide evolution produced evidence to the contrary [93]. In fact, far from inhibiting the activity of soil microorganisms, ethylene (and also acetylene and hydrocarbons of higher molecular weight) are consumed by them under aerobic conditions [61,125].

Concentration of low levels of unsaturated hydrocarbons evolved from incubated soil samples has been achieved by absorption in mercuric perchlorate solution, followed by liberation of the gases in a closed vessel and injection by syringe into the chromatograph [25]. Although Dimitriades et al. [126] have found mercuric perchlorate somewhat unsatisfactory for quantitative removal of ethylene from air samples containing vehicular exhaust gases, the technique is a valuable tool in the study of biological production of the gas at very low rates; and if several traps are connected in series and suitable flow rates are used, recoveries can be virtually complete.

An important area of soil science research in which gas chromatography has become a valuable analytical tool is the study of biological nitrogen fixation. Since the discovery that the nitrogenase enzyme which reduces atmospheric nitrogen to ammonia can also reduce acetylene to ethylene [127], the use of this reaction has become ever more widely used to detect the presence of nitrogen-fixing activity in soil or soil-plant systems. The relative ease with which the production of ethylene from acetylene can be determined by gas chromatography contrasts with the more elaborate techniques, and usually the greater expense, associated with the alternative method, namely, the use of ^{15}N tracer and its measurement either by mass spectrometry or optical emission spectrometry. However, considerable problems have been encountered by workers attempting to quantify nitrogen-fixing activity on the basis of results obtained with the acetylene reduction method. This subject is also discussed in Chaps. 8 and 9.

Porapak N, R, and T columns have been widely used for the separation of ethylene from acetylene [128-131] and have proved very suitable. The elution of ethylene before acetylene means that even a small trace of the former gas can be detected in the presence of a very large acetylene peak (concentrations of 0.1 ml ml^{-1} are commonly used in incubation procedures). Hardy et al. [132] used a 3 m × 3 mm column of 20% ethyl, N',N'-dimethyloxalamide on 100-200 mesh acid-washed firebrick, developed by Richmond [133]. This column was operated at 0°C, and was found to give an effective separation of acetylene and ethylene from other C_1-C_4 hydrocarbons and a linear response to ethylene over a 10^6-fold range of concentration, and to give reproducible results over a prolonged period. Propane has been used as an internal standard for the acetylene reduction assay, by injecting sufficient

---

Figure 9 Chromatograms of hydrocarbons in soil atmosphere samples taken on three occasions from the same probe at 300 mm depth. (a) April 26; (b) June 9; (c) June 21. (1) Methane (+ air), (2) ethane, (3) ethylene, (4) propane, (5) propene, (6) isobutane, (7) n-butane, (8) butene (mixture of isomers). (From Ref. 15, reproduced from the *Journal of Chromatographic Science*, by permission of Preston Publications, Inc.)

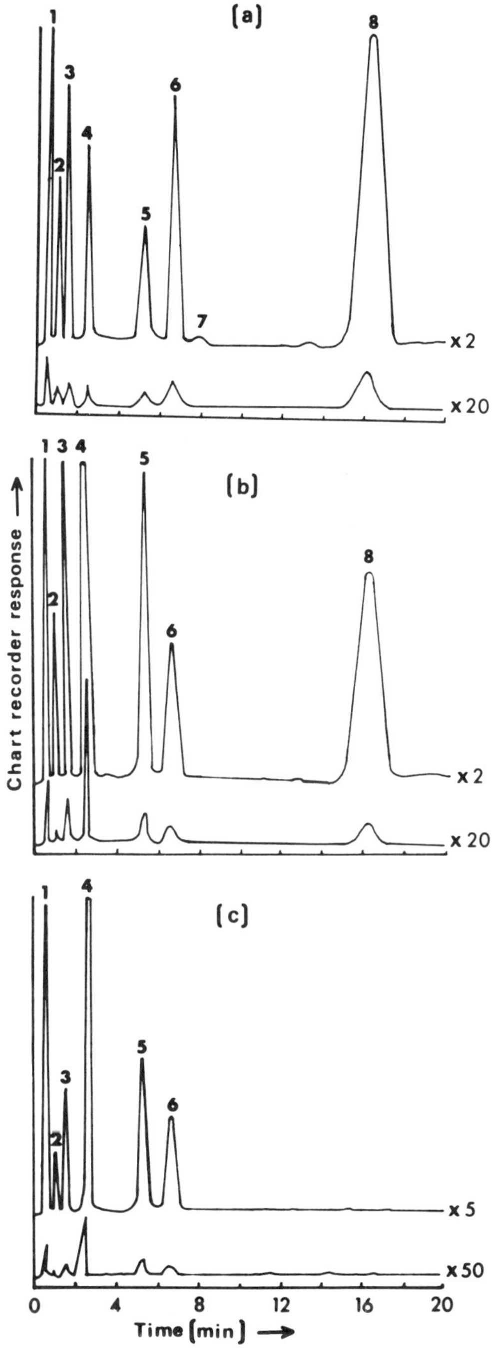
(a)
1
2
3
4
5
6
7
8
x 2
x 20
(b)
1 3 4
2
5
6
8
x 2
x 20
(c)
1
2
3
4
5
6
x 5
x 50
Chart recorder response →
0 4 8 12 16 20
Time (min) →

into the incubation chamber to produce a concentration of about 10^{-4} ml ml^{-1} [131]. Measurement of the propane peak, following the elution of ethylene and acetylene, provided a check on any leaks in the system.

E. Other Carbon Compounds

Biochemical processes taking place in the soil associated with microbial decomposition of plant residues, the germination of seeds, and the growth of plant roots, result in the liberation of a number of volatile organic compounds in addition to those already discussed. For example, the decomposition of carbohydrates under anaerobic conditions results in the formation of volatile fatty acids (formic, acetic, propionic, n- and iso-butyric, n- and iso-valeric). Several different columns have been found satisfactory for their determination, e.g., 6% FFAP on Porapak QS, temperature programmed from 150 to 210°C [134,135]; 33% dioctylsabacinate and 15% sabacic acid on Chromosorb W, at 150°C [136]: and 20% Tween 80 on Chromosorb W, at 110°C [137,138]. Separations carried out on Porapak QS and Chromosorb 101 [139] and Carbowax 400 [140] have shown that alcohols, aldehydes, ketones, and esters are also formed under these conditions.

The germination of the seeds of some plant species is thought to be controlled by the formation of volatile growth inhibitors within the seeds, which diffuse out, into the soil atmosphere. Ethanol, acetone, and acetaldehyde have been identified as behaving in this way, following their separation on a 1.8 m × 3 mm column of Porapak Q at 150°C [141].

Volatile compounds emanating from the decomposition of plant residues have been found to affect the growth of soil fungi and bacteria. Owens et al. [142] separated about 20 compounds on columns of Polypak-1 and 15% Carbowax 20M on Anakrom SD support, which were temperature programmed between 75 and 130°C. Of these 20 volatiles, acetaldehyde, isobutyraldehyde, isovaleraldehyde, 2-methylbutanal (and/or valeraldehyde), methanol, and ethanol were identified as the biologically active compounds (Fig. 10).

In the course of studies of microbial uptake of carbon monoxide, Smith et al. [61] found that this gas is also evolved by soils, by a nonbiological process, presumably due to chemical decomposition of soil organic matter. In this work, separation of carbon monoxide was achieved on a molecular sieve 5A column at 70°C and a helium ionization detector was used.

Brominated organic compounds are widely used as soil fumigants to control plant pests. Gas chromatographic methods have been used to study the distribution and the persistence of several of these substances in the soil. For example, methyl bromide has been determined on Chromosorb W coated with 20% Dow Corning high vacuum grease, at 85°C [143], ethylene dibromide on a column of 20% Castorwax at

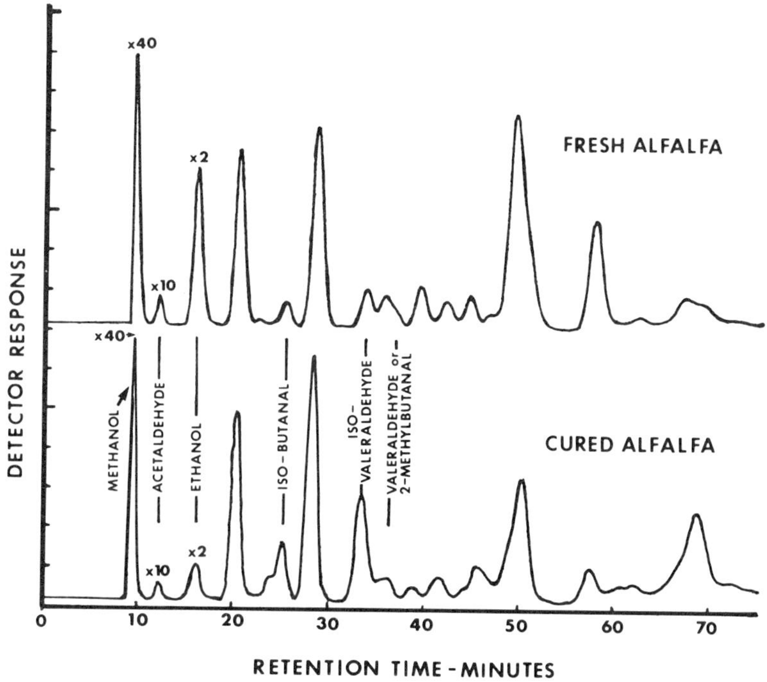

Figure 10 Gas chromatograms of volatile organic compounds from alfalfa, separated on Polypak-1. (From Ref. 142.)

101-105°C [144], and 1,2-dibromo-3-chloropropane (DBCP) on a column of 5% DC-11 on Chromosorb-W at 105°C, following hexane extraction of the soil [145]. Less than 1 ng of compounds such as DBCP can be detected with an electron capture detector.

V. SUMMARY AND CONCLUSIONS

The development of gas chromatographic methods over the last 20 years has made possible the rapid and accurate measurement of permanent gases and volatile compounds present as major constituents or only as trace components of the soil atmsophere. For several years the improvement in sensitivity over classical methods of analysis was vastly greater for organic compounds than for inorganic, because of the availability of the flame ionization detector. The development of the helium ionization and ultrasonic detectors has now led to comparable sensitivity for all substances, and detectors such as the electron capture and flame photometric combine great sensitivity with selectivity for certain types of molecule, so further extending the capabilities of the method.

It is possible, with currently available column packing materials, to separate all the gases of interest in the soil atmosphere, providing multiple-column systems or temperature programming procedures are used. One highly desirable development would be a column on which oxygen, nitrogen and carbon dioxide could be separated isothermally in a reasonably short time. Another would be the capability of separating argon from oxygen easily at ambient or elevated temperatures.

Naturally occurring soil gases which have been measured by gas chromatography include:

1. Oxygen, nitrogen and argon, originating from the earth's atmosphere .
2. Carbon dioxide produced by respiration of plant roots and microorganisms.
3. Nitrous oxide, nitric oxide, nitrogen dioxide, and molecular nitrogen evolved by microbiological and chemical processes in the soil.
4. Hydrogen sulfide, carbonyl sulfide, carbon disulfide, and organo-sulfur compounds produced by microbial decomposition of sulfur-amino acids and other sulfur compounds.
5. Methane, ethylene, and other C_2-C_4 hydrocarbons, produced by microbial activity in the soil, and also possibly present as a result of diffusion from underlying oil or gas-bearing strata.
6. Carbon monoxide, from chemical decomposition of soil organic matter; hydrogen, resulting from microbial decomposition of organic matter under severely reducing cnnditions.
7. The vapors of volatile fatty acids, aldehydes, alcohols, and ketones produced by microbial decomposition of plant residues.

In addition, the fate of gases which have been artificially introduced into the soil may be monitored by gas chromatography. Applications include: measurement of residual quantities of gaseous fumigants such as brominated hydrocarbons; estimation of biological nitrogen fixation rates by the reduction of acetylene to ethylene; identification of leaks of natural and manufactured gas from underground pipes; measurement of the rate of sorption of atmospheric pollutant gases by soils, and measurement of gaseous diffusion rates.

The study of physiologically active gases and volatiles which affect the growth of microorganisms and higher plants is only in its infancy, and it is most unlikely that the list of volatile organic compounds so far identified as having biological effects in the soil represents anything but a fraction of the total. The further application of gas chromatographic analysis to this area of research may be expected to provide much new information in the future.

REFERENCES

1. J. B. Boussingault and M. B. Lewy, Ann. Chim. Phys., *37*:5 (1853).
2. A. I. M. Keulemans, *Gas Chromatography*, Reinhold, New York, 1959.
3. P. G. Jeffery and P. J. Kipping, *Gas Analysis by Gas Chromatography*, Pergamon, Oxford, 1964.
4. J. C. Giddings, *Dynamics of Chromatography*, Marcel Dekker, New York, 1965.
5. L. Szepesy, *Gas Chromatography*, Iliffe, London, 1970.
6. A. J. P. Martin and R. L. M. Synge, Biochem. J., *35*:1358 (1941).
7. A. T. James and A. J. P. Martin, Biochem. J., *50*:679 (1952).
8. E. Glueckauf, Trans. Faraday Soc., *51*:34 (1955).
9. J. J. van Deemter, F. J. Zuiderweg, and A. Klinkenberg, Chem. Eng. Sci., *5*:271 (1956).
10. J. C. Giddings, Anal. Chem., *35*:439 (1963).
11. E. Grushka, J. Chromatogr. Sci., *13*:25 (1975).
12. H. M. McNair, J. Chromatogr. Sci., *16*:578 (1978).
13. O. L. Hollis, Anal. Chem., *38*:309 (1966).
14. C. G. Scott, J. Inst. Petroleum, *45*:118 (1959).
15. K. A. Smith and R. J. Dowdell, J. Chromatogr. Sci., *11*:655 (1973).
16. R. Kaiser, Chromatographia, *2*:453 (1969).
17. R. Kaiser, Chromatographia, *3*:38 (1970).
18. A. Zlatkis, H. R. Kaufman, and D. E. Durbin, J. Chromatogr. Sci., *8*:416 (1970).
19. D. H. Bollman and D. M. Mortimore, J. Chromatogr. Sci., *10*:523 (1972).
20. F. Bruner, A. Liberti, M. Possanzine, and I. Allegrine, Anal. Chem., *44*:2070 (1972).
21. E. W. Lord and R. C. Horn, Anal. Chem., *32*:878 (1960).
22. E. L. Obermiller and G. O. Charlier, J. Gas Chromatogr., *6*:446 (1968).
23. R. G. Bell, Soil Sci., *105*:78 (1968).
24. J. R. Burford and J. M. Bremner, Soil Biol. Biochem., *4*:191 (1972).
25. K. A. Smith and S. W. F. Restall, J. Soil Sci., *22*:430 (1971).
26. R. J. Dowdell and R. Crees, Lab. Practice, *23*:488 (1974).
27. S. Ghoshal and B. Larsson, Acta Agric. Scand., *25*:275 (1975).
28. K. C. Hall and J. R. Burford, *U. K. Agricultural Research Council Letcombe Laboratory Annual Report for 1975,* HMSO, London, 1976, p. 55.

29. D. H. Smith and F. E. Clark, Soil Sci. Soc. Am. Proc., *24*:111 (1960).
30. C. W. Quinlan and J. R. Kittrel, J. Chromatogr. Sci., *10*:691 (1972).
31. F. J. Stevenson and R. M. Harrison, Soil Sci. Soc. Am. Proc., *30*:609 (1966).
32. J. M. Trowell, J. Chromatogr. Sci., *9*:253 (1971).
33. S. A. Greene and H. Pust, Anal. Chem., *30*:1039 (1958).
34. D. H. Smith, F. S. Nakayama, and F. E. Clark, Soil Sci. Soc. Am. Proc., *24*:145 (1960).
35. D. W. Nelson and J. M. Bremner, Soil Biol. Biochem., *2*:203 (1970).
36. J. R. Burford and J. M. Bremner, Soil Biol. Biochem., *4*:489 (1972).
37. R. K. Stevens, J. D. Mulik, A. E. O'Keefe and K. J. Krost, Anal. Chem., *43*:827 (1971).
38. T. L. C. de Souza, D. C. Lane, and S. P. Bhatia, Anal. Chem., *47*:543 (1975).
39. T. Johns and A. C. Stapp, J. Chromatogr., Sci., *11*:234 (1973).
40. W. E. Wentworth and R. R. Freeman, J. Chromatogr., *79*:322 (1973).
41. W. E. Wentworth, E. Chen, and J. E. Lovelock, J. Phys. Chem., *70*:445 (1966).
42. D. C. Fenimore and C. M. Davis, J. Chromatogr. Sci., *8*:519 (1970).
43. S. O. Farwell and R. A. Rasmussen, J. Chromatogr. Sci., *14*:224 (1976).
44. K. A. Smith and W. Harris, J. Chromatogr., *53*:358 (1970).
45. K. A. Smith, R. J. Dowdell, and K. C. Hall, in *Measurement of Oxygen* (H. Degn, II. Balslev, and R. Brook, Eds.), Elsevier, Amsterdam, 1976, p. 226.
46. K. A. Smith, in *Advances in Chromatography*, vol. 15 (J. C. Giddings, E. Grushka, J. Cazes, and P. R. Brown, Eds.) Marcel Dekker, New York, 1977, p. 197.
47. S. Neglia and L. Favretto, in *Advances in Organic Geochemistry* (U. Colombo and G. D. Hobson, Eds.), Pergamon, Oxford, 1964, p. 285.
48. E. J. Russell and A. Appleyard, J. Agric. Sci. Camb., *7*:1 (1915).
49. J. L. Tackett, Soil Sci. Soc. Am. Proc., *32*:346 (1968).
50. O. A. De Camargo, F. Grohmann, E. Salati, and E. Matsui, Soil Sci., *117*:173 (1974).
51. J. R. Burford and R. J. Millington, *Transactions of the 9th International Comgress of Soil Science*, Alelaide, 1968, vol. 2, p. 505.
52. M. Yamaguchi, F. D. Howard, D. L. Hughes, and W. J. Flocker, Soil Sci. Soc. Am. Proc., *26*:512 (1962)

53. R. J. Dowdell, K. A. Smith, R. Crees, and S. W. F. Restall, Soil Biol. Biochem., *4*:325 (1972).
54. J. R. Burford and R. C. Stefanson, Soil Biol. Biochem., *5*:133 (1973).
55. J. M. Bremner and A. M. Blackmer, *Proceedings of Denitrification Seminar*, Fert. Inst., Washington, D.C., 1977.
56. J. R. Burford, J. Chromatogr. Sci., *7*:760 (1969).
57. M. E. Bednas and D. S. Russell, J. Gas Chromatogr., *5*:592 (1967).
58. J. S. Boyce and T. M. McCalla, Soil Sci., *108*:241 (1969).
59. G. H. Smith and M. M. Ellis, Bull. Am. Assoc. Petroleum Geol., *47*:1897 (1963).
60. E. P. Kavanagh and J. R. Postgate, Lab. Practice, *19*:159 (1970).
61. K. A. Smith, J. M. Bremner, and M. A. Tabatabai, Soil Sci., *116*:313 (1973).
62. W. L. Banwart and J. M. Bremner, Soil Biol. Biochem., *6*:113 (1974).
63. L. F. Elliott and T. M. McCalla, Soil Sci. Soc. Am. Proc., *36*:68 (1972).
64. M. Vilain and J.-P. Druelle, Ann. Argon. (N.S.), *18*:507 (1967).
65. A. M. Blackmer, J. H. Baker, and M. E. Weeks, Soil Sci. Soc. Am. Proc., *38*:689 (1974).
66. A. M. Blackmer and J. M. Bremner, Soil Sci. Soc. Am. J., *41*:908 (1977).
67. J. R. Burford, J. Sci. Fd. Agric., *27*:115 (1976).
68. J. M. Bollag and S. T. Czlonkowski, Soil Biol. Biochem., *5*:673 (1973).
69. I. S. Cornforth and R. J. Stevens, Plant Soil, *38*:581 (1973).
70. M. J. R. Bell and J. C. Holmes, unpublished work, 1973.
71. D. Laskowski and J. T. Moraghan, Plant Soil, *27*:357 (1967).
72. O. van Cleemput, J. Chromatogr., *45*:315 (1969).
73. B. T. Bunting and J. A. Campbell, Can. J. Soil Sci., *55*:69 (1975).
74. K. A. Smith and R. S. Russell, Nature (Lond.), *222*:769 (1969).
75. J. W. Swinnerton, V. J. Linnenbom, and C. H. Cheek, Anal. Chem., *34*:483 (1962).
76. K. A. Smith and R. J. Dowdell, J. Soil Sci., *25*:217 (1974).
77. R. J. Dowdell and K. A. Smith, J. Soil Sci., *25*:231 (1974).
78. K. G. Ikels, J. Gas Chromatogr., *3*:359 (1965).
79. E. Bjergbakke, in *Measurement of Oxygen* (H. Degn, I. Balslev and R. Brook, Eds.), Elsevier, Amsterdam, 1976, p. 1.
80. L. V. Petkovic, M. M. Kosanic, and J. G. Draganic, Bull. Boris Kidric Inst. Nucl. Sci., Belgrade, *15*:9 (1964).
81. H. A. C. Montgomery and C. Quarmby, Lab. Practice, *15*:538 (1966).

82. M. Yamaguchi and Y. Komatsu, Plant Soil, *47*:265 (1977).
83. K. C. Hall, J. Chromatogr. Sci., *16*:311 (1978).
84. D. M. Maharajh and J. Walkley, Nature Phys. Sci., *236*:165 (1972).
85. P. S. Liss and P. G. Slater, Nature Phys. Sci., *238*:64 (1972).
86. T. R. S. Wilson, Nature Phys. Sci., *239*:31 (1972).
87. A. L. Myers and J. A. Quinn, Nature Phys. Sci., *239*:32 (1972).
88. R. J. Wilcock and R. Battino, Nature (Lond.), *252*:614 (1974).
89. R. J. Dowdell and R. Crees, *U. K. Agricultural Research Council Letcombe Laboratory Annual Report for 1975*, HMSO, London, 1976, p. 52.
90. K. A. Smith, R. J. Dowdell, K. C. Hall, and R. J. Crees, *U. K. Agricultural Research Council Letcombe Laboratory Annual Report for 1973*, HMSO, London, 1974, p. 35.
91. K. A. Smith, *Forestry Commission Report on Forest Research 1976*, HMSO, London, 1976, p. 56.
92. K. A. Smith, *Forestry Commission Report on Forest Research 1977*, HMSO, London, 1977, p. 52.
93. K. A. Smith, Soil Biol. Biochem., *10*:269 (1978).
94. J. M. Lynch and S. H. T. Harper, J. Gen. Microbiol., *80*:187 (1974).
95. G. Goodlass and K. A. Smith, Soil Biol. Biochem., *10*:193 (1978).
96. D. S. McIntyre, Adv. Agron., *22*:235 (1970).
97. W. H. Patrick, Soil Sci. Soc. Am. J., *41*:651 (1977).
98. P. J. Crutzen, J. Geophys. Res., *76*:7311 (1971).
99. H. Johnston, Science, *173*:517 (1971).
100. D. D. Focht, in *Nitrogen in the Enviroment*, vol. 2 (D. R. Nielsen and J. C. MacDonald, Eds.), Academic Press, New York, 1978, p. 433.
101. D. E. Rolston, M. Fried, and D. A. Goldhamer, Soil Sci. Soc. Am. J. *40*:259 (1976).
102. C. C. Delwiche and D. E. Roston, Soil Sci. Soc. Am. J., *40*:324 (1976).
103. J. R. Burford, R. J. Dowdell, and R. Crees, *U. K. Agricultural Research Council Letcombe Laboratory Annual Report for 1977*, HMSO, London, 1978, p. 71.
104. H. Flühler, M. S. Ardakani, T. E. Szuszkiewicz, and L. H. Stolzy, Soil Sci., *122*:107-114 (1976).
105. J. M. Bremner and A. M. Blackmer, Science, *199*:295 (1978).
106. J. R. Freney, O. T. Denmead, and J. R. Simpson, *Transactions of the 11th International Comgress of Soil Science*, Edmonton, 1978, vol. 1, p. 300.
107. F. B. Cady and W. V. Bartholomew, Soil Sci. Soc. Am. Proc., *24*:477 (1960).
108. O. van Cleemput and L. Baert, Mededlingen Fakulteit Landbouwwetenschappen Gent, *37*:1177 (1972).

109. J. A. Lewis and G. C. Papavizas, Soil Biol. Biochem., *2*:239 (1970).
110. W. L. Banwart and J. M. Bremner, Soil Biol. Biochem., *6*:113 (1974).
111. W. L. Banwart and J. M. Bremner, Soil Biol. Biochem., *7*:359 (1975).
112. W. L. Banwart and J. M. Bremner, Soil Biol. Biochem., *8*:439 (1976).
113. W. L. Banwart and J. M. Bremner, Soil Biol. Biochem., *8*:19 (1976).
114. L. F. Elliott and T. A. Travis, Soil Sci. Soc. Am. Proc., *37*:700 (1973).
115. J. E. Lovelock, R. J. Maggs, and R. A. Rasmussen, Nature (Lond.), *237*:452 (1972).
116. J. M. Bremner and W. L. Banwart, Soil Biol. Biochem., *8*:79 (1976).
117. P. P. Dehérain, Ann. Agron., *10*:385 (1884).
118. W. Nielson Jones, Nature (Lond.), *136*:554 (1935).
119. T. Koyama, J. Geophys. Res., *68*:3971 (1963).
120. J. Juranek, Czech. Inst. Petroleum Res. Trans., *9*:57 (1958).
121. A. D. Rovira and M. Vendrell, Soil Biol. Biochem., *4*:63 (1972).
122. J. M. Lynch, Nature (Lond.), *240*:45 (1972).
123. G. Goodlass and K. A. Smith, Soil Biol. Biochem., *10*:201 (1978).
124. A. M. Smith and R. J. Cook, Nature (Lond.), *252*:703 (1974).
125. F. B. Abeles, L. E. Craker, L. E. Forrence, and G. R. Leather, Science *173*:914 (1971).
126. D. Dimitriades, C. F. Ellis, and D. E. Siezinger, in *Advances in Chromatography*, vol. 8 (J. C. Giddings and R. A. Keller, Eds.), Marcel Dekker, New York, 1969, p. 327.
127. M. J. Dilworth, Biochim. Biophys. Acta, *127*:285 (1966).
128. R. J. Roughley and P. J. Dart, Arch. Mikrobiol., *69*:171 (1969).
129. E. D. Spiff and C. T. I. Odu, Soil Biol. Biochem., *4*:71 (1972).
130. F. J. Bergersen, Aust. J. Biol. Sci., *23*:1015 (1970).
131. P. J. Dart, J. M. Day, and D. Harris, in *Grain Legume Production Technical Booklet*, FAO/IAEA, Vienna, 1972.
132. R. F. W. Hardy, R. C. Burns, and R. D. Holsten, Soil Biol. Biochem., *5*:47 (1973).
133. A. B. Richmond, J. Chromatogr. Sci., *7*;321 (1969).
134. D. A. Laskowski and F. E. Broadbent, Soil Sci. Soc. Am. Proc., *34*:72 (1970).
135. S. Chandrasekaran and T. Yoshida, Soil Sci. Plant Nutr., Tokyo, *19*:39 (1973).

136. T. S. C. Wang, S. -Y. Cheng, and T. Tung, Soil Sci., *103*:360 (1967).
137. E. M. Emery and W. E. Koerner, Anal. Chem., *33*:146 (1961).
138. H. W. Ford, Proc. Am. Soc. Hort. Sci., *86*:205 (1965).
139. J. A. Adamson, A. J. Francis, J. M. Duxbury, and M. Alexander, Soil Biol. Biochem., 7:45 (1975).
140. J. M. Lynch, J. Appl. Bacteriol., *42*:81 (1977).
141. R. E. Holm, Plant Physiol., *50*:293 (1972).
142. L. D. Owens, R. G. Gilbert, G. E. Gribel and J. D. Menzies, Phytopathology, *59*:1468 (1969).
143. M. J. Kolbezen and F. J. Abu-El-Haj, Pesticide Sci., *3*:73 (1972).
144. J. J. Jurinak and T. S. Inouye, Soil Sci. Soc. Am. Proc., *27*:602 (1963).
145. D. E. Johnson and B. Lear, J. Chromatogr. Sci., 7:384 (1969).
146. R. J. Dowdell, J. R. Burford and R. Crees, Nature (Lond.), *278*:342 (1979).
147. D. R. Keeney, I. R. Fillery, and G. P. Marx, Soil Sci. Soc. Am. J., *43*:1124 (1979).
148. J. R. Freney, O. T. Denmead, and J. R. Simpson, Soil Biol. Biochem., *11*:167 (1979).
149. A. R. Mosier and L. Meek, Soil Sci. Soc. Am. J., *44*:1121 (1980).
150. J. C. Ryden, L. J. Lund, and D. D. Focht, Soil Sci. Soc. Am. J., *42*:731 (1978).
151. J. C. Ryden, L. J. Lund, J. Letey, and D. D. Focht, Soil Sci. Soc. Am. J., *43*:110 (1978).
152. G. L. Hutchinson and A. R. Moiser, Soil Sci. Soc. Am. J., *45*:311 (1981).
153. G. A. Breitenbeck, A. M. Blackmer, and J. M. Bremner, Geophys. Res. Lett., 7:85 (1980).
154. V. L. Cochran, L. F. Elliott, and R. I. Papendrick, Soil Sci. Soc. Am. J., *45*:307 (1981).

11

Determination of Pesticides by Gas Chromatography and High Pressure Liquid Chromatography

DAVID J. EAGLE, JOHN L. O. JONES, and EDWARD J. JEWELL
Ministry of Agriculture, Fisheries and Food, Cambridge, England

I. INTRODUCTION

The rapid development in recent decades of pesticidal chemicals which are biologically active in minute concentrations has posed a formidable challenge to the analytical chemist, that of identifying and quantifying a wide variety of substances at nanogram or even picogram level in a highly complex organic matrix. The problems of the analyst are often complicated by an initial lack of information as to the pesticide he is to determine. Although screening methods are available for the main groups of chemicals, careful casework to obtain the maximum circumstantial evidence should always precede the work in the laboratory. Although no analytical method will detect all pesticides, the challenge of determining these materials has been met very largely by the advances which have been made in the technology of gas chromatography (GC). Of the hundreds of pesticides currently in use, the majority can be determined by GC, and for most of these it is the preferred method both for sensitivity and selectivity. More recently, high pressure liquid chromatography (HPLC) has joined with GC as a valuable technique in the methodology of the residue analyst. It is rapidly being adopted for the determination of compounds for which GC is unsuitable and it is challenging established GC methods for others. It has the potential to equal or exceed GC in its importance.

The history of chromatography has been well documented in texts on the subject (e.g., [1]) and the major landmarks [2-7] are well known. Many books (e.g. [8-14]) are available which treat the theory and practice in greater depth than is possible here, but an outline of the basic theory is given in Chap. 10. Although a few pesticide substances have been separated by gas-solid chromatography (GSC), the overwhelming majority of GC separations have involved the use of columns containing liquid-coated supports as stationary phase, i.e., gas-liquid chromatography (GLC). In this chapter the principles involved in the application of GLC to pesticide analysis are discussed, together with developments in the use of HPLC, and selected applications to pesticide residue analysis in soil are given.

II. SAMPLING AND PREPARATION FOR CHROMATOGRAPHY

A. Sampling

Procedures for taking representative samples of soil are given in texts of agricultural chemistry and in a leaflet issued by the Agricultural Development and Advisory Service in England and Wales (ADAS) [15]. Sampling methods for pesticide residue analysis have been reported by Ford et al. [16]. Where symptoms of crop damage are seen, samples from areas of differing severity can be helpful. Where a sample can be obtained from an area of comparable soil known to be free of pesticide treatment it can be valuable to the analyst in assessing the validity of the analysis. The depth is determined by the behavior of the compound to be determined. Many herbicides are adsorbed in the surface layer; sampling to greater than 50-75 mm dilutes the sample with uncontaminated material, raising the detection limit. For valid comparison of results the depth of sampling must always be known.

Unless metabolism of rapidly degraded chemicals is being studied, no special precautions are needed in transit of the sample to the laboratory other than protection from contamination. It should be noted that some packing materials, e.g., polythene [17] and cloth bags [18] may release interfering materials.

B. Sample Storage

It is often convenient to be able to store samples prior to analysis. However, few data are available on the stability of pesticides in stored samples and the analyst would be well advised to make his own investigation before undertaking a major program of work. Kawar et al. [19] examined the storage stability of 35 pesticides and found little evidence of breakdown under deep-freeze conditions other than for diazinon, malathion, parathion, captan, and dichlorvos, but at refrigerator temperatures a number of others showed some degradation. Webster and Reimer [20] have reported degradation of metribuzin in cold storage. Where evidence is lacking, storage at −20°C and retention for no more than 1-2 months is a wise precaution.

C. Preparation and Extraction

Samples should be thoroughly mixed and stones and plant debris removed. Where the pesticide is known to be stable during drying the sample should be dried, as the presence of water may be undesirable in the extraction and cleanup. Where the extractant is miscible with water and an aqueous phase is used in the first stage of cleanup, drying may be avoidable, but it should be remembered that even in these cases dilution by sample water may affect extraction efficiency. Drying at <30°C in a current of air is acceptable where stability data relating to these conditions indicate no problem of degradation. Where data are lacking or preclude drying in this way, removal of moisture may be achieved by blending with anhydrous sodium sulfate. Freeze-drying is also worthy of consideration. An alternative approach is extraction with a succession of solvents. A typical scheme used acetone for the first extraction, then the residue is re-extracted with acetone/dichloromethane mixture and finally with chloroform. On combination of the extracts and dilution, if necessary, with water the pesticide is obtained in the chloroform phase. With judicious choice of solvents and volumes such a sequence may simultaneously achieve extraction, cleanup, and concentration. In all operations it should be noted that contamination can be introduced from the laboratory environment [17,21,22]; a blank run without sample to check the apparatus and chemicals should always be carried out.

Extraction may be by shaking or Soxhlet extraction. The latter is to be preferred where the pesticide is adsorbed by the soil or is of limited solubility, but its use may be restricted by the thermal stability of the compound. In specifying a procedure the extraction time and rate of reflux should be stated. Condensers should allow the solvent to fall directly into the thimble. Hill and Stobbe [23] have reported ultrasonic extraction of soil, and this technique is currently being tried in the authors' laboratory with encouraging results.

The first consideration in the choice of a solvent is solubility. While residue levels are unlikely to exceed the solubility in all but a few solvents, use of a reagent in which macro-quantities are soluble favors rapid and complete extraction. The next consideration is the ability to release adsorbed material. There are no generally applicable rules, but polar solvents tend to be more effective. A further consideration is the solubility of unwanted organic matter. Miscibility with water is advantageous for the extraction of undried samples and acetone is widely used for this purpose. It is often more effective than other polar compounds, e.g., methanol, for adsorbed material but coextracts relatively large amounts of organic matter. Acetonitrile is less widely used on soils than on plants but the properties which make it popular for plant material are equally applicable to soil. Miscibility with water in most proportions makes it an efficient extractant for moist samples; it can be separated from water by dilution and addition of a salt, it is immiscible with a number of other solvents giving opportunities for partition cleanup, and it extracts relatively little organic matter. Where the compound ionizes, an aqueous extractant may be advantageous, the pH being chosen to ensure it is entirely in the acid or the salt form.

D. Cleanup

The extent of cleanup, i.e., the removal of unwanted constituents of the extract which may interfere with the subsequent analysis, is determined by the ability of the chromatographic column to separate the pesticide from these coextractives and by the specificity of the detector. With selective detectors many compounds may be determined with nothing more than concentration to small volume. For sensitive, nonspecific detectors, e.g., electron capture, cleanup is invariably needed.

Partition

Partition between immiscible solvents affords a rapid means of removing gross contamination. Apart from considerations of miscibility, a wide difference in polarity between the solvents is necessary to obtain a sufficiently favorable partition coefficient for the pesticide to be completely transferred to one phase in two to three extractions. Partition between aqueous and organic phases is most used but a system of polar and nonpolar organic phases, e.g., acetone and hexane, can be very useful, hexane being valuable for the removal of lipids. When the compound ionizes, adjustment of the pH will effect transfer between aqueous and organic phases, but this can lead to troublesome emulsification with some solvents.

Column Chromatography

Various forms of silica, alumina, and carbon are widely used for column chromatographic cleanup. Their characteristics are dependent on their method of preparation and some batch to batch variation may be encountered. With silica and alumina the moisture content is critical to their separation characteristics and may change with storage or exposure to the atmosphere. It is always wise to check the performance of the material by passing a standard solution of the pesticide at the time of analysis rather than relying on published elution data or past experience. For ionizable compounds, ion exchange resins may provide an effective cleanup material. The coarser HPLC packings have found some use as cleanup materials with relatively reproducible properties. HPLC itself, though primarily a determinative technique, should not be neglected as a possible preliminary to GC. Most detectors are nondestructive and with suitable fraction collecting arrangements the eluate may be recovered, giving a very clean solution for GC and providing valuable confirmatory evidence of peak identity.

Derivatization

Derivatization is normally looked on as a confirmatory technique or as a means of making the pesticide suitable in stability, retention characteristics or detectability for GC. However, this alteration of the physical and chemical properties may also be exploited for cleanup. Few, if any, of the coextractives are similarly altered, so the pesticide is likely to become separable by means which were not applicable to the parent compound. Conversely but less commonly, a troublesome coextractive may be derivatized to enable its separation from the pesticide.

Sweep Codistillation

Sweep codistillation [24-26], in which the sample extract is distilled through a glass wool column in which the coextractives are retained on the glass wool and the pesticide recovered in the distillate, has found some favor as a means of cleanup and concentration.

Some degree of concentration may be achieved in the course of cleanup but it is usually necessary to remove excess solvent to obtain a solution which will give an acceptable detection limit when injected into the chromatograph. Rotary evaporation under reduced pressure is effective in most cases. Small amounts of volatile solvent may be removed by impinging a stream of nitrogen. Where the solution has to be taken to dryness to be redissolved in another solvent, addition of a few drops of a nonvolatile liquid as a "keeper" is advised, as many compounds are degraded or difficult to redissolve if held dry more than momentarily. Chiba and Morley [27] reported that in the preparation and cleanup of soil extracts the greatest losses are at the evaporation stage. Certain GC detectors require rigorous elimination of halogenated solvent, which may entail repeated evaporation to dryness.

The residue is finally diluted with or redissolved in a measured volume of the solvent in which it is to be injected into the chromatograph. For HPLC this will always be the eluent. For GC any volatile solvent that is not retained by the stationary phase and does not saturate the detector may be used.

III. GAS-LIQUID CHROMATOGRAPHY

A. Columns

Solid Supports

Although sold under a variety of names, the majority of solid supports are derived from firebrick or diatomaceous earth, the latter being more common. Firebrick is preferred for strongly nonpolar compounds. An ideal support is inert; it serves only to keep the liquid in place and give it a large surface area. The base materials are therefore washed and treated to inactivate surface hydroxyl groups. Ashby [28] has listed the main materials and their pretreatments. Particle size is important—the smaller the particles the more efficient the column, the limit being set by resistance to gas flow. Efficiency depends equally on close packing so uniformity of size and resistance to fracture are essential. For further discussion of support materials the reviews by Ottenstein [29] and Palframan and Walker [30] will be found useful.

Liquid Phases

Many hundreds of compounds have been used as stationary phases and most suppliers offer a wide range. It is impracticable to find the most suitable by trial and error. Treatments of the factors governing their selection may be found in textbooks of GC but a few simple guidelines are helpful to narrow down the choice. The most basic is the concept of polarity—a polar phase will best separate polar compounds while a nonpolar phase is preferable for nonpolar analytes. Polarity is not directly quantifiable. Our knowledge of it is gained from the behavior of various compounds when chromatographed on different phases. From such studies comes the concept of retention index, based on the ability of a phase to retain benzene compared with the least polar—squalene. This concept has been reviewed by Ettre [31]. Rohrschneider [32] used the retention times of benzene, ethanol, methyl ethyl ketone, nitromethane, and pyridine to determine numerical indexes, and McReynolds [33], developing this work further, determined the behavior of 10 compounds on over 200 phases. Within a group of phases of similar polarity the Rohrschneider index may be used to predict separations from the structure of the compounds of interest, e.g., a high index number for methyl ethyl ketone relative to that for ethanol would be chosen if it were desired to elute an electron attractor compound (e.g., alcohols or acids) before an electron donor (e.g., ketones or ethers). An extensive list of McReynolds constants has been published by Supina [34]. The latter also discussed the effect of temperature and stationary phase loading on column efficiency and the effect of column packing.

In listing the properties of over 200 stationary phases, McReynolds sought to show the close similarity between many of them. His view that the number of phases had unnecessarily proliferated is shared by many chromatographers. A list has been published [35] of six preferred phases for pesticide analysis, with a further 24 to be tried if none in the short list is suitable. The chromatographer would do well to make an initial choice from these lists, exploring others only if the separation is unsatisfactory. Some analysts go further and use one column almost exclusively for a wide range of determinations. While this approach is unlikely to succeed for multiresidue analysis, the analyst who determines one particular compound in a sample, and has only to separate it from coextractives, may achieve considerable success with this approach and avoid the unproductive use of time which is a consequence of frequent column changing.

Coating and Packing

The efficiency of a column depends on the care and skill of its preparation. The preparation is not difficult and with a little practice it should be possible to equal or better the performance of commercially packed columns. The essential requirements are uniformity of coating, avoidance of particle damage, and no chemical degradation.

About 7 g of material is required to pack a 1.5 m × 4 mm column. Many liquid phases are viscous liquids or gums at ambient temperature and difficult to weigh out exactly. It is, therefore, best to weight the coating material first and then the appropriate amount of support to give the required loading. Suppliers' lists usually indicate a suitable solvent. Sufficient should be used to wet the support and form a slurry. The slurry (in a shallow basin) is placed over a gentle source of heat and kept continuously stirred. Thorough stirring is especially important as the slurry reaches a pasty state and particular attention should be given to the side of the basin where, if material is left undisturbed, there is a tendency for the coating to be deposited on the basin rather than the support. Once a friable state is reached, stirring may be relaxed and the basin put aside in a well-ventilated place for the last traces of solvent to evaporate.

Some chromatographers favor the use of a rotary evaporator to keep the slurry in motion. However, this is more liable to damage the particles. Another method is to prepare a liquid suspension of the support in a solution of the stationary phase. The procedure of Cook et al. [36] is typical. The suspension is poured into a Buchner funnel or wide-bore tube and the excess solution drained away. The wetted support is removed and spread in a thin layer to dry. The loading achieved is calculated from the difference in volume between the initial solution and that drained off. Some practice is needed to judge the concentration of solution required. The support may be expected to retain about its own volume of solution but this can vary by a factor of two. This method is the best where uniformity of coating and avoidance of particle damage are critical.

It is advisable to acid-wash and silanize glass columns and glass wool plugs. Pressure or vacuum may be used to assist close packing. The latter is more convenient but with columns exceeding 1 m in length the flow becomes barely sufficient in the later stages. The packing is introduced a little at a time with constant vibration and each lot allowed to pack down until there is no further movement before the next is added. Final settlement may conveniently be achieved by connecting the column to the injection port of the chromatograph and tapping the column throughout its length with maximum carrier pressure applied. The space remaining at the inlet should be such that a syringe needle just reaches the top of the packing. This space is filled with silanized glass beads and finally a small porous plug.

To condition the column it is placed in the chromatograph with carrier flow but no connection to the detector. The temperature is set 20-30°C higher than that at which the column will be used, but not exceeding the maximum for the liquid phase. This temperature is maintained at least overnight to remove last traces of solvent and volatile impurities and ensure stable behavior when the column is put into use.

Column Efficiency

It is important to be able to measure the efficiency of a column, so that if the separation is unsatisfactory it will be known whether the choice of liquid phase or the preparation of the column is at fault. The number of theoretical plates, n, is given by

$$n = \frac{16V_R}{w} \tag{1}$$

where V_R is the retention volume (or, more conveniently, retention time) of an eluted peak and w is the base width of a triangle formed by extrapolating the sides of the peak to the baseline.

It should be noted that V_R should properly be measured from the elution of an unretained peak, not from the injection. If an air peak is discernible this may be taken as the reference point, otherwise for maximum accuracy it is necessary to measure two peaks and solve the equation

$$\frac{V_{R_1} - V_a}{w_1} = \frac{V_{R_2} - V_a}{w_2} \tag{2}$$

to obtain the retention volume, V_a, of the air peak and thence the true values of the retention volume for the measured peaks. The number of theoretical plates is a measure of the ability of the column to separate but not of analysis time. A poor 2 m column may have the same number as a good 1 m column but take twice as long to achieve the same separation. Efficiency is given by n/L, where L is the length of the column. Values of up to 3000 per meter are attainable; 1500 is satisfactory. Efficiency may also be expressed as the height equivalent to a theoretical plate (HETP), h, where h is the number of plates divided by the column length in millimeters.

Capillary Columns

Capillary columns are glass or metal tubing, <1 mm in diameter and 25-100 m in length. They contain no support materials, the liquid phase being coated on the wall of the tube. They are comparable with packed columns in efficiency per unit length but by virtue of the great length which is possible without undue resistance to gas flow they can achieve very high efficiencies. While 1000-2000 plates is usual for a packed column, 100,000 is normal for capillary columns and up to 1,000,000 has been achieved. The small amount of liquid phase which can be deposited on the wall of the capillary makes the sample size admissible without overloading very small. As 0.2 μl is the least which can be injected reproducibly, a sample splitter is needed to further reduce the amount reaching the column. It is difficult to design a splitter which gives a constant ratio for all components of a mixture, so careful calibration is needed. The development of support-coated capillaries, in which the tube is first coated with support material to which the liquid may be bonded more permanently and uniformly, has allowed greater loadings than were formerly possible and made the technique more attractive. Though costly and short lived, capillary columns are worthy of consideration where a rapid separation is required of a particularly complex mixture.

B. Mobile Phase

In GC the mobile phase, the gas, plays little part in the separation. The choice is governed by the characteristics of the detector. Thermal conductivity detectors require a gas with high heat conductivity and therefore low molecular weight; helium is usual. Argon or nitrogen may be used with electron capture detectors. With alkali ionization detectors nitrogen is usual, although some forms of this detector have been reported to give higher sensitivity with helium. The gas must always be purified by passage through a tube containing molecular sieve to remove moisture and impurities. Moisture degrades many liquid phases, particularly the more polar ones. Oxygen must be removed where an electron capture detector is used [37].

C. Detectors

The choice of detector is influenced by the compounds to be determined and the sensitivity required, with reliability and cost as further considerations. There is much to be gained by using a detector which is selective for a particular element. Cleanup may be much reduced and some contributory evidence is afforded as to the identity of the compound detected. Often, however, such detectors may be less sensitive and less reliable than a nonspecific method such as electron capture. A number of useful comparative reviews have been published [38-44] which will guide the chromatographer as to the best choice for particular needs.

Flame Ionization Detector

Frequent reference to the flame ionization detector will be found in early papers on GC of pesticides, but for analyses at residue level it has been superseded by more sensitive and selective methods. Its operation is based on a combustion reaction which may be summarized as:

$$\text{Organic sample} + H_2 + O_2 \rightarrow CO_2 + H_2O$$

$$+ \text{+ve ions} + \text{−ve ions} + e^-$$

It responds to all carbon compounds, so cleanup is invariably necessary. Where cleanup is not difficult and sensitivity not demanding, the detector has some merit for its reliability and wide linear range. A description of the principle and operation is given in Chap. 10.

Thermal Conductivity Detector

Like flame ionization, the thermal conductivity detector—also known as the katharometer or hot-wire detector—was much used in the early days of GC of pesticides but has been replaced by more sensitive and selective techniques. Dependent on a temperature effect, it is slow to stabilize and sensitive to fluctuation of the detector oven temperature and carrier gas flow. Interruption of carrier flow will damage the filaments by overheating. The detector and its operation are more fully described in Chap. 10.

Electron Capture Detector

The development of the electron capture detector (ECD) [45] opened up a new order of sensitivity in residue analysis and did much to establish GC as the method of choice in the field. It remains the most sensitive detector for a wide range of compounds. The detector contains a beta-emitting radioactive source, usually cylindrical, and a central collector electrode (Fig. 1). Early models used tritium contained in metal foil or strontium-90. Both are volatile, limiting the safe operating temperature to 225°C or less. They have largely been superseded by nickel-63, which allows temperatures up to 400°C. Collisions of beta particles with the carrier gas cause ionization:

$$N_2 + \beta \rightleftharpoons N_2^+ + e^-$$

A potential applied between the source and the collector causes current to flow which may be amplified and recorded. When an electrophilic compound, e.g., one containing a halogen, is introduced, one of two reactions may take place:

$$e^- + AB \rightarrow AB^-$$

or

$$e^- + AB \rightarrow A + B^-$$

In either case the fast-moving electron is "captured" and replaced by a slow-moving negative ion which has a much greater chance of recombining with a positive ion. Less negative charges reach the collector and the current is reduced.

The detector may be operated in one of three modes. In the DC mode a constant potential of 2-30 V is applied. This mode is the least sensitive, as electrons are constantly removed and negative ions are most likely to reach the anode and lead to a buildup of contamination. In the pulsed mode a potential of 50 V is applied in pulses of 0.5-1 μs at intervals of 50-500 μs. The interval allows a buildup of electrons available to be captured, leading to increased sensitivity. The migration of negative ions is greatly reduced, lessening contamination. One of the limitations of both modes is the very limited range of concentration in which the response is linear. Beyond the concentration which gives 20% reduction in standing current the change for a given concentration increment is progressively reduced. Increase of the pulse interval may be expected to improve sensitivity, but by reducing the

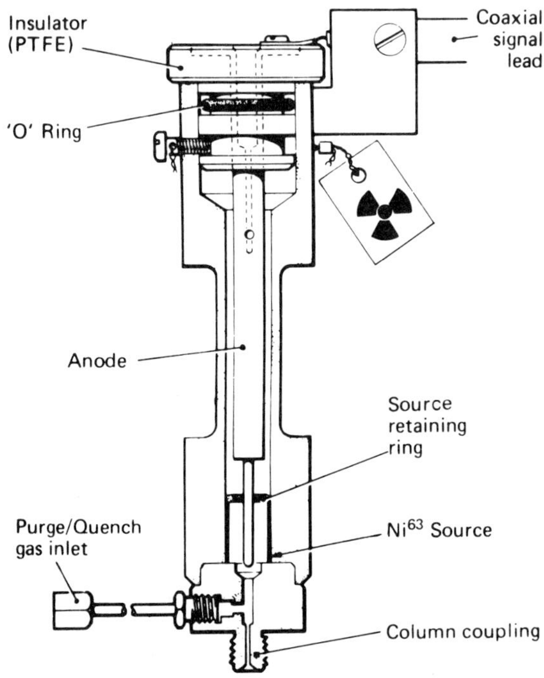

Figure 1 Electron capture detector (From Ref. 28.)

standing current it lowers the threshold of nonlinearity. Unless the linearizing amplifier circuits described in Chap. 10, Sec. II.D are used, the interval chosen must be a compromise between sensitivity and a wide working range. The pulse modulated mode [46] attempts to obtain the best compromise. The pulse interval is adjusted electronically to maintain constant detector current, the change of interval being monitored to give the measured sample signal. A linear range of four decades of concentration is attainable and it is claimed that contamination is reduced, but the authors have found it still to be a problem, particularly at higher column temperatures.

Although particularly sensitive to halogenated compounds, the ECD should not be regarded as specific to this class of compounds, let alone discriminating between them. The mistaken identification as DDT of peaks actually due to polychlorinated biphenyls is well known, but many less obvious pitfalls await the overconfident chromatographer. Cleanup is, therefore, invariably necessary. If the high sensitivity of the ECD is not essential, a more selective detector may obviate much time-consuming preparatory work. Where the object is screening to eliminate negative samples rather than identification of positive findings, the lack of specificity may be an advantage.

The detector is slow to stabilize and very slow to recover from overload. Thus halogenated solvents, which cause great problems in this respect, must be rigorously excluded from sample solutions. After a stable baseline has been achieved there may be further drift in sensitivity, and injection of a reference material at intervals after starting up or recovery is advisable. If not abused, the detector is one of the most stable and reliable, requiring little maintenance. Intervals between cleaning can be lengthened if it is operated so as to avoid condensation of contaminants. The temperature should be appreciably higher than that of the column but, although heating to the maximum working temperature is used as a means of cleaning, the use of maximum temperature under routine conditions may be disadvantageous, as some contaminants may bake onto the electrodes rather than volatilize. Some silicone stationary phases are troublesome in this respect. Cleaning by any other means than heating should not be attempted, except by personnel familiar with the regulations and precautions applicable to radioactive material. Otherwise, the detector should be returned to the manufacturer for cleaning. If the necessary expertise is available, some cleaning may be undertaken by the user, following the manufacturer's recommended procedure. Examination of the collector electrode will reveal the extent of cleaning necessary. Any discoloration is a sign that cleaning will be beneficial; a varnish-like deposit is indicative of gross contamination requiring action to prevent its recurrence.

Alkali Ionization Detector

Although not well reputed for its reliability, the alkali ionization detector can be a great asset to the residue analyst, enabling many analyses to be performed without cleanup. In its simplest form it resembles a flame ionization detector to which is added some means of introducing an alkali metal. It has been produced in many configurations and given a variety of names. Aue [47] proposed the term "alkali flame detector" to avoid mechanistic implications but subsequent developments have made "flame" a misnomer for some types.

The types may be considered in three groups: those in which the alkali is located around the flame and cannot be independently moved or polarized; those in which it is deposited on an electrode heated by the flame but which can be moved and polarized independently of the jet; and those in which the alkali is heated independently rather than by combustion. Some detectors may fall into more than one of these categories according to the mode in which they are used. The first type is reviewed by Aue [47], who emphasizes the importance of close control of hydrogen flow and electrode position for optimum sensitivity. Different selectivities may be obtained according to the alkali used. The selectivity relative to carbon is reported as 10^4-10^5 for phosphorus, 10^2-10^3 for halogens and nitrogen, and at least one order less for sulfur and arsenic.

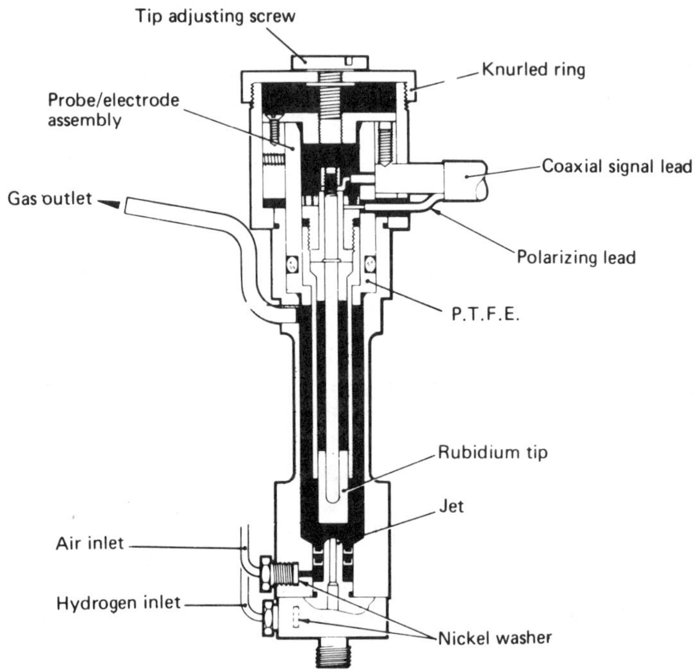

Figure 2 Alkali ionization detector (From Ref. 28).

The second type (Fig. 2) is exemplified by the Pye Unicam 104 and 204 detectors, with which the authors have extensive experience. The two models appear to have identical electrode configuration, but we have found the 204 to be more stable and reliable. We are currently investigating whether this performance is maintained when a 104 chromatograph is adapted to take the 204 detector. Exact positioning of the electrodes is critical to sensitivity. With some compounds a variation of 0.2 mm may make the difference between optimum sensitivity and virtually no response. Optimization of hydrogen flow is also important, and this parameter and electrode position are interdependent. Optimization is performed by injecting a nitrogen compound with a short retention time and observing the effect of adjustments on peak height. As a more rapid alternative for getting near the optimum, a syringe needle is available which gives a constant bleed of nitrogen compound into the column, enabling continuous adjustment to maximum recorder response. This approach can only succeed if the baseline remains relatively steady. In fact, it moves quite considerably but, fortunately, reaches a maximum near to the setting which is optimum for the detection of nitrogen. For the best performance we would recommend final adjustment with injections of the compound to be determined. While we have no conclusive evidence (and the evidence of the literature is conflicting), our subjective impression is that, within a narrow range, there may be some variation in the optimum electrode height according to the compound to be detected.

Frequent maintenance is important to good performance. There is a gradual deposit of alkali salt on the collector and "whiskers" of it form on the source. If a high noise level is encountered, cleaning should be the first remedy to be considered. Although mainly used as a nitrogen detector, it can be optimized for phosphorus, and the determination of sulfur has also been reported [48].

The commercial development of detectors in which the alkali source is in the form of an independently heated bead above the jet is a relatively new development. They were first described by Kolb et al. [49,50] in 1974 but have some similarities to much earlier dual flame detectors [51, 52]. Four variants are now marketed and further advances may be expected. This type of detector results from the recognition that the gas mixture which is optimal for conversion of the sample to the ions necessary to the reaction is not necessarily that which will heat the alkali sufficiently for vaporization. The position of the alkali relative to the jet is much less critical than with other types, allowing rapid differentiation between nitrogen and phosphorus by an alteration of operating conditions which does not involve mechanical adjustment.

In the form described by Kolb and co-workers, operation in the phosphorus mode uses a gas mixture sufficient to heat the alkali bead. The bead is negatively polarized and the jet grounded. Electrons from combustion are prevented from reaching the collector by the polarity of the bead, while those from the reaction of phosphorus at the bead surface are captured by the collector. In the phosphorus-nitrogen mode, bead and jet are polarized and hydrogen flow is reduced tenfold, the bead being heated electrically. It is postulated that a partial pyrolysis occurs, giving cyanide radicals. At the bead these capture electrons from alkali atoms, giving alkali ions which return to the bead and cyanide ions which either move to the collector or react in the oxidizing zone to give electrons, which are captured. Compounds containing nitrogen in the forms

$$—CO—NH—CO— \quad \text{or} \quad —CO—NH_2 \quad \text{or} \quad —O—NH_2$$

are not detected, as they are incapable of forming cyanide radicals unless alkylated. A recent paper by Kolb et al. [53] further describes the reaction mechanism and other workers have reviewed the performance of the detector [54-56]. Further commercial developments of the heated bead type of detector have been described by Burgett et al. [57] and by Patterson and Howe [58]. The detector of Burgett and co-workers is responsive to compounds not detected by the configuration of Kolb and co-workers, so a mechanism other than the formation of cyanide radicals must be postulated.

Flame Photometric Detector (FPD)

The first flame photometric detector (FPD) to be commercially developed was described by Brody and Chaney [59]. Eluted compounds are burnt in a hydrogen-rich flame and the emitted light observed through a narrow bandpass filter by a sensitive photomultiplier. Use of a 526 nm filter gives selectivity for phosphorus; 393 nm is selective for sulfur. Krost et al. [60] have reported flame photometric detection of nitrogen but with a detection limit two orders higher than for sulfur. Detectors are available which simultaneously monitor phosphorus and sulfur emission. Response to phosphorus is linear; mechanistic considerations suggest a sulfur response proportional to the square of the concentration. In practice, responses ranging from linear to square have been reported [61], exponents between 1.5 and 2.0 being usual. The variations are attributed to differences in burner design and gas flows [62], sample concentration, and the oxidation state of the sulfur. Grob [11] has cited reports of the ratio of the phosphorus response to the square root of the sulfur response being a good indicator of their relative amounts, independent of concentration, column temperature and retention time, and of the exponent of the sulfur response being used as an indicator of its oxidation state. In the sulfur mode, selectivity over phosphorus is 10^3-10^4; the phosphorus mode is much less discriminating against sulfur. Electronic devices for linearizing the sulfur response have been described (e.g., [63]) but Burnett et al. [64] have drawn attention to potential errors which may arise from their use.

Early designs of the detector were prone to extinction of the flame by elution of the solvent. Burgett and Green [65] developed a detector which gave improved performance and overcame extinction but there remained a major downward disturbance of the baseline which would interfere with quantitation of rapidly eluted peaks and could cause false baseline assignment by some types of integrator. Zehner and Simonaitis [66] attempted to offset this by addition of a volatile phosphorus compound. The response of the detector is quenched by simultaneous elution of non-sulfur compounds [67]. The selective response cannot, therefore, be used to quantify compounds not resolved by the column.

With response characteristics critically dependent on operating parameters, careful optimization is essential. Greenhalgh and Wilson [68] have reviewed the factors affecting optimization of the Pye Unicam detector. Recently Patterson, et al. [69] have described a dual flame detector in which a hydrogen-rich flame is used to decompose the sample and a second flame to produce light emission. A more uniform response to concentration and a high solvent tolerance is reported. Patterson [70] has compared its performance with that of single flame operation and found the dual flame mode to give less quenching but the single flame to be more sensitive.

Microcoulometric Detector

The microcoulometric detector described by Coulson et al. [71,72] in 1960 was the first successful approach to element-specific detection. They tackled the problem of detection of halogens by pyrolysis of the sample to halide ions and reaction with a solution containing electrochemically generated silver ions. The current passed to restore the silver ion concentration was measured. By using reducing conditions to form H_2S the principle was also applied to sulfur. More detailed papers on the theory and practice of the system and its application to pesticides subsequently appeared [73,74]. Burchfield et al. [75-77] reported on further developments in the system and extended its application to phosphorus. Martin [78] determined nitrogen by catalytic reduction to ammonia which was titrated with coulometrically generated hydrogen ions. Guiffrida and Ives [79] described a miniaturized furnace for use with the detector. Sensitivity tends to be poor—detection limits at the microgram level are usual though nanogram limits are attainable under favorable conditions. Careful optimization due to substantial changes in response with gas flow is generally considered to be necessary. However, Westlake [80] reported the Dohrman microcoulometric detector to be insensitive to gas flow, in contrast with earlier types. The later developments in the microcoulometric detector were paralleled by development of the electrolytic conductivity detector, which has to a considerable extent superseded it.

Electrolytic Conductivity Detector

Developed by the same workers as the microcoulometric detector, the electrolytic conductivity detector is also based on the oxidation or catalytic reduction of the column effluent and its dissolution. The ionic species is, however, measured by its conductivity rather than by titrimetry. The original detector was described by Coulson [81-83] and its performance compared with microcoulometric and electron capture detectors. A detailed description of its application to nitrogen has been given by Cassil et al. [84], and Westlake [85] has reported on its use for bromine and iodine. Patchett [86] described improvements to the detector and evaluated its performance. Hall [87] introduced an improved design which overcame many of the limitations of the Coulson detector and achieved substantially better sensitivity. The two models have been compared by Wilson and Cochrane [88].

Mass Spectrometry

The qualitative significance of data obtained from GC is discussed in Sec. VI. Suffice it to say here that the chromatographic behavior of a compound, while providing valuable evidence, is by no means conclusive proof of its identity. Almost the sole exception to this is mass spectrometric detection. The technique requires a large capital outlay

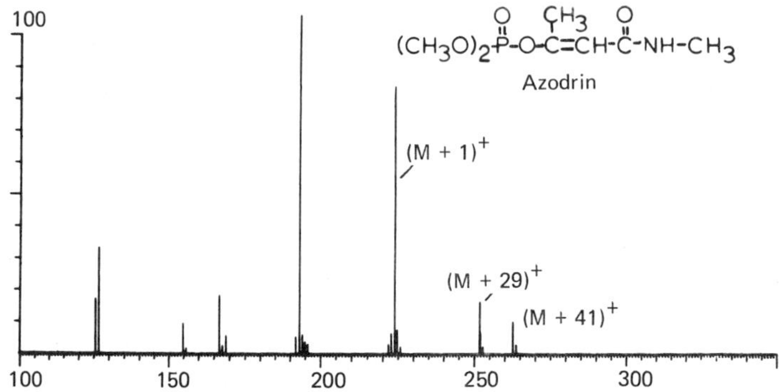

Figure 3 Chemical ionization (CH_4) mass spectrum of Azodrin (From Ref. 92, reproduced from the *Journal of Chromatographic Science*, by permission of Preston Publications Inc.)

and considerable operator skill. The reward is data which not only quantify a compound but identify it. The coincidence of retention time with the appearance of the characteristic ions with the appropriate relative intensities demonstrates the presence of the compound more conclusively than any other means available to the analyst.

In its most sophisticated form the complete mass spectrum of the compound is scanned and compared with the mass spectra of known compounds. Such a mode of operation requires a complex data system. It is also insensitive—on a par with flame ionization detection—as the time taken for a complete scan limits the number of possible in the time the compound is eluted. The more usual mode is to monitor only two or three of the most intense ions. Their relative intensities provide identification with sufficient certainty for most purposes and this mode is much more sensitive. Although capable of giving qualitative data, mass spectrometry should not be looked upon as a means of elucidating the composition of unresolved mixtures. Although in experienced hands it can be used to identify components of partially resolved peaks, it should not be treated as a subsitute for good chromatography. A typical mass spectrum of a pesticide is shown in Fig. 3, and mass chromatograms (showing peak height as a function of time) at selected mass/energy ratios are illustrated in Fig. 4.

Discussion of the types of mass spectrometer and the practical aspects of their operation is beyond the scope of this chapter. For further reading on the technique the works of Brooks and Middleditch [89], Jenden and Cho [90], and McFadden [91] will be found useful, and for its application to pesticides those by Vander Velde and Ryan [92] and Ryan [93] should be consulted.

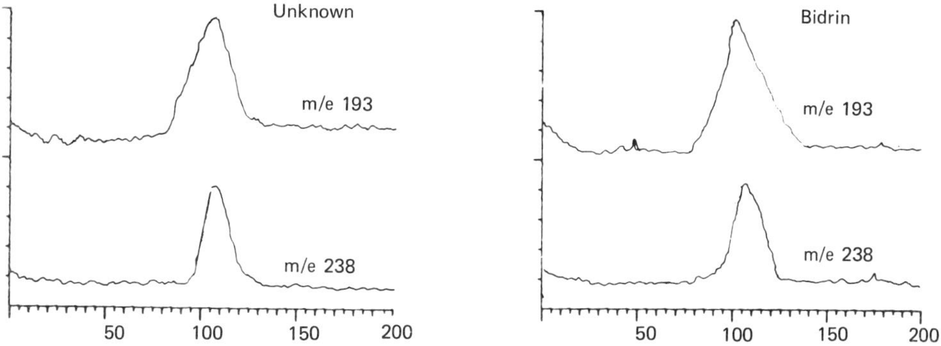

Figure 4 Mass chromatograms of m/e 193 and 238, Bidrin and unknown extract. (From Ref. 92, reproduced from the *Journal of Chromatographic Science* by permission of Preston Publications Inc.)

D. Operating Parameters and Technique

A number of matters of technique have already been referred to, e.g., optimization of detectors. Before concluding this section, some others should be mentioned which are of importance in ensuring a good chromatogram. Introduction of the sample is normally by syringe. With care good reproducibility is not difficult. Between samples the syringe should always be rinsed with solvent—it is good practice to do so at least 10 times—and finally with the sample to be injected. Checks should be made that there is no leakage; bubbles are indicative of a leak between needle and barrel, upward movement of the liquid as the septum is penetrated shows a loose fit of the plunger. Septa should be frequently checked for leakage—the most common cause of drift and noise. Injection heaters should be set a few tens of degrees above the column temperature to ensure rapid vaporization of the sample.

Column temperature is determined by the analysis time which is acceptable for the slowest eluting peak, by its effect on separation and, in some cases, by the susceptibility of the analyte to thermal degradation. As a general rule, increasing the temperature improves peak shape but tends to bunch rapidly eluting peaks. The increase of retention time as temperature is lowered is accompanied by peak broadening so the effect on resolution tends to be small, but where peaks are partially resolved a change of temperature can sometimes have a beneficial effect.

Drift, noise, and distorted peaks are problems encountered by all chromatographers. They can arise from a variety of causes; leaks and bleeding of the liquid phase are the most common. With experience much can be learned from symptoms revealed by the chromatogram. A helpful guide to the operating problems of gas chromatographs and their causes and cure has been published by one of the instrument manufacturers [94], and a booklet by Swan [95] gives flowcharts for problem solving with the ECD, alkali ionization detector and FPD, as well as concise theoretical and constructional descriptions of these detectors.

IV. HIGH PRESSURE LIQUID CHROMATOGRAPHY

While gas chromatographic apparatus is usually supplied as a complete unit, in liquid chromatography it is more usual for the individual parts—pump, injection system, column, detector, and data system—to be purchased separately. The absence of a column oven—most separations are performed at ambient temperature—facilitates the combination of components from more than one manufacturer, enabling a system to be assembled which meets the specific needs of the chromatographer at least cost. There is, however, a trend toward complete systems and this may be expected to increase as the equipment becomes more sophisticated and greater automation is introduced.

A. Pumps

Pumps are of two types: constant pressure, powered by pressure from a gas cylinder, or constant flow, electrically driven. In the former type the pressure may be applied directly to the liquid. Such pumps are simple and reliable; the pressure is pulse free and easily controlled. For routine repetitive analysis they are convenient but, where frequent changes of solvent are called for, other types are preferable. The pressure attainable is limited by the cylinder pressure. Most analytical separations can be achieved below 10 MPa (100 bar), so this limitation is not serious. It may be overcome by means of a pneumatic amplifier—a piston system in which air pressure applied to a large piston drives a small piston pressurizing the liquid. With reciprocating pistons a wide range of flow rates is attainable but pulsing is introduced. Constant pressure pumps can give constant flow only if the impedance of the column and plumbing remain constant. As detector signals are normally recorded in relation to time and retention volumes are assumed to be proportional, this is an important consideration, and again points to constant pressure pumps being most useful in routine analysis where operating parameters are infrequently altered.

Constant volume pumps employ either a large piston driven by a stepping electric motor or a small piston driven by a cam. The piston may be in contact with the liquid directly or via a diaphragm. Small pistons are normally mounted in pairs, the second maintaining the flow while the first is making its return stroke. Even with this arrange-

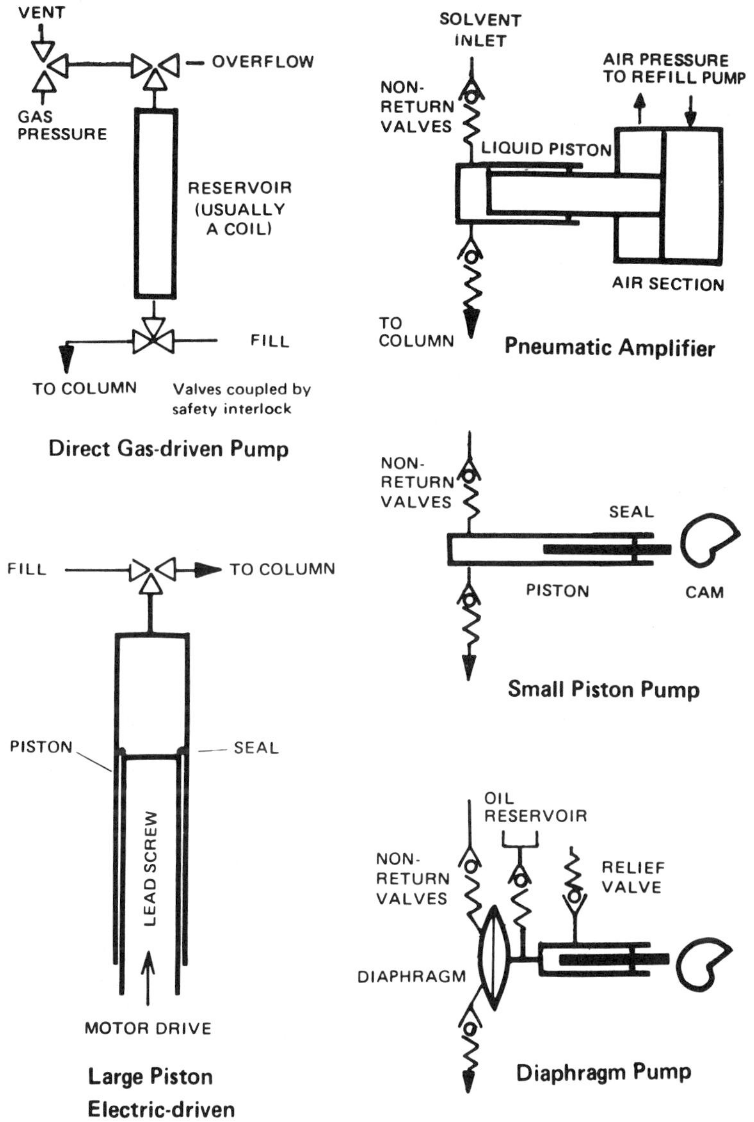

Figure 5 Principal types of HLPC pump. (From Ref. 13.)

ment they are subject to pulsing but with careful design of the timing of the strokes and, in some cases, with feedback to detect and compensate for pulses, the problem is very much reduced. Large piston pumps are pulse free but, as with direct gas driven pumps, their running time is limited by the capacity of the reservoir. Figure 5 illustrates schematically the main types of pump.

Where compounds of widely differing polarity have to be separated, a single eluent may not separate them in a reasonable time. A means of mixing the eluents in varying proportion as the run proceeds may be needed. These solvent gradients may be formed on either the low or high pressure side of the pump. The most simple but least versatile system is to form a step-wise gradient by changing reservoirs at intervals during the run. A sudden change of eluent is liable to disturb the baseline so the changes should be timed to avoid peaks of interest. A continuous gradient can be formed at low pressure by using a proportioning valve to feed from two reservoirs to a mixing chamber and thence to the pump. Time proportioning at high pressure is feasible but time accuracy and speed become critical as does efficient mixing. The more usual way to form a gradient at high pressure is to use two constant volume pumps. For separation development this may be the most satisfactory system but costly for routine applications. High accuracy of flow at low flow rates is necessary to ensure reproducibility at the extremes of the gradient.

B. Injection Devices

Injection with a syringe onto the column poses greater problems than in GC because of the pressure and the effect of solvents on septum materials. Syringes for HPLC have to be made to very close tolerances to prevent leakage and provided with a stop to prevent the plunger being expelled by back pressure. To overcome the problem of the effect of solvents on septa and of septum debris on flow, septumless injection systems have been designed [96,97] but tend to be complex. Not all versions allow the needle to reach the column bed, resulting in loss of efficiency due to edge effects.

The alternative to syringe injection is to use an injection valve. The sample solution is introduced into a loop in the valve and the valve then turned so that the flow to the column passes through the loop. Fixed volume loops give superior precision to syringe injection but the volume is difficult to change quickly. Some loops can be partially filled with a syringe. The volume is, therefore, variable but the precision of a fixed loop is lost and there is greater dilution of the sample before it reaches the column. In most valve injectors the column end fitting does not allow the sample to reach the top of the column coaxially, as it would from a syringe needle, so efficiency is impaired by edge effects.

C. Columns

Stainless steel is the usual material for columns. Unless the system can be operated in such a manner that the sample material does not reach the walls, the interior finish is of considerable importance. Glass has been used for its smoothness and resistance to corrosion but has serious pressure limitations. The lateral spread of the sample as it passes through the column can be calculated, and it has been shown that if it is introduced coaxially and is of negligible diameter at the point of entry, it will not reach the wall if $Ld_p/d_c < 0.4$, where L is the length of the column, d_p is the diameter of the particles, and d_c is the internal diameter of the column. If the column diameter satisfies this condition, the adsorptive properties of the wall and the effect of surface irregularities on the flow pattern are of no consequence. An appreciably greater diameter will needlessly dilute the sample.

Packing materials have not proliferated to the same extent as GC packings. Porous silica is widely used, a pore size of 5-10 nm being most useful. Its adsorptive properties result from surface hydroxyl groups; polar solutes are, therefore, most strongly adsorbed. The various silica materials differ in their particle size and shape. The trend is toward small (ca. 5 μm) spherical particles; uniformity of size is of equal or greater importance than shape in obtaining close packing. Alumina materials with properties similar to silica are available but fewer applications have been reported.

The strongly adsorptive properties of silica can produce excessive retention times for some solutes. To reduce the time taken for diffusion, packings have been developed in which nonporous glass beads are coated with the adsorbing material. This concept also allows a greater variety of active materials to be used. The disadvantage of these pelicular packings is that considerable column volume is occupied by inert material, leading to peak broadening, which reduces their resolving power and gives greater sample dilution. This, together with their very low sample capacity, has limited their application. Bonded-layer materials are now generally preferred.

Packings of lower polarity than silica have been produced by chemically bonding an organic molecule to the hydroxyl of the silica surface. The technique offers the possibility of an infinite variety of materials but the difficulties of preparing them reproducibly has kept their number sensibly restrained. Attachment of the organic molecule is usually achieved by reaction of di- or tri-chlorosilanes with the silica, the Si-O-Si-C linkage being more stable and easy to prepare than a direct Si-C bond. The most common radical is an alkyl group, notably octadecyl, producing a nonpolar stationary phase. Chromatography with such materials is commonly referred to as "reversed phase", as a polar solvent is used with a nonpolar packing, in contrast to the practice with silica. The bonded layer may be a single molecule thick, but if moisture is present in the preparation cross-linkage of the chlorosilanes gives a polymerized layer. Polymerized layers have greater capacity but are less efficient.

HPLC columns are often purchased ready packed, but with some practice and patience they can be filled by any capable user in a chromatography laboratory. Particles of 30 μm or over and spherical materials down to 10 μm may be packed dry, but packing as a slurry which is pumped into the column is generally preferred. Typical procedures are given by Bristow [13]. The solvent for preparing the slurry is chosen for low viscosity, flocculation or dispersion, and density to control settlement. There is conflicting evidence as to whether a flocculated or dispersed slurry is preferable. For silica, polar solvents should be used if flocculation is to be avoided; for nonpolar bonded phases a nonpolar solvent is needed. Acetone will give a dispersed suspension of both types.

After filling, the top 5 mm of packing is removed and a mesh disk is pushed in to allow removal of septum and sample debris without disturbing the packing. This is followed by glass beads and a PTFE sinter which is trimmed off about 1 mm proud of the column tube (Fig. 6). After assembling the end fittings the column may be attached to the chromatograph and eluting solvent pumped to wash out the slurry solvent. If the column is to be used in the "infinite diameter" mode, a top fitting which will introduce the eluent coaxially (Fig. 7) is desirable. Testing and subsequent care of the column is described in a booklet issued by one of the column suppliers [98].

D. Mobile Phase

In GC the mobile phase serves as a carrier and plays virtually no role in the separation. By contrast, in HPLC the choice of solvents is crucial. As with GC phases there have been many theoretical attempts to predict suitable solvents, with some degree of success, but in the present state of knowledge the most practical approach is still one of trial and error with generalized theoretical considerations pointing the direction. Snyder and Kirkland [99] have published a practical guide to solvent selection and Saunders [100] has provided practical conclusions from theoretical work. The principal points of their studies have been summarized by Bristow [13]. The complexities of solvent selection are intensified by the possibilities of using mixtures. A binary

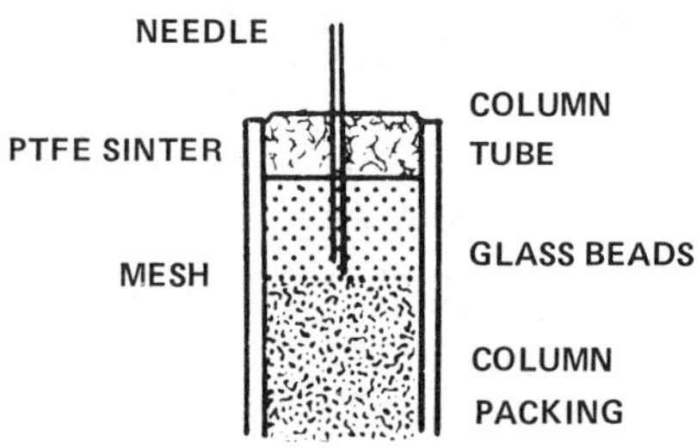

Figure 6 Finishing the top of a column for syringe injection. (From Rof. 13.)

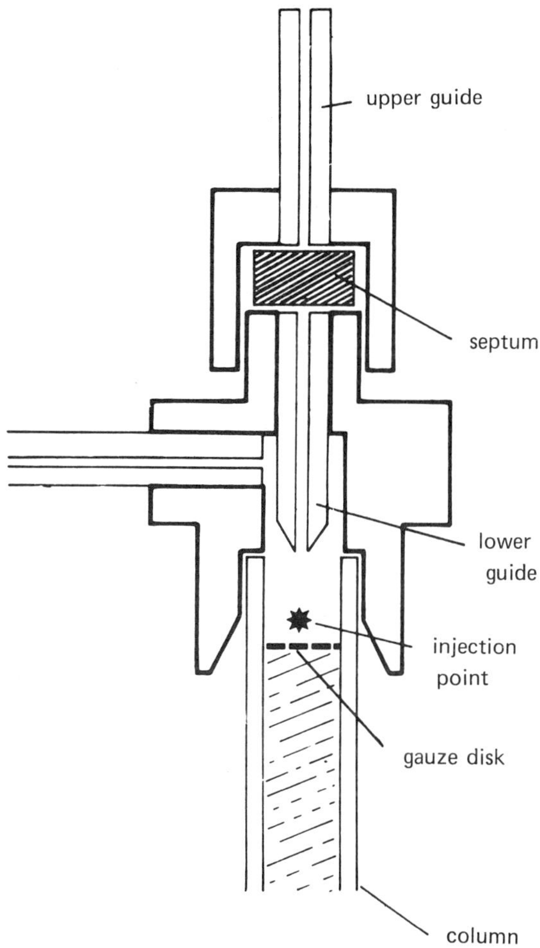

Figure 7 Outline design for syringe injection head. (From Ref. 14.)

mixture greatly aids the attainment of suitable polarity. Use of more than two major components is not often practiced but it is not unusual for minor amounts of other solvents to be added to achieve a particular selectivity. Where the solutes to be separated cover a wide range of polarity, gradient elution can be helpful in achieving a separation in reasonable time. Even where the compounds of interest can be separated isocratically, i.e., with a single eluent, a change of solvent after they have been eluted may be beneficial to clear strongly adsorbed coextractives from the column. The guidelines of Saunders [100] are helpful in assessing the properties of mixtures. For chromatography on silica a small concentration of a polar solvent may have a very marked effect compared with none. Doubling of the concentration approximately halves retention time. Alteration of the nonpolar component has relatively little effect.

E. Detectors

HPLC is demanding in detector design. Peak volumes as low as 100 µl may have to be detected so much of the resolution achieved by chromatography is lost unless the detection volume is small. A rapid response is required—on short columns of small particles peak duration may be as little as 1 sec. Photometric detectors are well capable of detecting such peaks, others less so. The recorder response may in some cases be limiting. The types of detector available for liquid chromatography have been described by Scott [101].

Photometric Detectors

Light absorption, usually in the ultraviolet, is by far the most widely used means of detection, both generally and in pesticide residue analysis. A choice has to be made between a fixed-wavelength instrument using a monochromatic source or a more versatile but more expensive variable-wavelength instrument. Fixed-wavelength instruments usually operate at the 254 nm mercury emission wavelength. Variable wavelength allows detection at the peak of the absorption spectrum, offering potentially greater sensitivity and selectivity. Where, however, a number of peaks with differing absorption maxima have to be quantified on a chromatogram, a compromise wavelength may have to be chosen to the detriment of sensitivity. Variable wavelength offers the possibility of obtaining some qualitative evidence by stopping the flow and observing the wavelength of maximum absorption. A good detector should have a cell volume not greater than a tenth of the smallest peak volume likely to be encountered and it should be so designed that any bubbles formed are cleared quickly. A restriction downstream helps to prevent their formation. The available types of photometric and fluorimetric detectors have been compared by Baker et al. [102].

Fluorimetric Detectors

Few applications have been reported in the pesticide residue field for fluorimetric detectors but the technique has considerable potential for sensitivity and selectivity if suitable fluorescent derivatives can be prepared. Derivatization may be carried out prior to chromatography or, if the conditions for the reaction can be satisfied, post-column. Suitable reactions have been described by Udenfriend et al. [103] and Katz and Pitt [104]. Methods specifically for pesticides have been reported by Duenges [105], Moye et al. [106] and Seiler and Demisch [107].

Instruments designed for HPLC are available but it is possible to fit suitable flowcells to existing fluorimeters. As the sensitivity depends on the cell volume irradiated, a larger cell is required than for UV detection. If peaks of short duration are to be detected, the problem of resolution must be overcome by using a larger diameter column to increase peak volume. Absorption by the fluorescent compound or any other eluate at the exciting wavelength seriously affects linearity of response, so it is prudent to monitor this absorption simultaneously; it should not exceed 0.1 absorbance unit.

Other Detectors

Refractometers have found application in HPLC, mainly in permeation chromatography. For pesticide residue analysis they do not have the necessary sensitivity and selectivity. They are highly sensitive to temperature and pressure.

Atomic absorption spectrometry has been applied to HPLC eluates. Its application to pesticides is limited, as few contain a metal. If the technique gains wider acceptance, extension of the range of application by derivatization may be worthy of consideration.

Dolphin et al. [108] have described the use of electron capture to detect organochlorine pesticides. A commercial instrument is available and the range of application is being extended to nitrogen and organophosphorus compounds. The range of solvents which are compatible is restricted and exclusion of dissolved oxygen is essential.

Attempts have been made (e.g. [109]) to separate the solvent from the eluate and transfer the solute to a flame ionization detector. At present, recovery of the solute presents formidable problems and response is too slow for adequate resolution with efficient columns. If the separation mechanism can be perfected, the concept holds promise for the future, particularly if linked to other types of GC flame detectors to give selectivity.

Highly specific identification information can be obtained by mass spectrometry of eluted peaks from HPLC. The technique has been assessed by Scott [110] in a review of various aspects of HPLC. With wide-bore columns it is not difficult to collect sufficient material. Linked HPLC-MS systems have been described, but as with HPLC-FID, a breakthrough in solvent/solute separation is needed to establish this technique. Like GC-MS it will always be costly and can only be justified where there is a high throughput and certainty of identification is paramount.

V. DATA HANDLING

The devices available for recording the output of chromatographs range from simple chart recorders to complex systems capable of quantifying the data, comparing it with standards, and calculating the final result.

If a recorder is used, a choice must be made between measurement of peak height or area. Much has been written on their relative merits. If the peak is of the ideal gaussian shape and well resolved, height measurement is legimate and easier; less perfectly shaped peaks may be measured accurately in this way if the shape is reproducible. It should always be the aim of chromatographer to produce such peaks. Where it cannot be done, area measurement may be more accurate. Areas may be measured manually by triangulation or by multiplying the height by the width at half height. Both are subject to some degree of error, which becomes significant and broad peaks and depends on regular peak shape. The shape problem can be overcome by cutting out the peak from the chart paper and weighing it but it is tedious, the assignment of the baseline remains subjective, and the permanent record of the separation is destroyed.

The simplest integrators provide a cumulative measure of all the area below the trace. Their use is, therefore, limited to chromatograms with a steady, level baseline. More sophisticated instruments assign the baseline and assess partially resolved peaks according to predetermined criteria, removing the subjective element. The accuracy of their assessment is only as good as the criteria programmed into them. The possibility of an unusual chromatogram being wrongly interpreted should be kept in mind and the recorder trace examined before results are accepted. An integrator is an adjunct to the chart recorder, not a replacement. Electronic data handling methods, including interfacing with computers, have been reviewed by Pierson and Steible [111].

VI. IDENTIFICATION AND CONFIRMATION

The coincidence of retention time of an unknown with that of a standard provides some evidence as to its identity. It is no more than a probability. If the chromatographer has extensive expierence of the compound and the type of sample, in which no "clean" samples have shown the peak and all positive findings have been confirmed, the balance of probability may be sufficient for some purposes. If the sample is rerun on a column of substantially different characteristics the probability is improved but it is still not conclusive. To be convincing a confirmatory technique must employ a different fundamental principle or modify the compound of interest so that its chromatographic behavior is different. Chemical derivatization is, therefore, a very valuable means of confirmation. A number of reviews of derivatization have been published [112-116]. On a routine basis it is convenient to derivatize an aliquot of the sample extract and chromatograph it immediately following the initial determination of the pesticide.

A. Triazines

The chloro-s-triazines can be derivatized by the following general reactions:

Methylation

The residue is heated with dimethyl sulfoxide, sodium hydride, and methyl iodide [117,118] to give the reaction

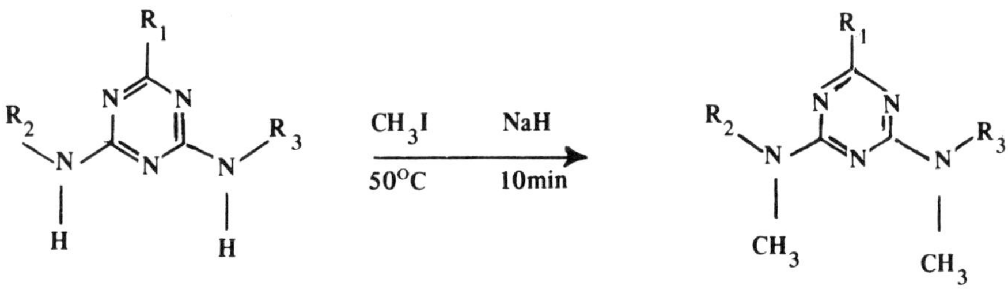

(I) (II)

$R_1 = Cl, OCH_3, SCH_3$

$R_2, R_3 = C_2H_5, CH_3.CH.CH_3$

Figure 8 [118] shows chromatograms of atrazine before and after alkylation.

Methoxylation

The reaction involves the replacement of the labile chlorine atom of the triazine with the methoxyl radical [117].

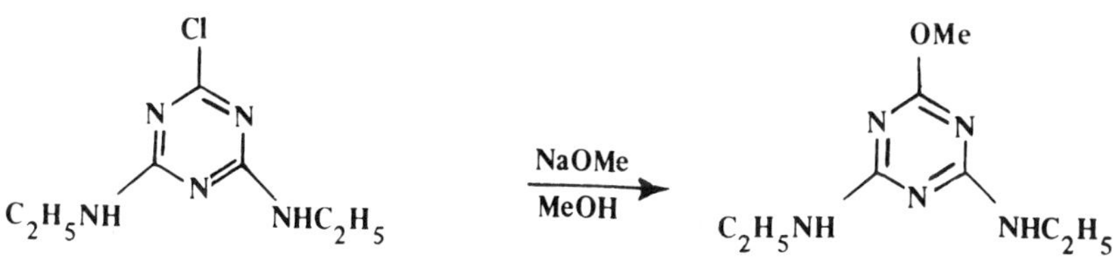

Simazine (III) Simetone (IV)

Hydrolysis and 2,4-Dinitrofluorobenzene

Hydrolysis with M HOl at 150°C is followed by conversion to the dinitrophenyl derivatives by reaction with 2,4-dinitrofluorobenzene in alkaline solution [117].

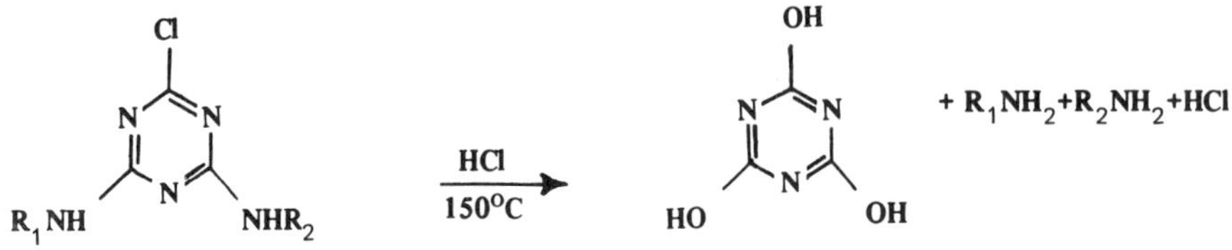

(V) (VI)

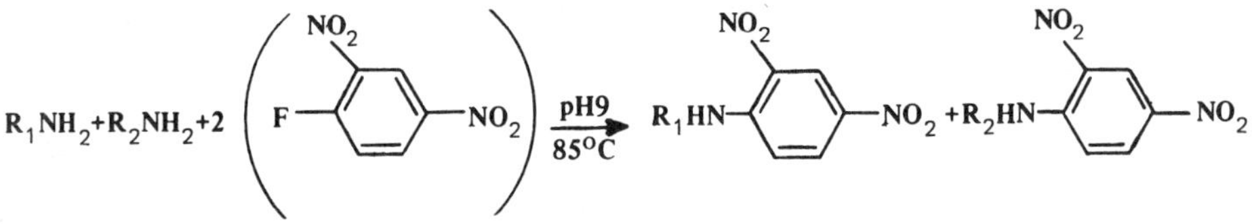

(VII) (VIII) (IX)

Lawrence [117] concluded that the methoxylation reaction is the simplest and fastest method. To confirm the presence of methoxy-s-triazines, methylation with methyl iodide should be used. The hydrolysis-DNFB reaction is useful for characterization of the triazines.

B. Phenoxyalkanoic Acids

Phenoxyalkanoic acid herbicides can be derivatized using the following reagents.

Pentafluorobenzyl Bromide

The acids are converted to pentafluorobenzyl bromide (PFB) esters by reacting with PFB in alkaline acetone solution overnight at room temperature and extracted into iso-octane [119]. The product is separated by column chromatography.

Dicyclohexylcarbodiimide Reagent

2-Chloro esters are prepared by reaction with dicyclohexylcarbodiimide (DCC) and 2-chloroethanol overnight at 60°C [119]. The product is separated from the reaction mixture by partition and column chromatography. The pentafluorobenzyl bromide reaction is simpler to perform than the dicyclocarbodiimide but less specific. The GC retention times

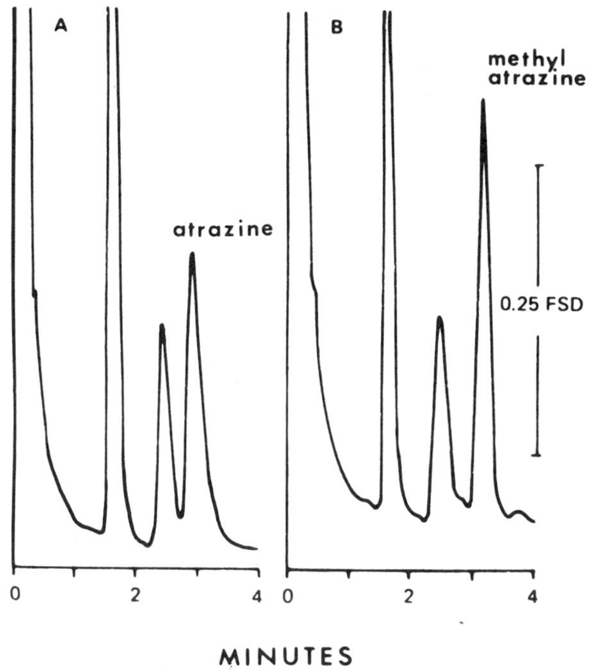

Figure 8 Gas chromatograms of crude methanol extract of organic soil (0.26 $\mu g\ g^{-1}$ of atrazine) before and after alkylation: (A) crude soil extract; (B) alkylated soil extract. [From Ref. 118, reprinted with permission from the *Journal of Agricultural and Food Chemistry*, *23*:325 (1975). Copyright by the American Chemical Society.]

are twice as long for the PFB esters as for the 2-chloro esters under the same conditions. The PFB esters have a greater sensitivity than the 2-chloro esters when determined by electron capture.

C. Organophosphorus Compounds

Hydrolysis and Pentafluorobenzyl Bromide Reagent

Organophosphorus pesticides are hydrolyzed with 10% methanolic potassium hydroxide and the product separated from the acidified reaction mixture by partition into benzene. Derivatization with pentafluorobenzyl bromide is carried out in alkaline solution and the product, partitioned into iso-octane, purified by column chromatography [120].

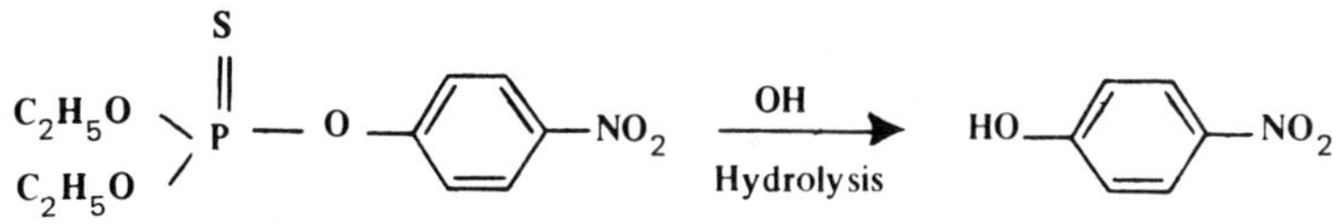

Parathion (X) p - Nitrophenol (XI)

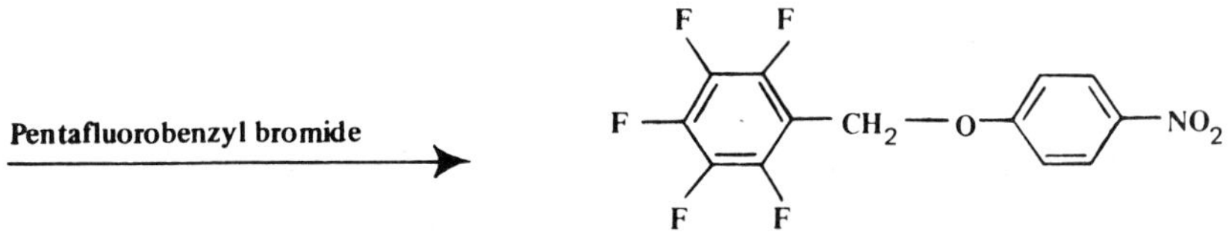

(XII)

Chemical Reduction

Metal chlorides, especially chromous chloride, have been used to confirm the identity of organophosphorus pesticides [121]. Their reactivity toward heptachlor was found to decrease in the order Cr^{2+}, Pd^{2+}, Ti^{2+}, Co^{2+}, and Fe^{2+}. Chromous chloride can be used to confirm organophosphorus pesticides containing a nitro group. The suggested pathway of the reduction of parathion in acetone to yield O,O-diethyl-O-(isopropylidineamino)-phosphorothioate is:

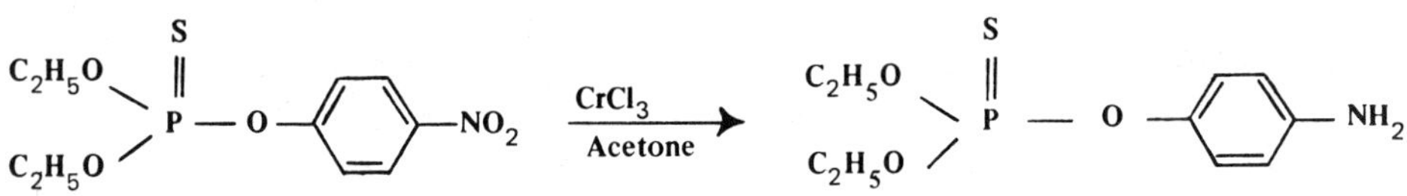

(XIII) (XIV)

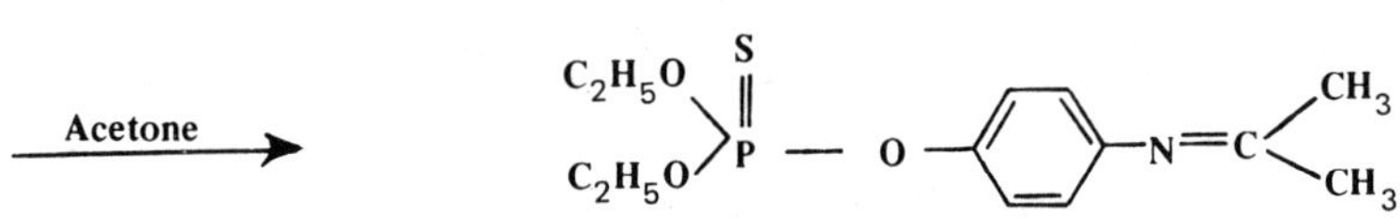

(XV)

Neutral Sodium Hypochlorite

Organothiophosphorus compounds are oxidized by neutral sodium hypochlorite solution [122]. Generally the compounds are oxidized to their corresponding oxons:

Alkylation (Base Catalyzed) Using Sodium Hydride-Methyl Iodide-Dimethyl Sulfoxide

This reaction is applicable to organophosphorus pesticides which have NH or NH_2 moieties [118]. This includes phosphoroamidothioates, $R_1R_2P(S)\text{-}NHR_3$, and phosphoroamidates, $R_1R_2(S)\text{-}NHR_3$.

It has been shown that these compounds in the presence of a strong base react with various alkylating agents to give the methyl derivative, for example:

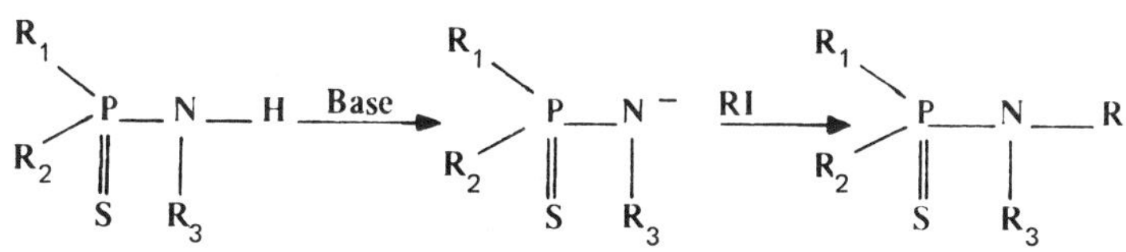

(XVI) (XVII) (XVIII)

The yield of this reaction is related to the polarity of the solvent.

Methylation Using Trimethylanilinium Hydroxide

Trimethylanilinium hydroxide (TMAH) reacts with most organophosphorus compounds to form short chain trialkyl phosphates which can be separated and quantified by GC [123]. A solution containing the organophosphate pesticide and TMAH is injected into a gas chromatograph equipped with a flame ionization detector. The efficiency of the on-column reaction varies from 61 to 100%, depending on the compound under test.

D. Carbamates and Ureas

These two classes will be considered together because of the similarity of their structure:

Carbamates (XIX)

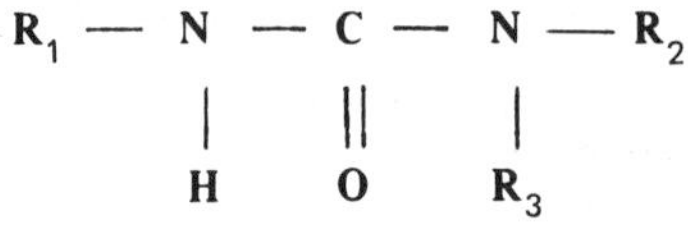

Ureas (XX)

Alkylation (Base Catalyzed)

This is applicable to pesticides which have NH or NH_2 moieties. The residue is heated in a sealed tube with sodium hydroxide, dimethyl sulfoxide, and methyl iodide [118]. The product is extracted from the reaction mixture with ether.

Carbamates decompose if heated at 50°C for 10 min at the alkylation stage. A single product is obtained if the reaction is carried out at room temperature for 5 min.

Esters of Sulfonic Acids: 2, 5-Dichloro-benzene Sulfonyl Chloride

Esters of sulfonyl chlorides can be prepared by the reaction of carbamate pesticides with halogenated benzene sulfonyl chlorides [124]. These derivatives respond well to electron capture detection. The reaction is carried out in a sealed tube at 80°C. Figure 9 shows a typical chromatogram of the derivatives.

Perfluoroacetylation

N-Methylcarbamate insecticides react rapidly and quantitatively with trifluoroacetic, pentafluoropropionic, and heptafluorobutyric anhydrides [125,126].

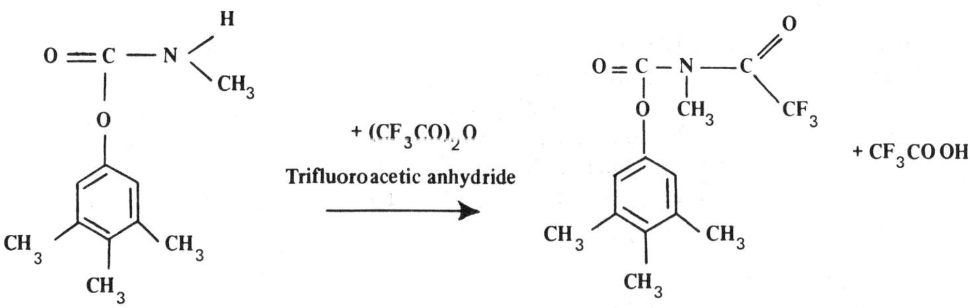

Methiocarb (XXI) (XXII)

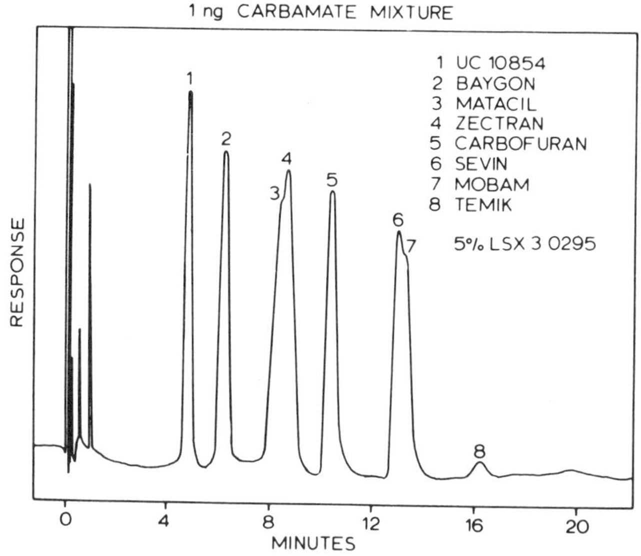

Figure 9 Chromatogram of 1 ng each of seven carbamates and two carbamoyl oximes after derivatization with 2,5-dichlorobenzenesulfonyl chloride. [From Ref. 124, reprinted with permission from the *Journal of Agricultural and Food Chemistry*, *23*:416 (1975). Copyright by the American Chemical Society.]

GC examinations of these derivatives are nearly identical. Relative electron capture responses increase, for example, twofold with carbaryl and fivefold for carbofuran on passing from the trifluoroacetyl to the heptafluorobutyryl derivatives. The use of pyridine as a catalyst enables the condensation of pentafluoropropionic anhydride to proceed at room temperature.

2,4-Dinitrophenyl Derivatives

Methylcarbamate residues are treated with 1-fluoro-2,4-dinitrobenzene by heating for 20 min at 100°C in alkaline sodium borate solution [127]. The ether derivative is extracted into iso-octane and analyzed by electron capture GC. The derivatives of carbamates are stable compounds.

2,4-Dinitroaniline Derivatives

Methyl- or dimethylcarbamates are hydrolyzed with alkali and the liberated amines are reacted with 1-fluoro-2,4-dinitrobenzene to form dinitroaniline derivatives which are determined using electron capture detection [128]. This procedure is not specific.

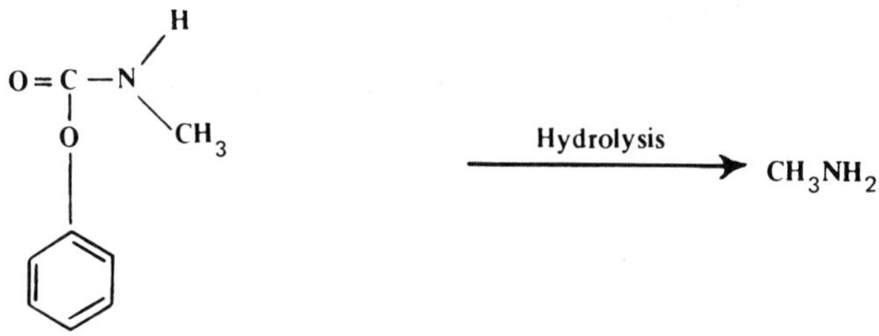

(XXIII)

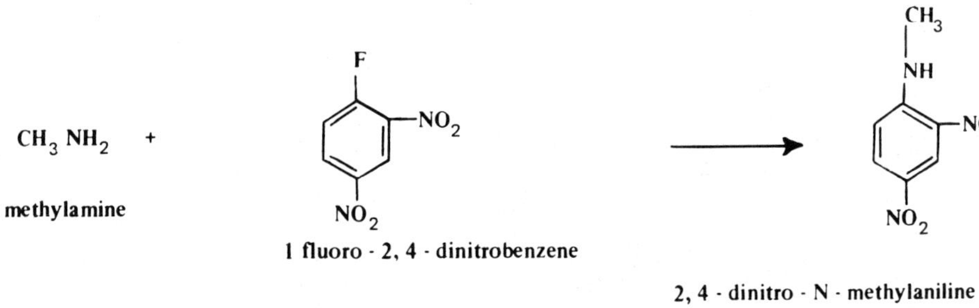

(XXIV)

Pentafluorobenzyl Bromide Reagent

N-Methylcarbamates are hydrolyzed in a 10% methanolic potassium hydroxide solution to form phenolic products. These are derivatized with pentafluorobenzyl bromide to produce ether derivatives which are cleaned up on a silica gel microcolumn [129]. They are determined by gas chromatography using an electron capture detector.

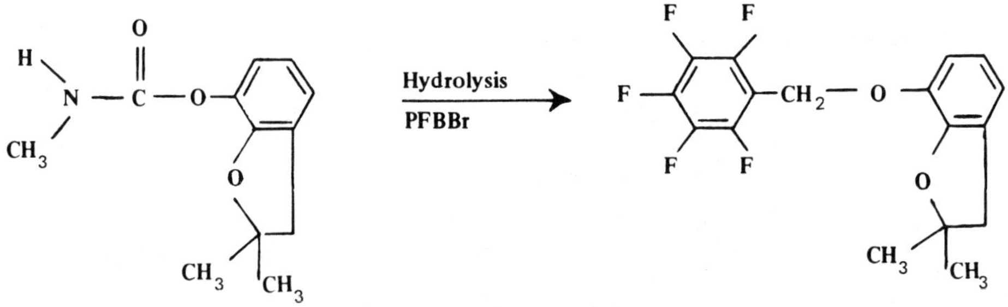

Carbofuran (XXV)

Pentafluorobenzyl derivative (XXVI)

VII. METHODS FOR THE DETERMINATION OF PESTICIDES IN SOIL

Gas-liquid chromatography is currently the main analytical technique used for the determination of pesticide residues in soil. High pressure liquid chromatography has been applied to the assay of a number of herbicides including the substituted ureas, triazines, and uracils. This technique is particularly applicable to thermally labile compounds which suffer some decomposition at the temperature necessary for gas-liquid chromatography or where problems occur in the preparation of a volatile stable derivative.

In general, the sensitivity and specificity of the detectors coupled to gas chromatographs are superior to the UV detector used with high pressure liquid chromatographs. The post-column fluorimetric labeling of some carbamate insecticides [130] has enabled nanogram quantities of these compounds to be detected. Chromatographic data published by Lawrence and Turton [131] details the application of high pressure liquid chromatography to the determination of 166 pesticides in standard solutions and a range of biological materials, including some soil samples. The methods of detection are based, in the main, upon measurement in the UV, but some pesticides are determined by electron capture or by the production of a fluorophore and monitored with a fluorescence detector.

Pesticides are extracted from soil by end-over-end shaking, Soxhlet refluxing, or by a leaching procedure. Comparative studies have been published for some residues in soil. Solvent systems commonly employed include hexane, acetone, methanol, acetonitrile, and chloroform, used either singly or in a mixture of two or more. Methods for the cleanup of soil extracts prior to chromatography involve the use of florisil, alumina, or liquid partition. This may not always be necessary if a selective detector is used and the soil extract is clean.

From the available solid support materials for gas chromatography, Chromosorb W-HP and Gas Chrom Q are widely used. The stationary phases SE-30, QF-1, Carbowax 20M, and the OV series are suitable for a range of residues. High pressure liquid chromatography columns for adsorption chromatography are available with 5 μm size silica or alumina particles. Bonded phase columns with both polar and nonpolar phases enable the operator to utilize their selectivity with high efficiency. Solvent programming may be used to decrease analysis time.

Confirmation of the result of a pesticide residue analysis should always be obtained by one of the standard procedures. The recovery of a pesticide added to a soil at residue level should generally be better than 75%.

A selection of published methods for the determination of the more commonly used pesticides in soil is given in Table 1. Figures 10 and 11 are typical of chromatograms obtainable by GC and HPLC, respectively.

Table 1 Methods for the Determination of Pesticides in Soil

Compound	Soil extractant/cleanup
A. Substituted ureas	
Chlortoluron	Methanol/silica gel
Diflubenzuron	Acetonitrile/water
Diflubenzuron	Acetonitrile/water Celite/Florisil-alumina-silica gel
Fenuron	Methanol
Linuron	
Linuron	Methanol/Florisil
11 substituted ureas	Acetone
8 substituted phenylureas	Methanol
B. Carbamates	
Asulam	$Ca(OH)_2$ solution
Carbofuran	Chloroform/diethyl ether (1:1)
Carbofuran	Chloroform

Method	Detector	Ref.
HPLC on Merckosorb SI-60; mobile phase n-hexane/2-propanol (85:15)	UV (240 nm)	132
HPLC, on Micro-Pak-CH; mobile phase acetonitrile-water (70:30)	UV (254 nm)	133
HPLC, on μ-Bondapak C-18 or μ-Porasil; mobile phases acetonitrile/water (60:40) or methanol/water (75:25), and isooctane/2-propanol (93:7), respectively	UV (254 nm)	134
GLC of methyl derivative; 4% SE-30 + 6% QF-1 on Gas Chrom Q	Alkali ionization (RbCl)	135
GLC of 3,4-dichloroaniline formed by alkaline hydrolysis; 7.5% SE-30 on silanized Chromosorb W	Electron capture	136
GLC, 1.5% XE-60 on Chromosorb W-HP	Electron capture	137
GLC of brominated anilines; 5% high-vacuum silicone grease on Gas Chrom Q. (Typical chromatogram-Fig. 10)	Electron capture	138
HPLC, on Spherisorb ODS; mobile phase methanol/water (60:40) + 0.6% (v/v) conc. NH_3 solution	UV (240 nm)	139
Colorimetric, reaction with sodium and N-(1-naphthyl)ethylenediamine dihydrochloride	Spectrophotometer (540 nm)	140
TLC, on silica gel G; mobile phase ether/hexane (3:1) Colorimetric, hydrolysis, and diazotization	Spectrophotometer (490 nm)	141
Colorimetric, reaction with diazotized aniline	Spectrophotometer (460 nm)	142

Table 1 (Continued)

Compound	Soil extractant/cleanup
Methomyl, carbofuran, and metabolite	Acidified ammonium acetate
N-methylcarbamates	Acidified ammonium acetate/silica gel
Carbamates (general)	Acetone
Carbamates (general)	
<u>C. Triazines</u>	
Atrazine and metabolites	Methanol/acidic alumina
Atrazine	
Atrazine and hydroxyatrazine	Methanol/acidic alumina
Simazine and prometryne	Methanol
7-s-triazines	Acetone/alumina
<u>D. Organochlorine insecticides</u>	
Aldrin	
Chlordane	Hexane/2-propanol (1:1)
Dieldrin metabolites	Methanol
Dieldrin and Heptachlor	Hexane/2-propanol (3:1)/alumina

Method	Detector	Ref.
GLC, on-column transesterification; Porapak P	Alkali ionization (Rb_2SO_4)	143
GLC of PFB derivative; 3% OV-225 or 3.6% OV-101 + 5.5% OV-210 on Chromosorb W-HP	Electron capture	129
GLC of dichlorobenzene sulfonyl chloride derivatives; 5% LSX-3-0295 or UCW98 on Chromosorb W-HP (typical chromatogram Fig. 9).	Electron capture or FPD (sulfur)	124
HPLC, on Permaphase ODS: mobile phase methanol/water (6:94) (typical chromatogram Fig. 11).	UV (254 nm)	144, 145
GLC of methyl derivative; 3% Carbowax 20M on Chromosorb W-HP	Alkali ionization (RbCl)	146
GLC of methyl derivative; 4% SE-30 and 6% QF-1 on Gas Chrom Q	Alkali ionization (RbCl)	135
GLC, 3% SE-30 or 3% Carbowax 20M on Chromosorb W-HP	Alkali ionization (CsBr)	147
GLC, 2% neopentyl glycol succinate on Chromosorb W-HP	Alkali ionization (RbCl)	148
GLC, 5% Reoplex or 3% Carbowax 20M or 3% OV-17 on Chromosorb W-HP	Alkali ionization (CsBr)	149
GLC, 10% Apiezon L on Chromosorb W AW-DCMS	Electron capture	150
GLC, 2% SE-30 + 6% QF-1 Chromsorb W-HP	Electron capture	151
GLC, 1% OV-1 on Chromosorb W AW-DCMS. TLC, on silica gel, mobile phase cyclohexane/acetone (4:1)	Electron capture and FID	152
GLC, 5% DC-200 and 7.5% QF-1 or 3% DC-200 on Gas Chrom Q	Electron capture	153

Table 1 (Continued)

Compound	Soil extractant/cleanup
Hexachlorobenzene	Hexane/acetone (1:1)/ Florisil
Hexachlorobenzene and Dichloran	Petroleum ether/acetone (1:1)/ Florisil
Isofenfos and metabolite	Ethyl acetate/Florisil, silica gel, alumina
11 organochlorine pesticides	Hexane/acetone (1:1)/Florisil
E. Organophosphorus insecticides	
Bromofenvinfos	Acetone/light petroleum (3:7)
Dichlorvos, Trichlorphon, and Isofenfos	Acetone/benzene (19:1); Nuchar C-190/low temperature
Fonofos	Hexane/acetone (1:1)
Methyl parathion	Incubation with sodium acetate, pH 7
Staufer N-2596 (O-ethyl-S-(4-chlorophenyl) ethanephosphorodithioate)	Benzene
Parathion, phorate	Acetone and ethyl acetate/Celite 545, MgO, Norit SG.
F. Halo-alkanoic acids	
Dalapon	Methanolic NaOH
G. Uracils	
Bromacil, Lenacil, and Terbacil	$Ca(OH)_2$ solution/chloroform
Terbacil and metabolites	Chloroform/liquid-liquid partition

Method	Detector	Ref.
GLC, 5% DC-200 and 7.5% QF-1 (support not stated)	Electron capture	154
GLC, 2% OV-225 or 3% mixture of OV-17 and OV-210 (3:22) on Gas Chrom Q	Electron capture	155
GLC, 5% DEGS on Gas Chrom Q	FPD (phosphorus)	156
GLC, 10% DC-200 or OV-101 on Chromosorb W-HP	Electron capture	157
GLC, 5% DC-200 on Varaport	Alkali ionization	158
GLC, 3% SE-30 on Gas Chrom Q	Alkali ionization (Rb_2SO_4)	143
GLC, 0.8% Carbowax 20M and 8% SE-30 on Gas Chrom Q	Alkali ionization (RbCl)	159
Colorimetric determination of the p-nitrophenol produced	Spectrophotometer (405 nm)	160
GLC, 10% OV-17 on Gas Chrom Q	FPD (phosphorus)	161
GLC, 3% OV-1 on Gas Chrom Q	FPD (phosphorus)	162
GLC of 1-butyl ester; 2% OV-17 on Chromosorb W-HP	Electron capture	163
GLC, 5% high vacuum silicone grease on Gas Chrom Q	Alkali ionization (RbCl)	164
GLC of silyl derivative; 5% XE-60 and 0.2% Epon 1001 on Gas Chrom Q	Microcoulometer	165

Table 1 (Continued)

Compound	Soil extractant/cleanup
H. Heterocyclic compounds	
Bentazone and metabolites	Leachate/methylene chloride
Metribuzin and metabolites	Acetonitrile/water/silica gel-Florisil
Metribuzin and metabolites	Water/methanol (1:9)/chloroform
Metribuzin and metabolites	20% aq. methanol/chloroform
Oxadiazon	Benzene/ethyl acetate/acetone (1:1:2) and 10% aq. methanol
Oxadiazon	10% aq. methanol
Picloram	10% aq. KOH, and $Ca(OH)_2$ solution
I. Aromatic acids	
2,4-D, Dicamba and Mecoprop	Acidified acetone/Florisil
J. Pyridines	
3,6-Dichloropicolinic acid	NaCl solution, pH 7-8; 5% ethanol/chloroform
3,6-Dichloropicolinic acid	Water
Cyperquat	H_2SO_4 solution; hydrogenation/methylene chloride
K. Anilides	
Benzoylprop ethyl and metabolites	Methanol/water/ethyl acetate/acetic acid (40:40:20:1)/alumina H

Method	Detector	Ref.
TLC, on silica gel coated sheets with a fluorescent indicator; mobile phase chloroform/methanol/ammonia (69:30:1)	UV light	166
GLC, 5% OV-225 on Chromosorb W-HP	Electron capture	167
GLC, 3% Silar 5 CP on Chromosorb W AW-DCMS	Alkali ionization	20
GLC, 3% OV-225 or Silar 5 CP on Chromosorb W-HP; 1% Reoplex 400 on Gas Chrom Q	Electron capture, Coulson conductivity, FPD (sulfur)	168
TLC, on silica gel plates (F-254); mobile phases hexane/acetone (60:40), methylene chloride/methanol (90:10 and 85:15), benzene	UV light	169
GLC of methyl derivative; 3% OV-17 on Gas Chrom Q	Electron capture	169
GLC of methyl ester; 1.5% XE-60 on Chromosorb W-HP	Electron capture	170
GLC of methyl ester or brominated derivative; 11% OV-17/QF-1 on Chromosorb W-HP	Electron capture	171
GLC of methyl ester; 3% OV-17 on Chromosorb W-HP	Electron capture	172
Colorimetric	UV spectrophotometer (280 nm)	173
GLC, 3% Carbowax 20M + KOH on Carbosorb W-HP	Alkali ionization (RbCl)	174
GLC of methyl derivatives; 2% OV-225 on Gas Chrom Q	Electron capture	175

Table 1 (Continued)

Compound	Soil extractant/cleanup
L. Nitroanilines	
Trifluralin	Methylene chloride
Cryzalin	Methanol/alumina
M. Nitrophenols	
Dinoterb	Ethanol
N. Amides and nitriles	
Dichlobenil	Ethyl acetate
Dichlobenil	20% acetone in hexane/TLC on alumina H
Diphenamid	Methylene chloride
Propyzamide	Sulfuric acid/methanol/Florisil
O. Oxime derivates	
Aldicarb	
Aldicarb, sulfoxide and sulfone	Aqueous leachate; chloroform, silica gel
Glyphosate and metabolites	Dilute NH_4OH
Oxamyl	Ethyl acetate
Oxamyl	Acetone/dichloromethane/Florisil

Method	Detector	Ref.
GLC, 10% DC-200 on Gas Chrom Q	Coulson conductivity (nitrogen mode)	176
GLC of methyl derivative; 5% XE-60 on Chromosorb W-HP	Electron capture	177
GLC of methyl derivative; 4% SE-30 on Gas Chrom Q	Electron capture	178
GLC, 10% DC-200 on Gas Chrom Q	Electron capture	179
GLC, 10% silicone oil and 0.5% Epikote resin 1001 on Celite; 2% NPGS or PDES on Celite	Electron capture	180
GLC, 10% DC-200 on Gas Chrom Q	Coulson conductivity (nitrogen mode)	172
GLC of methyldichlorobenzoate; 10% OV-17 on Gas Chrom Q	Electron capture	181
GLC, 5% Reoplex 400 on Gas Chrom Q	FPD (sulfur)	182
ABS, 8% Carbowax 20M on Anakrom	FPD (sulfur)	183
TLC, on microcrystalline cellulose plates, two-dimensional; mobile phases:phenol/water/acetic acid (84:16:1), isobutyric acid/water/ 1-propanol/conc. ammonia/2-propanol/1-butanol (500:95:70:20:15:15)	UV light/Hanes reagent, ninhydrin	184
GLC, after alkaline hydrolysis to oximino fragment, 10% SP-1200/1% H_3PO_4 on Chromosorb W-AW	FPD (sulfur)	185
GLC, on-column reaction with trimethylphenylammonium hydroxide; 0.5% Carbowax 20M and 5% SE-30 on Chromosorb W	FPD (sulfur)	186

Table 1 (Continued)

Compound	Soil extractant/cleanup
P. Fungicides (misc.)	
Benomyl	Acetone/M aq. NH_4Cl (1:1), solvent partition
Benomyl	Benzene, solvent partition
Quintozene	Petroleum ether/acetone (1:1)

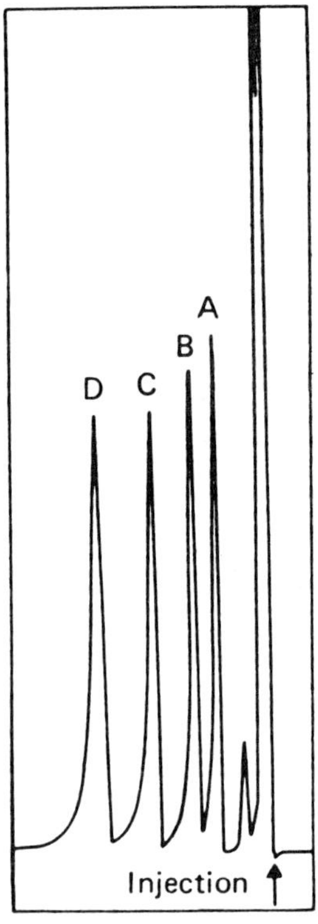

Figure 10 Gas chromatographic response of a standard mixture of brominated anilines derived from: A, monolinuron; B, fenuron; C, linuron; and D, chlorobromuron (R_t = 14 min). Amount applied 0.5 ng. (From Ref. 138.)

Method	Detector	Ref.
Colorimetric	UV spectrophotometer (282 nm)	187
GLC of trifluoroacetyl derivative; 5% SE-30 on Chromosorb R	Electron capture	188
GLC, 2% OV-225 or 3% of a mixture of OV-17 and OV-210 (3:22) on Gas Chrom Q	Electron capture	155

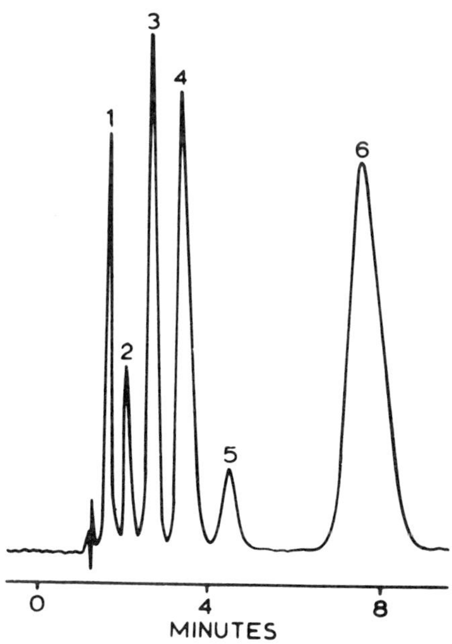

Figure 11 Chromatogram of carbamate pesticides on Permaphase ODS column, 1 m × 2 mm. Mobile phase 6% methanol, 94% water. (1) Furadan; (2) Matacil; (3) Sevin; (4) Landrin; (5) 1-napthol; (6) Mesurol. (From Refs. 144 and 145.)

ACKNOWLEDGMENTS

We thank Dr. Keith Smith for affording us the opportunity to contribute to this volume, and we acknowledge gratefully the permission of the Ministry of Agriculture, Fisheries and Food to draw upon experience gained in our employment. We owe a particular debt of gratitude to our wives for their tolerance and understanding of the many hours spent in study and at the typewriter, heedless of the competing demands of home and leisure activities. Without their forebearance this work, prepared in our own time, would not have been possible.

REFERENCES

1. R. L. Grob in *Modern Practice of Gas Chromatography* (R. L. Grob, Ed.), Wiley, New York, 1977, p. 3.
2. M. Tswett, Ber. Deut. Bot. Ges., *24*:316 (1906).
3. M. Tswett, Ber. Deut. Bot. Ges., *24*:384 (1906).
4. M. Tswett, Ber. Deut. Bot. Ges., *25*:71 (1907).
5. A. J. P. Martin and R. L. M. Synge, Biochem. J. *35*:1358 (1941).
6. A. T. James and A. J. P. Martin, Analyst (Lond.), *77*:915 (1952).
7. A. T. James and A. J. P. Martin, Biochem. J., *50*:679 (1952).
8. R. L. Grob (Ed.), *Chromatographic Analysis of the Environment*, Marcel Dekker, New York, 1975.
9. L. S. Ettre and A. Zlatkis (Eds.), *The Practice of Gas Chromatography*, Interscience, New York, 1967.
10. D. W. Grant, *Gas Liquid Chromatography*, Van Nostrand Reinhold, New York, 1971.
11. R. L. Grob (Ed.), *Modern Practice of Gas Chromatography*, Wiley, New York, 1977.
12. C. F. Simpson, *Practical High Performance Liquid Chromatography*, Heyden, London, 1976.
13. P. A. Bristow, *Liquid Chromatography in Practice*, HETP, Wilmslow, Cheshire, England, 1976.
14. J. H. Knox (Ed.), *High Performance Liquid Chromatography*, Edinburgh Univ. Press, Edinburgh, 1979.
15. ADAS, *Sampling Soil for Analysis*, Leaflet 655, MAFF, Pinner, Middlesex, England, 1979.
16. J. H. Ford, C. A. McDaniel, F. C. White, R. E. Vest, and R. E. Roberts, J. Chromatogr. Sci., *13*:291 (1975).
17. D. R. Rourke, W. F. Mueller, and S. H. Yang. J. Assoc. Off. Anal. Chem., *60*:233 (1977).
18. I. Levi and T. W. Nowicki, Bull. Environ. Contam. Toxic., *7*:133 (1972).
19. N. S. Kawar, G. C. deBatista, and F. A. Gunther, Residue Rev., *48*:45 (1973).

20. G. R. B. Webster and G. J. Reimer, Pestic. Sci., *7*:292 (1976).
21. I. Levi and T. W. Nowicki, Bull. Environ. Contam. Toxic., *7*:193 (1972).
22. J. A. Singmaster and D. G. Crosby, Bull. Environ. Contam. Toxic., *16*:291 (1976).
23. B. D. Hill and E. H. Stobbe, J. Agric. Fd. Chem., *22*:1143 (1974).
24. R. R. Watts and R. W. Storherr, J. Assoc. Off. Anal. Chem., *48*:1158 (1965).
25. R. W. Storherr and R. R. Watts, J. Assoc. Off. Anal. Chem., *51*:662 (1968).
26. R. W. Storherr and R. R. Watts, J. Assoc. Off. Anal. Chem., *48*:1154 (1965).
27. M. Chiba and H. V. Morley, J. Assoc. Off. Anal. Chem., *51*:55 (1968).
28. J. D. Ashby, *An Introduction to Gas Chromatography*, Pye Unicam Ltd., Cambridge, England, 1977, p. 25.
29. D. M. Ottenstein, *Advances in Chromatography*, vol. 3, Marcel Dekker, New York, 1966, p. 137.
30. J. F. Palframan and E. A. Walker, Analyst (Lond.), *92*:71 (1967).
31. L. Ettre, Anal. Chem., *36*:31A (1964).
32. L. Rohrschneider, in *Advances in Chromatography*, (J. C. Giddings and R. A. Keller, Eds.) vol. 4, Marcel Dekker, New York, 1967, p. 333.
33. W. O. McReynolds, J. Chromatogr. Sci., *8*:685 (1970).
34. W. R. Supina, *The Packed Column in Gas Chromatography*, Supelco, Belafonte, Pa., 1974.
35. Anonymous, J. Chromatogr. Sci., *13*:115 (1975).
36. R. F. Cook, R. P. Stanovick, and C. C. Cassil. J. Agric. Fd. Chem., *17*:277 (1969).
37. H. J. Van de Wiel and P. Tommassen, J. Chromatogr., *71*:1 (1972).
38. D. J. David, *Gas Chromatographic Detectors*, Wiley, New York, 1974.
39. J. Sevick, *Detectors in Gas Chromatography*, Elsevier, Amsterdam, 1975.
40. D. F. S. Natusch and T. M. Thorpe, Anal. Chem., *45*:1185A (1973).
41. W. A. Aue and S. Kapila, J. Chromatogr. Sci., *11*:255 (1973).
42. W. A. Aue, J. Chromatogr. Sci., *13*:329 (1975).
43. W. P. Cochrane and R. Greenhalgh, Chromatographia, *9*:255 (1976).
44. J. F. Taylor, Proc. Soc. Anal. Chem. *13*:168 (1976).
45. J. E. Lovelock and S. R. Lipsky, J. Am. Chem. Soc., *82*:431 (1960).

46. R. J. Maggs, P. L. Joynes, A. J. Davies, and J. E. Lovelock, Anal. Chem., *43*:1966 (1971).
47. W. A. Aue, in *Pesticide Identification at the Residue Level* (F. J. Biros, Ed.), Am. Chem. Soc., Washington, D.C., 1971, p. 43.
48. R. A. Hoodless, M. Sargent, and R. D. Treble, Analyst (Lond.), *101*:757 (1976).
49. B. Kolb and J. Bischoff, J. Chromatogr. Sci., *12*:625 (1974).
50. B. Kolb, M. Linder, and B. Kempen, Appl. Chromatogr. (Perkin Elmer), *21E*:1 (1974).
51. A. Karmen, Anal. Chem., *36*:1416 (1964).
52. K. Abel, K. Lannéau, and R. K. Stevens, J. Assoc. Off. Anal. Chem., *49*:1022 (1966).
53. B. Kolb, M. Auer, and P. Popisil, J. Chromatogr. *Sci.*, *15*:53 (1977).
54. J. A. Lubkowitz, J. L. Glajch, B. P. Semonian, and L. B. Rogers, J. Chromatogr., *133*:34 (1977).
55. B. P. Semonian, J. A. Lubkowitz, and L. B. Rogers, J. Chromatogr., *151*:1 (1978).
56. M. J. Hartigan, J. E. Purcell, M. L. McConnell, and M. L. Lee, J. Chromatogr., *99*:339 (1974).
57. C. A. Burgett, D. H. Smith, and H. B. Bente, J. Chromatogr., *134*:57 (1977).
58. P. L. Patterson and R. L. Howe, J. Chromatogr. *Sci.*, *16*:275 (1978).
59. S. S. Brody and J. E. Chaney, J. Gas Chromatogr., *4*:42 (1966).
60. K. J. Krost, J. A. Hodgeson, and R. K. Stevens, Anal. Chem., *45*:1800 (1973).
61. A. I. Mizany, J. Chromatogr. Sci., *8*:151 (1970).
62. S. O. Farwell and R. A. Rasmussen, J. Chromatogr. Sci., *14*:224 (1976).
63. J. G. Eckhardt, M. B. Denton, and J. L. Moyers, J. Chromatogr. Sci., *13*:133 (1975).
64. C. H. Burnett, D. F. Adams, and S. O. Farwell, J. Chromatogr. Sci., *15*:230 (1977).
65. C. A. Burgett and L. E. Green, J. Chromatogr. Sci., *12*:356 (1974).
66. J. M. Zehner and R. A. Simonaitis, Anal. Chem., *47*:2485 (1975).
67. T. Sugiyama, Y. Suzuki, and T. Takeuchi, J. Chromatogr., *80*:61 (1973).
68. R. Greenhalgh and M. A. Wilson, J. Chromatogr., *128*:157 (1976).
69. P. L. Patterson, R. L. Howe, and A. Abu-Shumays, Anal. Chem., *50*:339 (1978).
70. P. L. Patterson, Anal. Chem., *50*:345 (1978).

71. D. M. Coulson, L. A. Cavanagh, J. E. De Vries, and Walther, J. Agric. Fd. Chem., *8*:399 (1960).
72. D. M. Coulson and L. A. Cavanagh, Anal. Chem., *32*:1245 (1960).
73. D. M. Coulson, Advan. Pest Control Res., 5:153 (1962).
74. C. C. Cassil, Residue Rev., *1*:37 (1962).
75. H. P. Burchfield, D. E. Johnson, J. W. Rhoades, and R. J. Wheeler, J. Gas Chromatogr., *3*:28 (1965).
76. H. P. Burchfield, J. W. Rhoades, and R. J. Wheeler, J. Agric. Fd. Chem., *13*:511 (1965).
77. H. P. Burchfield, and R. J. Wheeler, J. Assoc. Off. Anal. Chem., *49*:651 (1966).
78. R. L. Martin, Anal. Chem., *38*:1209 (1966).
79. L. Guiffrida and N. F. Ives, J. Assoc. Off. Anal. Chem., *52*:541 (1969).
80. W. E. Westlake, in *Pesticide Identification at the Residue Level* (F. J. Biros, Ed.), Am. Chem. Soc., Washington, D.C., 1971, p. 78.
81. D. M. Coulson, J. Gas Chromatogr., *3*:134 (1965).
82. D. M. Coulson, J. Gas Chromatogr., *4*:285 (1966).
83. D. M. Coulson, in *Advances in Chromatography*, vol. 3, (J. C. Giddings and R. A Keller, Eds.) Marcel Dekker, New York, 1966, p. 197.
84. C. C. Cassil, R. P. Stanovick, and R. F. Cook, Residue Rev., *26*:63 (1969).
85. W. E. Westlake, in *Pesticide Identification at the Residue Level* (F. J. Biros, Ed.), Am. Chem. Soc., Washington, D.C., 1971, p. 79.
86. G. G. Patchett, J. Chromatogr. Sci., *8*:155 (1970).
87. R. C. Hall, J. Chromatogr. Sci., *12*:152 (1974).
88. B. P. Wilson and W. P. Cochrane, J. Chromatogr., *106*:174 (1975).
89. C. J. W. Brooks and B. S. Middleditch, Clin. Chim. Acta, *34*:145 (1971).
90. D. J. Jenden and A. K. Cho, Ann. Rev. Pharmacol., *13*:371 (1973).
91. W. McFadden, *Techniques of Combined Gas Chromatography/ Mass Spectrometry*, Wiley-Interscience, New York, 1973.
92. G. Vander Velde and J. F. Ryan, J. Chromatogr. Sci., *13*:322 (1975).
93. J. F. Ryan, in *Analytical Methods for Pesticides and Plant Growth Regulators*, vol. 9, *Spectroscopic Methods of Analysis* (G. Zweig and J. Sherma, Eds.), Academic Press, New York, 1978, p. 1.
94. Anonymous, *Logical Troubleshooting*, Hewlett-Packard, Inc., Avondale, Pa., 1970.

95. D. F. K. Swan, *Three Selective Detectors*, Pye Unicam Ltd., Cambridge, England, 1978.
96. R. M. Cassidy and R. W. Frei, Anal. Chem., *44*:2250 (1972).
97. H. Spaans, H. Terol, and A. Onderwater, J. Chromatogr. Sci., *14*:246 (1976).
98. Anonymous, *Care and Use of Partisil Columns for HPLC*, Whatman Inc., Clifton, N.J., 1975.
99. L. R. Snyder and J. J. Kirkland, *Introduction to Modern Liquid Chromatography*, Wiley, New York, 1974.
100. D. L. Saunders, Anal. Chem., *46*:470 (1974).
101. R. P. W. Scott, *Liquid Chromatography Detectors*, Journal of Chromatography Library, vol. 11, Elsevier, Amsterdam, 1977.
102. D. R. Baker, R. C. Williams, and J. C. Steichen, J. Chromatogr. Sci., *12*:499 (1974).
103. S. Udenfriend, S. Stein, P. Böhlen, W. Dairman, W. Leimgruber, and M. Weigele, Science, *178*:871 (1972).
104. S. Katz and W. W. Pitt, Anal. Lett., *5*:177 (1972).
105. W. Duenges, UV Spectrom. Group Bull., 1977 No. 5, p. 38.
106. H. A. Moye, S. J. Scherer, and P. A. St. John, Anal. Lett., *10*:1049 (1977).
107. N. Seiler and L. Demisch, in *Handbook of Derivatives for Chromatography* (K. Blau and G. S. King, Eds.), Heyden, London, 1977.
108. R. J. Dolphin, F. W. Willmott, A. D. Mills, and L. P. J. Hoogeveen, J. Chromatogr., *122*:259 (1976).
109. R. P. W. Scott and J. G. Lawrence, J. Chromatogr. Sci., *8*:65 (1970).
110. R. P. W. Scott, Analyst (Lond.), *103*:37 (1978).
111. H. L. Pierson and D. J. Steible, in Ref. 11.
112. J. C. Cavagnol and W. R. Betker, in Ref. 9.
113. M. Beroza and W. R. Betker, in Ref. 9.
114. W. P. Cochrane, J. Chromatogr. Sci., *13*:246 (1975).
115. Anonymous, *Handbook and General Catalog*, Pierce Chemical Co., Rockford, Ill., 1979.
116. K. Blau and G. S. King (Eds.), *Handbook of Derivatives for Chromatography*, Heyden, London, 1977.
117. J. F. Lawrence, J. Agric. Fd. Chem., *22*:936 (1974).
118. R. Greenhalgh and J. Kovacicova, J. Agric. Fd. Chem., *23*:325 (1975).
119. H. Agemian and A. S. Y. Chau, J. Assoc. Off. Anal. Chem., *60*:1070 (1977).
120. J. A. Coburn and A. S. Y. Chau, J. Assoc. Off. Anal. Chem., *57*:1272 (1974).
121. M. A. Forbes, B. P. Wilson, R. Greenhalgh, and W. P. Cochrane, Bull. Environ. Contam. Toxic., *13*:141 (1975).

122. J. Singh and M. R. Lapointe, J. Assoc. Off. Anal. Chem., *57*:1285 (1974).
123. J. W. Miles and W. E. Dale, J. Agric. Fd. Chem., *26*:480 (1978).
124. H. A. Moye, J. Agric. Fd. Chem., *23*:415 (1975).
125. J. N. Sieber, J. Agric. Fd. Chem., *20*:443 (1972).
126. S. C. Lau and R. L. Marxmiller, J. Agric. Fd. Chem., *18*:413 (1970).
127. E. R. Holden, J. Assoc. Off. Anal. Chem., *56*:713 (1973).
128. E. R. Holden, W. M. Jones, and M. Beroza, J. Agric. Fd. Chem., *17*:56 (1969).
129. J. A. Coburn, B. D. Ripley, and A. S. Y. Chau, J. Assoc. Off. Anal. Chem., *59*:188 (1976).
130. R. T. Krause, J. Chromatogr. Sci., *13*:281 (1978).
131. J. F. Lawrence and D. Turton, J. Chromatogr., *159*:207 (1978).
132. A. E. Smith and K. A. Lord, J. Chromatogr., *107*:407 (1975).
133. C. H. Schaefer and E. F. Dupras, J. Agric. Fd. Chem., *25*:1026 (1977).
134. S. J. Diprima, R. D. Cannizzaro, J. Roger, and C. D. Ferrell, J. Agric. Fd. Chem., *26*:968 (1978).
135. R. Greenhalgh and J. Kovacicova, Bull. Environ. Contam. Toxic., *14*:47 (1975).
136. L. B. Dmitriev and L. P. Yudina, Izv. Timiryazev. Sel-khoz Akad., *2*:215 (1974).
137. S. U. Khan, R. Greenhalgh, and W. P. Cochrane, Bull. Environ. Contam. Toxic., *13*:602 (1975).
138. D. J. Caverly and R. C. Denney, Analyst (Lond.), *103*:368 (1978).
139. D. S. Farrington, R. G. Hopkins, and J. H. A. Ruzicka, Analyst (Lond.), *102*:377 (1977).
140. A. E. Smith and A. Walker, Pestic. Sci., *8*:449 (1977).
141. K. Venkateswarlu, T. K. Siddarame Gowda, and N. Sethunathan, J. Agric. Fd. Chem., *25*:533 (1977).
142. J. R. Rangaswamy, Y. N. Vijayashankar, and S. Prakash, J. Assoc. Off. Anal. Chem., *59*:1276 (1976).
143. P. T. Holland, Pestic. Sci., *8*:354 (1977).
144. A. D. Thruston, Jr., US EPA Report R2-72-079, Corvallis, Ore., 1972.
145. H. A. Moye, J. Chromatogr. Sci., *13*:268 (1975).
146. S. U. Khan and P. B. Marriage, J. Agric. Fd. Chem., *25*:1408 (1977).
147. S. U. Khan, R. Greenhalgh, and W. P. Cochrane, J. Agric. Fd. Chem., *23*:430 (1975).
148. A. Walker, Pestic. Sci., *7*:41 (1976).

149. S. U. Khan and P. Purkayastha, J. Agric. Fd. Chem., *23*:311 (1975).
150. D. L. Struble, Bull. Environ. Contam. Toxic., *11*:231 (1974).
151. F. Tafuri, M. Businelli, L. Scarponi, and C. Marucchini, J. Agric. Fd. Chem., *25*:353 (1977).
152. I. Weisgerber, D. Bieniek, J. Kohli, and W. Klein, J. Agric. Fd. Chem., *23*:682 (1975).
153. H. P. Freeman, A. W. Taylor, and W. M. Edwards, J. Agric. Fd. Chem., *23*:1101 (1975).
154. A. R. Isensee, E. R. Holden, E. A. Woolson, and G. E. Jones, J. Agric. Fd. Chem., *24*:1210 (1976).
155. W. Dejonckheere, W. Steurbaut, and R. H. Kips, Bull. Environ. Contam. Toxic., *13*:720 (1975).
156. M. J. Brown and I. H. Williams, Pestic. Sci., *7*:545 (1976).
157. Anonymous, *Official Methods of Analysis*, 12th ed., Assoc. Off. Anal. Chem., Washington, D.C., Sec. 29.013.
158. Z. Chmiel, R. Knapek, and T. Utracki, Pr. Inst. Przem. org., *6*:147 (1974).
159. S. U. Khan, H. A. Hamilton, and E. J. Hogue, Pestic. Sci., *7*:553 (1976).
160. F. M. Kishk, T. El-Essawi, S. Abdel-Chafar, and M. B. Abou-Donia, J. Agric. Fd. Chem., *24*:305 (1976).
161. R. J. Bussey, M. A. Christenson, and M. S. O'Connor, J. Agric. Fd. Chem., *25*:993 (1977).
162. A. M. Kadoum and D. E. Moch, J. Agric. Fd. Chem., *26*:45 (1978).
163. E. G. Cotterill, J. Chromatogr., *106*:409 (1975).
164. D. J. Caverly and R. C. Denney, Analyst (Lond.), *102*:576 (1977).
165. R. F. Holt and H. L. Pease, J. Agric. Fd. Chem., *25*:373 (1977).
166. E. W. Stoller, L. W. Wax, L. C. Haderlie, and F. W. Slife J. Agric. Fd. Chem., *23*:682 (1975).
167. J. S. Thornton and C. W. Stanley, J. Agric. Fd. Chem., *25*:380 (1977).
168. G. R. B. Webster, S. R. Macdonald, and L. P. Sarna, J. Agric. Fd. Chem., *23*:74 (1975).
169. D. Ambrosi, P. C. Kearney, and J. A. Macchia, J. Agric. Fd. Chem., *25*:868 (1977).
170. C. E. McKone and E. G. Cotterill, Bull. Environ. Contam. Toxic., *11*:233 (1974).
171. S. U. Khan, J. Assoc. Off. Anal. Chem., *58*:1027 (1975).
172. A. J. Pik and G. W. Hodgson, J. Assoc. Off. Anal. Chem., *59*:264 (1976).
173. A. J. Pik, E. Peake, M. T. Strosher, and G. W. Hodgson, J. Agric. Fd. Chem., *25*:1054 (1977).

174. S. U. Khan and K. S. Lee, J. Agric. Fd. Chem., *24*:684 (1976).
175. A. N. Wright and B. L. Mathews, Pestic. Sci., *7*:339 (1976).
176. W. R. Payne, J. D. Pope, and J. E. Benner, J. Agric. Fd. Chem., *22*:79 (1974).
177. R. F. Sieck, W. S. Johnson, A. F. Cockerill, D. N. B. Mallen, D. J. Osborne, and S. J. Barton, J. Agric. Fd. Chem., *24*:617 (1976).
178. W. H. Dekker and H. A. Selling, J. Agric. Fd. Chem., *23*:1013 (1975).
179. G. G. Briggs and J. E. Dawson, J. Agric. Fd. Chem., *18*:97 (1970).
180. K. I. Benyon, L. Davies, K. Elgar, and A. N. Wright, J. Sci. Fd. Agric., *17*:151 (1966).
181. I. R. Adler, C. F. Gordon, L. D. Haines, and J. P. Wargo, J. Assoc. Off. Anal. Chem., *55*:802 (1972).
182. Y. Iwata, W. E. Westlake, J. H. Barkley, G. E. Carman, and F. A. Gunther, J. Agric. Fd. Chem., *25*:933 (1977).
183. M. Leistra, J. H. Smelt, and T. M. Lexmund, Pestic. Sci., *7*:471 (1976).
184. M. L. Rueppel, B. B. Brightwell, J. Schaefer, and J. T. Marvel, J. Agric. Fd. Chem., *25*:517 (1977).
185. R. F. Holt and H. L. Pease, J. Agric. Fd. Chem., *24*:263 (1976).
186. R. H. Bromilow, Analyst (Lond.), *101*:982 (1976).
187. D. J. Austin and G. G. Briggs, Pestic. Sci., *7*:201 (1976).
188. J. P. Rouchaud and J. R. Decallonne, J. Agric. Fd. Chem., *22*:259 (1974).

Author Index

Numbers in brackets are reference numbers and indicate that an author's work is referred to. Underlined numbers give the page on which the complete reference is listed.

Subject Index

Applications refer to soil and/or soil extracts except where indicated.